Heterogeneous Cellular Networks

This detailed, up-to-date introduction to heterogeneous cellular networking presents its characteristic features, the technology underpinning it, and the issues surrounding its use.

Comprehensive and in-depth coverage of core topics catalog the most advanced, innovative technologies used in designing and deploying heterogenous cellular networks, including system-level simulation and evaluation, self-organization, range expansion, cooperative relaying, network MIMO, network coding, and cognitive radio. Practical design considerations and engineering tradeoffs are also discussed in detail, including handover management, energy efficiency, and interference management techniques.

A range of real-world case studies, provided by industrial partners, illustrate the latest trends in heterogenous cellular network development. Written by leading figures from industry and academia, this is an invaluable resource for all researchers and practitioners working in the field of mobile communications.

Xiaoli Chu is a Lecturer in the Department of Electronic and Electrical Engineering at the University of Sheffield, UK.

David López-Pérez is a Member of Technical Staff at the Autonomous Networks and Systems Research Department of Bell Labs, Alcatel-Lucent, Dublin.

Yang Yang is a Professor at the CAS Shanghai Institute of Microsystem and Information Technology, People's Republic of China.

Fredrik Gunnarsson is a Senior Specialist at Wireless Access Networks, Ericsson Research, and an Associate Professor in the Division of Automatic Control, Linköping University, Sweden.

Heterogeneous Cellular Networks

Theory, Simulation and Deployment

Edited by

XIAOLI CHU
University of Sheffield

DAVID LÓPEZ-PÉREZ
Bell Labs, Alcatel-Lucent

YANG YANG
Shanghai Institute of Microsystem and
Information Technology

FREDRIK GUNNARSSON
Ericsson Research

CAMBRIDGE
UNIVERSITY PRESS

CAMBRIDGE UNIVERSITY PRESS
Cambridge, New York, Melbourne, Madrid, Cape Town,
Singapore, São Paulo, Delhi, Mexico City

Cambridge University Press
The Edinburgh Building, Cambridge CB2 8RU, UK

Published in the United States of America by Cambridge University Press, New York

www.cambridge.org
Information on this title: www.cambridge.org/9781107023093

Cover image designed by Kinga Koren.

First published 2013

Printed and bound in the United Kingdom by the MPG Books Group

A catalog record for this publication is available from the British Library

Library of Congress Cataloging in Publication data
Heterogeneous cellular networks : theory, simulation, and deployment / Xiaoli Chu, University
of Sheffield, David Lopez-Perez, Bell Labs, Alcatel-Lucent, Yang Yang, Shanghai Institute of
Microsystem and Information Technology, Fredrik Gunnarsson, Ericsson Research.
    pages   cm
Includes bibliographical references and index.
ISBN 978-1-107-02309-3 (hardback)
1. Cell phone systems.   2. Internetworking (Telecommunication)   I. Chu, Xiaoli.
TK5103.2.H485   2013
621.3845'6 – dc23   2013006208

ISBN 978-1-107-02309-3 Hardback

To those who have enlightened our lives
with love and knowledge

Contents

11   Network MIMO techniques   312

Gan Zheng, Yongming Huang and Kai-Kit Wong

Acknowledgments

We would like to thank our publishers, Phil Meyler, Elizabeth Horne and the rest of the staff at Cambridge University Press. We are grateful for their encouragement, enthusiasm and vision about this book, as well as for their professionalism. We learned a lot from them.

We would also like to thank our graphical designer, Kinga Koren, for her high-quality and on time job. We are thankful for her enthusiasm and professionalism.

We also express our thanks to the Engineering and Physical Science Research Council (EPSRC) and the project partners University of Bedfordshire and Alcatel-Lucent for supporting our research on femtocells and radio resource allocation.

We express our gratitude to all our families for their support throughout the years. We know that without their support we could not be what we are today.

We also express our gratitude to all our teachers/supervisors, mentors and friends who illuminated our way during our studies from primary school to PhD. In many ways, they lit up our dreams.

I, Xiaoli Chu, would like to sincerely thank my co-editors David López-Pérez, Fredrik Gunnarsson and Yang Yang, without whose expertise, hard work and commitment it would have been impossible to complete this book. I would like to thank all the authors for their dedication to this book. Without their devoted efforts, extensive knowledge and insights, the timely completion of this book would never have been possible. I am also grateful to all those who have directly or indirectly provided help with this book. Most important of all, I would like to thank my partner, Scott, who is the most loving, understanding and supportive companion anyone could ever ask for. I would like to express profound gratitude to my parents, Kaizhong and Jinju, for their constant love and support. I would also like to acknowledge my brother, Xiaoyan, who has always encouraged me in all my pursuits.

I, David López-Pérez, would like to dedicate this book to all those who love me, and those who have supported me through the years. They have made the adventure of living a wonderful story. I would like to specially thank my family, in particular my parents Juan Jose López and Antonia Pérez and my sister Charo for their loving care and support. Thank you! From the bottom of my heart, I am also very thankful to my good old friends from Zarzadilla de Totana as well as those good friends from Alicante and that I met abroad. Really, you are amazing, and in many ways, have changed my life and built my character and the person who I am now. Thank you all! I will probably never be able to pay off my debt with you guys, but do not forget that here you have a true friend.

I, Yang Yang, would like to thank my parents, wife and brother for their love and constant support. I would also like to thank my co-editors Xiaoli Chu, David López-Pérez and Fredrik Gunnarsson for their work.

I, Fredrik Gunnarsson, would like to thank my children Astrid, Kerstin and Ludvig, and my wife Sofia. Thanks for keeping my thoughts away from work and making every day cheerful in different ways. Thanks also to Ericsson Research and Linköping University for being such innovative and pleasant places to work at. I would also like to thank my co-editors for a very rewarding cooperation.

Forewords

Gordon Mansfield

Small Cell Forum

My name is Gordon Mansfield, and I currently serve as the elected chairman of the Small Cell Forum. The Forum is an industry body that promotes and drives the wide-scale adoption of small cell technologies to improve coverage, capacity and services delivered by mobile networks. I have many years of experience in the space, having previously served on the Femto Forum board from 2008–2010 and having led a tier one operators small cell effort since 2007. I consider it a great honor to be asked to write the foreword for this very informative book on small cells and heterogeneous networks. The authors are all highly respected researchers in academia and in industry, who have spent years working on the topics covered.

In recent years, small cells have become a very big topic when discussing mobile Internet and the tremendous data growth experienced over the past five years by operators around the globe. When we look at the recent history of data growth, some operators have experienced a 20,000 percent growth in data from 2007–2011. Combine that with the incredible forecast coming from all parts of the industry suggesting 10X and higher growth over the next four to five years, and it becomes clear that new ways to serve this data growth are necessary. We cannot continue to rely on new spectrum and advances in the air interface alone to sustain these types of data growth.

To be clear, small cells are not really new. They have been around for some time to provide capacity within limited amounts of spectrum. However, the solutions that were available up until a few years ago were limited in their functionality and, due to the massive growth in data, expensive to deploy on the scale required today. The high-power base stations in the macro network remain very important to maintain broad umbrella coverage, but, as demand in concentrated areas continue to rise, we will need not only traditional picocells and microcells that exist today but new innovative small cell solutions that allow for densification of networks in a cost-effective and aesthetically appealing way. These will range from femtocells, which have become a mature technology in the past couple of years, to new flavors of picocells and relays that are not yet productized in large numbers. Of course, to tie all of this together, we need software functionality to enable true heterogeneous networks and self-organizing network (SON) capabilities to manage the large increase in nodes within the mobile network of tomorrow.

The authors of this book effectively and insightfully break down the important factors of heterogeneous cellular networks and cover the various types of nodes that may be used. They lay a good technical foundation, and address several of the challenging areas that require focus for heterogeneous cellular networks to become a widespread success. They give the mobile operator several pieces of information that can be useful in helping them decide the best way to evolve their network, from the homogeneous network of today to the heterogeneous network of tomorrow.

As the topic of small cells and heterogeneous networks is at the forefront of many mobile operators' decision-making today, I am happy to see a book like this, which not only lays the foundation, but also provides many details on things that should be considered.

Gordon Mansfield
Chairman, Small Cell Forum

Alan Stidwell

Orange Labs, France Telecom

My name is Alan Stidwell, and I am Senior Technical Specialist within Wireless Technology Evolution at Orange Labs, France Telecom. My main role is to assess the performance benefits coming from future radio access technology evolutions (from two years +), to develop mid/long-term radio access network strategies, and to support work on radio access networks in 3GPP.

Until recently, most mobile network operators have been very focused on delivering customer performance needs mainly through macrocellular based deployments. This strategy is starting to be tested by the rapid increase in traffic volumes experienced during the last two to three years, which is expected to continue, and which are causing operators to deliver additional capacity at an accelerated rate compared to previous years. Additional capacity can be delivered by adding new technologies in additional spectrum on existing macrocell sites. This applies even for the early LTE deployments, primarily because this is the most cost-effective way to achieve coverage for a new technology whilst at the same time adding considerable capacity. Such deployments also bring big initial improvements in user experience because the early adopters will have the benefits of a largely unloaded network, thus amplifying the perceived performance compared to legacy 3G networks. However, it takes time for users to migrate their devices to LTE and in the meantime the traffic is still growing on the existing 3G network, which calls for a different strategy in order to avoid disappointing users on the legacy network. A primary way to achieve this additional capacity on legacy networks is through deployment of small cells, using legacy technologies that are in widespread use in existing devices. Furthermore, if the traffic predictions are correct, then it will not take too long before early LTE macrocell networks also become overloaded with a corresponding drop in user experience. Spectrum is also a finite and expensive resource, and therefore a continuing expansion of capacity in the macrocell domain alone may not be compatible with future capacity needs. In addition, macrocells are not necessarily the most cost- or energy-efficient way to deliver additional capacity in areas where there are predominantly non-uniformly distributed (or clustered) users. The need to maintain a reasonable user experience with this expected ongoing growth in traffic, whilst at the same time respecting the limits on spectrum, is therefore stimulating the whole industry to develop small cell solutions for both the new and legacy technologies.

These changes are happening very fast, and there is very little published on small cells and HetNets so far. This book is therefore very timely, and the way it comprehensively deals with all the relevant topics will appeal to many colleagues in all areas of the mobile industry who wish to get up to speed on this very important and fast-evolving technological change in network topology. Even experts and people with some prior knowledge in the field should find value in it, since it pulls together in one place a lot of valuable information and reference material.

Alan Stidwell
Senior Technical Specialist, Orange Labs, France Telecom

Preface

Driven by a new generation of wireless user equipments and the proliferation of bandwidth-intensive applications, mobile data traffic and network load are increasing in unexpected ways, and are straining current cellular networks to a breaking point. In this context, heterogeneous cellular networks, which are characterized by a large number of network nodes with different transmit power levels and radio frequency coverage areas, including macrocells, remote radio heads, microcells, picocells, femtocells and relay nodes, have attracted much momentum in the wireless industry and research community, and have also gained the attention of standardization bodies such as the *3rd Generation Partnership Project* (3GPP) LTE/LTE-Advanced and the *Institute of Electrical and Electronics Engineers* (IEEE) Mobile *Worldwide Interoperability for Microwave Access* (WiMAX).

The impending worldwide deployments of heterogeneous cellular networks bring about not only opportunities but also challenges. Major technical challenges include the co-existence of various neighboring and/or overlapping cells, intercell interference and mobility management, backhaul provisioning, and self-organization that is crucial for efficient roll-outs of user-deployed low-power nodes. These challenges need to be addressed urgently to make the best out of heterogeneous cellular networks. This asks for a thorough revisit of contemporary wireless network technologies, such as network architecture and protocol designs, spectrum allocation strategies, call management mechanisms, etc. There is also an urgent need in the wireless industry, academia and even end-users to better understand the technical details and performance gains that heterogeneous cellular networks would make possible.

Heterogeneous Cellular Networks – Theory, Simulation and Deployment provides a complete and thorough exposition of heterogeneous cellular networks, with a delicate balance between theory and practice. It contains cutting-edge tutorials on the technical and theoretical foundations upon which heterogeneous cellular networks are built, while also providing high-level overviews of standardization activities that are informative to readers of all backgrounds. The book is written by researchers currently leading the investigations on heterogeneous cellular networks, covering a wide spectrum of topics such as simulation and evaluation models, advanced radio resource management, self-organization, mobility and handover issues, cooperative communications, network *multiple-input multiple-output* (MIMO) techniques, network coding, cognitive radio, energy efficiency, etc.

The book is organized into 14 chapters. Chapter 1 provides the background information necessary for understanding the theoretical and technical foundations of heterogeneous cellular networks, as well as discussions of technical challenges in deploying heterogeneous cellular networks, and serves as a summary of the rest of the book.

Chapter 2 describes the process of building a radio propagation model, and provides a detailed review of different radio propagation modeling techniques, including empirical models, deterministic models, semi-deterministic models and hybrid models. The tradeoff between radio channel modeling accuracy and complexity is discussed. Moreover, this chapter identifies the need and requirements for new radio propagation modeling tools that are able to operate in heterogeneous cellular network scenarios, which may involve both indoor and outdoor environments.

In Chapter 3, we look into important issues in system-level simulations for heterogeneous cellular networks. We review different kinds of system-level simulation, and compare dynamic system-level simulations with static ones. Intricate modeling issues, such as wrap-around, shadowing, multi-path fading, antenna patterns, diversity combining and signal quality modeling, are discussed. Traffic and mobility modeling is also described. Simulation setups for heterogeneous cellular networks from the perspective of 3GPP are illustrated, taking into account features of heterogeneous cellular network nodes and channel modeling, and covering macro-pico, macro-femto and macro-relay scenarios.

Access control enables flexible installation and operation of low-power nodes for different deployment scenarios. Chapter 4 provides an in-depth description of access control in heterogeneous cellular networks from the perspectives of the core network, radio access network and user equipments. The benefits and tradeoff of different access control mechanisms are analyzed through numerical simulations using various performance metrics, such as the percentage of offloaded user equipments, as well as cell-average and cell-edge data throughputs. Moreover, the access control enhancements standardized during LTE Releases 8, 9 and 10 are discussed.

In Chapter 5, we first provide an overview of stochastic models where the locations of network nodes can be modeled following a *Poisson point process* (PPP) or a hardcore point process. Then, motivated by the accuracy and tractability of PPP network models, we present a stochastic-geometry-based two-tier network framework, and study the effects of spectrum allocation and different access schemes on the link reliability of each tier of a heterogeneous cellular network and on the network area spectral efficiency. We investigate both disjoint and joint subchannel allocation, where the two tiers are assigned disjoint sets of subchannels or share the whole spectrum, respectively. Moreover, we quantify the enhancement in a user equipment's link reliability when it can hand over from the macrocell to the closest femtocell, and analyze the resulting benefits to the network throughput. The proposed framework also allows us to generalize the results to multi-tier heterogeneous cellular networks.

Self-organizing networks (SONs) provide means to facilitate network and service management, while supporting radio network robustness and optimized coverage and capacity. SON functionalities are commonly divided into self-configuration, self-optimization and self-healing. These are especially important to heterogeneous cellular networks due to the increasing number of nodes that need to be managed. Self-configuration enables smooth introduction of low-power nodes into an existing macrocell deployment. Continuous self-optimization and self-healing ensure that the low-power nodes are able to provide the intended capacity without requiring significant human management and monitoring efforts. In Chapter 6, we provide relevant insights about SON for heterogeneous cellular networks, with a main focus on 3GPP LTE, and discuss key performance monitoring aspects. Different deployment options are described in relation to the 3GPP management architecture. The associated 3GPP SON feature requirements and signaling support are also described together with 3GPP specification references.

Chapter 7 reviews intercell interference problems and interference management techniques for heterogeneous cellular networks. Some simulation results are used to demonstrate the downlink/uplink coverage imbalance issues in heterogeneous deployments. Range expansion of small cells, cell-selection methods and intercell interference coordination techniques are discussed, including mechanisms recently proposed in the 3GPP LTE. The dedicated focus is on frequency-domain, power-based and time-domain interference management. Moreover, performance assessments and comparisons are provided through system-level simulations, indicating some design guidelines for dynamic interference management.

Interference management becomes even more challenging if there are restrictions in access mechanisms and internode signaling, for example in femtocells. Chapter 8 looks into uncoordinated femtocell deployments, and self-organizing femtocell networks are discussed, from the perspectives of the Small Cell Forum. With less strict backhaul requirements as compared with other heterogeneous cellular network nodes, femtocell backhaul time synchronization becomes especially critical and is paid special attention. Moreover, power-based, antenna-based and frequency-domain techniques for interference mitigation in uncoordinated femtocells are reviewed in detail.

Chapter 9 provides an overview of mobility robustness, handover management and *mobility load balancing* (MLB) in LTE/LTE-Advanced systems, and develops *mobility robustness optimization* (MRO) algorithms under both *Radio Resource Control* (RRC) connected and idle modes. Under the RRC connected mode, we introduce the hard handover procedure in LTE and the MRO use case, and propose an MRO scheme to assign different handover hysteresis values for user equipment traveling at different velocities. Under the RRC idle mode, we introduce cell selection/reselection procedures, and propose a negotiation mechanism to solve the *cell-specific reference symbol* (CRS) parameter mismatch between neighboring cells and the mismatch between CRS and handover parameters. Specifically for heterogeneous cellular networks, we identify the technical challenges in mobility management, evaluate the mobility performance under 3GPP Release 10 enhanced *intercell interference coordination* (ICIC) features, and propose a mobility-based ICIC scheme.

In Chapter 10, we first review and analyze the performance of *amplify-and-forward* (AF) relay, *demodulate-and-forward* (DMF) relay and *decode-and-forward* (DCF) relay, and propose a link adaptation scheme for the DMF relay. Then, the 3GPP LTE-Advanced relay system architectures in heterogeneous cellular networks are introduced. In order to fully explore the capacity of cooperative relaying, the emerging *estimate-and-forward* (EF) relay and joint network-channel coding are presented together with a new link adaptation scheme.

Network MIMO, also known as *multi-cell processing* (MCP) or *coordinated multi-point* (CoMP), combines MIMO and cooperation techniques to provide new means to manage interference more intelligently, while increasing the spectral efficiency. In Chapter 11, we first analyze the problems with existing *single-cell processing* (SCP), review different types of MCP and introduce application scenarios of MCP for heterogeneous cellular networks. Afterwards, we demonstrate how MCP together with MIMO transmit beamforming techniques and power control can mitigate *intercarrier interference* (ICI) to achieve a better tradeoff between system throughput and user fairness in heterogeneous cellular networks. Practical implementation issues such as backhaul requirements, base station clustering and *channel state information* (CSI) acquisition are discussed, and future research directions are outlined.

Network coding is a technique that smartly controls the interference from different source nodes to improve the overall network performance. Overlaying network coding techniques upon existing cellular networks provides a simple and economical way to tremendously enhance the network performance without too many hardware replacements. In Chapter 12, we first explain the fundamentals of network coding and its state-of-the-art developments. Then, we give coding-gain upper bounds with and without practical geometry considerations, as well as illustrative examples to show the performance gains and effectiveness of applying simple network coding solution to heterogeneous cellular networks. Following that, we analyze the efficiency and reliability performance of network coding using a two-way relay-aided X network. Finally, we introduce a low-complexity and distributed coding solution construction method that is ideal for heterogeneous cellular networks.

Dynamic spectrum access (DSA) or *cognitive radio* (CR) has the potential to provide significant benefits and simple solutions to overcome many of the challenges faced by heterogeneous cellular networks. Chapter 13 presents a detailed overview of the functionalities and techniques associated with DSA/CR technologies, including methods to acquire spectrum awareness, select the appropriate frequency of operation and share spectrum opportunities in a non-interfering manner, and explores the potential application of DSA/CR concepts and techniques in heterogeneous cellular networks. Several practical scenarios are explored. Various implementation alternatives along with the corresponding pros and cons are discussed. Related cross-layer issues and design tradeoffs are investigated. Finally, recent standardization efforts aimed at developing new systems based on DSA/CR principles and including DSA/CR capabilities in existing standards are reviewed.

Making cellular networks green would have a tangible positive impact on reducing the carbon footprint of *information communication technology* (ICT), and achieve a long-term profitability for mobile service operators. Chapter 14 focuses on heterogeneous cellular network architectures and their potential to deliver improved performance at reduced energy and cost consumption levels. In particular, the chapter examines the fundamental tradeoffs between performance and energy consumption, recently proposed heterogeneous cellular network architectures, dynamic base station designs, advanced transmission schemes, hardware improvements, and cross-layer integration solutions. We also present a comprehensive survey of latest green techniques for homogeneous and heterogeneous cellular networks with discussions of their merits and demerits. Moreover, recent research projects for green cellular networks are reviewed, and a taxonomy of green metrics is summarized.

Contributors

Ralf Bendlin
Texas Instruments Inc., USA

Vikram Chandrasekhar
Accelera Mobile Broadband Inc., USA

Min Chen
Huazhong University of Science and Technology, People's Republic of China

Huaxia Chen
Shanghai Institute of Microsystem and Information Technology, Chinese Academy of
Sciences, People's Republic of China

Xiaoli Chu
University of Sheffield, UK

Holger Claussen
Bell Labs, Alcatel-Lucent, Ireland

Anthony E. Ekpenyong
Texas Instruments Inc., USA

Mats Folke
Ericsson Research, Sweden

Fredrik Gunnarsson
Ericsson Research, Sweden

Weisi Guo
University of Sheffield, UK

Ismail Güvenç
Florida International University, USA

Honglin Hu
Shanghai Institute of Microsystem and Information Technology, Chinese Academy of Sciences, People's Republic of China

Yongming Huang
Southeast University, People's Republic of China

Shengyao Jin
Shanghai Research Center for Wireless Communications, People's Republic of China

Marios Kountouris
Supélec (École Supérieure d'Électricité), France

Zhihua Lai
Ranplan Wireless Network Design Ltd., Uk

Cong Ling
Imperial College London, UK

Miguel López-Benítez
University of Surrey, UK

David López-Pérez
Bell Labs, Alcatel-Lucent, Ireland

Meiling Luo
Ranplan Wireless Network Design Ltd., UK

Haishi Ning
Imperial College London, UK

Tony Q. S. Quek
Singapore University of Technology and Design and Institute for Infocomm Research, Singapore

Athanasios V. Vasilakos
University of Western Macedonia, Greece

Guillaume Villemaud
INSA Lyon, France

Jiang Wang
Shanghai Institute of Microsystem and Information Technology, Chinese Academy of Sciences, People's Republic of China

Kai-Kit Wong
University College London, UK

Jing Xu
Shanghai Institute of Microsystem and Information Technology, Chinese Academy of Sciences, People's Republic of China

Yang Yang
Shanghai Institute of Microsystem and Information Technology, Chinese Academy of Sciences, People's Republic of China

Jie Zhang
University of Sheffield, UK

Gan Zheng
University of Luxembourg, Luxembourg

Ting Zhou
Shanghai Research Center for Wireless Communications, People's Republic of China

Acronyms

3GPP	3rd Generation Partnership Project
4G	fourth generation
ABS	almost blank subframe
AF	amplify-and-forward
AM	acknowledged mode
AMC	adaptive modulation and coding
ANR	Automatic Neighbor Relation
AoA	angle of arrival
AoD	angle of departure
AP	access point
APP	a posteriori probability
ARCF	automatic radio configuration data handling function
ARQ	Automatic Repeat reQuest
AWGN	additive white Gaussian noise
BCJR	Bahl–Cocke–Jelinek–Raviv
BER	bit error rate
BLER	block error rate
BPSK	binary phase-shift keying
BS	base station
BSP	binary space partitioning
CA	closed access
CAPEX	capital expenditure
CAS	cluster angular spread
CCC	common control channel
CCO	coverage and capacity optimization
CDF	cumulative distribution function
CDMA	code division multiple-access
CDS	cluster delay spread
CFL	Courant–Friedrichs–Lewy
CIO	cell individual offset
CIR	channel impulse response
CM	configuration management
CN	core network

CoMP	coordinated multi-point
CPC	cognitive pilot channel
CPE	customer premises equipment
CQI	channel quality indicator
CR	cognitive radio
CRC	cyclic redundancy check
CRE	cell range expansion
C-RNTI	cell radio network temporary identifier
CRS	cell-specific reference symbol
CSCC	common spectrum coordination channel
CSG	closed subscriber group
CSG ID	CSG identity
CSI	channel state information
CSIR	receiver-side channel state information
CSO	cell selection offset
CTS	Clear-to-Send
DAB	Digital Audio Broadcasting
DAS	distributed antenna system
DCF	decode-and-forward
DCH	dedicated channel
DCM	directional channel model
DECT	Digital Enhanced Cordless Telecommunications
DeNB	donor eNB
DFS	Dynamic Frequency Selection
DHCP	Dynamic Host Configuration Protocol
DL	downlink
DM	domain manager
DMC	dense multi-path componens
DMF	demodulate-and-forward
DNS	Domain Name System
DMT	diversity and multiplexing tradeoff
DoA	direction-of-arrival
DoD	direction-of-departure
DPC	dirty-paper coding
D-QDCR	Distributed QoS-based Dynamic Channel Reservation
DS	downstream
DSA	dynamic spectrum access
DSL	digital subscriber line
DSP	digital signal processor
DVB	Digital Video Broadcasting
E2E	end-to-end
ECGI	Evolved Cell Global Identifier
EESM	exponential effective SINR mapping
EF	estimate-and-forward

EGC	equal gain combining
EM	element manager
eNB	evolved NodeB
EoA	elevation of arrival
EoD	elevation of departure
EPC	Evolved Packet Core
EPS	Evolved Packet System
E-RAB	E-UTRAN radio access bearer
ESF	even subframe
ETSI	European Telecommunications Standards Institute
E-UTRA	Evolved UTRA
E-UTRAN	Evolved UTRAN
FAP	femtocell access point
FCC	Federal Communications Commission
FCFS	first come first served
FCI	failure cell ID
FD	frequency domain
FDD	frequency division duplexing
FDTD	finite-difference time domain
FER	frame error rate
FGW	femto gateway
FIFO	first in first out
FM	fault management
FPGA	field-programmable gate array
FTP	File Transfer Protocol
FUE	femtocell user equipment
GERAN	GSM EDGE Radio Access Network
GNSS	global navigation satellite system
GPS	Global Positioning System
GSCM	geometry-based stochastic channel models
GSM	Global System for Mobile Communications
GTD	geometry theory of diffraction
GTP	GPRS Tunnel Protocol
HA	hybrid access
HARQ	hybrid automatic repeat request
HCN	heterogeneous cellular network
HDTV	high-definition television
HeNB	Home evolved NodeB
HGW	home gateway
HII	high-interference indicator
HMS	HNB/HeNB management system
HNB	Home NodeB
HO	handover
HOF	handover failure

HOM	handover hysteresis margin
HRD	horizontal reflection diffraction
HSDPA	High Speed Downlink Packet Access
HSPA	High Speed Packet Access
HUE	home user equipment
IC	interference cancellation
ICI	intercarrier interference
ICIC	intercell interference coordination
ICT	information communication technology
ID	identifier
IEEE	Institute of Electrical and Electronics Engineers
IETF	Internet Engineering Task Force
IIR	infinite impulse response
IMEI-TAC	International Mobile Equipment Identity Type Allocation Code
IMT	International Mobile Telecommunications
InH	indoor hotspot
IP	Internet Protocol
IRC	interference rejection combining
ISD	intersite distance
ITU	International Telecommunication Union
JFI	Jain's fairness index
KPI	key performance indicator
L1	layer one
L2	layer two
L3	layer three
LLR	log-likelihood ratio
LMDS	Local Multipoint Distribution Service
LMMSE	linear minimum mean square error
LOS	line of sight
LPN	low-power node
LR	likelihood ratio
LSP	large-scale parameter
LTE	Long Term Evolution
LTE/SAE	Long Term Evolution/System Architecture Evolution
LUT	lookup table
MAC	medium access control
MBS	macrocell base station
MCC	mobile country code
MCM	multi-carrier modulation
MCP	multi-cell processing
MCS	modulation and coding scheme
MDT	minimization of drive tests
MIB	Master Information Block
MIESM	mutual information effective SINR mapping

MIMO	multiple-input multiple-output
MISO	multiple input single output
ML	maximum likelihood
MLB	mobility load balancing
MME	Mobility Management Entity
MMSE	minimum mean square error
MNC	mobile network code
MPC	multi-path components
MR	measurement report
MRC	maximal ratio combining
MR-FDPF	multi-resolution frequency-domain ParFlow
MRO	mobility robustness optimization
MUE	macrocell user equipment
NAS	non-access stratum
NAV	Network Allocation Vector
NB	NodeB
NE	network element
NGMN	Next Generation Mobile Networks
NLOS	non-line-of-sight
NM	network management
NMS	network management system
NRT	neighbor relation table
NTP	Network Time Protocol
OA	open access
OAM	operation, administration, and maintenance
OCXO	oven controlled oscillator
OFDM	orthogonal frequency division multiplexing
OFDMA	orthogonal frequency division multiple access
OI	overload indicator
OPEX	operational expenditure
OSF	odd subframe
OSI	Open Systems Interconnection
P2MP	point to multi-point
P2P	point to point
PBCH	primary broadcast channel
PCI	physical layer cell identity
PDCP	Packet Data Convergence Protocol
PDF	probability density function
PDSCH	physical downlink shared channel
PeNB	pico evolved NodeB
PGW	Packet Data Network Gateway
PH	power headroom
PHY	physical
PLMN	public land mobile network

PML	perfectly matched layer
PPP	Poisson point process
PRB	physical resource block
PSC	primary scrambling code
PSS	primary synchronization signal
PTP	Precision Time Protocol
PUE	picocell user equipment
QAM	quadrature amplitude modulation
QCI	QoS class identifier
QoS	quality of service
QPSK	quadrature phase-shift keying
RAB	Radio Access Bearer
RACH	random access channel
RAN	radio access network
RAT	radio access technology
RAXN	relay-aided X network
RB	resource block
RCI	re-establish cell ID
RE	range expansion
REB	range expansion bias
RF	radio frequency
RFC	Request for Comments
RIM	RAN information management
RLC	Radio Link Control
RLF	radio link failure
RMSE	root mean square error
RN	relay node
RNC	Radio Network Controller
RNL	radio network layer
RNS	Radio Network Subsystem
RNTP	relative narrowband transmit power
RPSF	reduced-power subframe
RRC	Radio Resource Control
RRH	remote radio head
RRM	radio resource management
RS	reference signal symbol
RSC	Recursive Systematic Convolutional
RS-CS	resource-specific cell selection
RSQ	reference signal quality
RSRP	reference signal received power
RSRQ	reference signal received quality
RSS	reference signal strength
RSSI	reference signal strength indicator
RTS	Request-to-Send

S1AP	S1 Application Protocol
S1-MME	S1 for the control plane
S1-U	S1 for the user plane
SAE	System Architecture Evolution
SAEGW	System Architecture Evolution Gateway
SAS	spectrum allocation server
SBS	super base station
SCC	Standards Coordinating Committee
SCM	spatial channel model
SCP	single-cell processing
SCTP	Stream Control Transmission Protocol
SDMA	space-division multiple access
SDR	software defined radio
SE	spectral efficiency
SEF	soft symbol estimation and forward
SFBC	space frequency block coding
SGW	Serving Gateway
SI	system information
SIB	system information block
SIB1	System Information Block Type 1
SIB4	System Information Block Type 4
SIC	successive interference cancellation
SIM	subscriber identity module
SIMO	single input multiple output
SLA	service level agreement
SN	serial number
SNTP	Simple Network Time Protocol
SINR	signal to interference plus noise ratio
SIR	signal to interference ratio
SISO	single-input single-output
SLAC	stochastic local area channel
SLNR	signal to leakage interference and noise ratio
SNR	signal to noise ratio
SOCP	second-order cone programming
SON	self-organizing network
SoT	saving of transmissions
SPS	spectrum policy server
SSMA	spread spectrum multiple access
SSS	secondary synchronization signal
STBC	space-time block coding
TA	timing advance
TAC	tracking area code
TAI	tracking area identity
TCE	Trace Collection Entity

TCoSH	Triggering Condition of Self-Healing
TCP	Transmission Control Protocol
TCXO	temperature controlled oscillator
TDD	time division duplexing
TDM	time division multiplexing
TDMA	time division multiple access
TEID	tunnel endpoint identifier
TNL	transport network layer
TPC	Transmit Power Control
TR	transition region
TSG	Technical Specification Group
TTT	time-to-trigger
TU	Typical Urban
TV	television
TWXN	two-way exchange network
UE	user equipment
UL	uplink
UMa	urban macrocell
UMTS	Universal Mobile Telecommunication System
US	upstream
UTD	uniform theory of diffraction
UTRA	UMTS Terrestrial Radio Access
UTRAN	UMTS Terrestrial Radio Access Network
UWB	ultra-wide band
VD	vertical diffraction
VeNB	virtual eNB
VoIP	voice over IP
VR	visibility region
WCDMA	wideband code division multiple access
WG	working group
WiFi	Wireless Fidelity
WiMAX	Worldwide Interoperability for Microwave Access
WNC	wireless network coding
WLAN	wireless local area network
WRAN	wireless regional area network
X2AP	X2 Application Protocol

1 Introduction

Xiaoli Chu, David López-Pérez, Fredrik Gunnarsson and Yang Yang

Mobile broadband demands are increasing rapidly, driven by the popularity of various connected mobile devices with data services, such as smartphones, tablets, vehicles, machines and sensors. The notion of connected devices actually expands to encompass basically everything that can take benefits from a wireless connection. A true mobile broadband experience of high quality everywhere can be expected by consumers in the near future.

Mobile applications have become an indispensable part of people's everyday life, with requirements on seamless access to social media, video contents and cloud-based contents anytime, anywhere. To provide services that meet these requirements is of top priority for operators with ambitions to be a key wireless communications provider in the networked society. These requirements can only be met by mobile networks with sufficient capacity and coverage. Mobile broadband today is mainly provided via networks based on *UMTS Terrestrial Radio Access* (UTRA) or *Evolved UTRA* (E-UTRA), and solutions differ in the details. Mobile networks need to evolve through improving the existing mobile broadband networks and adding more cells in an optimal way to migrate to a *heterogeneous cellular network* (HCN). The migration path could be different for different operators. A thorough understanding of the various components involved is vital for a cost-efficient, spectrum-efficient and energy-efficient network evolution.

This chapter provides an introduction to the whole book. First, the need for more capacity and mobile broadband forecasts are discussed in Section 1.1. Then, Section 1.2 reviews different network evolution solutions, Section 1.3 describes the HCN nodes, and Section 1.4 provides a brief discussion about related standardization work. Finally, Section 1.5 addresses technical challenges associated with HCNs. These challenges will then be addressed in more detail by reviewing the remaining chapters of the book.

1.1 Mobile data explosion and capacity needs

Mobile data demands increase at exponential rates. Market analyses [1, 2] agree that this trend will continue as mobile data is becoming a larger and larger part of people's daily lives. Based on measurements over several years using a large base of live networks that cover almost all regions of the world, the statistics provide strong evidence of a mobile data explosion.

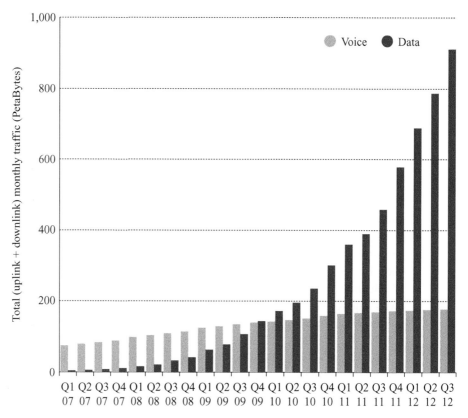

Figure 1.1 Global total traffic in mobile networks, 2007–2012.
Source: Ericsson (November 2012)

One key milestone was passed in the fourth quarter of 2009, when the volume of mobile data surpassed that of voice, as illustrated in Fig. 1.1. Not much later, in the first quarter of 2011, the volume of mobile data doubled that of voice.

The forecasts in Fig. 1.2 indicate that mobile data will continue to grow at exponential rates. Portable personal computers still dominate the traffic in mobile networks nowadays, but smartphones and tablets are increasing in popularity. The forecast for latter years even splits between mobile data via smartphones and via personal computers/tablets. It has been predicted [1] that mobile data will grow 12-fold between 2012 and 2018, corresponding to a compound annual growth rate of about 50%.

Moreover, users are increasingly aware of the connection speed, data rate, coverage and availability of their mobile broadband services. This has already led to disputes about whether the available capacity and coverage are sufficient to meet the increasing demands.

In order to meet or even exceed subscriber expectations, operators need to analyze the capacity and coverage challenges in their networks. Fig. 1.3 illustrates the capacity and coverage expansion needs that operators foresee to different extents. First, there is an overall need for enhanced capacity in the network, essentially across the entire service

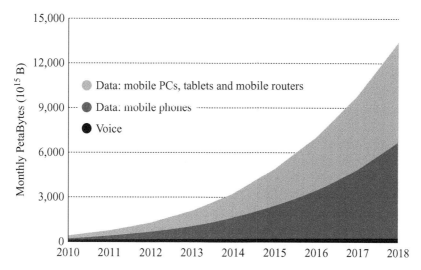

Figure 1.2 Global mobile traffic, voice and data, 2010–2018.
Source: Ericsson (November 2012)

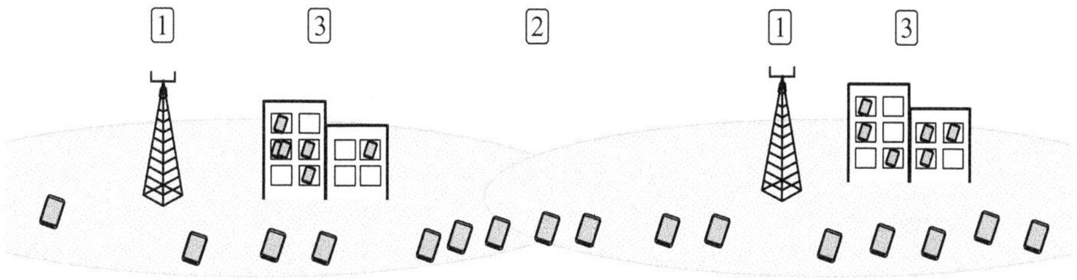

Figure 1.3 Different capacity and coverage challenges in mobile networks, characterized by (1) a need for an overall capacity increase, (2) a need for more favorable radio conditions far from current site installations and (3) adequate coverage and capacity indoors, where most of the mobile data is consumed.

area. Second, some of the capacity need comes far away from currently deployed site installations, resulting in poor user experiences over unfavorable radio links. Third, part of the capacity need comes from indoor environments with both coverage and capacity issues. For example, in most urban areas, most of the traffic is consumed indoors. There are different solution approaches to meet all these aspects, as will be discussed in the following section.

1.2   Capacity and coverage solutions

The challenges in terms of capacity and coverage illustrated by Fig. 1.3 can be addressed in different ways. In this section, we discuss some popular solutions.

1.2.1   Improving existing macrocell networks

To improve the existing macrocell network [3–5] is a cost-efficient way to address the overall capacity needs indicated by (1) in Fig. 1.3. Adding more spectrum to the existing sites together with interfrequency load sharing directly gives a capacity boost with limited efforts. Such interfrequency load sharing is facilitated by mobile terminals capable of supporting multiple carriers simultaneously, which is known as Multi-Carrier in UTRA and Carrier Aggregation in E-UTRA. Furthermore, cell sites can be upgraded with additional sectorization and improved radio link utilization, such as modulation and coding schemes corresponding to higher data rates. In fact, the capacity needs in one of the highest traffic density regions in the world have been met by improving the macrocell sites with additional spectrum [6].

Further improvement can be obtained via coordination between existing sites. Some signaling coordination is already available in the first releases of the radio technologies such as internode signaling in *Evolved UTRAN* (E-UTRAN) and *Radio Network Controller* (RNC) signaling in *UMTS Terrestrial Radio Access Network* (UTRAN). Additional benefits can be achieved through tighter coordination of data packet transmission and reception.

1.2.2   Network base station densification

The challenges indicated by (2) in Fig. 1.3 are difficult to address with improvements only to the existing cell sites, since the distances between *user equipments* (UEs) and cell sites cause the main problem. Accordingly, the alternative solution is to densify the network by installing new outdoor cell sites [3–5]. Large capacity gains in cellular networks are obtained through the shrinking of cell sizes and universal frequency reuse, and thereby the spatial frequency reuse is greatly improved by increasing the network node density.

The nature and properties of new site proposals depend on the radio propagation environment and installation limitations. The number of needed additional sites depends on the transmission power [7, 8], and is related to the deployment and maintenance costs. It is preferable to use node transmission power as high as possible to keep the number of needed sites small. Another important factor to consider is the spatial distribution of traffic. By identifying traffic hotspots and considering this information in the site selection process, the number of needed sites can be reduced [9].

When installing new cell sites, there are also advantages from intersite coordination. Some important coordination can be obtained via internode signaling in E-UTRAN and the RNC in UTRAN, and further gains can be achieved by introducing tighter coordination between sites.

1.2.3   Indoor capacity and coverage

Indoor areas are typically characterized by fairly good isolation from the macrocell network. *Low-power nodes* (LPNs) constitute a very attractive network expansion alternative to meet the indoor coverage and capacity needs [4, 5, 10], as indicated by (3) in Fig. 1.3, where the distances between the indoor LPNs and UEs are short and low transmission powers would be sufficient.

Indoor site installations should take benefit from considerations of the traffic density to keep the number of needed sites small [9]. Deployment considerations could be different if the LPNs are not intended to be used by all UEs but only a *closed subscriber group* (CSG) of UEs. There are also advantages from site coordination either between indoor nodes or between indoor and outdoor nodes. Such coordination can be obtained via internode signaling in E-UTRAN and the RNC in UTRAN, as well as via tighter coordination between nodes.

1.2.4   Heterogeneous cellular networks

Most likely, the optimal solution would be a combination of the approaches discussed above, and would also depend on specific network, propagation environment and traffic demand aspects. With the exponential traffic increase, uneven traffic distribution and limited spectrum availability, it will not be possible to address network evolution aspects only with *macrocell base stations* (MBSs) [11, 12]. More specifically, MBSs are inadequate to meeting the mobile data demands in certain scenarios, for example the following.

- Large outdoor areas with a high traffic density, for example urban commercial areas and town squares, where the network is already dense and the intercell interference could be high. Moreover, there can also be requirements of limited size to make the site equipment almost invisible.
- Large indoor areas with a high traffic density, such as enterprises, shopping malls, airports, subway stations and hotel lobbies, where it could be difficult to reach from an outdoor macrocell network.
- Small indoor areas and coverage holes, such as private apartments and houses, restaurants, small stores and offices, where the macrocell coverage is insufficient with respect to the mobile data demands. There may also be an interest to provide wireless access only to a CSG.

Hence, LPNs play an important role in the solution in combination with the existing macrocell network to meet future coverage and capacity needs. The most popular LPNs will be described in the next section.

1.3   Heterogeneous cellular network nodes

A *base station* or *node* is deployed at a geographical *site*, serving one or more *cell(s)*. When the discussion is specific for E-UTRAN a base station is referred to as an *evolved NodeB* (eNB), and for UTRAN a base station is denoted *NodeB* (NB).

HCN nodes are distinguished by their transmit power, coverage area, physical size, backhaul and radio propagation characteristics. MBSs will be operator-deployed with fairly large coverage areas, having transmit power levels per served macrocell that typically vary between 5 W and 40 W, plus large antenna gains. The transmit power of an LPN ranges from 250 mW to approximately 5 W if it is deployed outdoors and falls below 100 mW for indoor deployment [4, 5, 10, 12]. As will be discussed in the following subsections, LPNs include *remote radio heads* (RRHs), micro and pico base stations, femtocell access points and *relay nodes* (RNs).

1.3.1   Remote radio heads

RRHs are radio equipment typically connected to the MBS via a fiber optic cable. This enables a tight coordination between RRHs and the MBS, analogous to the coordination between cells served by the same *base station* (BS). RRHs can also be seen as a *distributed antenna system* (DAS) [13], even though the term more commonly refers to analog signal distribution over distributed antennas. Modern RRHs have the analog to digital converter in the RRH, which means that the interface to the RRH is digital, and that base band processing can be centralized. The transmission power levels depend on the deployment scenario, and can thus be in the range from macro- to picocells.

1.3.2   Micro base stations

Micro base stations or nodes, serving microcells, are regular BSs that provide standardized interfaces over the backhaul, but with lower transmit power than traditional macrocells. The transmit power of a microcell is typically of the order of 5–10 W for outdoor deployment. Micro base stations are deployed outdoors in an operator-planned fashion, particularly popular for providing outdoor hotspot coverage. They can be equipped with omnidirectional antennas, but may have antenna directivity as well, for example radiating outwards from a house wall deployment. Micro base stations can be coordinated with the macro network via the X2 interface in E-UTRAN and the RNCs in UTRAN, where tighter intersite coordination is possible.

1.3.3   Pico base stations

Pico base stations or nodes, serving picocells, are similar to micro nodes, but with even lower transmit power per picocell served, and possibly also of smaller size. This ranges from 250 mW to approximately 2 W for outdoor deployment, while it is typically 100 mW or less for indoor deployment. Pico base stations are deployed indoors or outdoors, often in an operator-planned fashion, particularly popular for providing hotspot coverage. They can be equipped with omnidirectional antennas or with some antenna directivity. Pico nodes can be coordinated with the macro network via the X2 interface in E-UTRAN and the RNCs in UTRAN, and tighter intersite coordination is also possible.

1.3.4   Femtocell access points

Femtocell access points (FAPs), base stations or nodes, serving femtocells, are small low-power BSs that are generally consumer deployed with a network backhaul facilitated by the consumer's own wired broadband connection [11]. They are typically provided by wireless operators as a managed service. Originally envisioned as a means to improve voice coverage in indoor environments, femtocells can also be viewed as one way to offload data traffic from macrocells. FAPs are typically equipped with omnidirectional antennas, and their transmit power is 100 mW or less. There are dedicated procedures for FAPs; in LTE, an FAP is denoted a *Home evolved NodeB* (HeNB), and in *Universal Mobile Telecommunication System* (UMTS) it is named a *Home NodeB* (HNB). The HeNBs or HNBs can be coordinated and controlled by a *Home NodeB/evolved NodeB Gateway* (HGW), which also can be used to interface the macro and core networks.

1.3.5   Relay nodes

In scenarios where a wired backhaul is not available, RNs can be deployed with the air interface used for both backhaul connection and access to UEs [14]. RNs can be deployed either indoors or outdoors. Their transmit power ranges from 250 mW to approximately 2 W for outdoor deployments, and is typically 100 mW or less for indoor deployments [12]. An RN can be considered as a full-fledged BS but without a wired backhaul. An RN appears as a UE to its *donor eNB* (DeNB) and as a regular BS to the UEs that it serves. The link between the RN and the DeNB is denoted the backhaul link, the RN to UE link is denoted the access link and the MBS to UE link (without relay participation) is denoted the direct link. RNs are typically equipped with directional antennas in the backhaul link (pointing to the DeNB) and may have omnidirectional or directional antennas in the access link. If the backhaul link uses the same frequency band as the access link, then the relay is an in-band relay; otherwise, the relay is an out-of-band relay.

1.4   3GPP LTE-Advanced heterogeneous cellular networks

The first release of *Long Term Evolution* (LTE), i.e., *3rd Generation Partnership Project* (3GPP) Release 8 [15], was published in December 2009. Subsequently, LTE evolved into LTE-Advanced (LTE-A), i.e., 3GPP Release 10 [16] and onwards, which are developed to meet the *International Mobile Telecommunications* (IMT)-Advanced requirements of the *International Telecommunication Union* (ITU). LTE-Advanced introduces carrier aggregation of up to five 20 MHz component carriers to provide a peak data rate of more than 1 Gb/s. Moreover, it features additional support for using multiple antennas in both the receiver and transmitter, enabling spatial multiplexing of data streams.

The physical layer is based on *Orthogonal Frequency Division Multiple Access* (OFDMA), which enables orthogonal intra-cell waveforms. However, interference from other cells can limit the performance. Therefore, LTE-Advanced features enhanced interference management and suppression techniques, in particular to address HCN interference scenarios. The *Self-Organizing Network* (SON) has also been part of LTE from the first release to facilitate deployment, installation, operation and maintenance – something that is important when the number of BSs increases. Throughout the releases, there has been increasing support for home eNBs, for example via dedicated mobility procedures and a specific architecture, as well as mechanisms to enable restricted access to CSGs. Later releases also have featured different interference management components, from intersite signaling mechanisms to interference suppression recievers, as well as mobility robustness support.

The book aims at a general treatment of HCN, but with an ambition to refer to related concepts in 3GPP. Concepts in 3GPP Release 11 and earlier releases are considered.

1.5   Heterogeneous cellular network challenges

HCNs have the potential to provide benefits over homogeneous network deployments [17]. These include the ability to place nodes close to the end users, enabling a dense node deployment, which in turn can provide unprecedented mobile network capacity to homes, enterprises and urban hotspots, where most mobile traffic is generated.

The deployment of LPNs also poses technical challenges, as will be discussed in this section. Most of the challenges are then addressed in the rest of this book, with the exception of backhaul, which is beyond the scope of the book. Another challenge not addressed in the book is interworking with *Wireless Fidelity* (WiFi), which is an important component in a mobile broadband offer.

1.5.1   Optimal network evolution path

As discussed before, the best network evolution path is individual to each network with its specific characteristics and traffic needs. It also depends on the cost structure for the investment in terms of *capital expenditure* (CAPEX), and the deployment, operation and maintenance costs in terms of *operational expenditure* (OPEX). The time needed for different kinds of site acquisition and planning activity is also a critical factor. Since the number of needed LPNs increases significantly with decreasing transmission power [9, 11], the total cost of ownership needs to match this. Clearly, this is not only about CAPEX, since a more expensive smart node would be more prepared to meet the other challenges discussed in this section, and can lead to lower OPEX. Overall, it is about evolving the network in a cost-efficient way.

One way to reach a better understanding of different deployment options is via simulations of HCN scenarios. Chapter 2 describes radio propagation models from empirical models with low computational complexity and reduced accuracy to fully deterministic models with high accuracy at the expense of large computational efforts. Semi-deterministic models act as a tradeoff between the two, and hybrid models combine two or more models. The ambition is to model indoor and urban radio propagation within an acceptable complexity. In addition to propagation models, antenna models, land-use models and detailed channel models are also central, especially for assessing *multiple-input multiple-output* (MIMO) components properly.

Chapter 3 describes static and dynamic system-level simulations, including the building blocks, link models, antenna models, shadowing and multi-path fading, with a focus on the models defined by ITU and 3GPP. Different 3GPP HCN deployment scenarios are described, together with a node placement discussion. Moreover, traffic models and mobility models are described as well. All models and scenarios are described with tabulated model parameters.

The propagation model can be tuned to match the characteristics of a particular network. The tuning can be based on radio signal strength and quality measurements provided by UEs as to be discussed in Section 6.6. These measurements may be geolocalized either based on location information provided by the UEs, or by network-based localization. With a geolocalized view of the network, it is possible to better analyze a particular network, and to propose suitable candidate sites for network evolution with LPNs.

1.5.2   Access control

Some deployments of LPNs are motivated by the need for restricted access to services. Such LPNs are not primarily for all subscribers but for the corresponding CSGs, such as enterprise employees or members of a household. Depending on whether an LPN allows access to all UEs or to a set of pre-registered UEs only, it can be classified into open access and closed access, respectively. With closed access, users on the CSG white list of the LPNs have exclusive rights to the LPN resources. Fewer resource allocation retrictions with open access enables a freedom that can be exploited to increase network capacity. An alternative is hybrid access, where all UEs can access the LPN but with a higher priority given to the UEs belonging to the CSG, while any other UEs are allowed with only a limited service grade.

Such access control mechanisms are supported by the 3GPP and are described in Chapter 4 together with details about the UMTS and LTE architectures. Both the core network and radio access network components are covered, including support for femtocells. The three different access modes (i.e., open, closed and hybrid access) are defined together with the CSG white list concept. The chapter reviews the different CSG components per 3GPP LTE release and describes the differences when it comes to UMTS.

1.5.3   Mobility and handover

Mobility is a challenge in the HCN, where the main difference from homogeneous networks is that the link connection to an LPN degrades quickly when moving out of the LPN coverage. Furthermore, very limited LPN coverage area may lead to short stays and frequent handovers by highly mobile users, which may increase both signaling and risk of handover failures. Efficient mobility is also a challenge with CSG LPNs in the HCN, since it is desirable to prevent unauthorized UEs from reporting CSG cells as handover candidates to avoid unnecessary signaling.

Chapter 9 describes the mobility and handover procedures in 3GPP LTE connected mode as well as in idle mode. The UE measurement report mechanisms are detailed, including the different configuration and report triggering options. Moreover, cooperation between LPNs is mandatory for any form of mobility management.

1.5.4   Self-organizing networks

As the number of LPNs is expected to be orders of magnitude greater than that of MBS, the costs associated with planning, deploying, configuring, optimizing and managing the LPNs become a critical issue. Manual deployment and maintenance associated with the LPNs should be reduced to a minimum. Especially for FAPs, which are likely installed by customers or private enterprises without traditional *radio frequency* (RF) planning, site selection or maintenance by the operator, plug-and-play operation becomes essential. Therefore, there are strong requirements on smart nodes, which self-organize in the network and simplify the network operations and maintenance.

Chapter 6 describes the network management architecture and the area of SONs, with a focus on the SON requirements and mechanisms in E-UTRAN, but also with respect to UTRAN and the use cases defined by *Next Generation Mobile Networkss* (NGMNs). SON functionalities can be divided into self-configuration, self-optimization and self-healing. Self-configuration concerns tasks that are carried out to introduce new or re-plan existing site installations, and automates installation and configuration procedures. While in operation, self-optimization tunes parameters that dictate algorithmic behaviors based on empirical network observations. Self-healing concerns tasks that are carried out to detect, and if possible compensate for, failures and disruptive events (e.g., malfunctioning equipment). Most SON aspects are equally applicable to macrocell networks. However, mobility and load sharing imply some aspects specific to HCNs. Moreover, there are needs for dedicated mechanisms to ensure mobility robustness and the sufficient utilization of LPNs through efficient load sharing. SON also includes aspects concerning spectrum and interference management. Specific SON aspects for femtocells are also discussed in Section 8.4.

As stated before, mobility is a critical issue in HCNs. Section 9.4 will address the tuning of mobility procedures for improving mobility robustness, and discuss the impact of interference management mechanisms.

1.5.5   Intercell interference

When densifying the radio networks they may become increasingly limited by interference. Orthogonal waveform multiplexing such as via OFDMA avoids intracell interference, which means that the primary source of interference would be intercell interference. The massive deployment of LPNs overlaying macrocells will create a large number of new cell boundaries, where UEs may suffer from strong intercell interference if LPNs and MBSs use the same operational carriers. In addition to the large number of newly created cell boundaries, the interference problem in HCNs is especially challenging due to the fact that UEs are not always served by the most favorable BS, radio propagation-wise in *downlink* (DL) or *uplink* (UL). This is due to the transmit power differences between MBS and LPN as well as the configuration of access control and handover mechanisms.

One way to address the intercell interference challenge is to solve the spectrum assignment problem. Chapter 5 describes carrier and subchannel assignments in a two-tier network with MBSs and LPNs in two separate tiers, where the former are regularly deployed and the latter are deployed at random locations. The spectrum resources are assigned in order to maximize the DL throughput.

Cognitive radio in Chapter 13 is related to spectrum management, where dynamic spectrum access is used to disclose whether a specific spectrum is available for use. Such spectrum aware methods are based on spectrum sensing, geolocation database lookups or broadcast information via beacon signals. Applications include rural broadband, dynamic backhaul, capacity extension, mobile to mobile communications, cognitive femtocells etc.

Chapter 7 addresses dynamic interference management and *intercell interference coordination* (ICIC) in LTE, where the considered mechanisms reside in the time or frequency domain, or are based on transmission power or antenna adjustments. In the frequency domain, intercell interference management via subchannel resolution actions or carrier resolution actions is reviewed. Time-domain mechanisms include *almost blank subframe* (ABS) with either zero or reduced power data transmissions in an aggressor cell to facilitate reception at the victim UEs. These mechanisms are illustrated and evaluated via extensive simulations.

Intercell interference management becomes even more challenging if there are restrictions in access mechanisms and intersite signaling. Examples include CSG femtocell operations and uncoordinated LPNs, as will be addressed in Chapter 8. Chapter 8 presents the femtocell network architecture and some deployment scenarios. With less strict backhaul requirements compared with other nodes, femtocell backhaul time synchronization becomes critical, especially for *time division duplexing* (TDD) systems. A femtocell often features a UE component, which is capable of measuring the surrounding environment even before the node is operational. Alternatively, connected UEs may provide the node with similar information. Such information supports carrier selection, transmission power control, antenna tuning and cell-selection configurations.

1.5.6   Intersite coordination

Coordination between MBSs and LPNs has the potential to provide additional benefits to the overall network performance. With intersite coordination, the unbalance between nodes due to different transmission power would no longer be a problem, since the most favorable UL receivers and DL transmitters can be selected. The highest coordination gain is achieved when using a dedicated high-bandwidth low-latency link among several radios provided by the same baseband, in which case the challenge is very much related to the backhaul quality.

Diversity is considered as a means to improve reliability and capacity of wireless links. One popular way to create such diversity is via cooperative relaying, which will be presented in Chapter 10. Chapter 10 reviews different types of data forwarding via relays, and analyzes their impact on *signal to interference plus noise ratio* (SINR) and throughput, for both a normal relaying function and via cooperative relaying. The relay architecture in 3GPP LTE is also described.

Chapter 11 addresses cooperation between multiple cells via joint processing. Soft handover via the RNC in UTRAN is one early example. With refined joint processing, the received signals at multiple nodes can be jointly processed, and multiple antennas at different nodes can beamform jointly. The impact on capacity is analyzed and different backhaul aspects are considered, for example those that mandate distributed implementations.

Another form of cooperation is in the coding domain. Chapter 12 describes coordination motivated by the observation that there could be significant overlaps between cells, especially in HCNs, where LPNs are typically more or less completely covered by a macrocell. The coding gain in such scenarios is analyzed and exemplified. The important aspects of efficiency, reliability and distributed coding are considered.

1.5.7   Energy efficiency

When adding a multitude of new nodes to a network, it is important that these nodes are using energy efficiently. It is not acceptable for an idle mode node to cause high power consumption, because the total power will then sum up to a significant amount. It is essential for the nodes to both consume very little power when idle and transmit data efficiently when active. HCNs bring network infrastructure closer to UEs. As a result, transmit power can be greatly reduced, leading to prolonged battery life and improved SINR. Large-scale deployments of LPNs are only ecologically worthwhile if new energy-efficient protocols and power saving mechanisms are adopted. Moreover, LPNs can be made independent from external power supplies, for example through energy harvesting mechanisms such as solar panels.

Chapter 14 reviews energy-efficient network architectures and techniques, and analyzes cellular networks from an energy-efficiency point of view. There is a fundamental tradeoff between capacity, energy and cost, which is described and elaborated upon. Moreover, green cellular network architectures as well as transmission techniques are presented.

1.5.8   Backhaul

Even if the spectrum resources are efficiently managed, the performance of an LPN could still be insufficient due to a backhaul bottleneck. From a network hierarchy perspective, the cellular network can be broadly divided into three parts: last mile, regional network and core network [18]. The core network is usually a nationwide optical-fiber-based network connecting main switching nodes [18]. The backhaul network covers the last mile and regional network. The cost, availability of leased line infrastructure, environment, and regulations could influence the decision to lease or build backhaul links.

Backhaul traffic characteristics are determined by the combination of end-user services and *radio access network* (RAN) protocols [19]. Like other wide-area networks, a cellular backhaul network requires *operation, administration, and maintenance* (OAM) functions to initially provision and then monitor the connections that it provides [19]. Any failure or degradation must be detected and corrected in a timely manner.

Backhaul access can use microwave radios, copper, optical fiber or other transmission technologies [19]. The specific technology to be used depends on transmission distance, terrain characteristics, local availability and relative costs [18].

Even though backhaul is outside the scope of this book, there are some aspects considered, namely self-backhauling, where the spectrum resources are used both for a wireless backhaul link between a DeNB and a RN and for an access link between the RN and a UE. This will be briefly discussed in Chapters 6 and 10.

References

[1] Ericsson, *Traffic and Market Data Report* (report Nov. 2012 and interim report Feb. 2013).

[2] Cisco, *Global Mobile Data Traffic Forecast Update 2010–2015*, White Paper (Feb. 2011).

[3] S. Landström, A. Furuskär, K. Johansson, L. Falconetti and F. Kronestedt, Heterogeneous networks – increasing cellular capacity. *Ericsson Review*, no. 1 (Feb. 2011).

[4] Ericsson, *Heterogeneous Networks*, White Paper (Feb. 2012).

[5] Nokia Siemens Networks, *Deployment Strategies for Heterogeneous Networks*, White Paper (2012).

[6] H. Beijner, J. Högberg and H. Marutani, The smartphone challenge is surmountable. *Ericsson Review*, no. 2 (Dec. 2011).

[7] K. Hiltunen, Comparisons of different network densification alternatives from the LTE uplink point of view. In *Proceedings of the IEEE International Symposium on Personal, Indoor and Mobile Radio Communications* (2011).

[8] K. Hiltunen, Comparisons of different network densification alternatives from the LTE downlink performance point of view. In *Proceedings of the IEEE Vehicular Technology Conference (VTC) Fall* (2011).

[9] K. Hiltunen, The gain of a targeted introduction of OSG femtocells into a LTE macro network. In *Proceedings of the IEEE Vehicular Technology Conference (VTC) Spring* (May 2012).

[10] Alcatel-Lucent, *Small Cells Create New Opportunities in the Enterprise*, White Paper (Feb. 2012).

[11] J. G. Andrews, H. Claussen, M. Dohler, S. Rangan and M. C. Reed, Femtocells: past, present, and future. *IEEE Journal on Selected Areas in Communications*, **30**:3 (2012), 497–508.

[12] A. Damnjanovic, J. Montojo, Y. Wei, T. Ji, T. Luo, M. Vajapeyam, T. Yoo, O. Song and D. Malladi, A survey on 3GPP heterogeneous networks. *IEEE Wireless Communication*, **18**:3 (2011), 10–21.

[13] A. Ghosh, J. G. Andrews, N. Mangalvedhe, R. Ratasuk, B. Mondal, M. Cudak, E. Visotsky, T. A. Thomas, P. Xia, H. S. Jo, H. S. Dhillon and T. D. Novlan, Heterogeneous cellular networks: from theory to practice. *IEEE Communication Magazine*, **50**:6 (2012), 54–64.

[14] B. R. A. Bou Saleh, S. Redana and J. Hamalainen, Comparison of relay and pico eNB deployments in LTE-Advanced. In *IEEE Vehicular Technology Conference Fall* (2009), vol. 1, pp. 412–415.

[15] 3GPP, *3GPP Evolved Universal Terrestrial Radio Access (E-UTRA) and Evolved Univeral Terrestrial Radio Access Network (E-UTRAN); Overall Description; Stage 2 (Release 10)* (2010).

[16] E. Dahlman, S. Parkvall and J. Sköld, *4G LTE/LTE-Advanced for Mobile Broadband* (Academic, 2011).

[17] J. Hoydis and M. Debbah, Green, cost-effective, flexible, small Cell networks, *IEEE ComSoc MMTC E-Letter Special Issue on Multimedia over Femtocells* (2010).

[18] S. Chia, M. Gasparroni and P. Brick, The next challenge for cellular networks: Backhaul. *IEEE Microwave Magazine,* **10**:5 (2009), 54–66.

[19] P. Briggs, R. Chundury and J. Olsson, Carrier ethernet for mobile backhaul. *IEEE Communication Magazine,* **48**:10 (2010), 94–100.

2   Radio propagation modeling

Zhihua Lai, Guillaume Villemaud, Meiling Luo and Jie Zhang

2.1   Introduction

In the past decade, radio wave propagation modeling has attracted a great deal of interest from both academia and industry, because it facilitates efficient computation of path loss between a transmitter and a receiver in a given scenario, and plays an important role in, e.g., radio link planning and optimization processes.

Radio waves are electromagnetic waves, which can be decomposed into electric and magnetic fields. Along the propagation direction, these two fields are perpendicular to each other, creating the effect of polarization. Radio propagation is affected by the environment. For example, radio waves diffract on the edges of objects, reflections occur on an object when the wavelength is much smaller than the dimension of the object, scattering occurs if the object surface is not smooth, and attenuation losses depend on the material and size of an object.

The radio spectrum is the most precious resource in wireless communications. Different frequencies are typically allocated for use in different systems. The frequency of a radio wave also has a great impact on its propagation. A higher frequency yields a smaller wavelength and vice versa. On the one hand, radio waves with a higher frequency generally experience a higher attenuation loss, and thus propagate a shorter distance before the carried signal strength falls below a threshold. On the other hand, radio waves with a lower frequency have a larger wavelength, and can bypass obstructions more easily. Moreover, radio waves at adjacent frequencies may interfere with each other, hence it is necessary to carefully allocate frequency bands to avoid significant interference between different communication systems or even between radio links within the same system. For example, the frequency bands of 700 MHz, 2300 MHz and 2600 MHz are allocated for LTE networks, while 900 MHz and 1800 MHz bands are allocated for *Global System for Mobile Communications* (GSM) networks. In the characterization, planning and optimization processes for a *heterogeneous cellular network* (HCN), where the radio propagation environment becomes even more complicated than in conventional cellular networks, radio wave propagation modeling is essential.

This chapter reviews different kinds of radio wave propagation model, discusses their trade-offs in terms of modeling accuracy and computational complexity, and describes the process of building a radio wave propagation model. Moreover, this chapter identifies the need and requirements for new radio propagation modeling tools that are able to operate in HCN scenarios involving both indoor and outdoor environments.

2.2   Different types of propagation model

The instantaneous electromagnetic field strength at a particular location can be expressed as the sum strength of radio waves coming from all possible directions, also known as *multi-path components* (MPCs). Since radio waves may reach the receiver either in phase (with phase difference in multiples of 180 degrees) or out of phase (with phase difference in multiples of 90 degrees), the contribution of each multi-path component could be constructive or destructive. Let us denote the phase shift between two multi-path components as θ. If $\theta = 90°$, then the two sine waves will interfere with each other in a destructive manner, while if $\theta = 180°$, then they will add up in a constructive manner [1]. The *stochastic local area channel* (SLAC) model [2] computes the complex scalar electric field strength E as follows:

$$E(r, f) = \sum_{n=1}^{N} g_n e^{-j2\pi f \tau_n} e^{-j\langle k_n, r \rangle} + \omega(r, f), \tag{2.1}$$

where τ_n, g_n and k_n ($n = 1, \ldots, N$) are the delay, amplitude and arrival direction of the nth MPC, respectively, f ($f \in B$) is the emitting frequency, B is the signal bandwidth, $\mathbf{r}$ is the radius vector, $\langle k_n, \mathbf{r} \rangle$ is the product of k_n and $\mathbf{r}$, and $\omega(\mathbf{r}, f)$ approximates the diffuse wave component following the zero-mean Gaussian distribution with standard deviation σ.

The instantaneous electromagnetic field strength at a particular receiver location may fluctuate from time to time due to the multi-path propagation and uncertainties in the environment, e.g., wind blowing trees, which are usually difficult to capture in the channel model.

The path loss PL represents the difference between the transmitted power level P_t and the received power level P_r, and can be expressed in decibel units as follows:

$$PL = P_t - P_r, \tag{2.2}$$

where PL is in dB, and P_t and P_r are in dBm. If PL is computed based on a propagation model, then the received power level P_r can be obtained for given P_t as $P_r = P_t - PL$. This means that changing the transmit power level does not require re-invoking the propagation model to obtain the changed received power level. In practice, antenna gains of both the transmitter and the receiver have to be considered in the field strength computation. Accordingly, (2.2) can be rewritten as

$$PL = P_t - P_r + G_t + G_r, \tag{2.3}$$

where G_t and G_r are the antenna gain (usually in dBi) of the transmitter and the receiver, respectively.

The instantaneous path loss PL can be decomposed into three parts (in decibel units) as follows:

$$PL = L(d, f) + X_\sigma + F, \tag{2.4}$$

where X_σ denotes shadow fading and F denotes small scale fading, while the mean path loss $L(d, f)$ is given by

$$L(d, f) = 10n \log(d) + 20 \log(f) + 32.45, \tag{2.5}$$

where d (in km) is the distance between the transmitter and the receiver, f (in MHz) is the emitting frequency, 32.45 dB is the path loss at the distance of 1 km and the frequency of 1 MHz, and n is the path loss exponent that varies with different scenarios. The shadow fading X_σ is created by obstructions between the transmitter and the receivers. The contribution to the field strength in the shadowed area comes from the multiple reflections or diffractions at the obstructions. If site-specific information is not available, a zero-mean Gaussian distributed random variable X_σ (in dB) with standard deviation σ can be employed to model the shadow fading. In site-specific scenarios, the shadow fading can be estimated in a deterministic manner. For example, the reflections and diffractions can be computed by the law of reflections and *uniform theory of diffraction* (UTD) [3], respectively. The small scale fading F refers to the signal variation over a short period of time, reflecting the rapid fluctuation of field strength due to environmental changes such as moving vehicles. In *line of sight* (LOS) scenarios where a strong MPC (e.g., the direct ray or reflection on the building surface) exists, F can be modeled as a Ricean distribution, while in *non-line-of-sight* (NLOS) scenarios, e.g., dense urban scenarios where the direct ray is usually blocked by buildings, F can be modeled using a Rayleigh distribution.

Radio wave propagation models can be classified into two types: large-scale propagation models, and small-scale propagation models. Small-scale propagation models are also referred to as channel models, which study signal variations in a few wavelengths over a short period of time. In contrast, large-scale propagation models are used to characterize signal variations over a long period of time, and are usually used for link budget computation, wireless network planning and optimization. For example, computation of path loss is usually the most time-consuming part in the process of network planning and optimization, and large-scale propagation models are expected to accurately predict path losses within reasonably short times.

There are typically four kinds of large-scale propagation model: empirical models, semi-empirical (or semi-deterinistic) models, deterministic models, and hybrid models. The accuracy of a large-scale propagation model mainly depends on the level of details of the model input, e.g., information about a specific building. A higher accuracy requires a higher computational complexity. Hence, there is commonly a trade-off between the accuracy and computational complexity of a propagation model. In the following sections, we shall first focus on large-scale propagation models. The advantages and drawbacks of these models will be discussed in detail, and their implementation will be described.

2.2.1 Empirical models

Empirical propagation models are often built from site-specific measurement campaigns, based on which relatively simple formulas with only a few parameters, such as the emitting frequency and the distance between emitter and receiver, are then constructed.

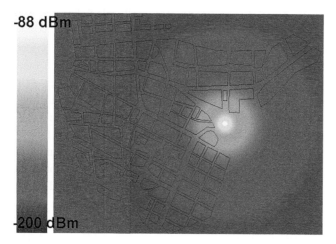

Figure 2.1 Outdoor one-slope-model coverage prediction using [5].

Instead of a high accuracy, empirical propagation models have the advantages of simple implementation and fast execution. In [4], the authors construct a set of empirical propagation models for residential environments from 0.9 GHz to 3.5 GHz. Empirical models have been widely used for the purposes of initial estimation of coverage margins and performance analysis, e.g., in 3GPP. More information on 3GPP empirical models for HCN scenarios will be provided in Chapter 3.

The COST 231 One Slope Model is a typical empirical propagation model, which is formulated using the path loss exponent as follows:

$$\text{PL} = L_0 + 10n \log(d), \qquad (2.6)$$

where d is the transmitter–receiver (T–R) separation distance, L_0 is a parameter that accounts for the system losses, and n is the path loss exponent, which is an empirical site-specific factor. For example, $n = 2$ in free space; in dense urban scenarios, most direct components are blocked by obstructions between the emitter and the receiver, i.e., shadow fading, and n is usually larger than 2; while in street canyons with wave-guiding effects, the field strength is typically strong due to the combined multiple reflections and/or diffractions in the near field, and n could be smaller than 2.

An example coverage prediction provided by the COST 231 One Slope Model is plotted in Fig. 2.1, where the transmitter is equipped with an omnidirectional antenna of 10 dBi gain, transmits at the GSM 947 MHz frequency with a power level of 3 dBm, and the receivers are 1.5 m above the ground. It can be seen that the path loss transition between any two pixels in the figure tends to be smooth regardless of the buildings, indicating that building structures do not have an influence on the received signal strength.

2.2.1.1   COST 231-Hata

The Okumura model (or Okumura–Hata model) was first developed by Japanese scholar Yoshihisa Okumura in 1968, and later simplified by Masaharu Hata in 1980, based on a series of measurements and observations in Tokyo, Japan. The Okumura–Hata model integrates empirically the clutter effect and terrain information into computation, which is one of its advantages over other empirical models. The Okumura–Hata model was further extended by COST 231 [6] into the well-known COST 231 Hata model. The COST 231 Hata model is formulated as follows [7]:

$$L = 46.3 + 33.9 \log f - 13.82 \log h_B - a(h_R) + (44.9 - 6.55 \log h_B) \log d + C,$$

$$(2.7)$$

where L is the median path loss (dB), f is the carrier frequency (MHz), h_B is the base station antenna effective height (m) ranging from 30 m to 200 m, h_R is the mobile station antenna effective height (m) ranging from 1 m to 10 m, d is the distance between the base station and the mobile station (km), typically less than 20 km, C is set to 0 dB for medium-sized cities (or suburban areas) and 3 dB for metropolitan areas, and $a(h_R)$ is the mobile station antenna height correction factor, which is formulated as follows for suburban or rural environments:

$$a(h_R) = (1.1 \log f - 0.7)h_R - (1.56 \log f - 0.8).$$

$$(2.8)$$

As an empirical model, the COST231 Hata model is applicable in the frequency range between 150 MHz and 2000 MHz, e.g., for GSM systems. Other empirical propagation models developed based on the Okumura–Hata model include ITU-R P.529-2 [8], ITU-R P.529-3 [9], and ERC 68 [10] models. The Okumura–Hata model requires that the base station antenna is higher than all nearby rooftops, and would be more accurate for a higher transmitter antenna height. For example, the Okumura–Hata model is suitable for base stations located on top of hills and with only a few obstacles between transmitter and receiver. Some semi-empirical models are also based on the Okumura–Hata model, but they differ in the methods for modeling diffraction loss, effective antenna height and land use class.

2.2.2   Deterministic models

Deterministic models take into account site-specific information in the modeling of reflections and diffractions, so as to provide accurate prediction of radio wave attenuation. However, such deterministic and accurate characterization of radio wave propagation is obtained at the cost of huge computational efforts. Deterministic models are generally categorized into two groups: *finite-difference time-domain* (FDTD) models, and geometry models.

2.2.2.1   Finite-difference models

FDTD

Maxwell's equations [11] provide an elegant and accurate description of electromagnetic wave propagation, but they are in the form of partial differential equations, which are not easy to solve. In 1966, Kane Yee proposed to replace this set of partial differential equations with a set of finite-difference equations (through central-difference approximations of the space and time partial derivatives) [12]. As a result, numerical solutions

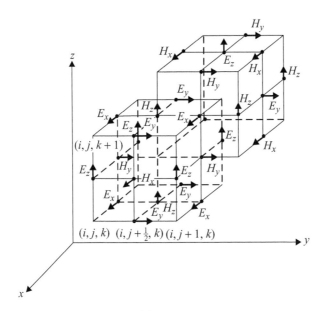

Figure 2.2 Staggered Yee grid.

to Maxwell's equations can be obtained at lower complexity. The finite-difference equations were further developed by Allen Taflove into the FDTD method [13], which takes all the propagation phenomena into account and provides a high modeling accuracy.

In FDTD simulations of electromagnetic wave propagation, the scenario of interest is first discretized and then constructed into a grid-based structure. The grid points for the electric field (E-field) and the magnetic field (H-field) are staggered, as shown in Fig. 2.2, where the E-field and the H-field components are marked. We can see that E-field components are orthogonal to H-field components, following the electromagnetic wave propagation rule in space. Besides, in the implementation of FDTD, the E-field and H-field are also interleaved in time [14]. For example, if the H-field is calculated at the discrete time instance n (where n is the abbreviation for the time instance $n\Delta t$ with time step Δt), then the E-field is calculated at time instance $n + 0.5$, followed by the calculations of H-field at $n + 1$ and E-field at $n + 1.5$, and so on, until the desired transient or steady-state electromagnetic field behavior is captured. This interleave in time is usually called the leapfrog scheme. Accordingly, the set of finite-difference equations of Maxwell's equations are solved iteratively: the H-field and E-field values of the current iteration are obtained by using the H-field values of the previous iteration and the adjacent E-field grid points of the previous iteration, respectively and conversely.

The electric field accurately predicted by FDTD can be directly used to analyze a wide variety of fading statistics. Moreover, different materials can be easily modeled in FDTD by specifying the values of electric permittivity, magnetic permeability and electrical conductivity. Since FDTD is a time-domain method, when a wideband pulse is used as the source to excite the grid, the FDTD solution obtained from a single simulation can cover a wide range of frequencies, which would be very useful for wideband system analysis.

However, FDTD models are also facing challenges.

- Maxwell's equations are discretized in both the time and space domains in FDTD models. The time-domain leapfrog scheme imposes an upper bound on the time step to ensure numerical stability [15], i.e., the time step should satisfy the *Courant–Friedrichs–Lewy* (CFL) stability condition [16]. Moreover, the space step is subject to an insignificant numerical dispersion [17]. The upper bounds on the time step and space step result in very large memory requirements and simulation times, especially when the simulated scenario is geographically large and the simulated frequency is high. Therefore, FDTD is usually used for radio wave propagation modeling in small environments.
- Due to limited computer memory, the size of the simulated scenario cannot be infinite. In order to simulate an unbounded scenario using FDTD, an artificial absorbing boundary should be inserted at the border of the simulated scenario so as to minimize errors. Most modern FDTD models use a *perfectly matched layer* (PML) to implement the absorbing boundary [18].

The ParFlow model

FDTD models possess high accuracy at the expense of computational load. Thus, simplified finite-difference models that can reach a better tradeoff between accuracy and computational load are of great interest. In this context, the ParFlow model was proposed by B. Chopard *et al.* [19] to simulate the electromagnetic wave propagation phenomenon in terms of fictitious flows on a grid. The ParFlow model provides discrete solutions to Maxwell's equations, and can be applied to the electric field, the magnetic field, or both [20].

The ParFlow model was originally proposed in the time domain, based on the cellular automaton formalism [19, 20]. In the time-domain ParFlow model, the scenario of interest should be first discretized into a two-dimensional (2D) grid-based structure, and the scalar electric field strength at a grid point r is the sum of the fictitious flows traveling along the connecting lines between the grid point r and its four neighboring grid points. From the discrete time instance $n\Delta t$ to $(n + 1)\Delta t$ (where Δt is the time step), as shown in Fig. 2.3, the fictitious flows may travel to neighboring grid points like the flows $\{f_E, f_W, f_S, f_N\}$ or remain as a stationary flow like $\tilde{f}_0$. The flows coming into a grid point are called inward flows, whereas the flows leaving a grid point are called outward flows. Note that the inward flows of a grid point r are the outward flows of its neighboring grid points and vice versa, i.e.,

$$\overleftarrow{f}_d(r + \Delta r, t) = \vec{f}_d(r, t - \Delta t), \quad d \in E, W, S, N \tag{2.9}$$

where Δr is the space step between any two adjacent grid points and satisfies the following conditions:

$$\Delta r = c_0\sqrt{2}\Delta t, \quad \Delta r \ll \lambda \tag{2.10}$$

where c_0 is the speed of electromagnetic wave in the propagation medium, and λ is the wavelength.

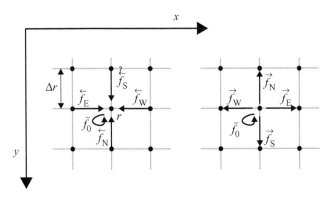

Figure 2.3 Inward and outward flows.

In accordance with Maxwell's equations, the discrete time evolution of the flows should obey the local scattering equation

$$\vec{F}(r, t) = W(r)\overleftarrow{F}(r, t - \Delta t) + \vec{S}(r, t), \tag{2.11}$$

where the inward flows vector $\overleftarrow{F}(r, t)$, the outward flows vector $\vec{F}(r, t)$, and the source flows vector $\vec{S}(r, t)$ are defined, respectively, as follows:

$$
\begin{aligned}
\overleftarrow{F}(r, t) &= [\overleftarrow{f}_E(r, t) \ \overleftarrow{f}_W(r, t) \ \overleftarrow{f}_S(r, t) \ \overleftarrow{f}_N(r, t) \ \breve{f}_0(r, t)]^T \\
\vec{F}(r, t) &= [\vec{f}_E(r, t) \ \vec{f}_W(r, t) \ \vec{f}_S(r, t) \ \vec{f}_N(r, t) \ \breve{f}_0(r, t)]^T \\
\vec{S}(r, t) &= [\vec{s}_E(r, t) \ \vec{s}_W(r, t) \ \vec{s}_S(r, t) \ \vec{s}_N(r, t) \ 0]^T
\end{aligned}
\tag{2.12}
$$

where the operator $[\cdot]^T$ denotes matrix transposition, the stationary flow $\breve{f}_0(r, t)$ models the dielectric media with the relative permittivities $\varepsilon_r \neq 1$, and the local scattering matrix $W(r)$ in (2.11) is defined as

$$
W(r) = \frac{1}{2n_r^2}
\begin{bmatrix}
1 & \alpha_r & 1 & 1 & Y_r \\
\alpha_r & 1 & 1 & 1 & Y_r \\
1 & 1 & 1 & \alpha_r & Y_r \\
1 & 1 & \alpha_r & 1 & Y_r \\
1 & 1 & 1 & 1 & \beta_r
\end{bmatrix}
\tag{2.13}
$$

where n_r is the refraction index, $\alpha_r = 1 - 2n_r^2$, $\beta_r = 2n_r^2 - 4$, and $Y_r = 4n_r^2 - 4$.

The local scattering equation (2.11) can be solved by a cellular automaton, and then the electric field is given by

$$\Psi(r, t) = \frac{1}{n_r^2} \left(\sum_{d=E,W,S,N} \vec{f}_d(r, t) + Y_r \breve{f}_0 \right). \tag{2.14}$$

The multi-resolution frequency-domain ParFlow (MR-FDPF) model
The MR-FDPF model was proposed by Gorce *et al.* [21, 22], and is the frequency-domain counterpart of the time-domain ParFlow model [19, 20]. The MR-FDPF model takes Fourier transform on both sides of the local scattering equation (2.11), makes use of the

relationship between the inward flows and outward flows in (2.9), and then efficiently combines them in the local scattering matrix $W(r)$ defined in (2.13). Finally, the local scattering equation becomes a linear equation in the frequency domain as follows:

$$\vec{\underline{F}}(\upsilon) = \tilde{\underline{W}}\, e^{-j2\pi\upsilon\Delta t}\, \vec{\underline{F}}(\upsilon) + \vec{\underline{S}}(\upsilon) \tag{2.15}$$

where υ is the frequency, and $\vec{\underline{F}}(\upsilon)$, $\vec{\underline{S}}(\upsilon)$, and $\tilde{\underline{W}}$ are the global outward flow vector, global source flow vector, and global scattering matrix, respectively.

Therefore, in the frequency domain, it is a linear matrix inverse problem:

$$\vec{\underline{F}}(\upsilon) = \left(I - \tilde{\underline{W}}\, e^{-j2\pi\upsilon\Delta t}\right)^{-1} \vec{\underline{S}}(\upsilon), \tag{2.16}$$

where I is the identity matrix.

However, the huge sizes of matrices corresponding to very large simulated scenarios make the inverse matrix calculation prohibitive. A multi-resolution approach was thus proposed to provide a computationally efficient way to solve this problem [21]. This multi-resolution approach constructs a binary tree structure for the simulated scenario in the preprocessing stage, which consumes most of the computational load and should run just once even if there are multiple signal sources. Such preprocessing greatly accelerates the MR-FDPF predictions and makes it outperform the time-domain ParFlow model [21, 23]. Furthermore, the MR-FDPF model has recently been improved to provide not only coverage prediction, but also information about fading statistics, such as Rice K factor, path loss component, and standard deviation of shadowing [24]–[26], which are useful for the evaluation of system-level performance.

An intuitive interpretation of ParFlow models is the discrete version of Huygens' principle, which states that each point of a wavefront may be considered as the source of a secondary wavelet. It is important to note that ParFlow models take into account all the radio propagation effects, such as reflections, diffractions, and refractions, and the computational load is independent from the number of reflections. The space discretization step Δr should be at least six times smaller than the wavelength in order to obtain good propagation predictions [27]. However, the smaller the space discretization step, the higher the computational load. Therefore, the main drawback of ParFlow models is the relatively high computational load, which may be addressed by using a lower fake frequency method [19, 27].

2.2.2.2 Geometry models

Radio wave propagation can also be modeled through geometry rays which interact with the environment when they travel through various media. The rays may be reflected, diffracted, or scattered, or penetrate obstructions before reaching their destination. The signal strength of a ray is attenuated along the propagation path, and the total received signal strength is determined by all the rays arriving at the destination and their phase shifts. Geometry models compute the MPCs between the transmitter and receiver depending on path searching algorithms, and are also referred to as ray-based (or ray optical) models in the literature.

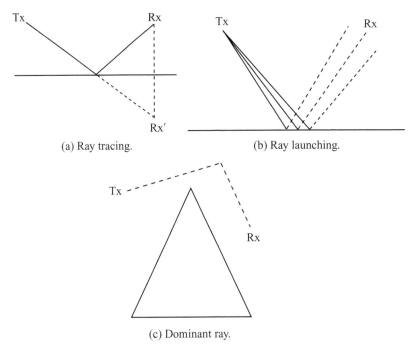

(a) Ray tracing.   (b) Ray launching.

(c) Dominant ray.

Figure 2.4 Path search algorithms (Tx – transmitter, Rx – receiver).

Similar to the SLAC model given in (2.1), the received signal strength modeled using geometry models can be expressed as follows:

$$P_r = P_t - \sum_{i=1}^{N} \omega(\tau_i, L(f, d_i)) + G_t + G_r, \qquad (2.17)$$

$$L(f, d_i) = F(f, d_i) + \sum_{i=1}^{R} \varphi(M_i, t_i), \qquad (2.18)$$

where P_r is the received signal power in dBm, P_t is the transmit power level in dBm, G_t and G_r are the antenna gains (in dBi) of the transmitter and the receiver, respectively, N is the number of MPCs arriving at the receiver, τ_i is the delay of the ith MPC, ω sums up the power carried by the MPCs, $L(f; d_i)$ is the path loss of the ith MPC, d_i is the distance that the ith MPC propagates along its path, f is the carrier frequency, $F(f; d_i)$ is the path loss as a function of d_i and f, e.g., (2.5), and $\varphi(M_i, t_i)$ is the interaction loss as a function of M_i and t_i, which denote the material type and the type of interaction (e.g., reflection or diffraction) of the ith MPC, respectively.

Geometry models compute the exact path that a ray travels based on the Fermat principle, which states that in optics the path taken by a ray between two points is the path that can be traversed in the least time [28]. For example, the reflected rays can be computed as in Fig. 2.4, where the transmitter location Tx is first mirrored with respect to the reflecting surface into point Tx′; if there is an intersection O between ray Tx′ ⟶ Rx and the reflecting object, then the intersection point will be considered as the reflection

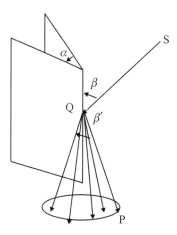

Figure 2.5 Keller cone diffraction principle [32, 33].

point and the reflection path is Tx$\longrightarrow$O$\longrightarrow$Rx. In this way, the computed rays satisfy the law of reflection, where $\theta_i = \theta_r$, and θ_i and θ_r are the incident ray angle and the reflected ray angle, respectively. The reflection loss can be formulated as follows:

$$L_r = 10\log_{10}\frac{(Z_1 - Z_2)^2}{(Z_1 + Z_2)^2}, \tag{2.19}$$

where L_r is the reflection loss in dB, and Z_1 and Z_2 are the impedances of medium 1 and medium 2, respectively. Scattering is often considered as multiple reflections on a rough surface, and is usually not computed deterministically but empirically. Many ray-based modeling tools offer the computation of scattering as an add-on, by configuring an empirical factor to compensate the scattering loss.

Diffraction occurs at the edge of an obstruction when the size of the object is smaller than the radio wavelength, and the radio wave bypasses the obstruction resulting in many secondary rays in new directions. The *geometry theory of diffraction* (GTD) [29, 30] was first proposed to approximate the electromagnetic near field of diffraction and determine diffraction coefficients, which are required to compute the field strength and phase for each diffracted ray of a geometry model. The GTD was later extended into the UTD [3]. Many existing ray tracing tools, such as WinProp by AWE Communications [31], are built together with the GTD or UTD. The Keller Cone [32, 33], as illustrated in Fig. 2.5, is often used to find diffracted rays in geometry models. In Fig. 2.5, Q is the diffraction point and is considered as a secondary source that virtually emits many diffracted rays in a cone, and the incident ray angle β is equal to the diffracted ray angle β'. The diffraction loss is also dependent on the angle between facets α.

In geometry models, how rays refract is described by Snell's law of refraction, which follows Fermat's principle. Snell's law of refraction is illustrated in Fig. 2.6 and is modeled as follows:

$$\frac{\sin\theta_1}{\sin\theta_2} = \frac{v_1}{v_2} = \frac{n_2}{n_1}, \tag{2.20}$$

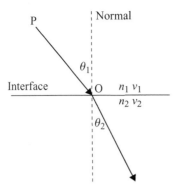

Figure 2.6 Snell's law of refraction.

where θ_1 is the incident ray angle with respect to the normal of the interface, θ_2 is the refracted ray angle, n_1 and n_2 are the refractive indices of the two media, and v_1 and v_2 are the phase velocities in the two media.

Geometry models can be generally divided into three categories: ray tracing, ray launching (casting), and dominant path models, which are illustrated in Fig. 2.4.

Ray tracing computes the exact paths going through reflections, diffractions and/or refraction from the transmitter to the receiver backwards. Ray tracing models provide the highest accuracy among all geometry models at the cost of high computational requirements. The ray tracing computations for every receiver location are independent of one another. Accordingly, the complexity (or the total amount of time) of ray tracing increases linearly with the number N of receiver locations required in the computation, i.e., $C = \sum_{i=1}^{N} c_i$, or approximately $C \propto c_i$, $\forall i \in \{1, \ldots, N\}$. This indicates that ray tracing models are precise and suitable for point-to-point field strength computation, but not suitable for coverage prediction due to the computational complexity. Different techniques have been proposed to accelerate ray tracing computations in order to adopt ray tracing tools in planning tools. For example, in [34], the authors proposed a reduced-complexity ray tracing model based on two techniques: (1) computing received signal strengths at neighboring locations and interpolating those skipped, based on the observation that the ray paths for adjacent locations are similar; (2) decomposing the building structure into a few small tiles according to the simulation resolution, then computing the visibility relationships between the tiles and storing them in a tree structure, thus the visibility relationships between the objects can be easily traced based on the preprocessed tree and rays can be traced accordingly. The advantage of this enhancement is that the visibility tree does not depend on the locations of transmitters, and therefore it can be computed only once and stored for use in the ray tracing model. However, it has been reported that the preprocessing technique in [34] consumes much time and memory in order to build geometrically large scenarios. Since ray–object intersections are the most time-consuming part of many geometry models, a fast ray-triangle algorithm was proposed in [35], where the *binary space partitioning* (BSP) recursively subdivides a space into convex sets by hyperplanes, so as to cut off unnecessary ray–object intersections. In addition, parallel execution of ray tracing models has been implemented in [36].

Ray launching is also known as ray casting or ray sampling, where the rays transmitted from the transmitter are traced as they propagate towards the receiver. In ray launching models, interactions between rays and obstructions can be modeled by the law of reflection, GTD/UTD, Snell's law of refraction etc. In practice, more ray iterations or a lower received signal strength threshold would lead to a longer running time of ray launching (or ray tracing) models and vice versa. In order to perform ray launching (or ray tracing) simulations within a reasonable time, some launched rays could be abandoned through two approaches: (1) upper bound the maximum number of ray interactions with the scenario; (2) lower bound the minimum received signal strength to ignore those rays that fall below a predefined power level. The launched rays are usually separated by an angle and disperse further away from one another as they propagate on. In this case, pixels that are far from the transmitter are likely to be abandoned by the aforementioned two approaches to speed up ray launching computation. Therefore, ray launching is generally less accurate than ray tracing. Meanwhile, ray launching is more suitable for coverage prediction than ray tracing, because the computation of a pixel can in most cases be easily obtained according to its precedent pixels along the same path. In contrast to ray tracing, the total number of pixels required for ray launching to perform a prediction does not grow linearly with the size of the scenario. However, if only the received signal strengths at a few known locations are of interest, then ray launching has to trace all the launched rays in order to compute the corresponding path losses, which is less computationally efficient than ray tracing. In practice, ray tracing and ray launching can be combined to offer a reasonable tradeoff between modeling complexity and accuracy. For example, the Volcano tool from Siradel implements ray tracing in combination with ray launching.

Both ray tracing and ray launching models can be implemented in vector and/or a grid-based discrete structure. The vector implementation ensures accurate ray–object intersection tests, but the execution is much slower than the grid-based discrete implementation where the ray–object intersection tests can be accelerated (e.g., by using bounding boxes built based on the uniformed grid structure). However, discrete implementations are inherently less accurate than vector implementations, because different rays from the same source reaching the same discrete pixel are treated as duplicated rays in the grid-based discrete structure. Vector and grid-based discrete implementations can be intelligently combined for ray launching to achieve a reasonable level of accuracy within a short period of time [5].

In most cases, the received signal strength is dominated by only one or a few strong rays [37]. Accordingly, the *dominant path* model [37] is constructed by analyzing the scenario and then computing the few dominant rays, as depicted in Fig. 2.4(c). If dominant rays do not exist, then the sum of many rays can be consider as a virtual dominant ray. The path loss along a dominant path can be expressed as follows:

$$L = 20 \log \left(\frac{4\pi}{\lambda/m} \right) + 20p \log(d/m) + \sum_{i=0}^{n} \alpha(\phi, i) - \frac{1}{c} \sum_{k=0}^{c} w_k, \qquad (2.21)$$

where λ is the wavelength of the transmitted signal, p is the visibility state between the transmitter and receiver, which can be either LOS or NLOS, $\alpha(\phi, i)$ denotes the direction of a ray, and w_k is the wave guiding gain. The terrain and clutter information, which affects the visibility p, is also considered in the dominant path model. The dominant path model has the following advantages. First, the computation of a few dominant rays is less time consuming than the ray tracing or ray launching model. Second, the dominant path model is less sensitive to the accuracy of the scenario database than the ray tracing or ray launching model, i.e., a slight change of the position or angle of an obstruction would not lead to much different prediction of the dominant path.

Intelligent ray launching algorithm

An intelligent 3D ray launching algorithm was proposed in [38] for wireless network planning and optimization. Its high accuracy and computational speed have been verified using the COST-Munich scenario [39]. The ray launching model comprises three components: LOS, *vertical diffraction* (VD), and *horizontal reflection diffraction* (HRD). The LOS component denotes the direct path from the transmitter to the receiver, and also identifies secondary cubes to launch reflections and/or diffractions. The VD component contains vertical diffractions. The HRD component emits all 3D rays that comprise reflections and diffractions. This ray launching model is based on a discrete data set of size (N_x, N_y, N_z), which represent the number of cubes along x, y, and z dimensions, respectively. The total number of pixels is given by $N_{\text{total}} = N_x N_y N_z$, while the total number of discrete rays launched from the transmitter to the cubes in all directions is calculated as $N_{\text{rays}} = 2N_x N_y + 2(N_z - 2)(N_x + N_y - 2)$, where $N_z > 1$. The obtained MPCs are then used to calculate the field strengths for all pixels in the three dimensions, with the delay of each ray considered to yield a high modeling accuracy. Compared with other geometry models, the intelligent ray launching model does not require a preprocessing stage, and it can be executed efficiently in wireless network planning tools.

The intelligent ray launching model has also been extended to indoor scenarios [40, 41], while advanced models that include outdoor, indoor, outdoor-to-indoor, and indoor-to-outdoor propagation predictions have also been proposed [42–45]. In the following, intelligent ray launching model predictions for an indoor scenario are illustrated. The indoor scenario is comprised of a building that has five floors, as shown in Fig. 2.7. Each floor has deployed four *Home evolved NodeBs* (HeNBs) using 10 MHz bandwidth and 10 dBm transmit power each, and serving 10 *user equipments* (UEs) each. A full buffer traffic model is used. Fig. 2.8 plots the *reference signal received power* (RSRP) coverage prediction map obtained. The propagation prediction considers the cross-floor penetration losses. The best RSRP map on each floor can be observed in Fig. 2.9, which shows that sufficient coverage can be obtained.

Diffraction

A radio wave propagates along a straight line if it is unobstructed; but if a radio wave is obstructed or there are obstructions near its LOS path, then the reflected rays may have different phases with respect to the direct ray, either enhancing or weakening the received signal strength at the receiver [46]. Fresnel's wave theory provides a method to determine whether reflected rays will add to the LOS component in a constructive or destructive manner, and to estimate the received signal strength as a result the reflected

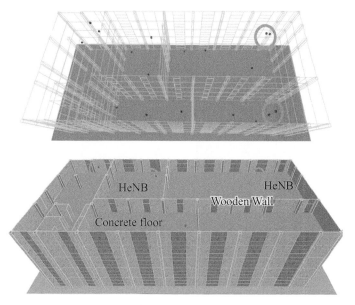

Figure 2.7 Case study: scenario.

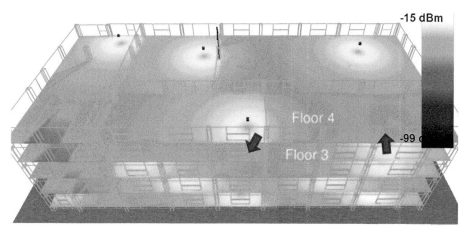

Figure 2.8 Case study: RSRP coverage map.

rays. At any point P as shown in Fig. 2.10, the Fresnel zones can be formulated as follows [47]:

$$r_n = \sqrt{\frac{n\lambda d_1 d_2}{d_1 + d_2}},$$ (2.22)

where r_n is the radius of the nth Fresnel zone in meters, λ is the wavelength of the transmitted signal in meters, d_1 is the distance from point P to the transmitter in meters, and d_2 is the distance from point P to the receiver in meters.

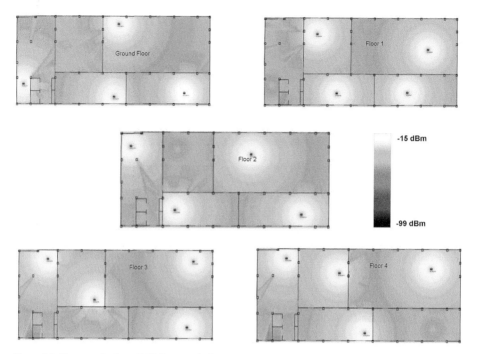

Figure 2.9 Case study: best RSRP on each floor.

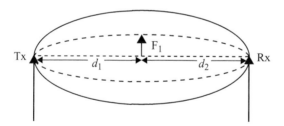

Figure 2.10 Fresnel Zones [48].

It has been observed that odd numbered Fresnel zones are constructive, while even numbered zones are destructive [47, 49]. If 60% of the first Fresnel zone is free of obstructions [47], then the radio link can be considered as LOS. The calculation of Fresnel zones is thus useful in determining the antenna heights to avoid significant obstructions.

In outdoor urban scenarios, multiple rooftop diffractions [47, 49] are typically considered as dominant contributions to the received signal strength, especially for receivers at distant locations. Multiple rooftop diffractions can be mathematically calculated using knife-edge models, which consider the effect of Fresnel zones. In knife-edge models, the knife edge refers to a perfectly absorbing edge, where no transmission or reflection occurs. It has been shown that the exact calculation of knife-edge diffractions involves

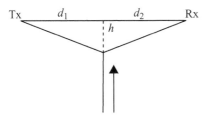

Figure 2.11 Single edge [47].

complex calculus, which is difficult to be efficiently solved within a reasonable time. Accordingly, knife-edge models have been simplified to different extents.

Single-edge diffraction

For the single-edge geometry illustrated in Fig. 2.11, h denotes the perpendicular distance from the obstacle edge to the straight line connecting transmitter and receiver, where $h \ll d_1$, $h \ll d_2$, $\lambda \ll d_1$, and $\lambda \ll d_2$, and the diffraction parameter v [7] can be calculated as follows:

$$v = h\sqrt{\frac{2(d_1 + d_2)}{\lambda d_1 d_2}}.$$
(2.23)

The diffraction loss can be approximated by a Cornu spiral, which is an estimation of the combined Fresnel sine and cosine integrals of the diffraction parameter v. An estimation of diffraction loss can be written as follows [47], and is depicted in Fig. 2.12.

$$L = \begin{cases} 6.9 + 20 \log_{10}(\sqrt{(v - 0.1)^2 + 1} + v - 0.1) & \text{if } v \geq -0.7 \\ 0 & \text{otherwise.} \end{cases}$$
(2.24)

Fig. 2.12 shows that the diffraction loss is around 6 dB when $h = 0$ and $v = 0$ (i.e., a gazing knife edge). If 60% of the first Fresnel zone is clear of obstacles (i.e., $v < -1.4$), then the diffraction loss is zero.

In addition to single-edge diffraction, there are several multiple-edge models [47], such as Burlington, Epstein–Peterson, and Deygout models, which simplify the calculation of multiple-edge diffraction. Given requirements of accuracy and computational complexity, these models may need to be modified for practical scenarios where there is no so-called knife-edge obstacle.

Burlington model

The Burlington model constructs an equivalent single knife edge at the intersection of the line connecting the transmitter and its nearby major edge and the line connecting the receiver and its nearby major edge (as shown in Fig. 2.13), and calculates the diffraction loss based on the equivalent single knife edge [47]. The Burlington model provides a simple approximation of multiple-edge diffraction that is easy to implement, but tends to underestimate diffraction losses because one single knife edge might not be able to represent the effects of all major edges.

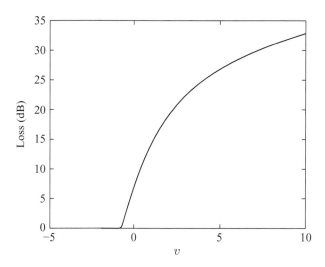

Figure 2.12 Loss in relation to parameter v.

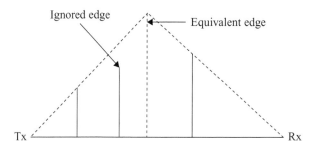

Figure 2.13 Burlington model [47].

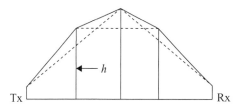

Figure 2.14 Epstein–Peterson model [47].

Epstein–Peterson model

The Epstein–Peterson model computes the diffraction loss at each single knife edge and sums them up. The effective height of each edge is computed using a geometric approach, e.g., the height h_A of edge A above the dotted line connecting the transmitter and edge B in Fig. 2.14. The Epstein–Peterson model tends to overestimate diffraction losses, especially in the case when there are many closely spaced edges.

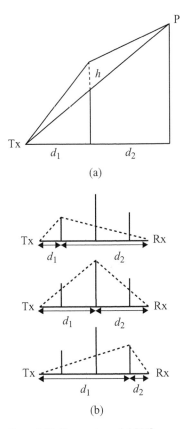

Figure 2.15 Deygout model [47].

Deygout model

The Deygout model calculates $v = h\sqrt{\frac{2(d_1+d_2)}{\lambda d_1 d_2}}$ of each edge, as shown in Fig. 2.15(a). The edge with the largest value of v is considered as the 'principal edge,' and the diffraction loss caused by the principal edge is calculated following single-edge diffraction. Then, the two lines that respectively connect the transmitter and the receiver to the principal edge are assumed as two virtual paths. For each of the two virtual paths, a principal edge is found if there is one, and the diffraction loss caused by it is calculated, e.g., in Fig. 2.15(b). Each previously found principal edge can further divide the corresponding virtual path into two new virtual paths. The process of searching for a principal edge and calculating its diffraction loss for each virtual path can be carried on until all edges between the transmitter and receiver have been considered, but with all edges in the first Fresnel zone ignored. Finally, the total diffraction loss is the sum of all computed intermediate diffraction losses.

2.2.3   Semi-deterministic models

Empirical models require low computational complexity at the cost of reduced accuracy, while deterministic models offer high accuracy at the expense of large computational efforts. Semi-deterministic models, also known as semi-empirical models, provide a tradeoff between empirical and deterministic models. Semi-deterministic models are usually developed for particular scenarios, but are more accurate than empirical models, because they still consider some deterministic factors, e.g., environmental influences on signal strength variation. In the the following, we review typical semi-deterministic models for heterogeneous network scenarios that involve both indoor and outdoor environments.

2.2.3.1 COST 231 Walfisch–Ikegami model

The empirical model developed by Walfisch and Ikegami was enhanced in the COST 231 Project for urban environments, resulting in the semi-empirical COST 231 Walfisch–Ikegami model [6], which takes into account diffractions in the vertical plane, but ignores reflections. The COST 231 Walfisch–Ikegami model is often adopted for outdoor field strength prediction and offers an acceptable level of accuracy, because antennas are usually placed above rooftops in outdoor urban scenarios, and thus diffractions that propagate on rooftops dominate the radio propagation. The modeling accuracy can be improved by deploying the COST 231 Walfisch–Ikegami model jointly with some deterministic approaches to include horizontal diffractions and/or reflections. Compared with the Hata model, the COST 231 Walfisch–Ikegami model considers more environment factors, e.g., the width of the street, the space between buildings, and the height difference between transmitter and receiver.

The COST 231 Walfisch–Ikegami model computes the path loss taking into account both LOS and NLOS components. The LOS path loss is given by

$$L_{\text{dB}} = 42.6 + 26 \log_{10}(d) + 20 \log_{10}(f), \tag{2.25}$$

where d (km) is the distance from the transmitter to the receiver, and f (MHz) is the transmitted frequency. The NLOS path loss is composed of contributions from rooftop diffractions and waves that go around buildings arriving at the receiver horizontally, i.e.,

$$L_{\text{NLOS}} = \begin{cases} L_0 + L_{\text{rts}} + L_{\text{msd}} & \text{if } L_{\text{rts}} + L_{\text{msd}} > 0 \\ L_0 & \text{if } L_{\text{rts}} + L_{\text{msd}} \leq 0 \end{cases} \tag{2.26}$$

where L_{msd} is the multi-screen path loss, L_0 is the free space path loss given as [47]

$$L_0 = 32.44 + 20 \log_{10}(d) + 20 \log_{10}(f), \tag{2.27}$$

and L_{rts} is the rooftop path loss based on the Walfisch–Ikegami model [47], which can be expressed as

$$L_{\text{rts}} = -16.9 - 10 \log_{10}(w) + 10 \log_{10}(f) + 20 \log_{10} \Delta h_{\text{mobile}} + L_{\text{ori}}, \tag{2.28}$$

where d (km) is the transmitter–receiver distance, f (MHz) is the transmit frequency, w (km) is the street width, L_{ori} is the street orientation, and $\Delta h_{\text{mobile}} = h_{\text{roof}} - h_{\text{mobile}}$ (see Fig. 2.16).

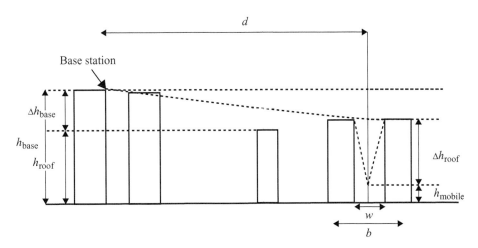

Figure 2.16 COST 231 Walfisch–Ikegami model [6].

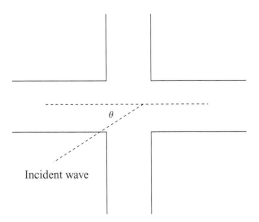

Figure 2.17 COST 231 Walfisch–Ikegami model street orientation [47].

In (2.28), the COST 231 Walfisch–Ikegami model may take different functions for L_{ori} (see Fig. 2.17), which can be expressed as

$$
L_{ori} = \begin{cases} -10 + 0.354\theta & 0° \leq \theta < 35° \\ 2.5 + 0.075(\theta - 35) & 35° \leq \theta < 55° \\ 4.0 - 0.114(\theta - 55) & 55° \leq \theta \leq 90°. \end{cases}
$$

In (2.26), L_{msd} can be formulated [47] as

$$
L_{msd} = L_{bsh} + k_a + k_d \log_{10}(d) + k_f \log_{10}(f) - 9 \log_{10}(b),
$$

where

$$L_{bsh} = \begin{cases} -18\log_{10}(1 + \Delta h_{base}) & \text{for } h_{base} > h_{roof} \\ 0 & \text{for } h_{base} \leq h_{roof} \end{cases}$$

$$\Delta h_{base} = h_{base} - h_{roof} \text{ (Fig. 2.16)}$$

$$k_a = \begin{cases} 54 & \text{for } h_{base} > h_{roof} \\ 54 - 0.8\Delta h_{base} & \text{for } d \geq 0.5 \text{ km and } h_{base} \leq h_{roof} \\ 54 - 0.8\Delta h_{base}\frac{d}{0.5} & \text{for } d < 0.5 \text{ km and } h_{base} \leq h_{roof} \end{cases}$$

$$k_d = \begin{cases} 18 & \text{for } h_{base} > h_{roof} \\ 18 - 15\frac{\Delta h_{base}}{h_{roof}} & \text{for } h_{base} \leq h_{roof} \end{cases}$$

$$k_f = -4 + \begin{cases} 0.7\left(\frac{f}{925} - 1\right) & \text{for medium sized city and} \\ & \text{suburban centres with medium tree density} \\ 1.5\left(\frac{f}{925} - 1\right) & \text{for metropolitan centres.} \end{cases}$$

The COST 231 Walfisch–Ikegami model has an applicable range in terms of f from 800 MHz to 1500 MHz, h_b from 4 m to 50 m, h_m from 1 m to 3 m, and d from 0.02 km to 5 km. However, this model does not consider multi-path propagation, and a good accuracy requires that the antenna is above average building rooftop and the terrain is flat, because L_{ori}, w and some other parameters are estimated empirically. As a result, the COST 231 Walfisch–Ikegami model might not be applicable to complex environments.

2.2.3.2   Multiwall model

The multiwall model [50] is a semi-empirical propagation model for indoor scenarios. The losses of transmissions through walls and/or floors are calculated as follows:

$$L_{multiwall} = L_p + \sum_{i=1}^{N_{walls}} \alpha_i W_i + \sum_{j=1}^{N_{floors}} \beta_i F_i + C \tag{2.29}$$

where $L_{multiwall}$ is the path loss (in dB), L_p is the free space path loss given in (2.5), N_{walls} and N_{floors} are the numbers of walls and floors between the transmitter and receiver, respectively, α and β denote the transmission coefficients through a wall and a floor, respectively, W_i and F_i are the losses caused by each wall and each floor, respectively, and C is an adjustment constant that accounts for system losses.

The computational complexity of the multiwall model depends only on the number of walls and/or floors involved. An example coverage prediction using the multiwall model is presented in Fig. 2.18 [5], where there are three 2.4 GHz emitters. The running time on a standard PC (Intel i7 CPU, 8GB RAM) is less than 3 s. The losses are accumulated for each transmission through walls. For example, a constant 10 dB penetration loss is set for each wall in this scenario. However, the multiwall model may give incorrect

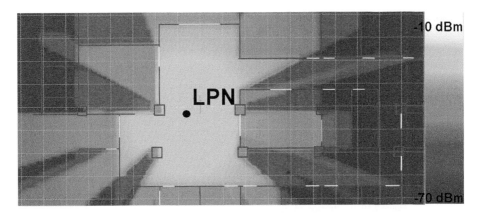

Figure 2.18 The indoor multiwall coverage prediction using [5], where LPN denotes low-power node.

predictions under certain circumstances. For example, multiwall predictions tend to be pessimistic in the far field where the direct ray between the transmitter and the receiver cross many walls and/or floors, because the multiwall model is not able to predict the multiple reflections and/or diffractions that dominate the received signal strength. Note that transmission coefficients α and β decrease with the propagation distance, and should be appropriately calibrated to improve the accuracy of the model.

2.2.3.3   COST 231 penetration model

Wall penetration attenuation can be predicted either statistically or deterministically. The statistical method is based on empirical losses, while the deterministic method traces rays when they penetrate through building rooftops or sidewalls and interact with interior walls inside the building [51].

The COST 231 Project [6] proposed an outdoor to indoor penetration model based on geometry, assuming isotropic antennas, *base station* (BS) heights less than 30 m, and a frequency range from 900 MHz to 1800 MHz. A typical outdoor to indoor ray propagation scenario is depicted in Fig. 2.19. The COST 231 penetration model can be formulated as follows:

$$L_{\text{total}} = L_{\text{out}} + L_{\text{in}} + L_{\text{tw}}, \tag{2.30}$$

where L_{total} is the total path loss from the outdoor transmitter to the indoor receiver, L_{out} is the path loss along the outdoor propagation path, L_{in} is the path loss of the indoor propagation path, and L_{tw} is the through-wall propagation loss.

In (2.30), L_{out} can be obtained from different propagation models, e.g., the free space path loss model [6],

$$L_{\text{out}} = 32.45 + 20 \log_{10}(S + d_{\text{in}}) + 20 \log_{10}(f), \tag{2.31}$$

where S is the outdoor distance, d_{in} is the indoor distance (see Fig. 2.19), and f (MHz) is the transmitting frequency.

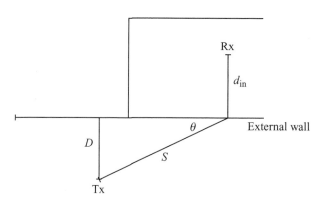

Figure 2.19 Illustration of outdoor to indoor propagation scenario.

The indoor path loss L_{in} in (2.30) can be determined as follows:

$$L_{in} = \max\{pW_i, \alpha(d_{in} - 2)(1 - \sin(\theta))^2\}, \tag{2.32}$$

where p is the number of interior walls penetrated by the ray, W_i is the penetration loss caused by each wall, α is the indoor path loss coefficient, and θ is the incident angle to the outer wall.

The COST 231 outdoor-to-indoor penetration model was extended in [52] and [53] to allow for a frequency range from 800 MHz to 2000 MHz, BS heights from 4 m to 50 m, and transmitter–receiver distances from 0.02 km to 5 km. More specifically, L_{out} and L_{in} in (2.30) are replaced by the COST 231 Walfisch–Ikegami model, i.e.,

$$L_{out} = 42.6 + 26\log_{10}(S) + 20\log_{10}(f), \tag{2.33}$$
$$L_{in} = \alpha d_{in}, \tag{2.34}$$

where it is assumed that there is one internal wall every 10 m, f is the transmitting frequency in MHz, α is the indoor path loss coefficient, and it is not necessary to know the locations of the receivers. Note that (2.33) computes the path loss between the outdoor BS and the external wall, while the indoor propagation path is included in (2.31).

Additional transmission losses, such as building losses due to insufficient resolution of the building data, can be computed empirically in the modeling of outdoor to indoor propagation. For example, based on the parameters in Fig. 2.19, the LOS path loss consists of two parts: the direct ray and the indoor losses [54, 55], and can be formulated as follows:

$$L_{stat,\theta} = \alpha(d_{in} - 2)\left(1 - \frac{D}{S}\right)^2, \tag{2.35}$$

$$L_{part.wall} = \beta d_{in}, \tag{2.36}$$

$$L_T = 32.4 + 20\log(f) + 20\log(S + d_{in}) + L_1$$
$$+ L_2\left(1 - \frac{D}{S}\right)^2 + \max(L_{part.wall}, L_{stat,\theta}), \tag{2.37}$$

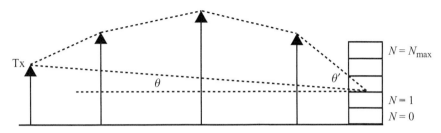

Figure 2.20 Outdoor-to-indoor NLOS illustration.

where L_1 is the external wall loss and L_2 is the path loss coefficient. Typical values for L_1, L_2 and α are 40 dB, 4–10 dB and 0.6 dB/m, respectively.

Based on the NLOS geometry in Fig. 2.20, a height gain model for NLOS path loss was proposed in [56] and [55], i.e.,

$$L_T = \min(L_{\text{outdoor}}) + L_1 + L_2 + \max(L_{\text{part.wall}}, L_{\text{state}}) - G_{\text{FH}}, \qquad (2.38)$$

$$L_{\text{state}} = \alpha d, \qquad (2.39)$$

$$G_{\text{FH}} = G_{\text{h}} h, \qquad (2.40)$$

where L_T is the total outdoor-to-indoor transmission loss, L_{outdoor} is the outdoor path loss, which takes Fresnel zones into account, d is the penetration distance α is the penetration distance coefficient, h is the floor height, and G_{h} is the floor height gain, which ranges from 1.1 dB/m to 1.6 dB/m when the floor height is between 4 m and 5 m.

Moreover, empirical computation of $L_1 + L_2 + \max(L_{\text{part.wall}}, L_{\text{state}})$ was suggested in [54], which transforms (2.38) into

$$L_T = \min(L_{\text{outdoor}}) + L_{\text{emp}} - G_{\text{FH}}. \qquad (2.41)$$

The propagation loss in the above empirical height model is approximately 2.9 dB per floor. The model was validated through measurements in GSM frequencies [57].

2.2.4   Hybrid models

Two or more propagation models can be combined into a hybrid model for complex scenarios, e.g., hybrid outdoor to indoor scenarios that are common in HCNs. Hybrid models often involve passing the outputs of one propagation model to another as inputs, in a way that combines the advantages of individual propagation models and usually avoids their drawbacks. For example, FDTD models are known to have limitations in geometrically large scenarios, such as outdoor urban environments, due to requirements of huge memory and complex computation; on the other hand, geometry models are able to provide a high accuracy in 3D scenarios, but only when the information of obstruction geometry and materials is accurately known. An FDTD model and a geometry model were combined into hybrid models for outdoor-to-indoor coverage predictions in [42] and [44] and for indoor-to-outdoor propagation scenarios in [45].

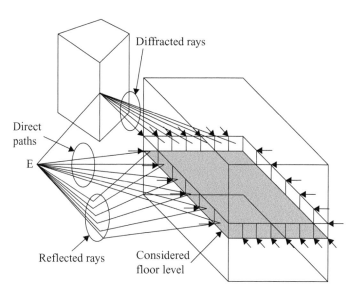

Figure 2.21 Hybrid outdoor-to-indoor model [44].

2.2.4.1 Hybrid outdoor-to-indoor models

Note that it is better to use geometry models for outdoor predictions, since outdoor scenarios are usually not flat and 3D predictions would be required, for which the computational complexity would be too high for FDTD models. In indoor scenarios, predictions based on 2D FDTD models are usually sufficient. The hybrid outdoor-to-indoor model in [44] uses the intelligent ray launching algorithm proposed in [38] for outdoor prediction, and the MR-FDPF model in [58] for indoor prediction. These two models were connected through the transformation of outdoor rays into indoor ParFlow. As depicted in Fig. 2.21, the outdoor prediction is first performed using the ray launching model. The rays hitting the cubes on the targeted building are accumulated for each cube and summed up considering their amplitudes, delays, and phases, and converted into ParFlow for use in the MR-FDPF model. For example, the converted ParFlow is given by $V = \sum_{i=1}^{N} v_i g_i \tau_i$, where N is the number of rays hitting a particular cube, v_i, g_i and τ_i are the amplitude, phase shift, and delay of the ith ray, respectively.

2.2.4.2 Hybrid indoor-to-outdoor models

The ray launching model and the MR-FDPF model can also be jointly used for indoor-to-outdoor scenarios, i.e., the MR-FDPF model is used for indoor prediction, the flows hitting the boundary of the building are stored and converted to outdoor rays, and the outdoor propagation is then computed using the ray launching model. Such a hybrid model is able to compute a considerably large scenario with a relatively high level of accuracy.

2.3 Clutter and terrain

The topography data of a typical outdoor scenario consists of terrain height information [7] and clutter data [59]. Building data is usually obtained from satellite and/or aerial imagery, assuming flat rooftops for the buildings. For outdoor predictions, terrain and clutter data has a huge impact on predictions of the received signal strengths, especially if the transmitted frequency is high or local environmental effects are significant. For example, an additional loss should be considered if a high-frequency signal propagates through a forest. There are three options to consider clutter data in propagation predictions [59].

- Use an empirical loss for each clutter type. This method does not incur extra complexity in the propagation model and is fast, but it is not very accurate.
- Assume that most clutters are 'non-transparent' blocks [59]. This approach has been proved to be the most effective for high-frequency signals, which see high attenuation in propagation [59].
- Assume the propagation mechanism of signals in a clutter as 'pass-through attenuation' [59]. For example, the signal strength degrades by a ratio in dB/m after passing through a clutter.

2.4   Antenna radiation pattern

In HCNs, UEs are typically equipped with omnidirectional antennas with, e.g., a 2 dBi gain. Outdoor macrocell BS antennas are much more directive and have a higher gain such as 18 dBi, while *low-power node* (LPN) antennas have a lower gain, e.g., picocell antenna gains are around 5 dBi. The antenna gains at both the transmitter and the receiver should be included in the path loss computation. Accordingly, the signal strength (in dBm) can be formulated as follows:

$$P_{receiver} = P_{emitter} - \text{PL} + G_{emitter} + G_{receiver} , \qquad (2.42)$$

where $P_{receiver}$ is the signal strength in dBm at the receiver, $P_{emitter}$ is the transmitted power level in dBm at the transmitter, PL is the path loss in dB between the transmitter and receiver, and $G_{emitter}$ and $G_{receiver}$ are the antenna gains in dBi for the transmitter and the receiver, respectively.

Antenna gains can be added to the obtained path loss predictions to form the propagation model, e.g., $P_{receiver} = P_{emitter} - \text{PL} + G(\omega, \theta)$, where $G(\omega, \theta)$ is the antenna gain in the horizontal direction of ω and vertical direction of θ. Antenna gains can be adjusted dynamically in geometry models, e.g., ray tracing or ray launching, where antenna gains should be added to each ray launched or traced.

Antenna patterns are typically given in the form of a $2 \times 2\text{D}$ plane, where the horizontal plane is of 360 degrees while the vertical plane is of 180 degrees. Many methods have been proposed to reconstruct the full 3D pattern from the $2 \times 2\text{D}$ antenna pattern, such as the non-linear interpolation method. Antenna manufacturers usually deliver antenna radiation patterns by means of horizontal and vertical cuts. A classical method to interpolate the 3D pattern is to add the antenna gains in the horizontal and vertical cuts, respectively, although comparisons with field measurements have shown the limitations of this method, especially for directive antennas that are tilted either electrically or mechanically. More information on 3GPP antenna gain models for HCN scenarios will be provided in Chapter 3.

2.5   Calibration

It is not possible to include all propagation phenomena in one model, and there are always uncertainties in a propagation model. For instance, it is not always possible to know exactly the materials of buildings, and some factors, such as unpredictable moving objects in the scenario, might be better ignored. Calibration is thus required for simulations to closely match the reality. Calibration is usually based on measurement campaigns to tune important parameters such as material properties and path loss coefficients. Given a solution domain, either finite or infinite, the calibration process needs to find the optimal set of parameters that make the simulation match the measurements as closely as possible. The calibration process can be formulated as the minimization of the *root mean square error* (RMSE) function Q as follows

$$Q = \sqrt{\frac{1}{m} \sum_{k=1}^{m} (M_k - P_k)^2}, \qquad (2.43)$$

where m is the number of measurement points, and M_k and P_k are the measured signal strength and the predicted signal strength of the kth measurement point, respectively. A constant offset between M_k and P_k can be obtained by computing the mean error as follows:

$$\Delta\Psi = \frac{1}{m} \sum_{k=1}^{m} (M_k - P_k), \qquad (2.44)$$

and then $P_k = M_k + \Delta\Psi$.

The calibration process can be slightly different for different propagation models. For example, the ray launching model in [38] and [40] employs a meta-heuristic simulated-annealing approach for calibration, as illustrated in Fig. 2.22 [40]. We can see that the RMSE decreases with the time evolution of simulated annealing.

2.6   MIMO channel models

Multi-antenna techniques play an important role in enabling HCN deployments [60]. Huge research efforts have been focused on *multiple-input multiple-output* (MIMO) technologies in order to meet the growing demands for faster and more reliable transmissions over harsh wireless channels. In MIMO systems, multiple collocated or distributed antennas are used to exploit the spatial dimension of radio channels, for capacity enhancement through spatial multiplexing, multi-path fading mitigation through spatial diversity, and/or *signal to interference plus noise ratio* (SINR) improvement through directional transmission (i.e., beamforming). MIMO technologies are included as a key component in latest wireless standards such as LTE and *Worldwide Interoperability for*

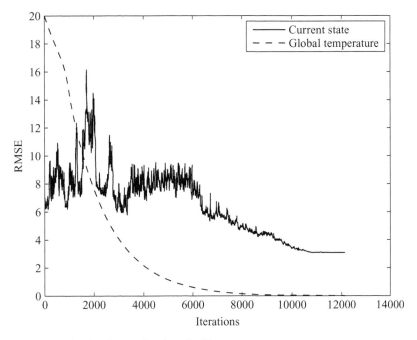

Figure 2.22 Simulated-annealing-based calibration.

Microwave Access (WiMAX). More information on MIMO technologies in HCNs will be provided in Chapter 11.

Channel models needed in the development of MIMO systems have to consider signal dispersions in the angular, delay, and Doppler domains. The emergence of multi-user MIMO and cooperative communications techniques also call for realistic multi-link channel models. Deterministic models, such as MR-FPDF and ray tracing models, can be used to model such complex multi-link channels, but even a few nodes of the network would lead to very high computational load. Moreover, deterministic models need a lot of improvements in order to address mobility and diffuse scattering.

Stochastic MIMO channel models can use a limited number of parameters to characterize the channel statistics. The computational complexity of the stochastic model depends on the scope of the system. On the one hand, analytical (non-physical) models mathematically characterize the MIMO channel matrices, including the antenna effects. Examples include the IEEE 802.11n tapped angular-delay line model, the correlation-based Kronecker model [60, 61], and the eigenspace-based Weichselberger model [62]. On the other hand, physical models characterize the radio waves by their delays, *direction-of-departures* (DoDs), *direction-of-arrivals* (DoAs), and complex path weights for different polarizations. Physical models are antenna independent, but can be directly combined with antenna array responses to synthesize the MIMO channel matrices.

The 3GPP *spatial channel model* (SCM) [61] was initially developed for evaluating different MIMO concepts in outdoor environments at a center frequency of 2 GHz and with a system bandwidth of 5 MHz. The WINNER I and II models [63] were based

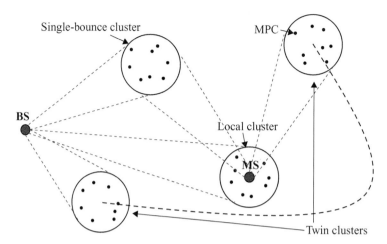

Figure 2.23 Example of the general structure of the COST 2100 model, containing three kinds of cluster: local clusters, single-bounce clusters and twin clusters.

on the COST 259 model [64] and the 3GPP SCM, with extended frequency range and bandwidth. *Geometry-based stochastic channel models* (GSCMs) constitute a group of advanced stochastic physical MIMO channel models that describe the channel statistics based on explicit geometries of the scatterers. The COST 259 channel model [64] was the first GSCM considering multi-antenna BSs, while full MIMO systems were later targeted by the COST 273 model [65]. The COST 2100 MIMO channel model extends the COST 273 model to cover MIMO systems at large, taking into account multi-user, multi-cellular and cooperative aspects, but without requiring a fundamental shift in the original modeling philosophy.

2.6.1   Geometry-based stochastic channel models

The principle of GSCMs is to model the stochastic properties of wireless channels by analyzing the geometric distributions of the interacting objects (or scatterers) in the environment [66], based on the concept that a radio channel can be modeled as the superposition of MPCs. MPCs are generated by the interactions between the radio waves and the objects in the environment, where each MPC is characterized by a geometric description in terms of delay, DoD and DoA. In an effort to reduce the number of model parameters, MPCs with similar delays and directions are sometimes grouped into a cluster (see Fig. 2.23). Different GSCMs differ mainly in the proposed distribution of scatterers. The simplest GSCM assumes that scatterers are uniformly distributed over a certain propagation region, and ignores low-power contributions from distant scatterers.

An important criterion to evaluate the accuracy of a GSCM is that the channel *large-scale parameters* (LSPs) synthesized from the model, such as the global delay and angular spread, should be statistically reliable and consistent with experimental observations. This means that cluster-based GSCMs should keep their accuracy at both the cluster and system levels. The 3GPP SCM [61] and the WINNER II model [63] are system-level GSCMs, while the COST channel models are cluster-level GSCMs.

In system-level GSCMs (taking the WINNER II model as an example), the modeling process for each instance of the channel between a BS and a UE is specified by

- defining the LSPs by their stochastic distributions for each channel instance,
- generating the clusters and MPCs according to the LSPs for any given locations of the BS and the UE.

The advantage of a system-level GSCM is that the LSP statistics in a specific scenario are always guaranteed at each channel instance. However, forcing the statistical consistency of LSPs brings two limitations:

- the rigid structure of the system-level GSCM does not spontaneously support continuous channel descriptions over intervals larger than the autocorrelation distance, hence hampering the simulation of large UE velocities;
- since the propagation environment is described based on the LSPs only, adding new LSPs (e.g., interlink correlation) into the model requires redefining the entire initialization of the environment, thereby hindering a straightforward extension of the model.

In cluster-level GSCMs, such as the COST 2100 model, the modeling process is specified once for the entire environment by

- defining a large quantity of clusters with consistent stochastic parameters throughout the simulation environment based on the BS location (yet, not all clusters will be visible at any time instance),
- defining the UE's location and determining the scattering from the visible clusters at each channel instance,
- synthesizing the LSPs based on the cluster scattering.

Cluster-level GSCMs are not constrained by the LSPs, and are independent of UE locations in the description of the propagation environment. This allows for smooth modeling of time-variant channels. Moreover, new LSPs can be included to explore advanced properties of clusters in a flexible way (i.e., without modifying the model structure).

In any single realization of the channel, a cluster-level GSCM is expected to exhibit larger deviations of LSP statistics than a system-level GSCM, because LSP statistics are not expressly forced by the cluster-level GSCM. Nevertheless, the LSP statistics of a cluster-level GSCM remain consistent with measurements on average. System-level and cluster-level GSCMs have similar simulation complexities, as they are both based on summing up contributions from a number of MPCs. The cluster-level COST 2100 model uses a relatively more complex representation of clusters, making the identification, estimation, and parametrization of clusters from measurements a critical task.

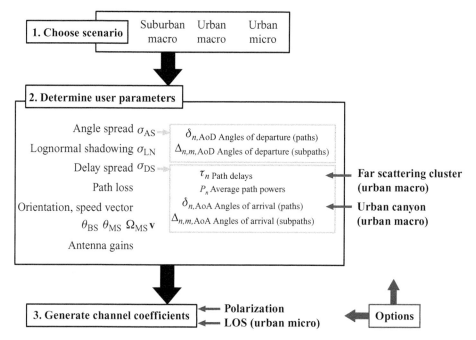

Figure 2.24 Channel model overview for simulations.

2.6.2   3GPP SCM and WINNER I model

The 3GPP/3GPP2 SCM [61] was established specifically for the simulation of 5 MHz bandwidth 3G networks in urban and suburban macrocells as well as in urban microcells. The SCM prescribes a specific discrete implementation, e.g., with fixed values for DoDs, DoAs, and azimuth spreads, and incorporates correlations between different LSPs. Accordingly, the SCM models different possible segments of the UE motion within a cell, instead of continuous and large-scale movements of the UE.

The 3GPP/3GPP2 SCM can be implemented as a tapped delay line model. Each tap consists of several subpaths that share the same delay, but have different DoDs and DoAs. Several options have been defined to include specific channel properties such as polarized antennas, far scatterer clusters, LOS, and urban canyons (Fig. 2.24). Strong interferers are modeled as spatially correlated, while weak interferers are taken as spatially white. Shadowing correlation is set as 0 for multiple UEs connected to the same BS, and 0.5 for a single UE connected to multiple BSs.

Because the bandwidth of the 3GPP/3GPP2 SCM is only 5 MHz, even with wideband extension, it was not adequate for advanced WINNER I simulations. Therefore, the 3GPP/3GPP2 SCM was extended into the WINNER I model [63], based on channel measurements performed at the 2 GHz and 5 GHz bands. The WINNER I model is a ray-based double-directional multi-link model, which is antenna independent, scalable, and capable of modeling channels for MIMO systems. In particular, it includes intra-cluster delay spreads, a LOS and K-factor model for all scenarios, time-variant shadowing, as

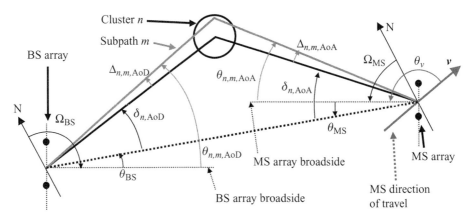

Figure 2.25 BS and UE angle parameters.

well as path angles and delays, as depicted in Fig. 2.25. It can accommodate statistical distributions and channel parameters extracted from measurements in the following propagation scenarios: indoor, typical urban microcell, typical urban macrocell, suburban macrocell, rural macrocell, and stationary feeder link. The WINNER I model can be implemented by defining a number of LSPs: the shadowing standard deviation, Ricean K-factor, delay spread, and spreads of DoDs and DoAs.

For a given link, the model fixes the LSPs according to prescribed distributions. This implies that only short segments of successive channel matrices can be generated. Different segments (i.e., different time periods of a given link) are related by correlating the LSPs as a function of the inter-segment distance, but the clusters of each segment are generated for that segment only. In other words, clusters (or scatterers) of any two segments are generated independently no matter how close they are. Two segments may share highly correlated LSPs, but see totally different clusters. For any segment, the WINNER I model then generates MPCs using clusters of scatterers.

2.6.3 WINNER II model

The WINNER I channel model was updated in 2007 to become the WINNER II channel model [63], which is a typical multi-link system-level GSCM. In the WINNER II multi-link model, the propagation environment is specified for each realization of all radio links between BSs and UEs. Whereas multiple links between multiple BSs, UEs or relays can be simulated simultaneously, each simulation is run for each link separately, similar to the 3GPP/3GPP2 SCM. The correlation between multiple links is introduced by correlating the LSPs as a function of distance. The shadowing correlation is set to zero for links from a UE to multiple BSs, which is different from the SCM shadowing correlation.

The advantage of the WINNER II model is that the large-scale statistics in a specific scenario are always guaranteed by any channel realization. However, the independent initialization of the propagation environment for each realization prohibits the connection between different realizations, which is important for characterizing the time variations caused by the motion of UEs. Typically, adding new LSPs in the analysis would require redefining the entire initialization of the propagation environment, thus deterring the straightforward extension of the model.

2.6.4   COST 259/273/2100 MIMO channel models

The COST 259 *directional channel model* (DCM) [64] provides a model for the delay and angle distributions on both sides of a radio link for 13 generic radio environments. Each radio environment is described by global parameters (e.g., BS location, frequency, heights of BS and UE) and their distributions. Values of those global parameters are determined through a mixed geometric and stochastic approach. Each radio environment contains a number of propagation environments, which are defined as an area where the local parameters are random realizations of the global parameters, representing instantaneous channel behaviors. The COST 259 DCM accounts for the relationship between several main parameters, such as BS–UE distance, delay dispersion, and angular spread. Its major restrictions include the assumption of stationary scatterers and the requirement of many scatterers to obtain an exponential power delay profile.

The COST 273 model [65] is a double-directional extension of the COST 259 model, allowing significantly different radio environments for the two sides of a radio link. For example, scattering clusters are modeled with two separate angular distributions at the transmitter and the receiver. Moreover, it is able to cover numerous new scenarios, such as peer-to-peer communications.

The COST 2100 model [67] further extends the COST 273 model by including

- a polarization model of multi-path contributions,
- the addition of dense MPCs to the specular contributions,
- the extension to multi-link (e.g., multi-cell, multi-UE) MIMO scenarios.

COST 2100 single-link MIMO channel model
The COST 2100 single-link MIMO channel model is illustrated in Fig. 2.23, where the radio waves travel from a BS to a UE through the physical environment via different MPCs, as a result of interactions with scatterers. An MPC is characterized by its delay (τ), *angle of departure* (AoD), *elevation of departure* (EoD), *angle of arrival* (AoA) and *elevation of arrival* (EoA). MPCs with similar delays and angles are grouped into a cluster, in order to reduce the number of channel parameters. The COST 2100 model calculates the time-varying *channel impulse response* (CIR) as follows:

$$h(t, \tau, \Omega^{BS}, \Omega^{UE}) = \sum_{n \in C} \sum_p \alpha_{n,p} \delta \left(\Omega^{BS} - \Omega^{BS}_{n,p} \right) \delta \left(\Omega^{UE} - \Omega^{UE}_{n,p} \right), \qquad (2.45)$$

where C is the index set of visible clusters, and $\alpha_{n,p}$, $\Omega^{BS}_{n,p}$, and $\Omega^{UE}_{n,p}$ are the complex amplitude, the direction of departure (AoD and EoD), and the direction of arrival (AoA and EoA) of the pth MPC in the nth cluster, respectively.

For a MIMO system employing V and U antennas at the BS and the UE, respectively, under the assumptions of plane waves and balanced narrowband arrays, the $U \times V$

MIMO channel matrix $H(t, \tau)$ can be expressed as follows:

$$
\begin{aligned}
H(t, \tau) &= \int \int h(t, \tau, \Omega^{\mathrm{BS}}, \Omega^{\mathrm{UE}}) s_{\mathrm{UE}}(\Omega^{\mathrm{UE}}) s_{\mathrm{BS}}^{\mathrm{T}}(\Omega^{\mathrm{BS}}) d\Omega^{\mathrm{UE}} d\Omega^{\mathrm{BS}} \\
&= \sum_{n \in C} \sum_{p} \alpha_{n,p} s_{\mathrm{UE}}(\Omega^{\mathrm{UE}}) s_{\mathrm{BS}}^{\mathrm{T}}(\Omega^{\mathrm{BS}}).
\end{aligned}
\tag{2.46}
$$

Clusters

A cluster is modeled as an ellipsoid in space. The size of a cluster is determined by its axes along different directions, corresponding to the maximum *cluster delay spread* (CDS) and *cluster angular spread* (CAS) in both the azimuth and elevation planes. There are two major kinds of cluster.

- Local cluster: located around the BS or the UE, and contains single-bounce MPCs uniformly distributed in the azimuth plane. The spatial spread of a local cluster is determined by its delay spread and elevation spread.
- Far cluster (or remote cluster): located away from both the BS and the UE, and could be either a single-bounce cluster or a multiple-bounce cluster.
  - Single-bounce cluster: a rotated ellipsoid with the spatial spread determined by the delay and angular spreads viewed from the BS, and the delay spread and the azimuth spread are independent.
  - Multiple-bounce cluster (or twin cluster), as shown in Fig. 2.23: described by two observations of an identical ellipsoid made at the BS and the UE, respectively.

The positions and spread of clusters need to be specified on both the BS and UE sides. For local and single-bounce clusters, such specifications are identical on both sides since they only need one representation in space. For twin clusters, a specification should be provided for each side of the radio link. All CDS and CAS parameters as well as the shadowing factor of a given cluster are considered as correlated non-negative random variables. The actual correlation coefficients depend on the scenario of interest.

Visibility region

The visibility of a far cluster is determined by its *visibility region* (VR), which decides the number of active far clusters for a radio link. Local clusters are visible by default. The VR is a circular region in the azimuth plane. A connection between the BS and a far cluster is established only when the UE is inside the VR of the far cluster. Each VR determines the visibility of one single cluster. The visibility level of a cluster is controlled by the visibility gain, A_{VR}, which increases as the UE approaches the VR center as follows:

$$
A_{\mathrm{VR}}(\gamma_{\mathrm{UE}}) = \frac{1}{2} - \frac{1}{\pi} \arctan \left(\frac{2\sqrt{2}(L_{\mathrm{C}} + d_{\mathrm{UE,VR}} - R_{\mathrm{C}})}{\sqrt{\lambda L_{\mathrm{C}}}} \right),
\tag{2.47}
$$

where $d_{\mathrm{UE,VR}}$ is the distance from the UE to the center of the VR, R_{C} is the VR radius, and L_{C} is the *transition region* (TR) width.

VRs are uniformly distributed in a cell, with a spatial density given by

$$\rho_C = \frac{N_C}{\pi (R_C - L_C)^2}[1/m^2],\tag{2.48}$$

where N_C is the average number of visible far clusters in the cell, and can be modeled as a Poisson distributed random variable. Different VRs may overlap, allowing UEs in the overlapping region to observe multiple visible clusters.

Multi-path and LOS components

The MPCs inside a cluster are spatially Gaussian distributed. The distribution of MPCs in a twin cluster is identical at the BS and the UE, in order to guarantee consistent delay and angular spreads on both sides. Rayleigh fading is assumed for all MPCs in each cluster. In the COST 2100 channel model, the LOS component is a special cluster containing a single MPC, with its VR determined by a cut-off distance from the BS. The power of the LOS component is determined by a log-normally distributed factor over the total power of all other visible clusters.

Time evolution

The COST 2100 channel model assumes a deterministic environment with all MPCs maintaining flat fading at any time instance. Therefore, time variation of the channel is caused by movements of UEs. The visibility of a cluster to the channel may change as the UE enters or leaves the corresponding VR. A moving UE will also cause the movement of local clusters around it, necessitating the update of local MPCs inside the local clusters for each new position of the UE.

Polarization

The COST 2100 channel model structure is designed for easy incorporation of polarization characteristics. More specifically, for each radio propagation path, four channels are generated and are then projected onto the transmit and receive antenna systems, with each of the four channels representing a particular combination of horizontal and/or vertical polarizations between the transmitter and the receiver. The phases of the four channels of each path are defined as independent and identically distributed uniform random variables taking value between 0 and 2π. The four-channel path matrix is then normalized and multiplied by a complex path amplitude coefficient containing path loss, shadowing, cluster attenuation, etc. The three cross-polar ratios are defined per path and per cluster. The cross-polar ratios (i.e. ratio between horizontal and vertical polarization) of different paths in one cluster are log-normally distributed.

Diffuse multi-paths

The COST 2100 model relies on specular scattering, which assumes that the interaction between the electromagnetic wave and a scatterer produces only one propagation path. In reality, rough surface reflections, corner diffractions, and reflections on different layers of a scatterer can all contribute to a large amount of diffuse scattering, which may form a significant component in both delay and angular domains. Such complex scattering mechanisms cannot be fully captured by a few specular MPCs with explicit geometry descriptions.

There are two approaches to characterize diffuse scattering:

- extending the propagation path with a continuous dispersion in delay and angular domains to include diffuse scattering characteristics;
- the superposition of a large number of specular MPCs with modified delays, angles, and amplitudes, which are also known as *dense multi-path componens* (DMCs) [68].

The first approach relies on the quality of the path dispersion modeling, i.e., the model mismatch might create a significant number of artifacts. The second method is able to capture the residual channel spectrum as long as the number of DMCs is sufficiently large, but it increases the total number of model parameters. The COST 2100 model considers the second approach as a direct extension of the MPC concept. In principle, a DMC is a modified MPC with its own delay, direction, fading, and power attenuation. The modeling complexity can be kept at a reasonable level by characterizing DMCs at the cluster level. The DMCs of a cluster share the large-scale properties of the cluster, such as shadowing and cluster power attenuation.

COST 2100 multi-link MIMO channel model
Measurements have shown that links could be correlated even if their BSs and/or UEs are well separated. Accordingly, large-scale correlation is introduced to characterize the correlation between two links separated by a distance larger than the coherence distance on either the BS side or the UE side. One possible explanation for such correlation is the existence of correlated clustering, i.e., clusters in different links show correlated fading or LSPs. However, using correlated clusters would contradict the uncorrelated-clustering assumption adopted in the COST 2100 model, hence requiring substantial modifications of the modeling approach. Another possible explanation is that some clusters are common in multiple links, i.e., clusters are simultaneously visible in different links. This approach requires the characterization of cluster visibility in different links, without altering the other physical properties of the clusters. Consequently, this approach would guarantee a compatible extension of the COST 2100 model upon the existing structure in single-link scenarios. In multi-link scenarios, a cluster might be associated with multiple VRs, while each VR is associated with a single cluster and determines to which BSs the cluster is visible. Therefore, the VRs define the cluster visibility to multiple BSs.

2.6.5   Perspectives of channel modeling

Depending on the considered scenario, the scale of the network, and the heterogeneity of the system, each channel model has its advantages and drawbacks. In HCNs, it is essential for channel models to be precise and flexible, taking into account the heterogeneity, mobility, and large numbers of links. A good tradeoff between realism and complexity has to be found. The COST 2100 model offers many advantages and is constantly enhanced thanks to the continuous work of the COST community, with particular focuses on non-BS-centric cases and on multi-scale scenarios commonly found in HCNs. Any evolution of existing channel models must be extracted from adequate measurement campaigns, particularly responding to new network scenarios.

However, measurement campaigns are technically challenging and time consuming. A promising way forward seems to be the combination of measurements and deterministic simulations of a large number of environments and scenarios. For instance, ray tracing or discrete methods can be well calibrated with some measurements, and the calibrated ray tracing model can be applied to a wide range of configurations afterwards without the need of more measurement campaigns.

2.7   Summary and conclusions

This chapter has provided a detailed description of radio propagation modeling. There are several different types of channel model: empirical models, deterministic models, semi-deterministic models, and hybrid models, which can be used for network planning and optimization, etc. Generally speaking, empirical models are easy and fast to implement, but have the disadvantage of a low level of accuracy. In contrast, deterministic models achieve a much higher level of accuracy than empirical models, but at the cost of high complexity. Thus there exists a tradeoff between accuracy and complexity in the channel models. In the last part of the chapter, MIMO channel models are included since MIMO technology plays an important role in the modern wireless communication standards such as LTE, WiMAX etc. In this chapter, the case studies and simulation results are presented using Ranplan iBuildNet tool.

The authors of this chapter wish to thank all contributors to the COST 2100 working group on channel modeling and particularly the chairman Claude Oestges. This work was supported by the EU-FP7 iPLAN (Project Number 230745) and EU-FP7 IAPP@RANPLAN (Project Number 218309). Also, acknowledgments have to be extended to the iBuildNet® tool from Ranplan Wireless Network Design Ltd., UK.

References

[1] F. Fontan and P. Espineira, *Modeling the Wireless Propagation Channel* (Wiley, 2008).

[2] G. D. Durgin, Theory of stochastic local area channel modeling for wireless communications. *Technology*, Dec. (2000), 1–207.

[3] D. Mcnamara, C. Pistorius and J. Malherbe, *Introduction to the Uniform Geometrical Theory of Diffraction* (Artech, 1990).

[4] A. Valcarce and J. Zhang, Empirical indoor-to-outdoor propagation model for residential areas at 0.9–3.5 GHz. *IEEE Transactions on Antennas and Propagation*, **9** (2010), 682–685.

[5] U. Ranplan Wireless Network Design Ltd., ranplan iBuildNet and RRPS. http://www.ranplan.co.uk

[6] European Cooperation in the Field of Scientific and Technical Research, *Digital Mobile Radio Towards Future Generation Systems, COST231 Final Report* (1999) http://www.lx.it.pt/cost231/

[7] D. John, *Introduction to Radio Propagation for Fixed and Mobile Communications* (Boston, MA: Artech, 1996).

[8] *Recommendation itu-r p.529-2* (1995) Prediction Methods for the Terrestrial Land Mobile Service in the VHF and UHF Bands.

[9] *Recommendation itu-r p.529-3* (1999) Prediction Methods for the Terrestrial Land Mobile Service in the VHF and UHF Bands.

[10] *Erc Report 68* (Naples, 2000) Monte Carlo Radio Simulation Methodology.

[11] J. Maxwell, On physical lines of force. *The London, Edinburgh, and Dublin Philosophical Magazine and Journal of Science*, **21**:139 (1861), 161–175.

[12] K. Yee, Numerical solution of initial boundary value problems involving Maxwell's equations in isotropic media. *IEEE Transactions on Antennas and Propagations*, **14**:3 (1966), 302–307.

[13] A. Taflove, Application of the finite-difference time-domain method to sinusoidal steady-state electromagnetic-penetration problems. *IEEE Transactions on Electromagnetic Compatibility*, **EMC-22**:3 (1980), 191–202.

[14] A. Rial, Applying the finite-difference time-domain to the modelling of large-scale radio channels. Ph.D. dissertation (2010).

[15] A. Taflove and M. Brodwin, Numerical solution of steady-state electromagnetic scattering problems using the time-dependent Maxwell's equations. *IEEE Transactions on Microwave Theory and Techniques*, **23**:8 (1975), 623–630.

[16] A. Taflove and S. Hagness, *Computational Electrodynamics*. (Boston, MA: Artech, 1995).

[17] J. Schneider and C. Wagner, FDTD dispersion revisited: faster-than-light propagation. *IEEE Microwave and Guided Wave Letters*, **9**:2 (1999), 54–56.

[18] J. Berenger, A perfectly matched layer for the absorption of electromagnetic waves. *Journal of Computational Physics*, **114**:2 (1994), 185–200.

[19] B. Chopard, P. Luthi and J. Wagen, Lattice Boltzmann method for wave propagation in urban microcells. *IEE Proceedings – Microwaves, Antennas and Propagation*, **144**:4 (1997), 251–255.

[20] B. Chopard and M. Droz, *Cellular Automata Modeling of Physical Systems* (Cambridge: Cambridge University Press, 1998).

[21] J. Gorce, K. Jaffres-Runser and G. De La Roche, Deterministic approach for fast simulations of indoor radio wave propagation. *IEEE Transactions on Antennas and Propagation*, **55**:3 (2007), 938–948.

[22] K. Runser and J. Gorce, Assessment of a new indoor propagation prediction method based on a multi-resolution algorithm. In *IEEE 61st Vehicular Technology Conference, VTC 2005–Spring* (2005), vol. 1, pp. 35–38.

[23] G. De La Roche, J. Wagen, G. Villemaud, J. Gorce and J. Zhang, Comparison between two implementations of parflow for simulating femtocell networks. In *2011 Proceedings of 20th International Conference on Computer Communications and Networks (ICCCN)* (IEEE, 2011), pp. 1–4.

[24] M. Luo, D. Umansky, G. Villemaud, M. Lafort and J. Gorce, Estimating channel fading statistics based on radio wave propagation predicted with deterministic MRFDPF method. In *Fifth European Conference on Antennas and Propagation (EuCAP 2011)* Rome (2011).

[25] M. Luo, G. De La Roche, G. Villemaud, J. Gorce, D. Umansky and J. Zhang, Simulation of wide band multipath fast fading based on finite difference method. In *IEEE Vehicular Technology Conference (VTC Fall)* (2011).

[26] M. Luo, N. Lebedev, G. Villemaud, G. De La Roche, J. Zhang, and J. Gorce, On predicting large scale fading characteristics with the MR-FDPF method. In *Sixth European Conference on Antennas and Propagation (EuCAP 2012)*, Prague (2012).

[27] P. Luthi, Lattice wave automata: from radio wave to fracture propagation. Ph.D. thesis, Computer Science Department, University of Geneva.

[28] A. Schuster, *An Introduction to the Theory of Optics*. London: Arnold (1904).

[29] V. Borovikov and B. Kinber, *Geometrical Theory of Diffraction*. (Institution of Electrical Engineers 1994).

[30] G. James, *Geometrical Theory of Diffraction for Electromagnetic Waves*. (Peregrinus, 1976).

[31] WinProp – wave propagation tool – ProMan (integrated in Atoll) – user reference. AWE Communication Online Documents (2003).

[32] J. Keller, Diffraction by an aperture. *Journal of Applied Physics*, **28**:4 (1957).

[33] J. Keller, One hundred years of diffraction theory. *IEEE Transactions on Antennas and Propagation*, **33** (1985), 123–126.

[34] G. Wolfle, B. Gschwendtner and F. Landstorfer, Intelligent ray tracing – a new approach for the field strength prediction in microcells. In *IEEE Vehicular Technology Conference*, Phoenix, AZ (1997), 790–794.

[35] T. Möller and B. Trumbore, Fast, minimum storage ray-triangle intersection. *Journal of Graphics, GPU, and Game Tools*, **2**:1 (1997), 21–28.

[36] A. Cavalcante, M. De Sousa, J. Costa, C. Frances and G. Dos Santos Cavalcante, A parallel approach for 3D ray-tracing techniques in the radio propagation prediction. *Journal of Microwaves and Optoelectronics*, **5**:5 (2007), 271–279.

[37] R. Wahl, G. Wolfle, P. Wertz, P. Wildbolz and F. Landstorfer, Dominant path prediction model for urban scenarios. In *14th IST Mobile and Wireless Communications Summit*, Dresden (2005).

[38] Z. Lai, N. Bessis, G. De La Roche, H. Song, J. Zhang and G. Clapworthy, An intelligent ray launching for urban propagation prediction. In *The Third European Conference On Antennas and Propagation (EUCAP)*, Berlin (2009), pp. 2867–2871.

[39] *COST231 Urban Micro Cell Measurements and Building Data,* http://www2.ihe.uni-karlsruhe.de/forschung/cost231/cost231.en.html

[40] Z. Lai, N. Bessis, G. De La Roche, P. Kuonen, J. Zhang and G. Clapworthy, On the use of an intelligent ray launching for indoor scenarios. In *The Fourth European Conference on Antennas and Propagation EUCAP*, Barcelona (2010).

[41] Z. Lai, G. De La Roche, N. Bessis, P. Kuonen, G. Clapworthy, D. Zhou and J. Zhang, Intelligent ray launching algorithm for indoor scenarios. *Radioengineering*, **20**:2 (2011), 398–408.

[42] G. De La Roche, P. Flipo, Z. Lai, G. Villemaud, J. Zhang and J. Gorce, Combination of geometric and finite difference models for radio wave propagation in outdoor to indoor scenarios. In *The Fourth European Conference On Antennas and Propagation (EUCAP)*, Barcelona (2010).

[43] G. De La Roche, P. Flipo, Z. Lai, G. Villemaud, J. Zhang and J. Gorce, Combined model for outdoor to indoor radio propagation. In *10th COST2100 Management Meeting, TD(10)10045*, Athens (2010).

[44] G. D. Roche, P. Flipo, Z. Lai, G. Villemaud, J. Zhang and J. Gorce, Implementation and validation of a new combined model for outdoor to indoor radio coverage predictions. *EURASIP Journal on Wireless Communications and Networking*, (2010), 215–352.

[45] D. Umansky, G. D. Roche, Z. Lai, G. Villemaud, J. Gorce and J. Zhang, A new determin-istic hybrid model for indoor-to-outdoor radio coverage prediction. In *The Fifth European Conference on Antennas and Propagation (EUCAP)*, Rome (2011).

[46] H. Holma and A. Toskala, *WCDMA for UMTS, Radio Access for 3G Mobile Communications*, 3rd edn (Wiley, 2004).

[47] C. Haslett, *Essentials of Radio Wave Propagation* (Cambridge University Press, 2008).

[48] M. Klepal, Novel approach to indoor electromagnetic wave propagation modeling. Ph.D. dissertation, Czech Technical University in Prague (2003).

[49] Y. Corre, Y. Lostanlen and Y. Helloco, A new approach for radio propagation modeling in urban environment: knife-edge diffraction combined with 2D ray-tracing. In *IEEE Vehicular Technology Conference* (2002).

[50] M. Lott and I. Forkel, A multi-wall-and-floor model for indoor radio propagation. In *IEEE Vehicular Technology Conference*, Rhodes (2001).

[51] Y. Lostanlen and Y. Corre, Indoor coverage maps over large urban areas: an enhanced ray-tracing method. In *The First International Conference On Wireless and Mobile Communications*, Nice (2006).

[52] E. Suikkanen, A. Tolli and M. Latva-aho, Characterization of propagation in an outdoor-to-indoor scenario at 780 MHz. In *2010 IEEE 21st International Symposium on Personal Indoor and Mobile Radio Communications*, Istanbul (2010).

[53] M. Alatossava, E. Suikkanen, J. Meinila, V. Holappa and J. Ylitalo, Extension of COST 231 path loss model in outdoor-to-indoor environment to 3.7 GHz and 5.25 GHz. In *International Symposium on Wireless Personal Multimedia Communication*, Sarriselka, Finland (2008).

[54] T. Kurner and A. Meier, Prediction of outdoor and outdoor-to-indoor coverage in urban areas at 1.8 GHz. *IEEE Journal on Selected Areas in Communications*, **20**:3 (2002), 496–506.

[55] J. Berg, *Digital Mobile Radio Toward Future Generation Systems* (European Commission, 1997).

[56] J. Berg, 4.6 building penetration: digital mobile radio toward future generation systems. In *COST Telecom Secretariat, Commission of the European Communities* (Brussels, 1999), pp. 167–174.

[57] R. Visbrot, A. Kozinskya, A. Freedman, A. Reichman and N. Blaunstein, Measurement campaign to determine and validate outdoor to indoor penetration models for GSM signals in various environments. In *2011 IEEE International Conference on Microwaves, Communications, Antennas and Electronics Systems (COMCAS)*, Tel Aviv (2011).

[58] G. De La Roche, J. Gorce and J. Zhang, Optimized implementation of the 3D MR-FDPF method for indoor radio propagation predictions. In *The Third European Conference on Antennas and Propagation (EUCAP)*, Berlin (2009).

[59] H. Anderson, T. Hicks and J. Kirtner, The application of land use/land cover (clutter) data to wireless communication system design. Technology White Paper, EDX Wireless LLC. http://www.edx.com (2008).

[60] S. Primak and V. Kontorovich, *Wireless Multi-Antenna Channels: Modeling and Simulation*. Wireless Communications and Mobile Computing Series (Wiley, 2011).

[61] 3GPP, Tr 25.996: spatial channel model for multiple input multiple output (mimo) simulations. *Change*, Release 10, 2011.

[62] N. Costa and S. Haykin, *Multiple-Input, Multiple-Output Channel Models: Theory and Practice*. Adaptive and Learning Systems for Signal Processing, Communications and Control Series (Wiley, 2010).

[63] IST-4-027756 winner II, D1.1.2. *Change*, **0** (2008).

[64] A. Molisch, H. Asplund, R. Heddergott, M. Steinbauer and T. Zwick, The COST259 directional channel model – part I: Overview and methodology. *IEEE Transactions on Wireless Communications*, **5**:12 (2006), 3421–3433.

[65] L. Correia, *Mobile Broadband Multimedia Networks: Techniques, Models and Tools for 4G, European Cooperation in the Field of Scientific and Technical Research (Organization)* (Elsevier–Academic, 2006).

[66] M. Steinbauer, A. Molisch and E. Bonek, The double-directional radio channel. *IEEE Antennas and Propagation Magazine*, **43**:4 (2001), 51–63.

[67] R. Verdone and A. Zanella, *Pervasive Mobile and Ambient Wireless Communications: Cost Action 2100*. Signals and Communication Technology Series (Springer, 2012).

[68] J. Poutanen, Geometry-based radio channel modeling: propagation analysis and concept development. Ph.D. dissertation, Aalto University (2011).

3 System-level simulation and evaluation models

David López-Pérez and Mats Folke

3.1 Introduction

In order to aid vendors and operators in the development and deployment of new *heterogeneous cellular networks* (HCNs), and the refinement of existing procedures such as *handover* (HO) and *radio resource management* (RRM), network simulation, planning and optimization tools that are able to evaluate the overall performance of complex cellular networks are highly regarded. In this context, system-level simulations have become a widely adopted methodology. In system-level simulations, the elements and operations of a cellular network are modeled by computer software. This approach is usually simpler and cheaper than real implementation, and is more accurate and reliable than analytical modeling. The number of assumptions and simplifications made in system-level simulations is up to the software designer, but it is usually less than that of analytical modeling. System-level simulations can also model more complex cellular networks than analytical modeling. However, system-level simulations for modeling intricate procedures or getting statistically representative results of network performance usually require significant computing capabilities. As a result, in order to avoid prohibitive computational costs, a tradeoff between accuracy and complexity should be reached. In this line, the *3rd Generation Partnership Project* (3GPP) provides guidelines on network simulations, and defines simulation procedures and parameters.

In this chapter, we look into important issues in system-level simulations for HCNs. In Section 3.2, we introduce different kinds of system-level simulation, including both static and dynamic simulations. In Section 3.3, we compare static system-level simulations versus dynamic ones. In Section 3.4, we discuss intricate modeling issues such as wraparound, shadowing, multi-path fading, antenna patterns and signal quality modeling. In Section 3.5, simulation setups for HCNs from the perspective of 3GPP are illustrated, taking into account the different features and channel models of small cells, and covering macro–pico, macro–femto and macro–relay scenarios. In Section 3.6, node and UE placement are also illustrated. In Section 3.7 and Section 3.8, traffic and mobility modeling are considered, respectively. For more detail on the simulation and evaluation models considered in this chapter, as well as models used for other specific scenarios, readers are referred to [1] for simulation of homogeneous networks, [2] for simulation of heterogeneous networks, and [3] for mobility modeling.

3.2   System-level simulation

Since simulating the transmissions of all bits between all *base stations* (BSs) and *user equipments* (UEs) in a cellular network could be prohibitive due to large computational costs, the simulation of a cellular network is usually divided into two distinct levels: *link-level simulation* and *system-level simulation*.

At the *link level*, the behavior of a radio link between a single transmitter and a single receiver is studied, considering all relevant aspects/procedures of both the *physical* (PHY) and *medium access control* (MAC) layers and radio propagation phenomenas at small temporal scales (e.g., per symbol duration).

- In the PHY layer, functions such as modulation, coding and interleaving should be carefully modeled according to the corresponding standard.
- In the MAC layer, the impact of different procedures/parameters, e.g., *hybrid automatic repeat request* (HARQ), should also be considered since they may significantly affect network performance.

In *Orthogonal Frequency Division Multiple Access* (OFDMA) systems, link-level simulations have been widely used, e.g., to investigate the performance of *adaptive modulation and coding* (AMC) and sub-carrier mapping schemes [4].

At the *system level*, in contrast, the behavior of a cellular network as a set of BSs and UEs is studied, considering issues such as HO, RRM etc. In this case, the target is not to analyze the performance of a specific radio link over small temporal scales, but to evaluate the performance of the entire network (in terms of *quality of service* (QoS), coverage, capacity, delay etc.) over long periods of time (e.g., minutes, hours and even days).

Link-level and system-level simulations are usually independently executed due to their different time scales and computational costs. However, in order to achieve an accurate and reliable network evaluation, they may still interact through static interfaces called *lookup tables* (LUTs). LUTs are generated by link-level simulations, and present link performance results in a simplified manner. For example, radio link quality in terms of *signal to interference plus noise ratio* (SINR) can be mapped to a radio link performance indicator, such as *block error rate* (BLER), as shown in Fig. 3.1 for different modulation and coding schemes. LUTs are usually queried during system-level simulations to assess radio link performance, resulting in a simple but efficient approach for including the effects of both the MAC and PHY layers as well as the fluctuations of the radio channel, which occur in small time scales, into system-level simulations, which are targeted at larger time scales.

Since the link-level simulation of different building blocks of a cellular network has been well defined by current cellular standards, e.g., 3GPP *Long Term Evolution* (LTE) and LTE-Advanced, in the rest of this chapter we will focus on system-level simulations. In more detail, we will describe the main building blocks and challenges of HCN system-level simulations, and explain the models and parameters suggested by 3GPP to simulate macrocell networks overlaid with *relay nodes* (RNs), picocells and femtocells.

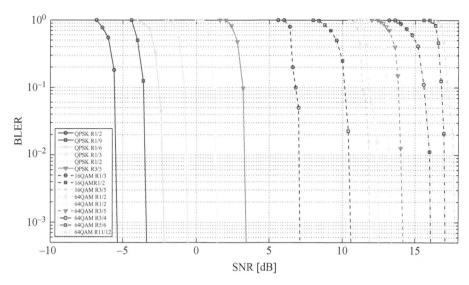

Figure 3.1 Block error rate lookup table for LTE system-level simulations.

3.3   Static versus dynamic system-level simulations

Although the general target of system-level simulations is to characterize the overall performance of a cellular network, they may be used to address a very wide range of issues, e.g., to analyze the benefits and drawbacks of deploying a large number of *low-power nodes* (LPNs) over existing macrocell networks, or to evaluate the impact of LPN self-organization techniques on the overall network performance.

Although there exist general-purpose system-level simulation tools, different system-level simulations tailored to specific targets are typically used by vendors and operators to investigate possible performance enhancements. In this chapter, we classify system-level simulations into *static* and *dynamic* system-level simulations [5].

3.3.1   Static snapshot-based approaches

In a static system-level simulation, the aim of the simulation is to study the average performance of a network over large areas and for long periods of time. In this case, simulations are typically based on Monte Carlo approaches, where multiple independent snapshots are used to evaluate network performance in a statistical manner [6]. Monte Carlo simulations have been widely used in network planning and optimization [7, 8]. In each snapshot, the time domain is neglected, and a random realization of the network is generated following appropriate distributions. The performance of the overall network is then analyzed in terms of coverage and capacity, for which cell/UE throughput and UE outage probability are widely used as *key performance indicators* (KPIs). Since the time domain is neglected, Monte Carlo simulations are easier to implement, and usually run much faster than dynamic simulations at the expense of reduced accuracy. In order to get a statistically representative result for the average network performance, Monte Carlo simulations require an adequate number of snapshots. In an HCN setting, a significant large number of snapshots may be needed to generate many different random drops of LPNs. Therefore, fast system-level simulation algorithms able to support hundreds of snapshots within a limited period of time are necessitated to statistically assess the overall performance of an HCN.

3.3.2   Dynamic event-driven approaches

In a dynamic system-level simulation, the target of the simulation is to accurately model the functioning of the network with a high degree of detail. In this case, the evolution of the network over time is taken into account, and the simulation allows the network to live as a function of time or series of events. In order to capture the end-to-end behavior of a network, the dynamic features of UEs (mobility models) and traffic (traffic models), as well as the fluctuations of the radio channel over frequency and time (e.g., correlated shadowing, speed-dependent fast fading) should be simulated [9]. In this way, the behavior of different techniques, such as HO procedures, power control, and subchannel scheduling, can be analyzed over time in terms of convergence, signalling, delay, etc. Accordingly, the running time of a dynamic system-level simulation significantly increases with the size of the network scenario considered, and thus only small areas are usually analyzed for short periods of time. In dynamic system-level simulations, network performance is typically assessed by measures such as cell/UE throughput, call block and drop rates, end-to-end delay and jitter, packets losses and/or retransmission ratios.

3.4   Building blocks

In this section, we will introduce system-level simulation building blocks, detailed description of which may be difficult to find in the current literature, and we will mainly consider 3GPP system-level simulation models.

3.4.1   Wrap-around

Fig. 3.2 illustrates the homogeneous macrocell layout used in 3GPP for analyzing network performance in both suburban and urban environments, where BSs are deployed in a hexagonal grid. In this case, tri-sectored BSs are considered. The number of BSs is 19, while the number of sectors is 57. This scenario is fully characterized by the *intersite distance* (ISD) D, which defines the regular distance between any two neighboring BSs.

Since the central BS is surrounded by a larger number of neighboring cells than any first- or second-tier BS, it is obvious that the the central BS will suffer from the highest intercell interference among all BSs in Fig. 3.2. In order to analyze network performance without incurring border effects, it would be necessary to either collect data only from the central BS, or consider a cellular network with an infinite number of tiers. The former decreases statistical sampling, while the latter increases computational complexity of system-level simulations. An alternative approach that allows simulating

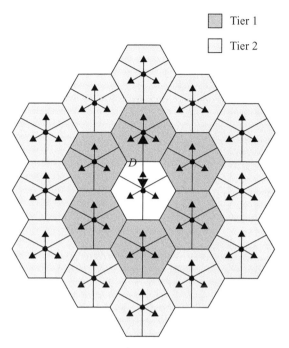

Figure 3.2 Homogeneous hexagonal-grid macrocell scenario.

a continuous network using a finite model is wrap-around, where an infinite planar cellular network, defined by a infinite hexagonal lattice, can be embedded on a torus to obtain a symmetric regular continuous finite network sample region for performance analysis. More specifically, the finite region is obtained by drawing a rhombus over the infinite hexagonal lattice, as shown in Fig. 3.3, and then joining the opposing edges of the rhombus to build a torus [10]. In Fig. 3.3, a cluster of 19 BSs, similar to that in Fig. 3.2, is repeated eight times at rhombus lattice vertices. The original cluster remains in the center, while the eight repeated clusters evenly surround the central one. By drawing a rhombus (solid lines in Fig. 3.3) over the hexagonal lattice and joining the opposing edges of the rhombus, a toroidal and continuous surface is obtained. In [11], the authors show that a hexagonal wrap toroidal embedding preserves translational and rotational symmetry on all three axes of a hexagonal lattice and is preferable to a cartesian wrap embedding.

In a wrap-around model, defining a coordinate system with the central BS at $(0,0)$ m, the distance d from a BS at (a, b) m to a UE at (x, y) m is given by

$$
d = \min \left\{
\begin{array}{c}
\sqrt{(x - a)^2 + (y - b)^2}, \\
\sqrt{(x - (a + 3D/3))^2 + (y - (b + 4D))^2}, \\
\sqrt{(x - (a - 3D/3))^2 + (y - (b - 4D))^2}, \\
\sqrt{(x - (a + 4.5D/3))^2 + (y - (b - 7D/2))^2}, \\
\sqrt{(x - (a - 4.5D/3))^2 + (y - (b + 7D/2))^2}, \\
\sqrt{(x - (a + 7.5D))^2 + (y - (b + D/2))^2}, \\
\sqrt{(x - (a - 7.5D))^2 + (y - (b - D/2))^2}.
\end{array}
\right\}
\tag{3.1}
$$

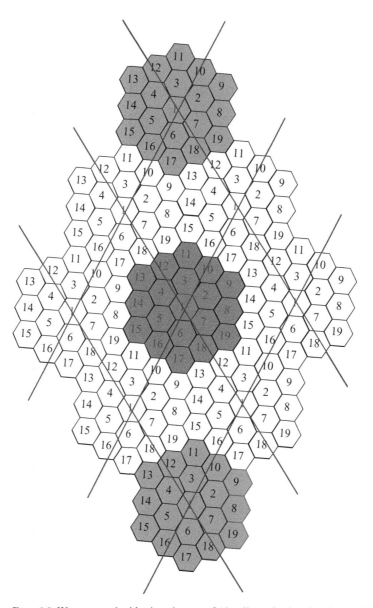

Figure 3.3 Wrap-around with nine clusters of 19 cells each, showing the toroidal nature of the wrap-around surface [10] (reproduced with permission from CEPT).

3.4.2   Shadow fading: auto- and cross-correlation

In wireless communications, fading characterizes the varying attenuation that a radio signal experiences through a certain propagation medium. Fading may vary with the geographical location, frequency and/or time, and is often modeled as a random process. In cellular networks, fading caused by obstacles affecting the radio signal propagation is referred to as shadow fading, while fading caused by multi-path propagation is known as multi-path fading [12]. In this section, we will focus on the modeling of shadow fading, while in the next section, we will discuss the modeling of multi-path fading.

A radio signal will typically experience obstructions caused by objects in its propagation path, thus generating random fluctuations of the received signal power at the receiver side. The locations, sizes and dielectric properties of the obstructing objects, as well as the reflecting surfaces and scattering obstacles are usually unknown. Due to such uncertainties, statistical models are generally used to model shadow fading. A widely used model is the log-normal shadowing model [12], which has been demonstrated to be able to accurately predict shadow fading attenuations in both outdoor and indoor environments.

Accordingly, the shadow fading in the path between a BS and a UE can be a priori modeled using a log-normal random variable, $L_s \sim N(\mu, \sigma^2)$, where μ and σ are the mean and standard deviation in dB, respectively. However, the modeling of shadow fading when considering multiple BSs and UEs is more intricate due to the spatial auto-correlation and cross-correlation properties.

- The shadowing process is auto-correlated, i.e., a moving UE may see similar shadow fading attenuations from a given BS at different but nearby locations (neighborhood of the UE), whereas the fading attenuations may significantly differ at distant positions.
- The shadowing process is cross-correlated, i.e., a UE at a given location may see similar shadow fading attenuations from different BSs, depending on the surrounding environment.

Deterministic radio propagation tools presented in Chapter 2 can be used to study auto-correlation and cross-correlation shadowing properties, and extract models. In the following, we introduce a simple but useful model to comprehend both auto- and cross-correlation properties, which are widely used in the literature.

Auto-correlation
A widely adopted auto-correlation model for shadowing is the Gudmundson model [13], which defines the auto-correlation coefficient as follows:

$$\rho_a(\Delta x) = \exp\left(-\frac{|\Delta x|}{d_{cor}} \ln 2\right), \tag{3.2}$$

where Δx is the distance between two positions, and d_{cor} is the decorrelation distance (which is defined here as the distance at which the correlation coefficient $\rho_a(d_{cor})$ falls to 0.5).

In simulations, the auto-correlation of shadowing can be implemented as follows. If L_s^1 is the log-normal shadowing in dB at position P_1 and L_s^2 is the log-normal shadowing in dB at position P_2, which is Δx away from P_1, then L_s^2 can be modeled as a normally distributed random variable in dB with mean $\mu' = \rho_a(\Delta x)L_s^1$ and standard deviation $\sigma' = \sqrt{(1 - \rho_a^2(\Delta x))\sigma^2}$ [14].

Cross-correlation

Multiple links from different BSs to one UE may observe a highly cross-correlated shadow fading depending on the network scenario. For example, if two BSs are very close to each other, a UE may observe similar shadow fading from the two BSs. Since the cross-correlation coefficient ρ_c among BSs and sectors is usually given in simulation assumptions (e.g., $\rho_c = 1$ for collated sectors and $\rho_c = 0.5$ for neighboring cells), cross-correlated shadowing can be modeled using Cholesky decomposition [14].

In simulations, the cross-correlation of shadowing can be implemented as follows.

1. Define N independent normally distributed random variables $\mathbf{x} = [x_s^1, \ldots, x_s^n, \ldots, x_s^N]$, where $x_s^n \sim N(\mu, \sigma^2)$ with both μ and σ in dB, to represent the shadow fading in the links from N different BSs to one UE at a given time.
2. Define the cross-correlation coefficient matrix $\mathbf{G}$ as follows:

$$\mathbf{G} = \begin{pmatrix} \rho_c^{1,1} & \rho_c^{1,2} & \cdots & \rho_c^{1,n} \\ \rho_c^{2,1} & \rho_c^{2,2} & \cdots & \rho_c^{2,n} \\ \vdots & \vdots & \ddots & \vdots \\ \rho_c^{N,1} & \rho_c^{N,2} & \cdots & \rho_c^{n,n} \end{pmatrix} \tag{3.3}$$

where $\rho_c^{1,2} = \rho_c^{2,1}$ is the cross-correlation coefficient of shadow fading between links from BS 1 and BS 2 to the UE. Since $\mathbf{G}$ is a symmetric and positive definite matrix, it can be decomposed into a lower or upper triangular matrix by using Cholesky decomposition, i.e., $\mathbf{G} = \mathbf{C}^T\mathbf{C}$, where $\mathbf{C}$ is an upper triangular matrix.
3. Then, the cross-correlated shadow fading attenuations for the N links are given by $\mathbf{y} = \mathbf{xC} = [y_s^1, \ldots, y_s^n, \ldots, y_s^N]$.

In order to consider both auto- and cross-correlated shadow fading, we can combine the auto-correlation and cross-correlation implementation procedures together as follows. During the initialization, the cross-correlation procedure described above is performed for each UE. Thereafter, when the UE locations are updated according to a mobility model, the cross-correlation procedure runs again. However, in step 1 the N independent normally distributed random variables $\mathbf{x} = [x_s^1, \ldots, x_s^n, \ldots, x_s^N]$ should be updated using the auto-correlation procedure depicted earlier.

Due to complexity issues, in scenarios with a large number of BSs and UEs, computing both auto- and cross-correlated shadow fading attenuations in real time through Cholesky decomposition may significantly increase the running time and memory requirements of system-level simulations. In order to solve this issue, pre-computed shadowing maps obtained through low-complexity methods may be used. Similar to LUTs, shadowing maps can be imported in a system-level simulation and queried during its execution to model the shadow fading attenuations of UEs at given locations. Fig. 3.4 illustrates a shadowing map (where path losses are also included) for a BS at coordinates (100 m, 100 m) and with correlation distance $d_{cor} = 20$ m. This map was computed using the low-complexity method described in [15], which also makes use of Cholesky decomposition.

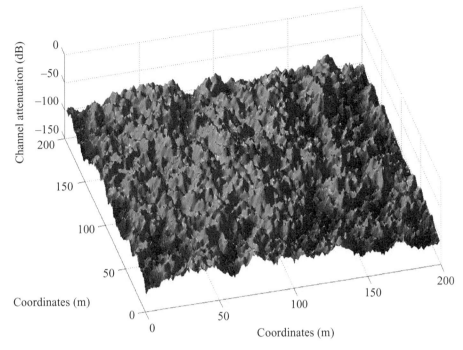

Figure 3.4 A generated channel attenuation map including path loss and shadow fading for a BS at coordinates (100 m, 100 m) and correlation distance $d_{cor} = 20\,\text{m}$ [15] (© 2005 IEEE).

3.4.3 Multi-path fading: International Telecommunication Union (ITU) and Typical Urban (TU) models

The objects in the propagation path from the transmitter to the receiver may not only cause obstruction, but they may also produce reflected, diffracted and scattered copies of the signal, resulting in *multi-path components* (MPCs). The MPCs may arrive at the receiver attenuated in power, delayed in time and shifted in frequency (and/or phase) with respect to the *line of sight* (LOS) component, thus adding up constructively or destructively at the receiver. As a result, the received power may vary significantly (e.g., in tens of dB) over very small distances in the order of a few wavelengths [12]. In other words, since different MPCs travel over different paths of different lengths, a single impulse sent from the transmitter will result in multiple copies received at different times. As a result, the impulse response of a multi-path channel is often modeled using a tapped delay line as follows:

$$h(t, \tau) = \sum_{i=0}^{\nu-1} h_i(t)\delta(\tau - \tau_i), \tag{3.4}$$

where t and τ capture the time and delay variability of the channel impulse response, respectively, ν is the total number of resolvable MPCs and $h_i(t)$ and τ_i are the fading coefficient at time t and delay of the ith path, respectively.

Table 3.1 ITU multi-path channel models (reproduced with permission from ITU).

Tap	Delay (ns)	Relative power (dB)	Delay (ns)	Relative power (dB)
		Pedestrian ($\leq$3 km/h)		
	Channel A		Channel B	
1	0	0	0	0
2	110	−9.7	200	−0.9
3	190	−19.2	800	−4.9
4	410	−22.8	1 200	−8.0
5	−	−	2 300	−7.8
6	−	−	3 700	−23.9
		Vehicular (30, 60, 120 km/h)		
	Channel A		Channel B	
1	0	0	0	−2.5
2	310	−1	300	0
3	710	−9	890	−12.8
4	1090	−10	12 900	−10.0
5	1730	−15	17 100	−25.2
6	2510	−20	20 000	−16.0
		Indoor		
	Channel A		Channel B	
1	0	0	0	0
2	50	−3	100	−3.6
3	110	−10	200	−7.2
4	170	−18	300	−10.8
5	290	−26	500	−18.0
6	310	−32	700	−25.2

Empirical multi-path channel models are often specified using the number of taps ν, and the relative average power and delay associated with each tap in the form of a power-delay profile. The most frequently used power-delay profiles are those suggested by the *International Telecommunication Union* (ITU) and COST 207. These types of model can also be created through the deterministic radio propagation tools presented in Chapter 2. ITU has specified two multi-path profiles, Channel A and Channel B, for vehicular, pedestrian and indoor channels [16] (see Table 3.1). Channel A is used for rural macrocell and microcell scenarios, where the cell radius is typically less than 500 m. Channel B has a much longer delay spread than Channel A, and is usually used for urban macrocell environments. In contrast, COST 207 has specified models for rural areas, hilly terrain and urban areas [17], where *Typical Urban* (TU) is widely used within 3GPP (see Table 3.2). These models are usually defined by 12 taps. However, due to complexity issues, it may not be possible to simulate the complete model, and a

Table 3.2 TU multi-path channel models (reproduced with permission from ITU).

Tap	Delay (ns)	Relative power (dB)	Delay (ns)	Relative power (dB)
	Alternative 1		Alternative 2	
1	0.0	0.0	−4.0	−4.0
2	0.1	0.2	−3.0	−3.0
3	0.3	0.4	0.0	0.0
4	0.5	0.6	−2.6	−2.0
5	0.8	0.8	−3.0	−3.0
6	1.1	1.2	−5.0	−5.0
7	1.3	1.4	−7.0	−7.0
8	1.7	1.8	−5.0	−5.0
9	2.3	2.4	−6.5	−6.0
10	3.1	3.0	−8.6	−9.0
11	3.2	3.2	−11.0	−11.0
12	5.0	5.0	−10.0	−10.0

reduced-complexity configuration of six taps has also been defined. Note that, for each model, two alternative tap settings are proposed, as denoted by Alternative 1 and Alternative 2 in Table 3.2.

Simulating different channel realizations using empirical multi-path channel models is usually time consuming. For a state-of-the-art personal computer, it may take hours to generate an adequate number of channel realizations. Moreover, we need to predict multi-path fading for the carrier signal and all interfering signals for each UE, and, thus, the number of required multi-path fading predictions scales with the number of BSs and UEs. Hence, pre-computed multi-path fading maps are commonly used to reduce complexity and speed up system-level simulations. Multipath fading maps have a matrix-like structure, in which each frequency and time resource is associated with a multi-path fading attenuation. Multi-path fading maps can be imported in the system-level simulation and queried during its execution to obtain multi-path fading attenuations of UEs at a given frequency and time. Fig. 3.4 illustrates a multi-path fading map predicted using the TU model with 12 taps for a UE moving at 3 km/h. Results are shown in a map of 50 *resource blocks* (RBs) and 100 subframes (i.e., 100 ms), where fading attenuations in dB were sampled every RB (180 kHz) and every time slot (0.5 ms).

Finally, let us note that, for fast fading channel modeling, 3GPP recommends that "if fast fading modeling is disabled in system-level simulations for relative evaluations, the impairment of frequency-selective fading channels shall be captured in the physical layer abstraction" [2]. This means that ordinary fast fading models such as ITU and TU may be disabled if the effects thereof are modeled somewhere else, e.g., in the PHY layer abstraction. The effect of fast fading has to be captured, but how to capture it is up to the implementation of the simulator. In 3GPP, such an agreement was the result of a compromise among different companies.

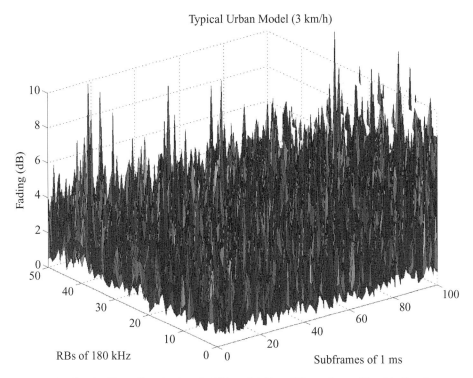

Figure 3.5 Fading in dB predicted using the TU model (for a UE moving at 3 km/h). Results are shown in a matrix of 50 RBs and 100 subframes (i.e., 100 ms), where fading was sampled every RB (180 kHz) and every time slot (0.5 ms).

3.4.4   Antenna patterns

If downtilt is not considered, omnidirectional antennas can be modeled using a constant antenna gain. However, for downtilt antennas, their horizontal and vertical antenna patterns should be simulated appropriately. The 3GPP has suggested the following antenna pattern models [1, 2].

- Horizontal antenna pattern (see Fig. 3.6):

$$A_H(\varphi) = -\min\left[12\left(\frac{\varphi}{\varphi_{3dB}}\right)^2, A_m\right],$$

$$\varphi_{3dB} = 70 \text{ degrees}, \quad A_m = 25 \text{ dB}. \tag{3.5}$$

- Vertical antenna pattern (see Fig. 3.7):

$$A_V(\theta) = -\min\left[12\left(\frac{\theta - \theta_{etilt}}{\theta_{3dB}}\right)^2, SLA_V\right], \tag{3.6}$$

$$\theta_{3dB} = 10 \text{ degrees}, \quad \theta_{etilt} = 15 \text{ degrees}, \quad SLA_V = 20 \text{ dB} \tag{3.7}$$

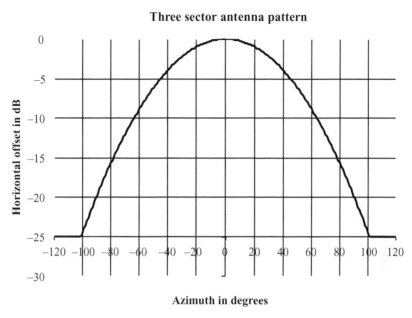

Figure 3.6 3GPP 25.814 horizontal antenna pattern.

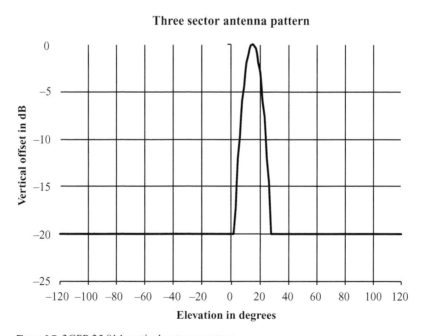

Figure 3.7 3GPP 25.814 vertical antenna pattern.

where $A_H(\varphi)$ and $A_V(\theta)$ are the attenuation offsets introduced by the horizontal and vertical antenna patterns with respect to the maximum antenna gain, respectively, φ and θ denote the angles of arrival in the horizontal and vertical planes with respect to the main beam positioning direction, respectively, φ_{3dB} and θ_{3dB} represent the main beam width of the horizontal and vertical antenna patterns within which at least half of the maximum transmit power is radiated, respectively, θ_{etilt} is the vertical angle (also known as tilt) between the main beam positioning directions of the vertical antenna pattern and the horizontal plane, and A_{m} and SLA_{V} are the maximum offsets of the horizontal and vertical antenna patterns, respectively.

In order to consider the joint effect of the horizontal and vertical antenna patterns, the 3GPP has proposed the following approach to create a 3D antenna pattern:

$$A(\varphi, \theta) = -\min\left\{-\left[A_{\text{H}}(\varphi) + A_{\text{V}}(\theta)\right], A_{\text{m}}\right\}. \tag{3.8}$$

3.4.5   Signal quality: maximal ratio combining (MRC) and exponential effective SINR mapping (EESM)

Under 3GPP system-level simulation assumptions, it is usual to find system models with two or more antennas at the receivers, which use *maximal ratio combining* (MRC) to combine the signals received by different antennas [1, 2]. Moreover, for estimating wideband SINRs, the 3GPP suggests the use of *exponential effective SINR mapping* (EESM) to compute the effective SINR across the entire bandwidth based on the individual SINRs of the corresponding frequency resources. In the following, we pay special attention to the implementation of MRC and EESM models.

Maximal ratio combining

Assuming perfect channel knowledge at the receiver, MRC is the optimum diversity combining technique over *additive white Gaussian noise* (AWGN) channels. Theoretically, when multiple independent copies of the same signal are combined through MRC, the instantaneous *signal to noise ratio* (SNR) at the output of the MRC combiner is maximized [18]. MRC is probably the most common approach adopted in 3GPP system-level simulations to realize the benefits of spatial diversity at receivers with two or more antennas [1].

In the case of a receiver having two antennas, assuming perfect channel knowledge at the receiver and independent channels observed at both antennas, the SINR γ_{MRC} at the output of the MRC combiner in a frequency resource can be modeled as

$$\gamma_{\text{MRC}} = \gamma_1 + \gamma_2, \tag{3.9}$$

where γ_1 and γ_2 are the SINRs of the received signals at antenna 1 and antenna 2 in the given frequency resource (e.g., subcarrier or RB), respectively, and SINR values are given in linear units.

The above model can be generalized to A antennas at the receiver as follows:

$$\gamma_{\text{MRC}} = \sum_{a=1}^{A} \gamma_{\text{a}}, \tag{3.10}$$

where γ_{a} is the SINR of the received signal at antenna a [18].

Table 3.3 EESM β values (reproduced with permission of 3GPP).

Modulation	Code rate	β
	1/3	1.49
	1/2	1.57
QPSK	2/3	1.69
	3/4	1.69
	4/5	1.65
	1/3	3.36
	1/2	4.56
16QAM	2/3	6.42
	3/4	7.33
	4/5	7.68

Exponential effective SINR mapping

In system-level simulations, there is a need for calculating the effective SINR of each RB based on the SINR of each subcarrier within the RB. There is also a need for calculating the effective SINR of the entire band (i.e., wideband SINR) based on the SINR of each RB within the considered spectrum. One of the most commonly used methods for this purpose is the EESM (one may also refer to *mutual information effective SINR mapping* (MIESM) models). The basic idea of EESM is to provide a mapping function that maps a set of SINRs to a single SINR value, which well represents the corresponding BLER [19].

The EESM mapping function is as follows:

$$\gamma_{\text{eff}} = \text{EESM}(\gamma, \beta) = -\beta \ln \left(\frac{1}{N} \sum_{i=1}^{N} \exp \left(-\frac{\gamma_i}{\beta} \right) \right), \tag{3.11}$$

where γ is the set of SINRs in linear units to be mapped (e.g., SINRs obtained from MRC of all RBs in a band), which are typically different in a frequency selective channel, N is the cardinality of set γ, γ_i is the ith element of set γ and β is a *modulation and coding scheme* (MCS)-dependent parameter that compensates the difference between the exact BLER and the predicted BLER.

Adequate values of β for different MCSs have been proposed in the literature. For example, Table 3.3 presents β values suggested by 3GPP [20], while Table 3.4 and Table 3.5 present results based on ITU models [21].

3.5   3GPP reference system deployments and evaluation assumptions

The 3GPP defines four reference cases for system-level simulations as summarized in Table 3.6. Case 1 and Case 3 are commonly used to simulate urban and suburban scenarios, as indicated by their relatively short and large ISD, respectively, and low UE velocities, while Case 2 and Case 4 are less frequently used. It is important to note that

Table 3.4 ITU (Pedestrian A)-based EESM β values [21].

Modulation	Code rate	β
QPSK	1/2	1.59
	3/4	1.70
16QAM	1/2	5.33
	3/4	8.45
64QAM	1/2	16.10
	3/4	27.21

Table 3.5 ITU (Pedestrian B)-based EESM β values [21].

Modulation	Code rate	β
QPSK	1/2	1.67
	3/4	1.75
16QAM	1/2	5.41
	3/4	7.94
64QAM	1/2	17.27
	3/4	32.33

Table 3.6 3GPP system simulation cases (reproduced with permission from 3GPP). CF is the carrier frequency, BW is the operating bandwidth, PLoss is the penetration loss and Speed is the speed at which a UE moves in the simulation.

Simulation cases	CF (GHz)	ISD (m)	BW (MHz)	PLoss (dB)	Speed (km/h)
1	2.0	500	10	20	3
2	2.0	500	10	10	30
3	2.0	1732	10	20	3
4	0.9	1000	1.25	10	3

these models represent idealized network models, which may differ from the realistic ones but are still good approximations.

3.5.1   Homogeneous deployments

In a homogeneous cellular network, all cells have similar characteristics, such as output power, *evolved NodeB* (eNB) height, antenna patterns etc., and the cell sites are placed in a regular pattern over an area as shown in Fig. 3.2. Such a homogeneous deployment is also called a "macro-only" deployment, as only macrocells are present in the deployment. Relevant 3GPP parameters can be found in Table A.2.1.1-3 of [1]. Table A.2.1.1-2 of [2] further extends the set of parameters. As a summary, Table 3.7 presents a complete view of the 3GPP system-level simulation baseline parameters.

Note that, regarding the antenna patterns, more detail can be found in Section 3.4.4, and in Section A.2.1.6 of [2].

Table 3.7 3GPP system simulation baseline parameters (Table A.2.1.1-3 [1] and Table A.2.1.1-2 [2], reproduced with permission from 3GPP).

Parameter	Assumption
Cellular layout	Hexagonal grid, 19 cell sites, 3 sectors per site.
Intersite distance	Depends on case, see Table 3.6.
Distance-dependent path loss	$L = I + 37.6 \log_{10} R$, R in km. $I = 128.1$ if carrier frequency is 2 GHz. $I = 120.9$ if carrier frequency is 900 MHz.
Lognormal shadow fading	Similar to Section B 1.4.1.4 [22] (see Section 3.4.2).
Shadow fading standard deviation	8 dB.
Shadow fading correlation distance	50 m (see Section D.4 [22]).
Shadow fading correlation	0.5 between cells, 1.0 between sectors.
Penetration loss	Depends on case, see Table 3.6.
Antenna pattern (horizontal)	$A_H(\varphi) = -\min\left[12\left(\frac{\varphi}{\varphi_{3dB}}\right)^2, A_m\right],$ $\varphi_{3dB} = 70°$, $A_m = 25$ dB.
Antenna pattern (vertical)	$A_V(\theta) = -\min\left[12\left(\frac{\theta-\theta_{etilt}}{\theta_{3dB}}\right)^2, SLA_V\right],$ $\theta_{3dB} = 10$, $SLA_V = 20$ dB, and θ_{etilt} is the electrical antenna downtilt. For calibration purposes, $\theta_{etilt} = 15°$ can be used for 3GPP Case 1 and $\theta_{etilt} = 6°$ for 3GPP Case 3. Antenna height at the BS is set to 32 m. Antenna height at the UE is set to 1.5 m.
Combining method in 3D antenna pattern	$A(\varphi, \theta) = -\min\{-[A_H(\varphi) + A_V(\theta)], A_m\}.$
Carrier frequency and bandwidth	Depends on case, see Table 3.6.
Channel model	3GPP SCM [TR 25.996]. For single transmit antenna evaluations, the TU channel model may be used.
UE speeds of interest	3 km/h, 30 km/h, 120 km/h, 350 km/h.
Total BS TX power (Ptotal)	43 dBm for 1.25 or 5 MHz bandwidth; 46/49 dBm for 10 or 20 MHz bandwidth. Some evaluations to exploit carrier aggregation techniques may use wider bandwidths, e.g., 60 or 80 MHz (FDD), where the total BS transmit power (49 dBm) should be used.
UE power class	23 dBm (200 mW), corresponding to the sum of power amplifier (PA) powers in the multiple-transmit-antenna case.
Intercell interference modeling	UL: Explicit modeling (all cells occupied by UEs). DL: Explicit modeling else cell power $= P_{total}$.

3.5.2 Heterogeneous deployments

In an HCN, the characteristics of cells can be different, e.g., different output powers. In a heterogeneous deployment, typically existing eNBs are overlaid with LPNs. The LPNs can be deployed on the same frequency as the macrocell network or on a separate frequency. LPNs include *remote radio heads* (RRHs), *pico evolved NodeBs* (PeNBs), *home evolved NodeBs* (HeNBs) and RNs. The main differences among these LPNs are summarized in Table 3.8.

It is important to understand the consequences of the differences among LPNs. Regarding the backhaul, low delay enables efficient coordination among network nodes. Hence,

Table 3.8 Categorization of LPNs (based on Table A.2.1.1.2-1 [2], reproduced with permission from 3GPP).

Type	Backhaul	Access	Notes
RRHs	Several μs latency to eNBs	Open to all UEs	Placed indoors or outdoors
PeNB (i.e., node for Hotzone cells)	X2	Open to all UEs	Placed indoors or outdoors. Typically planned deployment
HeNB (i.e., node for femto cells)	No X2 in baseline	CSG	Placed indoors. UE deployed
RNs	Through air interface with an eNB (for in-band RN)	Open to all UEs	Placed indoors or outdoors

Table 3.9 HCN deployment scenarios (based on Table A.2.1.1.2-2 [2], reproduced with permission from 3GPP).

Case	Environment	Deployment scenario	LPN node
5.1	Macro and indoor	Macrocell and indoor femtocell	HeNB
5.2	Macro and indoor	Macrocell and indoor relay	Indoor RN
5.3	Macro and indoor	Macrocell and indoor RRH or hotspot	e.g., PeNB
6.1	Macro and outdoor	Macrocell and outdoor relay	Outdoor RN
6.2	Macro and outdoor	Macrocell and outdoor RRH or hotspot	e.g., PeNB

coordination with RRHs is the most efficient, while coordination with HeNBs is the most challenging, especially due to the lack of an X2 interface. Regarding the access method, within 3GPP terminology, all nodes are *open access* to all UEs, except for the HeNBs, which are possibly closed access (see Chapter 4). Closed access, also referred to as CSG, may create coverage holes when the HeNB is deployed on the same frequency as the macrocell network, because UEs may not always be able to connect to the cells that provide the strongest signal. Note that open access HeNBs may solve this coverage problem, at the expense of more often handovers and higher security concerns.

LPNs can be combined to form various deployment scenarios. According to different combinations of LPNs, 3GPP has defined five baseline HCN deployment scenarios, as listed in Table 3.9. These scenarios will be further discussed later in this chapter. The selection of a scenario will primarily influence the simulation parameter settings for the radio environment.

3.5.2.1   Macrocells overlaid with indoor or outdoor picocells

Macrocells overlaid with indoor or outdoor picocells are the scenarios denoted by Case 5.3 and Case 6.2 in Table 3.9, respectively. In these two scenarios, the existing macrocells cannot provide the required capacity in hotspots, e.g., a shopping mall with many UEs, a train station, an airport etc. In order to increase the network capacity

Table 3.10 Simulation baseline parameters of Model 1 and Model 2 for picocells in a macro–pico scenario (based on Table A.2.1.1.2-3 [2], reproduced with permission from 3GPP).

Parameter	Value
Nodes per macro cell	1, 2, 4, or 10 nodes
Lognormal shadowing	Similar to Section B [22] (see Section 3.4.2)
Shadowing correlation between cells	Outdoor: 0.5. Indoor: 0
Shadowing correlation between sectors	N/A
Antenna pattern (horizontal)	$A(\theta) = 0$ dB
Carrier frequency	2 GHz for Cases 1 and 3
UE speeds	3 km/h for Cases 1 and 3
Total BS Tx power (Ptotal)	24 or 30 dBm for Case 1. 24, 30, or 37 dBm for Case 3. 37 dBm is for outdoor only. Both cases assume 10 MHz carrier bandwidth
UE power class	23 dBm (200 mW)
Intercell interference modeling	Uplink: Explicit modeling (all cells occupied by UEs). Downlink: Explicit modeling, or set cell power equal to P_{total}
Antenna configuration	2 Rx and 2 Tx antenna ports, or 4 Tx and 4 Rx antenna ports
Antenna gain + connector loss	5 dBi
Minimum distance from macro site to pico site	75 m
Minimum distance from macro site to UE	35 m
Minimum distance from pico site to UE	Outdoor: 10 m, Indoor: 3 m
Minimum distance among the pico sites	40 m

in hotspots, operators deploy PeNBs at the hotspots using the same frequency as the macrocells. The short distance between a PeNB and a UE implies that output power can be reduced for both the PeNB and the UE, leading to reduced interference and prolonged battery life. PeNBs can be deployed indoors or outdoors, depending on the nature of the targeted hotspots. The choice of indoor or outdoor deployment affects radio propagation parameters as shown in Table 3.10, where Model 1 is based on 3GPP TR25.814 [1] and has been updated according to *International Mobile Telecommunications* (IMT) Advance evaluations, while Model 2 is based on field measurements and ITU models.

For macrocells overlaid with outdoor picocells, Model 2 has more detail than Model 1, particularly in the modeling of LOS and *non-line-of-sight* (NLOS) path loss. Model 1 models both LOS and NLOS path loss together, while Model 2 models these two kinds of path loss separately with a probabilistic function selecting which one to use. Table 3.11 lists differences between Model 1 and Model 2 for outdoor PeNBs. The distance-dependent path loss is also dependent on the carrier frequency, which is assumed to be 2 GHz in Table 3.11, while R is measured in km.

For macrocells overlaid with indoor picocells, the channel models are more compli-cated, because UEs could be inside or outside the building containing the PeNB. UEs could also be in a building other than the one containing the PeNB. The different cases mostly affect the penetration loss. Moreover, there are also two channel models, Model 1 and Model 2, for the indoor scenario, where Model 2 is a simplified model, in which all UEs are assumed to be indoors. Table 3.12 shows the parameters of Model 1 for indoor

Table 3.11 Differences between Model 1 and Model 2 for outdoor picocells (based on Table A.2.1.1.2-3 [2], reproduced with permission from 3GPP).

Parameter	Model 1	Model 2
Distance-dependent path loss, macro to UE	$L(R) = 128.1 + 37.6 \log R$	$L_{\text{LOS}}(R) = 103.4 + 24.2 \log R$
Probability of LOS, macro to UE	Not applicable	Case 1: $P_{\text{LOS}}(R) =$ $\min(0.018/R, 1)(1 - e^{-R/0.063}) + e^{-R/0.063}$ Case 3: $P_{\text{LOS}} = e^{-(R-0.01)/0.2}$
Distance-dependent path loss, pico to UE	$L(R) = 140.7 + 36.7 \log R$	$L_{\text{LOS}}(R) = 103.8 + 20.9 \log R$ $L_{\text{NLOS}}(R) = 145.4 + 37.5 \log R$
Probability of LOS, pico to UE	Not applicable	Case 1: $P_{\text{LOS}} = 0.5 - \min(0.5, 5e^{-0.156/R}) + \min(0.5, 5e^{-R/0.03})$ Case 3: $P_{\text{LOS}} = 0.5 - \min(0.5, 3e^{-0.3/R}) + \min(0.5, 3e^{-R/0.095})$
Shadowing standard deviation	10 dB	LOS: 3 dB, NLOS: 4 dB
Penetration loss	20 dB	20 dB

Table 3.12 Channel Model 1 for indoor picocells (based on Table A.2.1.1.5-1 [2], reproduced with permission from 3GPP).

Case	Parameter	Value
Macro to UE	Distance-dependent path loss	$L_{\text{LOS}}(R) = 103.4 + 24.2 \log R$ $L_{\text{NLOS}}(R) = 131.1 + 42.8 \log R$
	Shadowing standard deviation	10 dB
	Penetration loss model 1	If UE is outside, 0 dB, if UE is inside, 20 dB
	Penetration loss model 2	20 dB
	Probability of LOS	Case 1: $P_{\text{LOS}} = \min(0.018/R, 1)(1 - e^{-R/0.063}) + e^{-R/0.063}$ Case 3: $P_{\text{LOS}} = e^{-(R-0.01)/1.0}$
Pico to UE[1]	Distance-dependent path loss	$L = \max(131.1 + 42.8 \log R, 147.4 + 43.3 \log R)$
	Shadowing standard deviation	10 dB
	Penetration loss model 1	If UE is outside: 20 dB. If UE is in another building: 40 dB
	Penetration loss model 2	20 dB
Pico to UE[2]	Distance-dependent path loss	$L_{\text{LOS}} = 89.5 + 16.9 \log R$ $L_{\text{NLOS}} = 147.4 + 43.3 \log R$
	Shadowing standard deviation	LOS, 3 dB, NLOS, 4 dB
	Penetration loss both models	0 dB
	Probability of LOS	$P_{\text{LOS}} = \begin{cases} 1 & R \le 0.018 \\ e^{\frac{-(R-0.018)}{0.027}} & 0.018 < R < 0.037 \\ 0.5 & R \ge 0.037 \end{cases}$

Table 3.13 Simulation baseline parameters of Model 1 for femtocells in a macro–femto scenario (based on Table A.2.1.1.2-3 [2], reproduced with permission from 3GPP).

Parameter	Value
Nodes per macro cell	1 node
Distance-dependent path loss	Femto to UEs links: $L(R) = 127 + 30 \log R$
	Other links: $L(R) = 128.1 + 37.6 \log R$
Lognormal shadowing	Similar to Section B [22]
Shadowing standard deviation	Femto to UEs 10 dB, other links 8 dB
Shadowing correlation between cells	0
Shadowing correlation between sectors	N/A
Penetration loss	Femto to UEs 0 dB, other links 20 dB
Antenna pattern (horizontal)	$A(\theta) = 0$ dB
Carrier frequency	2 GHz for Case 1 and 3
UE speeds	3 km/h for Case 1 and 3
Total BS Tx power (Ptotal)	20 dBm (assuming 10 MHz carrier bandwidth)
UE power class	23 dBm (200 mW)
Intercell interference modeling	Uplink: Explicit modeling (all cells occupied by UEs).
	Downlink: Explicit modeling, or set cell power equal to Ptotal
Antenna configuration	2 Rx and 2 Tx antenna ports, or 4 Tx and 4 Rx antenna ports
Antenna gain + connector loss	5 dBi
Minimum distance from macro site to femto site	75 m
Minimum distance from macro site to UE	35 m
Minimum distance from femto site to UE	3 m
Minimum distance among the pico sites	40 m

picocells, where the carrier frequency is assumed to be 2 GHz and R is measured in km. In Table 3.12, Pico to UE[1] applies when the UE is outdoors or in a building other than the one containing the PeNB, while Pico to UE[2] applies when the UE is in the building containing the PeNB.

3.5.2.2 Macrocells overlaid with indoor femtocells

Femtocells are likely to be deployed in residential homes and small enterprises, where coverage and capacity might be insufficient. Hence, it is also possible to overlay an existing macrocell network with indoor femtocells. An HeNB typically has a lower transmission power than a PeNB. Unlike picocells that are open to all UEs, CSG femtocells are only accessible to CSG UEs. In Case 5.1 of Table 3.9, HeNBs are always deployed indoors, with wall penetration losses increasing their "separation" from the outdoor macrocell network. Such separation is beneficial not only for macrocell UEs that are not allowed to connect to CSG femtocells, but also for CSG femtocell UEs, since they can be protected from macrocell transmissions.

For the simulation of a scenario with macrocells and femtocells, several models and configurations have been described in [2]: e.g., Model 1 assumes all UEs are on the same floor as the HeNB. Models for more complex indoor scenarios will be explained in Section 3.6. Table 3.13 presents a baseline parameter set for a macro–femto scenario, where "Femto to UEs links" are for communications between a femtocell and its UEs, while "Other links" are for communications between a UE and other femtocells, as well as macrocells. Note that some parameters in Table 3.13 correspond to the parameters in Table 3.10, and the note on how to model fast fading applies. It is also possible to use ITU/SCM models, in which *urban macrocell* (UMa) and *indoor hotspot* (InH) should be used for the link between a macrocell and a UE and that between a femtocell and a UE, respectively.

3.5.2.3   Macrocells overlaid with indoor or outdoor relays

Macrocells overlaid with indoor or outdoor RNs are the scenarios denoted by Case 5.2 and Case 6.1 of Table 3.9, respectively. RNs are usually overlaid on a macrocell network to improve its coverage. For example, an RN can be deployed in rural areas, where the eNB is far away from UEs to extend the coverage of the macrocell. In the case of in-band RNs, because RNs use the air interface as backhaul, the capacity of the macrocell network is not increased, but its coverage is extended using a much cheaper backhaul compared with microwave or fiber backhaul links.

For the simulation of a macrocell network overlaid with RNs, there are five different links to consider: eNB to UE, eNB to indoor relay, eNB to outdoor relay, indoor relay to UE and outdoor relay to UE. An outdoor relay shares many path loss properties with an outdoor picocell, and an indoor relay shares many path loss properties with an indoor femtocell. Baseline simulation parameters for relay deployments are presented in Table 3.14. These parameters are aligned with the femtocell model presented in Table 3.13. It is possible to explore more elaborate antenna configurations for the link between an eNB and a RN, depending mostly on whether there is LOS between them, as described in Table A.2.1.1.4-3 of [2].

3.6   Placing of low-power nodes and users

3.6.1   Macrocells overlaid with indoor or outdoor picocells or relays

In the simulation of macrocells overlaid with indoor or outdoor picocells or relays, 3GPP has suggested four different configurations for the placement of PeNBs, RNs and UEs, as listed in Table 3.15. The UE density is defined as the number of UEs in the geographic area of a macrocell coverage. The LPN density is proportional to the UE density in each macrocell. Note that PeNBs and RNs are often deployed by operators, and thus may not follow a uniform distribution within the macrocell coverage area. For example, PeNBs can be specifically deployed at the cell-edge for coverage and capacity enhancement, or in shopping malls or train stations to increase capacity. UEs may also be clustered in hotspots, where their positions could be modeled as uniformly distributed within an area around the LPN defined by the hotspot radius.

The following steps should be followed to deploy UEs in system-level simulations [2].

Table 3.14 Baseline simulation parameters for RNs in a macro–relay scenario (based on Table A.2.1.1.2-3 [2], reproduced with permission from 3GPP).

Parameter	Value
Nodes per macro cell	Outdoor relays: 1, 2, 4, or 10 nodes
	Indoor relays: 1 node
Distance-dependent path loss	Macro to UEs and macro to indoor relay: See path loss model for macro to UE, Model 2 in Table 3.11
	Indoor relay to UE: See path loss model in Table 3.13
	Outdoor relay to UE: See path loss model for pico to UE, Model 2 in Table 3.11
	Macro to outdoor relay:
	$L_{\mathrm{LOS}} = 100.7 + 23.5 \log R$
	$L_{\mathrm{NLOS}} = 125.2 + 36.3 \log R$
	$P_{\mathrm{LOS}} = \min(0.018/R,\, 1)(1 - \mathrm{e}^{-R/0.072}) + \mathrm{e}^{-R/0.072}$ (Case 1)
	$P_{\mathrm{LOS}} = \mathrm{e}^{-(R-0.01)/0.23}$ (Case 3)
Lognormal shadowing	Similar to Section B [22]
Shadowing standard deviation	Macro to relay: 6 dB (outdoor), 8 dB (indoor)
	Relay to UE: 10 dB (outdoor), 10 dB (for relay link indoor), 8 dB (other links indoor)
Shadowing correlation between cells	0.5 (outdoor) 0 (indoor)
Shadowing correlation between sectors	N/A
Penetration loss	Macro to UE: 20 dB
	Macro to relay: 0 dB (outdoor), 5 dB (indoor)
	Relay to UE: 20 dB (outdoor), 0 dB (for relay link indoor), 20 dB (other links indoor)
Antenna pattern (horizontal)	$A(\theta) = 0$ dB
Carrier frequency	2 GHz for Cases 1 and 3
UE speeds	3 km/h for Cases 1 and 3
Total BS Tx power (Ptotal)	Case 1: 30 dBm. Case 3: 30, 37 dBm. Both cases assume 10 MHz carrier bandwidth
UE power class	23 dBm (200 mW)
Intercell interference modeling	Uplink: Explicit modeling (all cells occupied by UEs)
	Downlink: Explicit modeling, or set cell power equal to P_{total}
Antenna configuration	2 Rx and 2 Tx antenna ports, or 4 Tx and 4 Rx antenna ports
Antenna gain + connector loss	5 dBi
Minimum distance from macro site to relay site	75 m
Minimum distance from macro site to UE	35 m
Minimum distance from relay site to UE	10 m (outdoor), 3 m (indoor)
Minimum distance among the relay sites	40 m

Table 3.15 Placing of LPNs and UEs (based on Table A.2.1.1.2-4 [2], reproduced with permission from 3GPP).

Config.	UE density across macros	UE distribution within a macro	LPN distribution within a macro	Comments
1	Uniform 25/macro	Uniform	Uncorrelated	Capacity enhancement
2	Non-uniform [10-100]/macro	Uniform	Uncorrelated	Sensitive to non-uniform distribution across macros
3	Non-uniform [10-100]/macro	Uniform	Correlated Planned	Cell-edge coverage enhancement
4a, 4b	Non-uniform Table 3.16	Clusters	Correlated Planned	Hotspot capacity enhancement

Table 3.16 Parameters of Configurations 4a and 4b for clustered UE dropping (based on Table A.2.1.1.2-5 [2], reproduced with permission from 3GPP).

Config.	N_{UEs}	N	$P_{hotspot}$
4a	30 or 60	1	1/15
		2	2/15
		4	4/15
		10	2/3
4b	30 or 60	1	2/3
		2	2/3
		4	2/3

1. Fix the total number N_{UEs} of UEs dropped within each macrocell coverage, where N_{UEs} is 30 or 60 in fading scenarios, and 60 in non-fading scenarios.
2. Randomly and uniformly drop the configured number N of LPNs within each macrocell coverage, where N may take values from $\{1, 2, 4, 10\}$.
3. Randomly and uniformly drop $N_{UEs_{lpn}}$ UEs within a 40 m radius of each LPN, where $N_{UEs_{lpn}} = P_{hotspot} \frac{N_{UEs}}{N}$, with $P_{hotspot}$ defined in Table 3.16 as the fraction of the number of all hotspot UEs over the total number of UEs in the network. Note that $N_{UEs_{lpn}} = 0$ for non-clustered scenarios, i.e., uniform UE distribution.
4. Randomly and uniformly drop the remaining $N_{UEs} - N_{UEs_{lpn}} N$ UEs over the entire macrocell coverage (including the LPN coverage).

3.6.2 Macrocells overlaid with indoor femtocells

For the case of macrocells overlaid with indoor femtocells, 3GPP has proposed a suburban model and two dense-urban models, i.e., the dual stripe model and the 5×5 grid model [23].

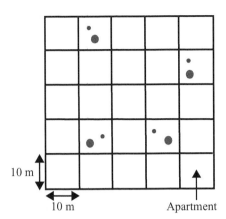

Figure 3.8 5 × 5 apartment grid [24].

3.6.2.1 Suburban model

The suburban model is considered when HeNBs are deployed inside residential houses located within the eNB coverage area. In the suburban model, residential houses are represented as 2D 12 m × 12 m squares, and are deployed following a uniform distribution within the eNB coverage area. The density of houses per eNB coverage area is variable. Houses are at least 35 m away from any eNB, and there should be no overlapping between any two houses. One HeNBs is uniformly distributed within each house. It is assumed that HeNBs are always "active," i.e., there is always at least one active call for each HeNB. In each house, femto UEs are randomly dropped following a uniform distribution, and with a minimum separation of 20 cm from the HeNB. Femto UEs have a 10% probability of being outdoors around the house. All UEs associated with eNBs are assumed to be uniformly dropped within the eNB coverage area, and hence they may be within a house containing an HeNB.

3.6.2.2 5×5 grid model

The 5×5 grid model is a simple model for HeNB urban deployment. A total of 25 apartments are distributed in a 5×5 grid, as shown in Fig. 3.8, where the number of floors per grid is variable. Each apartment is of size 10 m × 10 m. The density of grids per eNB coverage area is variable, grids are at least 35 m away from any eNB, and there should be no overlapping between any two grids. One HeNB with one associated UE are randomly dropped following a uniform distribution inside each apartment area. An occupation ratio, which is uniformly distributed in the range [0,1], determines the probability of an HeNB existing inside an apartment. Moreover, each HeNB has an activation parameter, which is also uniformly distributed in the range [0,1] and gives the probability of the HeNB being active.

3.6.2.3 Dual strip model

The dual strip model is a more complex model for urban HeNB deployments, since it considers multiple blocks of apartments. As shown in Fig. 3.9, each block of apartments

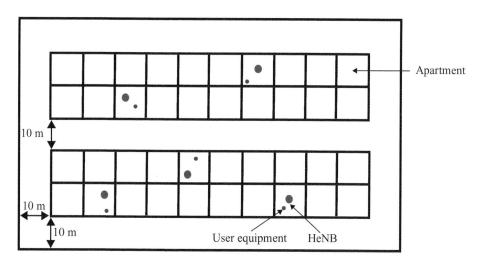

Figure 3.9 3GPP dual strip femtocell block model [24].

has two strips, separated by a 10 m wide street. Each strip has two rows of apartments, each row contains 10 apartments, and each apartment is of size 10 m×10 m, and the number of floors is uniformly distributed in the range [1,10]. Around the two strips there is also a 10 m wide street, thus the total block size is $10 \times (10 + 2) \times 70$ m². In each eNB coverage area, one or more dual-strip blocks may be randomly dropped following a uniform distribution, but without overlapping between any two blocks. An HeNB is randomly located in an apartment. It is assumed that for each HeNB there is only one UE, which is uniformly dropped inside the apartment with a minimum separation of 20 cm from the HeNB. All UEs associated with eNBs are assumed to be uniformly dropped within the eNB coverage area, and hence they may be within an apartment containing an HeNB. Similar to the 5×5 grid model, an occupation ratio and an activation parameter are also defined in the dual-strip model. In Fig. 3.9, active HeNBs are represented by larger solid circles, and femto UEs are represented by smaller solid circles.

3.7   Traffic modeling

It is important to consider the effects of traffic in system-level simulations. The bit rate and burstiness of traffic are the two major factors used to classify various traffic models. In Section A.2.1.3 of [2], three basic traffic models used in 3GPP are defined: full buffer model, *File Transfer Protocol* (FTP) model and *voice over IP* (VoIP) model.

3.7.1   Full buffer model

The full buffer model is a traffic model in which a UE tries to send and receive as much data as the air interface allows. This model is typically used to test the air interface in the sense of "worst case scenario." It is important to note that very few real applications exhibit the behavior of the full buffer traffic model. The full buffer traffic model represents a theoretical maximum of the system load, against which other traffic models may be compared, e.g., for evaluating the system throughput.

3.7.2   FTP model

The FTP model is a traffic model in which a UE downloads a single file of size of the order of 0.5–2 MB. UEs arrive according to a Poisson process, and thus the load of the system depends on the arrival UE rate. One problem with the single-file FTP model is that many UEs may need to be created over the course of a simulation to obtain meaningful statistical results. Depending on how the simulator is designed and implemented, this might lead to increased memory requirements. In Section A.2.1.3.1 of [2], the single-file FTP model is described in detail and an alternative model is also presented to deal with the mentioned drawback. In the alternative model, a UE downloads several files. The time between downloads, i.e., the reading time, is set following an exponential distribution. Thus, in this case, the load of the network depends on the arrival UE rate and the reading time. Which model to use for the system-level simulation is a matter of preference.

3.7.3   VoIP model

The VoIP model is a traffic model for real-time voice services, where two UEs take turns in sending data at a fixed rate, corresponding to the bit rate of a voice codec. The length of each talk spurt is typically exponentially distributed. The load of the system can be modeled analogously to the FTP model. Chapter A2 of [25] introduces a two-state Markov model for VoIP. For voice services, the average delay of packets is used to identify whether a UE would be satisfied or not. A UE is deemed satisfied if the radio interface delay is 50 ms or less for at least 98% of the packets. The voice capacity is typically defined as the number of UEs in the cell, when more than 95% of them are satisfied with voice services.

3.8   Mobility modeling

Mobility modeling attempts to describe the movements of UEs across a network in order to analyse how such UEs' motion affects network performance. 3GPP is currently studying HO-related metrics in HCNs with LPNs overlaid on an existing macrocell network. As the coverage areas of LPNs are smaller than those of existing macrocells, the number and frequency of HOs in an HCN will increase, thus leading to more signaling overhead and possibly more HO failures due to the nature of LPN coverage. New mechanisms for HOs in networks with LPNs may thus be necessary. Some recent results are reported in TR 36.839 [3]. The models and parameters used in [3] are based on [2], together with some newly added models, which are necessary to model and investigate HOs for LPNs.

In order to analyse mobility performance, how the LPNs are geographically placed relative to the macrocells is a factor to be considered. In [3], a PeNB is placed in the bore sight of the directional antenna of the eNB. The distance between the eNB and the PeNB is 0.3 ISD or 0.5 ISD, where the interference from the eNB to the PeNB would be higher for the 0.3 ISD case, resulting in a geographically smaller coverage area of the PeNB. For the 0.5 ISD case, the PeNB is located in the middle between two eNBs, and thus the HO scenarios might be more challenging in terms of mobility management because the picocell borders two macrocells. In [3], uniform distributions of PeNBs on a simulated scenario are also considered.

How UEs move around the scenario is also an important factor to be considered. In [3], two mobility models adopted in system-level simulations.

- *Hotspot simulations*: The hotspot has a radius of 100 m around the PeNB. A UE is first placed randomly at the edge of the hotspot area. Then, the UE moves towards the PeNB along a straight line in a randomly selected direction with a maximum deviation angle of 45° from the PeNB. When the UE reaches the edge of the hotspot, it bounces back towards the PeNB in another randomly selected direction, again with a maximum deviation angle of 45° from the PeNB.
- *Large-area simulations*: UEs are uniformly distributed over the entire simulation scenario, each with an independent random direction of motion. Each UE moves along a straight line in its independent direction of motion. When UEs hit the border of the simulation scenario, they bounce back inward the simulation scenario and move along newly randomly selected directions.

3.9   Summary and conclusions

This chapter has reviewed different kinds of system-level simulation, and has shed light on intricate modeling issues such as wrap-around, shadow fading, multipath fading, antenna patterns, MRC and EESM. Moreover, simulation setups for HCNs from the perspective of 3GPP have been illustrated, taking into account channel modeling and spatial distributions of LPNs and UEs for macro–pico, macro–femto and macro–relay scenarios. Furthermore, traffic and mobility models have also been discussed.

Copyright notices

therefore provided to you "as is" for information purposes only. Further use is strictly prohibited.

Tables 3.7, 3.8, 3.9, 3.10, 3.11, 3.12, 3.13, 3.14, 3.15 and 3.16 reproduced, with permission, from [2] © 3GPPM 2010. 3GPPTM TSs and TRs are property of ATIB, ATIS, CCSA, ETSI, TTA and TTC, who jointly own the copyright in them. They are subject to further modifications and are therefore provided to you "as is" for information purposes only. Further use is strictly prohibited.

Figure 3.4 © 2005 IEEE. Reprinted, with permission, from H. Claussen, Efficient modelling of channel maps with correlated shadow fading in mobile radio systems. *IEEE Proceedings*, 11–14 Sep. 2005.

References

[1] TR 25.814, *Physical Layer Aspects of Evolved Universal Territorial Radio Access UTRA*, 3GPP Technical Report (Sept. 2006).

[2] TR 36.814, *Evolved Universal Terrestrial Radio Access (E-UTRA); Further Advancements for E-UTRA Physical Layer Aspects*, 3GPP Technical Report (Mar. 2010).

[3] TR 36.839, *Mobility Enhancements in Heterogeneous Networks*, 3GPP Technical Report (June 2011).

[4] J. G. Andrews, A. Ghosh and R. Muhamed, *Fundamentals of WiMAX Understanding Broadband Wireless Networking* (Prentice-Hall, 2007).

[5] J. Zhang and G. de la Roche, *Femtocells: Technologies and Deployment* (Wiley, 2010).

[6] U. Türke, ed. Monte-Carlo snapshot analysis. In *Efficient Methods for WCDMA Radio Network Planning and Optimization* (DUV, Dec. 2007), pp. 45–66.

[7] Forsk, *Forsk Atoll – Global RF Planning Solution*.

[8] AWE Communications, *AWE Communications – Wave Propagation and Radio Network Planning*.

[9] T. Kwon, H. Lee, S. Choi, J. Kim and D.-H. Cho, Design and implementation of a simulator based on a cross-layer protocol between MAC and PHY layers in a WiBro compatible IEEE 802.16e OFDMA system. *IEEE Communications Magazine*, **43**:12 (2005), 136–146.

[10] J. Khun-Jush, *CDMA Uplink Power Control Methodology in SEAMCAT (VOICE ONLY)*, STG(03)13 r1 (Oct. 2003).

[11] M. Iridon and D. Matula, Symmetric cellular network embeddings on torus. In *Seventh International Conference on Computer Communications and Networks* (Oct. 1998), pp. 732–736.

[12] A. Goldsmith, *Wireless Communications* (Cambridge University Press, Sept. 2005).

[13] M. Gudmundson, Correlation model for shadow fading in mobile radio systems. *Electronics Letters*, **27**:23 (1991), 2145–2146.

[14] S802.16m-07/060, Correlation models for shadow fading simulation, *IEEE Technical Report* (Mar. 2007).

[15] H. Claussen, Efficient modelling of channel maps with correlated shadow fading in mobile radio systems. In *IEEE 18th International Symposium on Personal, Indoor and Mobile Radio Communications (PIMRC 2005)* (Sept. 2005), 512–516.

[16] ITU-R M.1225 1, *Guidelines for Evaluation of Radio Transmission Technologies for IMT-2000*, ITU Technical Report (1999).

[17] TS 05.05, *Technical Specification Group GSM/EDGE Radio Access Network; Radio Transmission and Reception*, 3GPP Technical Report (Nov. 2005).

[18] J. G. Proakis, *Digital Communications* (McGraw-Hill, 2000).

[19] E. Tuomaala and H. Wang, Effective SINR approach of link to system mapping in OFDM/multi-carrier mobile network. In *Second International Conference on Mobile Technology, Applications and Systems*, Guangzhou, China (Nov. 2005).

[20] TR 25.892, *Feasibility Study for OFDM for UTRAN Enhancement*, 3GPP Technical Report (Mar. 2004).

[21] F. Andrs-Quiroga, Link-to-system interfaces for system level simulations featuring hybrid ARQ. Master thesis, Technische Universität München, Munich (2008).

[22] TR 101 112, *Universal Mobile Telecommunications System (UMTS); Selection Procedures for The Choice of Radio Transmission Technologies of The UMTS*, ETSI Technical Report (April 1998).

[23] 3GPP R4-092042 TSG RAN WG4 (Radio) Meeting 51, *Simulation Assumptions and Parameters for FDD HeNB RF Requirements*, 3GPP Technical Report (May 2009).

[24] A. M. Galindo-Serrano, *Self-organized Femtocells: a Time Difference Learning Approach*, PhD Thesis, Univertat Politecnica de Catalunya, Barcelona (2012).

[25] ITU-R M.2135:, *Guidelines for Evaluation of Radio Interface Technologies for IMT-Advanced*, ITU Technical Report (Nov. 2008).

4 Access mechanisms

Vikram Chandrasekhar, Anthony E. Ekpenyong and Ralf Bendlin

4.1 Introduction

The objectives behind this chapter are twofold. One is to provide an in-depth description of access control in *heterogeneous cellular networks* (HCNs) from the perspectives of the *core network* (CN), the *radio access network* (RAN) and the *user equipment* (UE). The second objective is to provide benefits and tradeoff analysis of different access control schemes through numerical simulations with respect to various performance metrics such as the percentage of offloaded UEs as well as cell-average and cell-edge data throughput.

A rudimentary understanding of the *Universal Mobile Telecommunication System* (UMTS) and *Long Term Evolution/System Architecture Evolution* (LTE/SAE) cellular architectures is essential to get some intuition behind how access control is implemented in contemporary HCNs. Section 4.2 introduces the motivation for access control and describes available access methods. Sections 4.3 and 4.4, respectively, provide a basic overview of the UMTS and *Long Term Evolution* (LTE) cellular architectures. We describe two main components of the system architecture, namely the CN and the RAN. At a high level, the CN is responsible for overall control of the UE including packet processing, *quality of service* (QoS) enforcement and connection with the operator network. The RAN is responsible for the radio air interface functions to the UE. In UMTS, the RAN functions are split between a *radio network controller* (RNC) and a NodeB, in LTE, the RAN functions reside at a single logical node called the *evolved NodeB* (eNB).

Sections 4.5, 4.6 and 4.7 describe the access control enhancements standardized during LTE Releases 8, 9 and 10, respectively. We discuss the respective roles of the CN and the RAN, while implementing access control during inbound and outbound mobility to/from *closed subscriber group* (CSG) cells. Since access control in the UMTS standard is substantially similar to that in LTE, we only highlight their main differences in Section 4.8. Finally, the chapter concludes by analyzing the performance of access control in a representative LTE HCN.

4.2 Access control modes

The main motivation behind access control is to provide maximum deployment flexibility in a cellular network. As an example, in traffic hotspots such as malls, exhibitions, coffee shops and university areas, an operator may install either picocells or alternatively *relay nodes* (RNs) for offloading UEs from high-powered and potentially overloaded macrocells onto *low-power nodes* (LPNs). In a second scenario a user may deploy a femtocell to get a high capacity wireless link (by reserving exclusive access to its resources) and a prolonged battery life, thanks to the proximity to its femtocell access point. In order to address these scenarios, the *3rd Generation Partnership Project* (3GPP) provides three access mechanisms as follows.

Open access **(OA):** An OA cell allows unrestricted access to all UEs.

Closed access **(CA):** A CA cell provides privileged access to *user equipments* (UEs) belonging to a *closed subscriber group* (CSG) and denies access to other UEs. A cell implementing CA is henceforth referred to as a CSG cell. Whether or not a cell is CSG is determined by its broadcast parameter, namely, *CSG identity* (CSG ID). A CSG cell broadcasts a CSG ID, whose value is set to *TRUE*, and a unique CSG ID, if applicable. Example scenarios include residential and office buildings wherein the femtocells will inevitably limit CSG access to residents, employees and possibly the visitors to the building.

Hybrid access **(HA):** HA cells strike a compromise between completely closed and completely open access. Like a CSG cell, an HA cell broadcasts a CSG indicator and a specific CSG ID. However, unlike a CSG cell, the value of the CSG indicator from a HA cell is always set to *FALSE*. Thus, a cell employing HA appears as a CSG cell to UEs that are subscribed to the CSG, while it appears as a normal cell to other UEs.

The CSG whitelist concept

A UE maintains a whitelist of cells to whose CSG it belongs. Upon detecting a CSG cell, the UE checks suitability for access based on whether or not the broadcast CSG ID belongs to its CSG whitelist. The CSG whitelist is the union of a so-called allowed CSG list and the operator CSG list. Akin to Wi-Fi, if needed, the user may manually add/remove entries from its allowed CSG whitelist through a process called *manual selection*. The operator can modify both the aforementioned lists residing at the UE. One key difference of HA with respect to CA is that HA cells have the capability to provide differing levels of QoS to UEs based on whether they are subscribed (i.e., the broadcast CSG ID already belongs to their whitelist) or not.

4.3   Basics of the UMTS cellular architecture

The basic UMTS system consists of a set of logical network elements that together provide mobile telecommunication services to a UE. These logical elements are categorized under three main components: the CN, the *UMTS Terrestrial Radio Access Network* (UTRAN) and the UE itself. The CN is responsible for all call/packet switching and routing with external networks, whereas the UTRAN handles all radio-related functionalities (refer to [1] for more detail).

This logical framework for the UMTS network architecture facilitates interoperability between physical implementations of various network elements from different vendors. Towards this goal a set of open logical interfaces is standardized while the exact implementation of network elements is left to vendor implementation. The Iu interface connects the UTRAN to the CN, while the Uu interface connects between the UE and the UTRAN. This allows a cellular operator the flexibility of selecting a core network from one vendor, the UTRAN from another vendor and any number of UEs from several UE manufacturers.

4.3.1   Core network

The UMTS CN architecture is largely based on the preceding *Global System for Mobile Communications* (GSM) CN architecture, and has dedicated network elements for either circuit-switched connections or packet-switched data connections. This helped to simplify and reduce the cost incurred in evolving from *2G* to *3G* networks. As an example, similar to GSM networks, the UTRAN is connected to the Mobile Services Switching Center (MSC) and the Voice Location Register (VLR) for circuit-switched services such as voice calls.

4.3.2   Access network

The UTRAN consists of a set of *radio network subsystems* (RNSs), which are connected to the CN through the Iu interface. Each RNS consists of an RNC, which controls one or more radio access nodes referred to as NodeBs in UMTS nomenclature. The radio access functionality is split between the NodeB and the RNC. The RNC handles all *radio resource management* (RRM) functions, whereas the main function of a NodeB is the actual radio air interface processing to/from UEs within its cell. Note that the separation of network functionalities between the RNC and the NodeB is logical; whether or not the RNC and the NodeB are physically collocated is not standardized (left to network implementation).

A femtocell is referred to as a *Home NodeB* (HNB) in the UMTS specification. Because the HNB is likely user deployed, a separate architecture was specified for the HNB. The HNB access network consists of an HNB gateway and one or more HNBs. The HNB gateway appears as an RNC to the CN and is also a concentrator of multiple HNB connections towards the CN. The HNB gateway acts as the RNS to, and provides RAN connectivity for, the UE.

The RNC provides the following functions

Radio resource management: RRM comprises functions such as radio bearer control, radio admission control, connection mobility control and dynamic allocation (scheduling) of resources to UEs in both the *uplink* (UL) and the *downlink* (DL).

Header compression and encryption: Header compression and encryption of the UE packets are performed prior to transmission, which minimizes the radio overhead while transmitting small-sized packets, e.g., in *voice over IP* (VoIP).

CSG functionality: The RNC hosts the CSG functionality, which involves broadcasting its CSG ID and handling signaling with the CN with respect to CSG membership verification and access control.

Paging and broadcast signaling: The RNC ensures that all paging and broadcast information originated from the CN are communicated to the UE either via a scheduled transmission or via a broadcast transmission.

Measurement reporting configuration: The RNC configures the UEs to measure and report their radio link quality – to both serving and neighboring cells – which can be used to determine candidate target cells for *handovers* (HOs).

4.3.3   Radio protocol functions in UTRAN

The radio protocol functions in UTRAN can be categorized as the *user plane* and the *control plane*. The user plane functionality consists of the *packet data convergence protocol* (PDCP), the *radio link control* (RLC) and the *medium access control* (MAC) layers. The control plane handles the *radio resource* (RRC) protocol at the access stratum (between the UTRAN and the UE), which involves establishing radio bearers, handling mobility functions, handling measurement and reporting configuration, transmitting broadcast and paging information and configuring the lower layers through dedicated signaling from the UTRAN to the UE.

User plane

The UTRAN user plane protocol consists of the PDCP, RLC and the MAC protocols (also known as the Layer 2) each of which is terminated at the NodeB on the network side. Under the user plane, the functions of these protocols are described as follows.

Packet data convergence protocol (PDCP): The PDCP protocol processes the *internet protocol* (IP) traffic packets (radio bearers) on the user plane. This includes the encryption, header compression and ensuring security (via encryption) of the packets for the intended UE.

Radio link control (RLC): The RLC protocol segments and re-formats the upper layer packets so as to ensure they are suitable for transmission over the radio interface. For radio bearers requiring lossless transmission such as file downloads, the RLC layer performs reordering to compensate for potential out-of-order delivery as a result of *hybrid automatic repeat request* (HARQ) operation performed in the lower (MAC) layer.

Medium access control (MAC): The MAC protocol is the lowest sublayer in Layer 2. The MAC connects to its higher layer (RLC) via logical channels and connects to its lower layer (physical layer) via transport channels. The MAC is responsible for QoS management and multiplexing/demultiplexing data belonging to different radio bearers from/into transport blocks. Additionally, the MAC protocol handles HARQ so as to perform error correction via packet retransmission. Finally, the MAC layer establishes priority handling between different logical channels belonging to a UE as well as different UEs through dynamic assignment of frequency–time resources.

Control plane

The control plane of the access stratum – between the UTRAN and the UE – configures radio resources via the radio resource control (RRC) protocol of Layer 3. Some RRC functions include broadcasting *system information* (SI) messages, paging, control of radio bearers, transport and physical channels, mobility control, RRC connection establishment and release, and control of UE measurements for RRM. With respect to RRC the UE can either be in idle mode or RRC connected mode. The UE behavior in a cell depends on the RRC states as described below.

Idle mode: When a UE is first powered on, it searches for a cell of a *public land mobile network* (PLMN) on which it can receive communication services. Once a suitable cell is found, the UE tunes to the cell's control channel to receive SI and broadcast messages. This process is known as *camping on a cell* and the UE is said to be in idle mode. The UE performs cell re-selection, if it finds a *more suitable cell*, where suitability is based on a set of defined criteria. Prior to requesting services from the network, the UE registers its presence on the PLMN. This registration enables the network to notify the UE in the event of an incoming call by sending a paging message on the control channels of all the cells in the UE's registration area.

CELL_DCH: When an idle mode UE requires initial access to the UTRAN it initiates an RRC connection to the UTRAN. Upon successful RRC connection setup the UE is in the CELL_DCH state and is allocated a *dedicated channel* (DCH). In this mode the UE is also configured by the UTRAN to perform a set of measurements that facilitate RRM. The DCH is not released until the state transitions back to idle mode or to the CELL_FACH state. This circuit-switched mode of operation is not well suited for the bursty nature of data transmission due to the latency involved with frequent transitions between idle and connected modes. To reduce the latency between idle and connected modes three other RRC states are defined as follows.

CELL_FACH: In this state, the UE is not allocated a DCH; however, it can transmit signaling and small data packets on the random access channel (RACH) and the forward access channel (FACH). The UE also listens to the broadcast channel (BCH) for SI. State transitions can occur to/from all RRC states. The UE can also perform a cell re-selection and subsequent cell update procedure while in the CELL_FACH state. As networks evolved from voice-centric to data-centric applications, there was a need for an *on demand* RRC state, which could be set up and released in response to the bursty nature of data. However, the latency involved with transitions to/from CELL_DCH state was seen as unacceptable. Therefore, starting from Release 7 an Enhanced CELL_FACH feature was introduced, which supports *High Speed Packet Access* (HSPA) in the CELL_FACH state.

CELL_PCH and URA_PCH: In the absence of a voice or data call, the UE may transition back to idle mode. To reduce the connection setup time from idle to connected mode, the UE can transition to either the CELL_PCH or URA_PCH states. In these states, the UE is still known to the network at the serving RNC, but it can only be reached via a paging notification on the paging channel. If the UE performs a cell re-selection in the CELL_PCH state, it moves to the CELL_FACH state to update the network on its new location and then moves back to the CELL_PCH state if no other activity is required. On the other hand, in the URA_PCH state, the UE does not perform a cell update after a cell re-selection unless the new cell is in a different UTRAN registration area. This helps to reduce ping-ponging effects in the case of frequent cell re-selections, which can be observed in HCNs with many overlapping cells. Moreover, such frequent cell updates would drain battery life.

4.4　Basics of the LTE cellular architecture

The LTE/SAE network architecture was specifically designed for packed-switched services, as opposed to the circuit-switched connection-oriented nature of the preceding radio access technologies (RATs) such as GSM and *wideband code division multiple access* (WCDMA). A packetized interface aims to provide seamless IP connectivity between the packet data network and the UE. The connectivity is provided via the concept of *bearers*, which are packet flows with a predefined QoS between the packet data network and the UE.

A UE can engage in multiple IP services such as file downloads, web-browsing, streaming services and VoIP, each of which can be mapped to a bearer with an associated QoS. From a RAT perspective, LTE therefore exploits fast packet scheduling across frequency, time and space as well as link adaptation to the variations of the wireless channel. From a protocol perspective, bearers can be scheduled with a packet duration of 1 ms to exploit multi-user diversity and efficient resource allocation.

The two constituents of the LTE RAN architecture are called the *Evolved Packet Core* (EPC) and the *Evolved UTRAN* (E-UTRAN). The EPC refers to a set of logical nodes responsible for overall control of the UE and establishment of the radio bearers (refer to [2] for a more detailed description). The E-UTRAN consists of a single node, the eNB, which connects to and controls the behavior of the UEs. In LTE, the eNBs are normally connected with each other by means of a standardized interface called the X2 interface and to the EPC through the S1 interface.

The standardized nature of these interfaces allows for inter-operability across nodes manufactured by different vendors and provides operators the flexibility to choose a multi-vendor installation, e.g., based on geographical location and/or economic considerations.

4.4.1　Evolved Packet Core (EPC)

In contrast to the E-UTRAN, which comprises of a single logical node, namely, the eNB, the CN consists of multiple logical nodes. For the sake of brevity, we will confine our discussion to those logical nodes in the CN, which are directly connected to the access network: the *Mobility Management Entity* (MME), which connects to the E-UTRAN via the S1-MME interface; and the *Serving Gateway* (S-GW), which connects to the E-UTRAN via the S1-U interface. Other entities of the CN are beyond the scope of this chapter, for they are not directly connected to the E-UTRAN and use interfaces other than the S1 interface discussed here.

MME: The MME controls the *non-access stratum* (NAS) signaling, which refers to protocols between the CN and the UE. The functions of the MME include the creation and maintenance of the UE context (including CSG subscription data for connection authorization), and the establishment of radio bearers and paging. The MME also manages and secures the connection between the network and the UE, handles tracking area maintenance, P- and S-GW selection, roaming, authentication and HO procedures such as target MME selection and termination of interfaces. The functionalities of the MME are manifold and a complete description is omitted here for brevity (refer to [3] for a detailed description).

S-GW: The S-GW serves as a mobility anchor for the radio bearers associated with a given UE when it moves between eNBs. All the IP packets intended for the UE are routed through the S-GW, which is responsible for the buffering, marking and forwarding of packets. Similar to the MME, it hosts functionalities for both inter-RAT and intra-RAT HOs, including the termination of interfaces.

In addition to the aforementioned aspects, the MME hosts certain special functions for implementing access control during HOs to LPN cells. In the case of an HO to a CSG cell, the MME implements access control to determine whether the UE is a member of the CSG to which the target cell belongs. Access control is based on the CSG ID of the target CSG cell provided to the MME by the serving E-UTRAN. In case of an HO to an HA cell, the MME also performs membership verification based on the CSG ID of the target cell. Additionally, after membership verification, the MME signals the CSG membership status to the target eNB, so that the latter can decide the QoS for the UE accordingly. More details are discussed in the following sections.

4.4.2   Access network

Because the E-UTRAN was designed to be a flat architecture, there is no predefined hierarchy between macrocells and LPNs. In contrast to UMTS, LTE integrates the radio controller functionality into the eNB. This ensures tighter interaction across protocol layers and higher efficiency. A consequence of such an integrated functionality is that the UE context has to be transferred from one eNB to another during HO via the X2 and S1 interfaces. The *X2 Application Protocol* (X2AP) is used for inter-eNB communication, while the *S1 Application Protocol* (S1AP) is used between the eNB and the MME. Fig. 4.1 shows the LTE cellular architecture.

The E-UTRAN architecture supports a special gateway for interfacing *Home evolved NodeBs* (HeNBs) (used to refer to femtocells in LTE) to the CN. This network entity, called the HeNB Gateway, is logically placed between a HeNB and the MME. The motivation behind this gateway is to enable the S1 interface between each HeNB and the EPC to support a large number of HeNBs to be installed and moved within a given geographical area. The gateway appears as an MME to its connected HeNBs, and as an eNB to its connected MME. As a result, the HeNB gateway can simplify HOs to/from HeNBs connected to it.

Because the HeNB operates in licensed spectrum, the operator may desire to control the HeNB [4]. Since the gateway provides a single consistent interface to the EPC irrespective of the number of deployed HeNBs, it simplifies certain procedures such as activation/deactivation of a HeNB and configuration of the *tracking area code* (TAC) and PLMN identity.

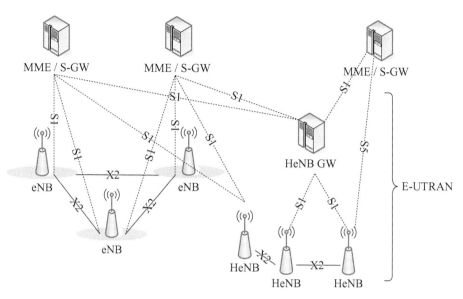

Figure 4.1. The LTE cellular architecture (reproduced with permission from 3GPP).

4.4.3   Radio protocol functions in Evolved-UTRAN (E-UTRAN)

Similar to UMTS, the LTE radio protocol functionalities at the eNB can be categorized as the user plane and the control plane. User plane functionality in LTE is as described earlier in Section 4.3.3. The following subsections examine the control plane in greater detail.

Control plane

As described in Section 4.3.3, the control plane for the access stratum handles radio-related functionalities via the RRC protocol. The control plane handling also involves the PDCP protocol (e.g., encryption via ciphering and integrity protection) as well as the RLC and MAC protocols (which perform the same functions as for the user plane). The RRC is responsible for mobility-related functions such as measurement configuration and reporting. The RRC is also responsible for establishing a radio link connection between the E-UTRAN and the UE, and for broadcasting paging and system information. Depending on the RRC state of the UE, the RRC protocol uses the following states.

RRC_IDLE: In the RRC_IDLE state, the UE autonomously decides a suitable cell to camp on during a cell search process. Cell search consists of identifying a set of suitable candidate cells and selecting the strongest cell from the set to camp on. Once camped on that cell, the UE continues to monitor neighboring cells for cell re-selection. Note that the mobility function in RRC_IDLE is UE controlled. The UE acquires critical SI by monitoring the broadcast channel. An RRC_IDLE UE also monitors the paging channel through which the CN informs the UE about incoming calls and/or updates to the SI.

RRC_CONNECTED: In the RRC_CONNECTED state, the UE has already established a connection to an eNB, and it can transmit and receive data. During each subframe, the UE monitors its control channels to see whether it has been scheduled during the current transmission, and if scheduled, it determines the frequency and time resources for transmitting/receiving its data. The network controls the mobility of an RRC_CONNECTED UE. To facilitate HO, the UE provides the network with measurement reports with regards to the DL channel quality to its serving cell and possibly a set of neighboring cells. When the network deems that the UE has a neighboring cell with a better DL channel quality relative to the current serving cell, the network can choose to trigger the UE for HO.

To aid UE mobility in RRC_IDLE and to control UE mobility in the RRC_CONNECTED state, the network transmits dedicated RRC messages. Three different types of system information can be provided to the UE, namely, the *Master Information Block* (MIB), the *System Information Block Type 1* (SIB1) and other SI messages. Note that in general, even though a UE may be camped on an LTE cell, it may continue to monitor neighboring cells including cells on a different carrier frequency or even on a different RAT (e.g., GSM, WCDMA, WiMAX). The eNB may configure the UE with measurement gaps for interfrequency monitoring wherein the UE ceases to transmit and receive data from its serving cell. A detailed description of UE mobility activities is outside the scope of this chapter; we refer the interested reader to [3] and Chapter 9.

In the following sub-sections, we discuss inbound and outbound mobility procedures for CA and HA cells, as introduced during LTE Releases 8, 9, 10 and 11 (as of the writing of this book), for both RRC_IDLE and RRC_CONNECTED states.

4.5   LTE Release 8: mobility management to CSG cells

An LTE UE determines the type of access (CA/OA/HA) to a certain cell depending on the CSG indicator and the CSG ID that is broadcast by the cell in the SIB1 message [3, 5]. The CSG ID identifies the cell as a member of a particular CSG within the PLMN. As discussed in Section 4.2, the UE checks suitability for access to CSG cells based on whether or not their broadcast CSG ID belongs to its whitelist. The final decision on whether the UE will connect to a CSG cell is controlled either by the UE (for a UE in RRC_IDLE state) or by the MME (for a UE in RRC_CONNECTED state).

4.5.1   Idle mode mobility to and from CSG cells

The basic principles of CSG cell association in RRC_IDLE mode were defined in LTE Release 8. For CSG cells, cell (re-)selection follows the same principles of idle mode mobility as macrocells. That is, the UE tries to camp on a CSG cell if such a cell provides the best DL channel quality and if the CSG ID of that cell belongs to the CSG whitelist of the UE. For the case of intrafrequency mobility, the UE ignores all CSG cells it is not allowed to connect to (i.e., whose CSG ID does not belong to its whitelist) and ranks the remaining ones by their radio link quality. The UE will disable the autonomous search function for CSG cells if the UE's CSG whitelist is empty [6].

Broadcasting CSG information

To aid cell detection of CSG cells at the UE, the network can configure cell re-selection parameters to prioritize camping on a CSG cell. Alternatively, an eNB can broadcast information corresponding to a set of neighboring (possibly CSG) cells. For example, via its *System Information Block Type 4* (SIB4) transmission, a cell could broadcast a set of *physical layer cell identity* (PCI) values, which are reserved for CSG cells, thereby triggering the UE to search for CSG cells (UEs without a CSG whitelist can skip reporting PCIs from this range). A UE shall associate the last received PCI range to be valid for a maximum of 24 hours since it was last received. This range of reserved PCIs is only applicable to the frequency of the PLMN in which it received such information. Optionally, the SIB4 information element also contains a set of *blacklisted* cells, which must not be considered by the UE during the intrafrequency cell re-selection procedure. Moreover, manual selection of CSG cells at the UE is possible as well. During cell re-selection, if the UE detects one or more suitable CSG cells on different frequencies, then the UE shall re-select to one of the detected cells irrespective of the frequency priority of the cell the UE is currently camped on, if the concerned CSG cell is the highest-ranked cell on that frequency [6].

Performance considerations

While specifying the detection procedure for CSG cells, care was taken to minimize the search complexity, while allowing for a non-standardized, autonomous cell-search implementation at the UE. It is especially important that subscribed UEs to a CSG cell give higher priority to camping on that cell, particularly when they are in its vicinity. For example, a user may deploy an HeNB since the indoor coverage of the macrocell layer may be weak at it premises. It could very well be the case that the billing rates applied for HeNB usage could differ with respect to the normal subscriber billing rates. It is thus important that, once a subscriber UE is within radio range of its HeNB cell, it immediately camps on it.

With increasing deployment of CSG cells, it is possible that the number of UE measurements that are likely to be wasted becomes large. This could lead to a loss in efficiency from a mobility perspective as well as a loss in battery life at the UE. Therefore, it is especially important that the network strives to minimize unnecessary measurements of a CSG cell when the UE is not subscribed to the cell.

Finally, the mobility performance of the UE should not be significantly impacted due to changes to either the carrier frequency or the PCI of a CSG cell. The PCI of an HeNB may change on a slow basis during normal operation, e.g., when high interference is detected. Similarly, it is possible that, following power-up, the HeNB selects a different carrier frequency out of a pre-configured set of carrier frequencies. When the carrier frequency at a CSG cell changes, the UE should detect the change autonomously by noticing the absence of the CSG cell on the PCI and/or carrier frequency where it was previously expected, and thereafter initiating a full scan of the CSG cell. By checking the CSG ID of detected cells, the UE can locate the CSG cell on a new PCI and/or carrier frequency.

4.5.2 Mobility to and from CSG cells in RRC_CONNECTED mode

In LTE Release 8, inbound mobility to CSG HeNB cells is not fully supported. The reasons are as follows: first, no access control is defined to ensure that a UE handed off to a CSG cell is a valid member of the CSG to which the HeNB belongs; second, due to the large number of HeNBs overlaying the macrocell layer, there may be ambiguity when determining the target HeNB for HO. This problem, referred to as the so-called *PCI confusion problem*, is further described in the following sub-section.

For outbound mobility from CSG cells to non-CSG cells (e.g., from a CSG HeNB to a macrocell eNB), the normal outbound mobility procedures as in mobility across non-CSG cells apply. For LTE Releases 8 and 9, because X2 connectivity to HeNBs is not allowed, the aforementioned HO procedures involve using the S1 interface between the HeNB and the CN.

4.5.3 PCI confusion

Due to the smaller cell sizes associated with LPNs, there could be a large number – hundreds or even thousands – of installed LPNs within the coverage overlay of the macrocell. However, the PCI assigned to an LPN can take only one of 504 different values. Making matters worse, the PCI for the LPN may change following power-up. A likely outcome is that there could be more than a single LPN sharing the same PCI within the set of reserved PCIs for CSG cells. One main reason is that the eNB may not maintain neighbor relations to LPN, especially closed access cells, such as closed access HeNB. As a result, the source eNB may be unable to determine the correct target LPN for HO based on the PCI included in the measurement reports transmitted from the UE. This ambiguity when determining the correct target eNB based on the PCI prior to HO is known as the *PCI confusion* problem. In such a scenario, the network might not be able to correctly HO the UE to the target eNB – leading to dropped calls during HO – due to the ambiguity at the MME, while distinguishing two cells with identical PCI values.

4.6 LTE Release 9: mobility enhancements to CSG cells and introduction of HA cells

In LTE Release 9, four enhancements were introduced to improve mobility support to LPN cells with closed access. First, the HA concept [5], described in Section 4.2, was introduced in addition to the other two levels of access, namely OA and CA, that were already present in LTE Release 8. Second, inbound mobility to CSG cells was introduced for UEs in RRC_CONNECTED mode by employing access control at the MME. Third, to enable the UE to perform efficient measurements in large-scale HeNB deployments,

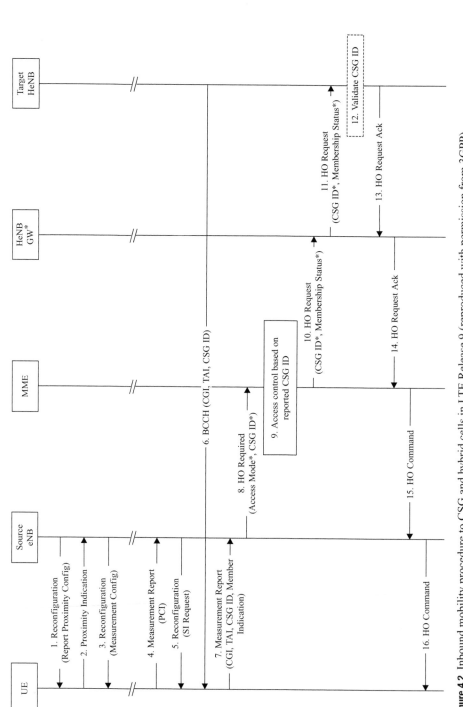

Figure 4.2 Inbound mobility procedure to CSG and hybrid cells in LTE Release 9 (reproduced with permission from 3GPP).

a UE reporting mechanism called proximity indication was introduced. Finally, the PCI confusion problem (Section 4.5.3) was resolved using a new UE assisted global cell identity reporting procedure. The following sub-sections survey these enhancements.

4.6.1   Hybrid access

Like a CSG cell, an HA cell broadcasts a CSG ID but it sets its CSG indicator to *FALSE*. In addition, the PCI values for HA cells are not contained within the reserved PCI range for CSG cells. If desired, the network can reserve a separate PCI range for HA cells. Manual selection of HA cells is also supported in the same way as for CSG cells.

The idle mode cell selection and reselection procedure in LTE had to be modified to account for the possibility that an HA cell will likely be accessed by both CSG UEs and non-CSG UEs. For a UE whose whitelist contains the CSG ID of the HA cell, that cell appears as a CSG cell. On the other hand, since the HA cell broadcasts its CSG indicator bit set as *FALSE*, a non-CSG UE can access an HA cell like a normal cell.

In RRC_CONNECTED, in the case of an HO to an HA cell, the MME performs membership verification to determine whether the UE belongs to the CSG advertised by the HA cell. Following membership verification, the MME informs the target eNB whether or not the UE is a member of the CSG associated with the HA cell. This partitioning of UEs based on their CSG membership status allows the eNB to provide differentiated QoS to CSG and non-CSG UEs. For example, a subscribed UE could be assigned a higher priority relative to a non-subscribed UE, which could only be assigned basic levels of access such as voice, or low-rate data. Similarly, an HA cell can accept and deny inbound HOs of CSG and non-CSG UEs differently. For example, as part of the load balancing function residing at an HA cell, a non-CSG UE could be denied inbound mobility to an HA cell or could be handed over to another cell depending on the load experienced at that HA cell.

4.6.2   Access control, PCI confusion resolution and proximity indication

The inbound mobility procedures to CSG and HA cells are illustrated in Fig. 4.2.

Proximity indication: Based on its autonomous neighbor cell search procedure, if a UE determines that it is either entering or leaving the radio range of one or more CSG or HA cells whose CSG ID lies in the whitelist maintained at the UE, the UE can inform the network that it is within the proximity of a CSG cell [3, 5]. Proximity indication provides the source eNB with improved mobility performance and prolongs battery life at the UE by minimizing the number of unwanted neighbor cell measurements and SI decoding that would otherwise be performed at the UE when in the presence of a large number of CSG cells. For example, the source eNB can configure the UE to perform measurements and acquire the SI of a CSG cell only upon reception of the proximity indication message. To control the number of proximity indication messages and limit their impact on the network load, the transmission of the proximity indication message is initiated only if at least 5 s have elapsed since a prior proximity indication (if applicable) was transmitted.

Resolving PCI confusion: Following proximity indication, the source eNB can reconfigure, using an *RRC Connection Reconfiguration* message, the UE to acquire the PCI of a target eNB and report the PCI back to the source eNB. Note, however, that the PCI of the target HeNB may be shared among multiple HeNBs. Following the reporting of the PCI of the target eNB, the source eNB requests the UE to acquire the SI broadcast at the target HeNB cell. Following SI acquisition, the UE transmits to the source eNB a measurement report comprising the *Evolved Cell Global Identifier* (ECGI), the *tracking area identity* (TAI), the CSG ID, the CSG indicator and its CSG membership status at the target CSG cell [5]. The ECGI is unique to each cell – even though multiple cells may share a common PCI – since the ECGI is derived based only on the PLMN to which the cell belongs and its cell identity within the PLMN. Knowledge of the ECGI of the target cell at the MME therefore resolves any ambiguity at the network due to PCI confusion. The source eNB passes on to the MME the ECGI, the CSG ID and optionally the CSG access mode of the target cell in case of HA.

Access control: Following the acquisition of SIB1 broadcast by the target CSG or HA cell, the UE sends a measurement report to its source eNB. The membership report includes the ECGI and CSG ID of the target eNB and the membership of the UE within the CSG of the target eNB. Note that the acquisition of SI of the target CSG or HA cell can be triggered by the reception of a proximity indication message at the source eNB. The source eNB includes the ECGI and the CSG ID of the target eNB (and, if applicable, the membership status) in the HO message sent to the MME, which enforces access control to the target eNB. If access control fails, the MME sends an HO preparation failure message to the source eNB, thereby terminating the HO. If access control is successful, and if the target cell is an HA cell, the MME also executes membership verification. If the UE is indeed allowed to access the target cell, the MME sends an HO request message to the target eNB, which includes the CSG ID as reported by the UE and the membership status in the case of HA cells. The target eNB verifies that the received CSG ID matches the CSG ID broadcast in the target cell. Following successful validation, the target eNB can allocate resources accordingly, i.e., prioritize CSG members over non-CSG members. The successful HO concludes with the target eNB sending an HO request acknowledge message to the MME, which, in turn, sends the HO command message to the UE via the source eNB.

Exceptions to access control and management of CSG status
From LTE Release 9 onwards, in certain exceptional circumstances such as when the UE has an emergency call, the MME allows access even if the access control procedure fails. In addition, a new feature was introduced to enable the MME to signal a change of CSG membership status to a CSG and/or HA cell. Specifically, if a change in membership status of a UE connected to a CSG cell occurs, such an event is signaled to that CSG cell by the MME. In such a case, the CSG cell could initiate HO to a non-CSG cell. On the other hand, if the serving eNB is an HA cell, the UE's QoS can thereafter be handled analogous to that of non-CSG members.

4.7 LTE Release 10 and beyond: introduction of X2 interface for HeNBs

The key mobility-related enhancement introduced in LTE Release 10 was with regard to the creation of the X2-based mobility directly between HeNBs. Such mobility is only allowed if access control is not required at the MME [3]. Applicable scenarios include those in which (a) the HO is between CSG/HA HeNBs with identical CSG IDs, or alternatively (b) the HO is to a target HeNB providing OA.

Until LTE Release 10, mobility from an eNB/HeNB to an HeNB CSG/hybrid cell used the S1 HO procedure [3]. For LTE Release 11, further mobility enhancements are currently being considered, such as the establishment of a so-called *X2-proxy* between eNBs and HeNBs. Architecturally, the X2-proxy, if deployed, logically resides at the HeNB gateway. The X2-proxy "appears" as an eNB to other eNBs and HeNBs. Consequently, the X2-proxy can provide direct X2 connectivity between itself and other eNBs/HeNBs connected to the HeNB gateway. Such X2 connectivity to an HeNB gateway was not specified up to LTE Release 10. During mobility, the X2-proxy processes and forwards all X2 messages, akin to a relay, between the HeNB and other eNBs for all UE-dedicated procedures. All non-UE-dedicated X2 application protocol procedures (such as X2 connection setup [7]) are terminated at the X2-proxy, and handled locally between the HeNB and the X2-proxy, and between the X2-proxy and other eNBs [8].

In dense HeNB deployments, the X2-proxy can also act as a concentrator to allow for a large number of HeNBs to interconnect via the proxy without requiring direct X2 links that would be needed otherwise. Since the number of X2 connections increases as the square of the number of deployed HeNB cells, the incurred overhead would be excessive. Hence, the main motivations behind the X2-proxy are to reduce the number of X2 backhaul links that would otherwise increase with increasing number of deployed HeNBs and to enable a more efficient operation of a large number of CSG cells over the X2 interface. Other enhancements, i.e., regarding a more efficient implementation of membership verification during an X2-based HO to HA cells, are also being considered for standardization in LTE Release 11.

4.8 Distinguishing features of UMTS access mechanisms

Mobility management techniques for LTE CSG cells were described in Sections 4.5, 4.6 and 4.7. To a large extent, these techniques are equally applicable to UMTS CSG cells. A few differences exist between UTRAN and E-UTRAN, where such differences are primarily due to the dissimilarity in air interfaces and network architectures. A few differences include the following.

Broadcast of CSG Information: At the physical layer, a UTRAN cell can be identified by its *primary scrambling code* (PSC), whereas an LTE cell is identified by its PCI. Therefore, to aid cell selection in UTRAN, CSG and non-CSG cells may broadcast a reserved range of PSCs in their SI. The reserved PSC range is only applicable to the carrier frequency of the PLMN on which it was received for a maximum duration of 24 hours.

PSC confusion: Similar to the PCI confusion that was described in Section 4.5.3, more than one HNB may share the same PSC. For such a scenario, the serving RNC cannot determine the desired target cell for HO. The mechanism for resolving the PSC confusion is the same as for PCI confusion resolution in LTE.

Soft HO: A UE may be in soft HO with multiple HNBs either through the HNB-GW or when the HNBs are directly connected via the Iurh interface. To support soft HO, the serving HNB receives a measurement report from the UE indicating that soft HO is possible. The serving HNB can then decide to set up a radio link to the target (also called the drift) HNB.

4.9   Case study of access control in LTE

In the previous sections, access schemes for LPNs were treated with emphasis on their underlying network architecture and mobility support. We now shift the focus to evaluating the numerical performance of different access control schemes over the DL in an LTE HCN with co-channel deployed macrocells and LPNs. The computer simulations model a tri-sectored macrocell network with 19 macrocell sites with an intersite distance equaling 500 m. Both a geographically regular LPN deployment (e.g., user deployment of HeNBs inside apartments) and a uniform random LPN deployment are considered.

For the sake of intuition and clean exposition, the DL performance analysis separately considers CA- and OA-based HCNs, although in reality an HCN deployment will invariably consist of a mix of CA, OA and HA cells. While such an approach may appear simplistic, there is no loss in generality; the performance trends presented in Sections 4.9.1 and 4.9.2 would be identical to those obtained from considering a mixed HCN deployment. The ensuing organization of this section is as follows. Section 4.9.1 studies the performance of an OA-based HCN. Two performance metrics are considered. First, performance gains of an OA HCN are evaluated relative to a homogeneous (macrocell only) network by obtaining their respective cell area spectral efficiencies (measured in bits/second/Hertz/sector). An increase in area spectral efficiency is indicative of a higher overall network throughput as a result of a higher number of concurrent transmissions per unit area. Second, we obtain the percentage of UEs associating with as a function of the number of deployed picocell LPNs within the umbrella coverage of each macrocell.

Thereafter, in Section 4.9.2, we change our focus to a CA-based HCN. Herein a given subset of UEs are assumed to maintain a CSG whitelist and hence associate with CSG cells. The remaining UEs are unsubscribed and hence are forced to associate with a macrocell, even though such UEs could potentially receive better coverage and DL throughputs if they were (hypothetically) not denied access by CSG cells. As a consequence, we shall show that, although deploying CA-based LPN cells results in considerably improved cell area spectral efficiencies for CSG UEs, the performance of non-subscribed UEs is significantly degraded due to severe intercell interference from CSG cells.

4.9.1 Open access heterogeneous cellular network

The main motivation behind LPNs is to harness the so-called *cell-splitting gain*, i.e., to increase the number of simultaneous transmissions in each macrocell area. In a co-channel deployment of macrocells and LPNs, as the number of deployed LPNs increases so does the spatial reuse of the available spectrum. The cell-splitting gains are produced at the expense of increased intercell interference, especially at UEs located at the cell border of a macrocell and an LPN.

Similarly, the transmit power of an LPN controls the cell-splitting gain, for it directly relates to the size of its coverage area and the amount of intercell interference it introduces into the network. The larger the coverage area of the LPN, i.e., the larger its transmit power, the greater is its radio footprint, and hence a larger number of UEs will connect to it. This frees up resources at the macrocell eNB, which in turn can allocate more resources to the remaining UEs not offloaded to LPNs. To capture these effects, we will use the cell area spectral efficiency as the fundamental performance metric. The cell area spectral efficiency is defined as the average normalized aggregate throughput (the normalization is performed with respect to the available channel bandwidth) delivered by each macrocell and LPNs deployed within the macrocell coverage area.

4.9.1.1 Methodology

In the urban macrocell network, we consider that three macrocell sectors form one cell site modeled as a hexagon. Within each cell site, the three cells transmit in same spectrum (universal frequency reuse) and are controlled by a single macrocell eNB. Thus, a total of 19 cell sites constitute 57 macrocell sectors. Each sector has on average 60 UEs. The minimum distance between two macrocell sites is 500 m and the minimum distance between a UE and a macrocell eNB is equal to 35 m. Each macrocell sector is overlaid with uniformly distributed LPNs, whose minimum distance from neighboring macrocell eNBs and LPNs is equal to 75 m and 40 m, respectively.

All transmitters are equipped with a single transmission antenna, while all receivers (UEs) are equipped with two receive antennas. Each picocell transmits at either 24 dBm (approximately 0.25 W) or at 30 dBm (1 W). The transmit power at the macrocell is significantly higher, 46 dBm (or nearly 40 W). UEs associate with their serving eNB based on LTE cell selection principles. Specifically, each UE measures the *reference signal received power* (RSRP) of a set of cells and connects to the eNB providing the highest DL channel quality (after averaging over channel fluctuations due to fast fading, path losses and shadowing).

The numerical evaluations model a clustered UE placement (to model traffic hotspots) with non-uniform UE densities across macrocells. For modeling traffic hotspots around the vicinity of LPNs, it is assumed that two-thirds of the UEs are clustered around an OA picocell LPN with a minimum (maximum) distance of 10 m (40 m), while the remaining UEs are uniformly and randomly scattered, as illustrated in Fig. 4.3. For example, given two LPN cells within the coverage area of each macrocell, this implies that 40 UEs are located within a 40 m radius of LPN cells. The remaining 20 UEs are scattered uniformly and randomly within each macrocell sector [9].

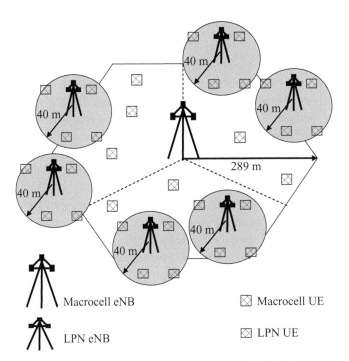

Figure 4.3 An OA-based HCN. Each hexagonal cell site is tri-sectored with the macrocell eNB controlling three sectors. Two LPNs are uniformly deployed within each macrocell sector. The UE density is non-uniform across the cell site, consisting of a mixture of randomly placed and LPN-clustered UEs.

4.9.1.2   Numerical results

Fig. 4.4 compares the DL cell spectral efficiencies of a homogeneous network (no LPNs are deployed) against an HCN with one, two and four LPNs deployed in each macrocell sector. Each LPN uses a transmit power equal to 24 dBm. The first bar in each cluster depicts the spectral efficiency of the homogeneous network. For comparison, the second and third bars correspond to the macrocell layer (second bar) and the LPN layer (third bar) of the HCN with identical UE distribution to the homogeneous network. For a fair comparison of cell spectral efficiencies, one needs to compare the first bar in each cluster (corresponding to a homogeneous network) with the sum of the second and third bars (corresponding to the normalized sum throughput delivered from both the macrocell layer and the LPN layer in an HCN).

The first observation is that the deployment of OA LPNs barely affects the performance of the existing macrocells. This is evident from Fig. 4.4, wherein the first two bars in each cluster are approximately of the same height. The third bar depicts the cell-splitting gain provided by deploying LPNs with OA in the same spectrum as the macrocell eNBs. In particular, it is observed that a single LPN can double the cell spectral efficiency with very little impairment of the macrocell layer. However, as can be seen from Fig. 4.4, the cell-splitting gain does not scale linearly with the number of LPNs. If we compare the

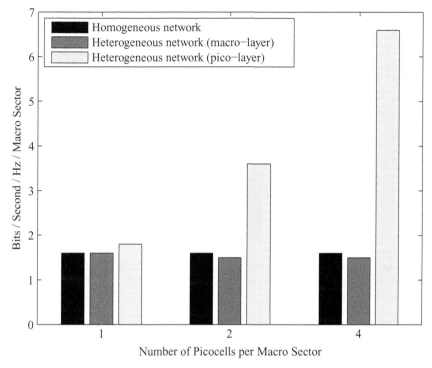

Figure 4.4 Cell spectral efficiencies in an OA-based HCN comprising one, two and four LPNs in each macrocell sector. The transmission power at each LPN is equal to 24 dBm (© 2011 IEEE).

last bars of the first and third clusters, we notice that the latter has less than four times the height of the former, due to the growing amount of intercell interference that each additional LPN creates in the network.

As shown in Fig. 4.5, a similar trend can be observed when we increase the transmit power of an LPN. For instance, with two LPNs per macrocell sector, as the LPN transmit power is increased from 24 dBm to 30 dBm, the cell spectral efficiency increases from 5.3 to 5.5 bits/second/Hertz/sector. The reason for this is that, with increasing LPN transmit power, more UEs are offloaded from the macrocell eNBs on to LPNs; consequently, the offloaded UEs obtain better throughputs due to the proximity to their serving LPN. From the computer simulations, when the LPN transmit power is equal to 30 dBm, on average, approximately 37% of UEs are offloaded on to LPNs. That is, assuming 60 UEs in each macrocell sector, on average only 38 UEs connect to their strongest macrocell eNB, while the remaining UEs are offloaded on to an LPN.

From the above results, it is evident that deploying OA cells provides network operators with a formidable tool to provide targeted coverage, improved offloading and capacity enhancements to existing macrocell networks. However, as analyzed in the following section, when the LPNs implement CA for access control the picture looks vastly different.

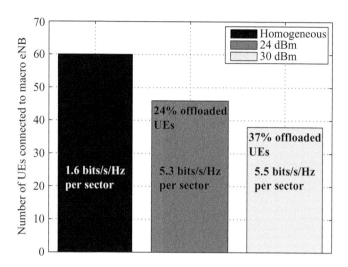

Figure 4.5 Cell spectral efficiencies and percentages of offloading obtained with different LPN transmission powers in an OA-based HCN (© 2011 IEEE). Two LPNs are deployed in each macrocell sector.

4.9.2   Closed access heterogeneous cellular network

The fundamental difference between OA and CA LPNs is the fact that the latter deployment could potentially create so-called *DL coverage holes* for certain important DL channels including broadcast, control and data. In OA, coverage holes do not typically occur, since the E-UTRAN can configure the UE to always connect to the cell for which it obtains the maximum RSRP.

However, this is not the case in a CA-based HCN. Specifically, if the CSG ID of a certain cell does not belong to the whitelist maintained at the UE, the UE cannot connect to the CSG cell, although that cell could very well offer the highest RSRP – and hence be the "best" for cell association – at the UE of interest. For instance, when an unsubscribed UE is close to a CSG HeNB but far away from a macrocell eNB, the UE would still be forced to connect to the macrocell HeNB, although the RSRP of the CSG HeNB could be higher than the RSRP of the macrocell eNB. The imbalance in RSRPs between the HeNB and the macrocell could occur even though the transmit power at an HeNB is significantly lower than that of the macrocell. Consequently, such a UE may experience deteriorated coverage on its physical DL channels since the received power from a dominant CSG interferer could be significantly higher than the received power from its serving cell. The size of the aforementioned DL coverage holes, experienced at macrocell UEs, is mainly determined by the transmit power of the HeNBs. The larger the HeNB transmit power, the larger is the concentric area around the HeNB in which the received signal from the HeNB masks the received signal from the nearest OA cell.

4.9.2.1   Methodology

To model the aforementioned impact of CA cells on network performance, we assume a single apartment complex containing 40 residents within each macrocell sector. The

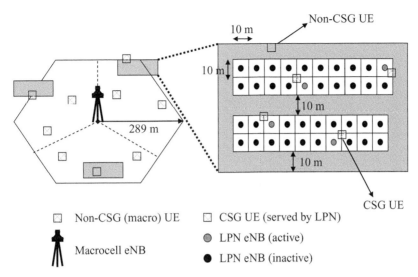

Figure 4.6 Dual-stripe apartment model: an apartment block consists of two buildings, each of which has 20 rooms of size 10 m × 10 m. In each room there can be either one or no HeNB. Each HeNB is assumed to have one UE (□) connected to it.

apartment complex consists of two buildings, with each building made up of 20 rooms of size 10 m × 10 m. A 10 m wide road separates the two buildings. The location of an apartment complex within the macrocell sector is chosen uniformly and randomly, subject to the minimum distance of 75 m between the HeNB and the macrocell eNB. Furthermore, any two adjoining apartment complexes are dropped at least 40 m apart. The complete topology of the simulation setup is illustrated in Fig. 4.6.

Within each room, an HeNB is dropped with a 10% probability of being active (i.e. being in use). Each HeNB is assumed to have one UE connected to it and there can be either one or no HeNB per room. The HeNB and the corresponding UE are uniformly distributed within the same room with a minimum distance of 3 m. Thirty-five percent of the macrocell UEs are clustered around the apartments both inside and outside the buildings. The remaining 65% of UEs are uniformly distributed within each macrocell sector.

HeNBs also have omnidirectional antennas in this model, but their transmit power is even further reduced as compared to the OA cells described in the previous section. In this evaluation, each HeNB, if active, has a transmit power of either 0 dBm or 10 dBm. The penetration loss of an exterior wall is 20 dB, whereas walls within a building have a penetration loss of 5 dB. The total penetration loss for each link is determined by ray tracing (see Chapter 2), while explicitly modeling propagation between multiple walls across different buildings.

4.9.2.2 Numerical results

Figs. 4.7 and 4.8 plot the cumulative distribution function (CDF) of the DL throughputs for UEs connected to macrocell eNBs and HeNBs, respectively. When comparing two with four single-antenna HeNBs per macrocell sector, a significant deterioration of the

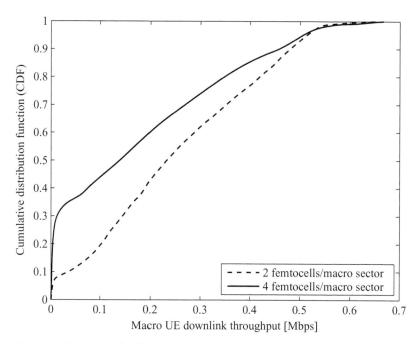

Figure 4.7 Cumulative distribution function plots of macrocell layer UEs in a CA-based HCN with HeNB transmission power equal to 0 dBm (© 2011 IEEE). The DL throughput of macrocell layer UEs is significantly deteriorated due to CSG HeNB interference.

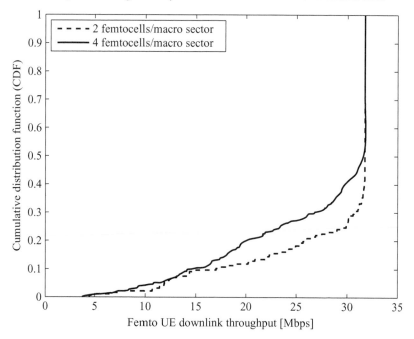

Figure 4.8 UEs connected to an HeNB (0 dBm transmission power) experience very good radio link quality and are thus scheduled with large modulation orders and coding rates (© 2011 IEEE).

DL throughput obtained at the macrocell layer UEs can be observed, visible by the gap between the two curves in Fig. 4.7.

In this example, since each HeNB restricts access to only a single UE, the remaining 35% of the UEs – clustered around the apartments – are forced to connect to a macrocell eNB. That is, such UEs cannot necessarily connect to their "best" cell, which could be a CSG HeNB, in terms of DL channel quality. In addition, because non-CSG UEs can be within the vicinity of a CSG HeNB, the relatively small HeNB transmit power and the in-building penetration loss cannot insulate the UEs from DL HeNB interference. Consequently, non-CSG UEs can potentially experience deteriorated DL reception. These observations are in stark contrast to the results in the previous section discussing the OA-based HCN, wherein it was observed that the introduction of additional LPNs does not deteriorate the performance of the macrocell UEs.

In contrast, as seen in Fig. 4.8, the radio link quality for UEs connected to HeNBs is generally excellent. Since each HeNB serves only a single UE, each CSG UE can be allocated the entire chunk of the bandwidth and benefit from improved DL link quality as a result of the proximity to its serving HeNB. In fact, nearly 50% of CSG UEs obtain the largest modulation and coding scheme such that their DL throughput saturates. While there is additional DL interference as a result of increasing the number of deployed HeNBs from two to four single-antenna HeNBs per macrocell sector, the performance of CSG UEs does not appear to be significantly deteriorated. Although each HeNB also creates additional interference for UEs connected to other HeNBs, their DL throughputs are still significantly higher compared to macrocell layer UEs.

To summarize, the focus of this section was to present detailed system-level simulation results for different access control schemes in an LTE HCN. In an OA-based HCN, overlaying the macrocell layer with just a single LPN is seen to double the cell area spectral efficiency without any impairment at the macrocell layer. The main reasons for such benefit are the relative proximity between the LPN to its serviced UEs and the offloading of UEs from the macrocell eNBs on to LPNs. In contrast, CSG-based HCNs will require additional interference avoidance and *intercell interference coordination* (ICIC) mechanisms (studied further in Chapter 7), for they considerably impair the performance of UEs that do not belong to the CSG.

4.10   Conclusions

Access control enables flexible installation and operation of LPNs for different deployment scenarios. This chapter has surveyed different access control mechanisms targeting HCNs in contemporary cellular standards. On the one hand, through OA, the operator can guarantee undifferentiated access and potentially improve coverage and DL throughputs in geographical areas experiencing high data traffic. On the other hand, through CA, users can install their LPN providing privileged access to a selected group of subscribed UEs. Finally, HA provides undifferentiated access akin to OA, while providing a level of service differentiation to UEs based on their CSG membership. A performance study revealed the impact of access control on the underlying network performance. For each of the access control schemes, this chapter has also presented their ramifications from the perspective of the CN, the RAN and the UE.

Copyright notices

References

[1] H. Holma and A. Toskala, *WCDMA for UMTS: Radio Access for Third Generation Mobile Communications*, 3rd edn (Wiley, 2004).

[2] S. Sesia, I. Toufik and M. Baker, *LTE, The UMTS Long Term Evolution: from Theory to Practice* (Wiley, 2009).

[3] 3GPP TS 36.300, *Evolved Universal Terrestrial Radio Access (E-UTRA) and Evolved Universal Terrestrial Radio Access Network (E-UTRAN): Overall Description*, Release 10, Technical Report (2011).

[4] Nomor Research, *LTE Home NodeBs and its Enhancements in Release 9*, Technical Report (2010).

[5] 3GPP TS 36.331, *Evolved Universal Terrestrial Radio Access (E-UTRA): Radio Resource Control; Protocol Specification*, Release 10, Technical Report (2011).

[6] 3GPP TS 36.304, *Evolved Universal Terrestrial Radio Access (E-UTRA): UE Procedures in Idle Mode*, Release 10, Technical Report (2011).

[7] 3GPP TS 36.423, *Technical Specification Group RAN; Evolved Universal Terrestrial Radio Access (E-UTRA): Network (E-UTRAN); X2 Application Protocol*, Release 10, Technical Report (2011).

[8] 3GPP TR 37.803, *Technical Specification Group RAN; UMTS and LTE; Mobility Enhancements for HeNB*, Release 11, Technical Report (2011).

[9] 3GPP TR 36.814, *Evolved Universal Terrestrial Radio Access (E-UTRA): Further Advancements for E-UTRA Physical Layer Aspects*, Release 9, Technical Report (2010).

5 Interference modeling and spectrum allocation in two-tier networks

Tony Q. S. Quek and Marios Kountouris

5.1 Introduction

Cellular networks have been undebatably a success story, which resulted in wide proliferation and demand for ubiquitous heterogeneous broadband mobile wireless services. With the exponential increase in high-rate traffic driven by a new generation of wireless devices, data is expected to overwhelm cellular network capacity in the near future. Multi-tier *heterogeneous cellular networks* (HCNs) have been recently proposed as an efficient and cost-effective approach to provide unprecedented levels of network capacity and coverage. Cellular operators have started integrating *small cells* as a means to provide dedicated additional capacity either where most data usage generally occurs (i.e., enterprises, households) or where *user equipments* (UEs) are likely to experience poor data rate performance (i.e., cell edges, subway stations and households). Small cells such as femtocells offer radio coverage through a given wireless technology while a broadband wired link connects them to the backhaul network of a cellular operator. In conventional single-tier networks, the *macrocell base stations* (MBSs) have to cater to the needs of both outdoor and indoor UEs, which leads to poor indoor coverage and the appearance of dead spots [1–3]. In contrast, in femtocell-aided cellular networks, indoor UEs can enjoy high-quality wireless service from their designated *femtocell access points* (FAPs) in close proximity and outdoor UEs can experience higher capacity gains due to traffic offload by FAPs through the backhaul. Moreover, FAPs have the economical advantage of being less costly to manufacture and maintain as compared with MBSs. Thus, femtocells appear as a very promising pathway to cater to the ever-increasing appetite for high-data-rate wireless applications in the near future.

Nevertheless, one of the major challenges in deploying femtocells is the incursion of cross-tier interference due to aggressive frequency reuse. When the two tiers share the whole spectrum, indoor UE communication is hindered by interference from undesignated FAPs as well as MBSs, and the same holds for outdoor UEs. This problem is further exacerbated by the random deployment of FAPs, which are often installed and controlled by their subscribed UEs, thus making centralized interference management not viable. This sparked off a significant amount of research on a variety of interference management techniques for two-tier networks such as power control [4–8], multiple antennas [8–10], adaptive FAP access schemes [11, 12, 14, 15], cognitive radio [16, 17], and spectrum allocation [13, 18–20].

In parallel, stochastic geometry has recently gathered considerable attention as a powerful and flexible interference modeling tool, which allows us to study the average behavior over many spatial realizations of a network with randomly distributed nodes [21–25]. The spatial distribution of MBSs in the network is traditionally modeled by lattices or hexagonal cells since their deployment is considered well planned, centralized and hence regular. Nevertheless, recent results for the case where MBS locations are points of a homogeneous *Poisson point process* (PPP) and associating macrocell UEs to their closest MBSs are shown to yield accurate analytical results on the distribution of *signal to interference plus noise ratio* (SINR) and the coverage performance [26]. For modeling of the macrocell UEs, the results do not change if we consider that the macrocell UEs are distributed as a stationary and homogeneous point process, which is true for large networks. On the other hand, due to FAPs' random nature, it is necessary for the network operators to analyze the worst-case scenario, namely when the users install their FAPs randomly and independently, without deliberately positioning their FAPs at the macrocell edges. This justifies the modeling of FAPs by PPP, whereby each point is distributed independently and uniformly at random. Indeed, PPP has been used to model FAPs' positions in numerous existing research papers, e.g., in [19, 27, 28]. For modeling the femtocell UEs, they are usually indoor and have close proximity to their FAPs, such that the ring model for UEs is still sufficiently accurate [29].

In this chapter, we first provide a very brief overview of stochastic models where the locations of the nodes in the network can be modeled as random, following, for example, a PPP or a hardcore point process. Then, motivated by the accuracy and tractability of PPP network models, we capitalize on [26, 27] and extend them to a two-tier HCN model.[1] Using tools from stochastic geometry, we study the effect of both spectrum allocation and closed/open access schemes on the link reliability of each tier and on the network area spectral efficiency. Specifically, the whole spectrum is partitioned into equal and consecutive orthogonal subchannels via *Orthogonal Frequency Division Multiple Access* (OFDMA), and different sets of subchannels can be assigned to different tiers. We focus on two special types of allocation, namely disjoint subchannel allocation, where the two tiers are assigned disjoint sets of subchannels, and joint subchannel allocation, where the two tiers share the whole spectrum. For each subchannel allocation scheme, we also investigate the effect of closed and open access FAPs. In the closed access configuration each FAP is only accessible by its authorized femtocell UEs, whereas in open access an FAP could be accessed by both its UEs and all co-channel macrocell UEs. Moreover, we quantify the enhancement in the macrocell UE's link reliability when it can hand over its communication from its closest MBS to its closest FAP for sufficiently close FAPs, and we also analyze the benefits of this scheme to the network throughput under the optimal joint allocation scheme.

Decentralized disjoint subchannel allocation and optimal spectrum partition have been investigated in [19, 28]. While disjoint subchannel allocation is shown to be sensible in dense networks, it is not clear whether disjoint schemes are required in sparse

---

[1] While our model of two-tier networks might lead to an FAP being unacceptably close to an MBS, our model can still serve as a lower bound on the two-tier femtocell networks in reality.

networks, whereby the cross-tier interference incurred through subchannel sharing can be tolerated. On the other hand, it was demonstrated by simulations that judiciously sharing a portion of subchannels between the two tiers is beneficial for maximizing throughput [30]. These works are our main motivation to investigate the optimal joint and disjoint schemes in two-tier networks with sparse and dense femtocell network deployment, respectively, capturing the spatial randomness of FAPs and without any restriction on the per-tier subchannel allocations. The implementation of open access and closed access in downlink two-tier networks is considered in [11, 12]. In [12], femtocell open access is considered in a stochastic geometric setting. Nevertheless, interference from other MBSs and interference between FAPs was ignored, which turns out to be of significant importance in our model for the case of sparse femtocell deployment. In [11], a handover policy of a UE from an FAP to an MBS based on the geographic positions is studied. Nevertheless, the spatial randomness of FAPs is not captured and other-cell interference is not taken into account. Here, we consider a simple femtocell open access policy, in which a macrocell UE hands over its link from its closest MBS to its closest FAP when the latter is sufficiently near.

In this chapter, we present a tractable stochastic two-tier network framework and investigate the effect of spectrum allocation and FAP closed/open access on mitigating cross-tier interference. Specifically, our main contributions are as follows.

- *Derivation of success probabilities under closed/open access FAPs*: We extend the tractable macrocell network proposed in [26] to a mathematically tractable two-tier network and propose a proximity-based FAP open access policy. For the case of closed access FAPs, we derive the success/coverage probability for each tier in an unshared or shared subchannel using stochastic geometric techniques. For the case of open access FAPs, we derive the enhanced success probability for a macrocell UE and show that the effect of cross-tier interference could be eliminated through careful adjustment of the FAP open access policy.
- *Optimizing throughput of various subchannel allocation schemes*: We analytically derive the optimal joint and disjoint subchannel allocation schemes under femtocell open or closed access. The optimization is performed subject to constraints on success probabilities and per-tier throughputs. Under femtocell closed access, we demonstrate that the optimal joint allocation achieves the highest throughput for sparse femtocell network deployments, while the optimal disjoint allocation achieves the highest throughput in dense femtocell networks. Under femtocell open access, we observe that the area spectral efficiency under optimal joint allocation is very close to the network throughput under optimal general allocation.

5.2   Interference modeling

The arising of HCNs has been the subject of intense investigation over the last decades. An analytical framework to study the coexistence issues in such networks has been proposed to analyze the effect of aggregate interference [21, 33, 35, 36]. A common theme of all these works is the use of a PPP for modeling the positions of the emitting nodes [32, 37, 38]. The PPP has been widely used in diverse fields such as astronomy [40], positron emission tomography [41], optical communications [42–45], ultra-wide-bandwidth communications [46–49] and wireless communications [50–54]. More recently, the PPP has been applied to other wireless networks such as ad hoc, sensor, relay, cognitive radio or femtocell networks [29, 55–73].

In wireless communications, a key performance measure is the received SINR. Without loss of generality, the SINR of a receiver located at the origin can be written as

$$\text{SINR} = \frac{p_0 h_0 \text{PL}(x_0)}{\sum_{i \in \Phi} p_i h_i \text{PL}(x_i) + N_0}, \qquad (5.1)$$

where h_0 is the random channel gain from the desired transmitter to the receiver at the origin, p_0 is the transmit power of the desired transmitter, N_0 is the noise power and Φ is the set of interfering nodes or a subset of all active transmitters. The desired transmitter is located at position x_0 at a distance $\|x_0\|$ from the desired receiver, and the position of the ith interferer is x_i at a distance $\|x_i\|$ away. For the set of interferers Φ, h_i is the random channel gain from the ith interferer to the receiver at the origin, and p_i is the transmit power of the ith interferer. The function $\text{PL}(x)$ accounts for the far-field path loss with distance x and is usually defined in the form of a power law given by $\|x\|^{-\alpha}$, $(1 + \|x\|^{\alpha})^{-1}$ or $\min(1, \|x\|^{-\alpha})$, where α is the path loss exponent.

Since the random locations of the interfering nodes, together with the random path loss, determine the interference to first order, the distribution of the interference can be essentially characterized using techniques from stochastic geometry and point process theory. Following [21, 33, 35, 36], we can model the spatial distribution of the interfering nodes according to some point process in a two-dimensional plane. Due to the analytical tractability, PPP has been by far the most popular spatial model. In a homogeneous PPP, defined in the infinite plane, the probability of n nodes being inside a region $\mathcal{A}$ is given by [32, 37, 38]

$$\mathbb{P}\{\Phi(\mathcal{A}) = n\} = \frac{(\lambda A_{\mathcal{R}})^n}{n!} e^{\lambda A_{\mathcal{R}}}, \quad n \geq 0 \qquad (5.2)$$

where $A_{\mathcal{R}} = |\mathcal{A}|$ is the total area of the region and λ is the spatial intensity of interfering nodes (in nodes per unit area). For notational convenience, we denote $I \triangleq \sum_{i \in \Phi} p h_i \text{PL}(x_i)$ assuming equal transmit power (i.e. $p_i = p, \forall i$) and we derive the Laplace transform of I as follows [21, 33, 35]:

$$\mathcal{L}_I(s) = \exp\left(-\lambda \int_{\mathbb{R}^2} [1 - \mathcal{L}_h(s\text{PL}(x))] \, dx\right) \qquad (5.3)$$

where $\mathcal{L}_X(s) \triangleq \mathbb{E}\{e^{-sX}\}$. Note that we can also use the characteristic function instead of the Laplace transform [36]. When a standard power-law path loss model is used, i.e., $\text{PL}(x) = \|x\|^{-\alpha}$, we obtain

$$\mathcal{L}_I(s) = \exp\left(-\lambda \pi \, \mathbb{E}h^{\delta} \Gamma(1 - \delta) s^{\delta}\right), \qquad (5.4)$$

where $\delta = 2/\alpha$ and $\Gamma(\cdot)$ is the Euler gamma function. We can observe from (5.4) that the interference follows a stable distribution with characteristic exponent $\delta < 1$, zero drift, skew parameter $\beta = 1$ and dispersion $\lambda \pi \mathbb{E}\{h^\delta\} \Gamma(1-\delta)$.[2] Using (5.3), we can further quantify other performance metrics such as outage probability, coverage probability, transmission capacity [21, 33, 35, 36], area spectral efficiency and average rate.

Instead of considering a random number of nodes, binomial point processes (BPPs) can be used to model the case when there is a fixed number of nodes within a bounded domain [33]. The locations of the interfering nodes in a BPP follow an independent and identically distributed (i.i.d.) distribution and the number of nodes in a given area is fixed. For a given number of nodes n, the probability that there are $k < n$ interfering nodes in $\mathcal{A} \subset \mathbb{R}^2$ of a BPP is given by

$$\mathbb{P}\{\Phi(\mathcal{A}) = k\} = \binom{n}{k} q^k (1-q)^{n-k}, \tag{5.5}$$

where $q = |\mathcal{A}|/|\mathcal{B}|$, and $|\mathcal{A}|$, $|\mathcal{B}|$ represent the Lebesgue measures of the sets $\mathcal{A}$, $\mathcal{B}$, respectively. For BPP, the Laplace transform of I becomes [33]

$$\mathcal{L}_I(s) = \prod_{i=1}^{n} (q\mathcal{L}_h(s\mathrm{PL}(x)) + 1 - q). \tag{5.6}$$

The assumption that the underlying node distribution follows a homogeneous point process may not be always true for certain wireless networks. For instance, some nodes tend to cluster due to geographical factors such as indoor femtocells in office buildings or high-rise apartment buildings. In addition, nodes may also cluster or are separated by a specified minimum distance due to some *medium access control* (MAC) protocols, resulting in non-homogeneous point processes. In particular, we present below two types of non-homogeneous point process, namely hard core processes and cluster processes [32, 38]. A hard core process is a point process in which no two points in a realization are closer together than a specified minimum distance $2r$. The points of such a process are the centers of non-overlapping m-dimensional spheres with radius r. The points of a hard core process are not independent, since positioning one point constrains the placement of the others. Thus, hard core models correspond to hard packing problems and have been applied in a number of engineering problems [38, 39]. Fig. 5.1 shows an example of the comparison between PPP and the hard core process. A cluster process is a finite point process generated from realizations of a parent process and a family of daughter processes. For each point generated from the parent process, there is a daughter process for this parent point and the set of points in a realization of the daughter process is called a cluster. In general, these point processes are finite point processes and are PPPs. In particular, we look at a more tractable cluster process known as the Poisson cluster processes (PCPs), where the parent points form a homogeneous PPP $\Phi_p = \{x_1, x_2, \ldots\}$ of intensity λ_p. The clusters are of the form $N_{x_i} = N_i + x_i$ for each $x_i \in \Phi_p$, and N_i is a

---

[2] Stable distributions with characteristic exponents less than one do not have any finite moments. In this case, the divergence of the mean interference is due to the singularity of the path loss law at the origin.

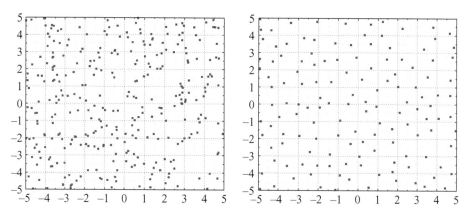

Figure 5.1 Comparison between original PPP with $\lambda = 3$ and hard core process with $\lambda = 3$ and minimum distance $h = 0.5$.

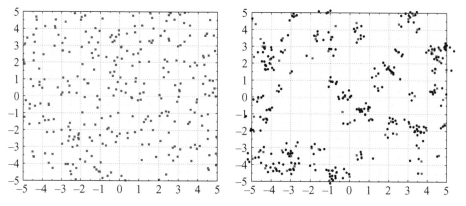

Figure 5.2 Thomas process with parents point intensity $\lambda_{\mathrm{p}} = 1$, children point intensity $\lambda = 3$ and $\sigma = 0.2$.

family of i.i.d. finite point sets with distribution independent of Φ_{p}. The complete PCP is then given by [32, 38]

$$\Phi = \bigcup_{x \in \Phi_{\mathrm{p}}} N_x. \tag{5.7}$$

Note that the parent points themselves are not included in the complete process. We also assume that the scattering density of the daughter process is isotropic and this makes the process isotropic. The intensity of the cluster process is $\lambda = \lambda_{\mathrm{p}}\bar{c}$, where $\bar{c}$ is the average number of points in each representative cluster, and the number of points is Poisson distributed. When each point is uniformly distributed in a ball of radius a around the origin, we have a Matern cluster process. When each point is distributed according to a normal distribution of zero mean and variance σ^2, we have a Thomas cluster process as illustrated in Fig. 5.2.

In terms of system performance evaluation, key performance metrics of interest include outage probability, spatial throughput, mean throughput, transmission capacity and area spectral efficiency [21, 33, 35, 36].

Outage probability: The outage probability is the probability that a transmission fails, i.e., the received SINR is below a specified threshold. Mathematically, the outage probability is given by

$$P_{\text{out}}(\lambda) = \mathbb{P}\{\text{SINR} < \gamma\} \tag{5.8}$$

where λ is the spatial density of the interfering nodes and γ denotes the minimum SINR threshold, which depends on physical layer parameters such as transmission rate, modulation, coding and type of traffic.[3]

Spatial throughput: The spatial throughput is the expected spatial density of successful transmissions and is given by

$$\tau(\lambda) = \lambda(1 - P_{\text{out}}(\lambda)). \tag{5.9}$$

Note that the spatial throughput does not give any guarantee on the outage probability, i.e., high spatial throughput can be obtained at the expense of unacceptably high outage.

Area spectral efficiency: The area spectral efficiency (ASE) is the product of the spatial throughput and the link spectral efficiency, which is given by

$$T(\lambda) = \tau(\lambda) \log_2(1 + \gamma). \tag{5.10}$$

Similarly to spatial throughput, a high mean throughput does not give any guarantee on the outage probability. The ASE is also called mean throughput.

Transmission capacity: The transmission capacity is the maximum throughput subject to a outage probability constraint and is given by

$$c(\epsilon) = \arg \max_{\lambda_\epsilon} \quad \lambda_\epsilon(1 - \epsilon) \\ \text{subject to} \quad P_{\text{out}}(\lambda_\epsilon) \leq \epsilon \tag{5.11}$$

where the optimal value of λ_ϵ in the above optimization problem is defined as the optimal contention density, which corresponds to the maximum spatial density subject to an outage probability constraint.

5.3   System model

5.3.1   Two-tier network model

We consider a macrocell network overlaid with FAPs. The first tier consists of MBSs arranged according to a homogeneous PPP Θ of intensity λ_{m} in the Euclidean plane, i.e., $\Theta \sim \text{PPP}(\lambda_{\text{m}})$. The spatial distribution of macrocell UEs is modeled as a stationary point process $\hat{\Theta}$ with constant density μ_{m}. Each MBS operates at a transmit power of P_{m}. The association of a macrocell UE to a base station depends on the access policy

---

[3] The transmission success probability is the complementary outage probability.

adopted by the femtocells, namely open or closed access; this will be discussed in a subsequent section.

The second tier consists of a set of randomly deployed FAPs Φ, whose locations are modeled according to a homogeneous PPP with intensity λ_f, i.e., $\Phi \sim \text{PPP}(\lambda_f)$. Each FAP operates at a constant transmit power P_f. Within the coverage of each FAP, there is a cluster of $|N| \sim \text{Poi}(\bar{c})$ femtocell UEs, where $\text{Poi}(\bar{c})$ denotes the Poisson distribution with mean $\bar{c}$. Each femtocell UE is distributed uniformly and independently in an infinitesimally thin ring of thickness Δ centered towards its designated FAP. The *probability density function* (PDF) of the distance from an FAP to its UE is given by $f_R(r) = \frac{2r}{((R_f+\Delta)^2 - R_f^2)} \mathbb{1} \, (R_f \le r \le R_f + \Delta)$, where $\Delta > 0$ (see Appendix 5.8.1). Therefore, the set of femtocell UEs $\hat{\Phi}$ forms a Neyman–Scott process (NSP) [32–34]. This ring model allows us to approximate the femtocell transmitter–receiver distance as R_f with arbitrarily small error for the computation of success probability (and have tractable expressions), while the set of femtocell UEs can still be modeled as an NSP. Note that $\Delta > 0$ is necessary, since for $\Delta = 0$, we no longer have an NSP as defined in [33, 34]. All point processes Θ, $\hat{\Theta}$, Φ and $\hat{\Phi}$ are stationary; moreover, Θ, $\hat{\Theta}$ and Φ are independent, and so are Θ, $\hat{\Theta}$ and $\hat{\Phi}$.

5.3.2   Spectrum allocation

The total available spectrum has a bandwidth of W Hz, and we consider localized channelization for grouping a consecutive fixed number of subcarriers to form a subchannel of B Hz for data transmission. Such a channelization technique can reduce the system overhead in terms of the required amount of feedback and control signaling. Thus, the total number of subchannels available is $\lfloor W/B \rfloor$ and we denote the subchannel index as j, where $j \in \mathcal{J} = \{1, 2, \ldots, \lfloor W/B \rfloor\}$. Without loss of generality, we assume that $B = 1$. We denote $\mathcal{J}_m$ and $\mathcal{J}_f$, where $\emptyset \neq \mathcal{J}_m, \mathcal{J}_f \subset \mathcal{J}$, as the sets of subchannels assigned to the macrocell and femtocell tiers, respectively. By denoting $\mathcal{J}(X)$ as the set of subchannels used by BS X, we see that $\mathcal{J}(X) \subseteq \mathcal{J}_m$ if X is an MBS, and $\mathcal{J}(X) \subseteq \mathcal{J}_f$ if X is an FAP. To enhance macrocell throughput, we assume that $\mathcal{J}(X) = \mathcal{J}_m$ for all MBSs X. On the other hand, to reduce interference from femtocells, for each FAP Y we assume that $|\mathcal{J}(Y)| = k \le |\mathcal{J}_f|$ and each subchannel in $\mathcal{J}(Y)$ is selected independently and uniformly from $\mathcal{J}_f$. Thus, for different FAPs Y and Z, $\mathcal{J}(Y)$ and $\mathcal{J}(Z)$ are not equal in general, though $|\mathcal{J}(Y)| = |\mathcal{J}(Z)| = k$. The probability of an FAP accessing a subchannel $j \in \mathcal{J}_f$ is $p = k/|\mathcal{J}_f|$, where $\mathcal{J}_f$ is assumed to be sufficiently large so that $p \in (0, 1]$. Throughout the chapter, the subchannel allocation is represented as $(p, |\mathcal{J}_m|, |\mathcal{J}_f|)$. Furthermore, we say that a subchannel j is *shared* if $j \in \mathcal{J}_m \cap \mathcal{J}_f$, and it is *unshared* if $j \in \mathcal{J}_m \backslash \mathcal{J}_f$ or $j \in \mathcal{J}_f \backslash \mathcal{J}_m$. Fig. 5.3 showcases the cross-tier and co-tier interference in a shared subchannel.

We investigate the following three spectrum access schemes

- *Disjoint subchannel allocation*, where femtocells are allocated with subchannels that are not being used by the MBSs, i.e., $\mathcal{J}_m \cap \mathcal{J}_f = \emptyset$.

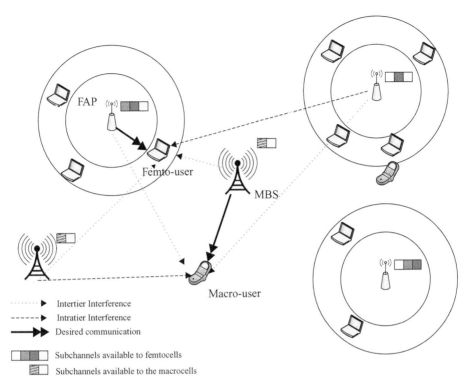

Figure 5.3 Two-tier network model (© 2012 IEEE).

- *Joint subchannel allocation*, where MBSs and FAPs can both access the entire spectrum, i.e. $\mathcal{J}_\mathrm{m} = \mathcal{J}_\mathrm{f} = \mathcal{J}$.
- *General subchannel allocation*, where there is no restriction on $\mathcal{J}_\mathrm{m}$ and $\mathcal{J}_\mathrm{f}$.

In subsequent sections, these subchannel allocations are abbreviated as *disjoint allocation*, *joint allocation* and *general allocation*, respectively. We focus on the disjoint and joint scheme cases as they are shown to be instrumental in granting insights on the behavior of the optimal general scheme, which will be defined later.

5.3.3   Femtocell access

In closed access femtocells, only registered femtocell UEs can communicate with their FAPs and macrocell UEs can only access their MBSs even in a shared subchannel. Following [26], every macrocell UE u is associated with its geographically closest MBS M. In open access femtocell operation, macrocell M is still designated as u's MBS for an unshared subchannel $j \in \mathcal{J}_\mathrm{m} \backslash \mathcal{J}_\mathrm{f}$. On the other hand, for a shared subchannel $j \in \mathcal{J}_\mathrm{m} \cap \mathcal{J}_\mathrm{f}$, a macrocell UE can access both MBS and FAP as long as it is within the femtocell coverage area. We have the following policy for handing over the macrocell UE's communication from M to its closest FAP F. As illustrated in Fig. 5.4, for a certain constant $0 \leq \kappa < 1$, we have the following open access policy:

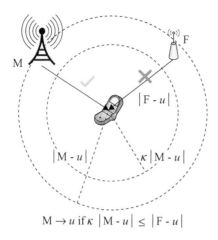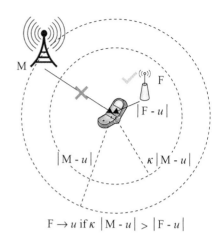

$$M \rightarrow u \text{ if } \kappa \, |M - u| \leq |F - u| \qquad\qquad F \rightarrow u \text{ if } \kappa \, |M - u| > |F - u|$$

Figure 5.4 Open access policy (© 2012 IEEE).

- M is UE u's designated BS if $\kappa \, \|M - u\| \leq \|F - u\|$,
- F is UE u's designated BS if $\kappa \, \|M - u\| > \|F - u\|$,

where $\|\cdot\|$ is the Euclidean norm. Here, we assume that $\kappa < 1$ since in general $P_f < P_m$ so that we only allow F to communicate with u if F is sufficiently closed to u compared to all MBSs. Note that increasing κ is equivalent to favoring the handover from M to F.

5.3.4   Signal-to-interference ratio

For notational convenience, we denote a BS or a UE by its location. For receiver u and transmitter x among the set of transmitters Ω, we define the *signal to interference ratio* (SIR) from x to u as follows:

$$\mathsf{SIR}(x \rightarrow u) = \frac{P_x h_x g_a(x - u)}{\sum_{y \in \Omega(x)} P_y h_y g_a(y - u)}, \tag{5.12}$$

where $\Omega(x)$ is the set of transmitters in Ω that interferes with x, P_y is the power of the transmitting node at y, and h_x is the exponential power fading coefficient from transmitter x to receiver u. Noise is ignored since we focus on the interference-limited regime [55]. We assume a standard power law path loss function $g_a(x) = \|x\|^{-\alpha}$, where α is the path loss exponent and $\|\cdot\|$ is the Euclidean norm. We denote α_i as the path loss exponent for an indoor–indoor link and α_o as the path loss exponent for an outdoor–indoor, indoor–outdoor or outdoor–outdoor link. The success probability from x to u is given by $\mathbb{P}\{\mathsf{SIR}(x \rightarrow u) > \gamma\}$, where γ is a prescribed *quality of service* (QoS) threshold. We denote γ_m and γ_f as the QoS thresholds of the macrocell UE and femtocell UE, respectively. Due to the stationarity of the point processes, the success probability for each Tx–Rx pair in each tier is the same.

5.4   Downlink success probability

In this section, we derive the success probability for each tier under different subchannel allocation schemes and FAP access control. In the following, we use $\mathbb{P}_{m,c}(j)$ and $\mathbb{P}_{m,o}(j, \kappa)$ to denote the success probability of macrocell Tx–Rx in subchannel j under closed access and open access FAP with parameter κ, respectively. On the other hand, we denote the success probability of femtocell Tx–Rx in subchannel j as $\mathbb{P}_f(j)$.

5.4.1   Success probabilities with closed access femtocells

First, we derive the downlink success probabilities in a shared or unshared subchannel when femtocells operate in closed access.

Lemma 5.1 *With closed access FAP, the macrocell success probability* $\mathbb{P}_{m,c}(j)$ *and the femtocell success probability* $\mathbb{P}_f(j)$ *on a subchannel* j *are given as follows.*

- *For* $j \in \mathcal{J}_m \backslash \mathcal{J}_f$,

$$\mathbb{P}_{m,c}(j) = \frac{1}{1 + \rho(\gamma_m, \alpha_o)}, \quad \mathbb{P}_f(j) = 0,$$

where $\rho(\gamma_m, \alpha_o) = \int_{\gamma_m^{-2/\alpha_o}}^{\infty} \frac{\gamma_m^{2/\alpha_o}}{1 + u^{\frac{\alpha_o}{2}}} du.$
- *For* $j \in \mathcal{J}_f \backslash \mathcal{J}_m$,

$$\mathbb{P}_{m,c}(j) = 0,$$

$$\mathbb{P}_f(j) = \exp\left(-p\lambda_f \pi R_f^{\frac{2\alpha_j}{\alpha_o}} C(\alpha_o) \gamma_f^{\frac{2}{\alpha_o}}\right) + O(\Delta),$$

where $C(\alpha_o) = \frac{2\pi/\alpha_o}{\sin(2\pi/\alpha_o)}$, *and the term* $O(\Delta) \to 0$ *as* $\Delta \to 0$ *for all choices of parameters.*
- *For* $j \in \mathcal{J}_m \cap \mathcal{J}_f$,

$$\mathbb{P}_{m,c}(j) = \frac{1}{1 + p\frac{\lambda_f}{\lambda_m} C(\alpha_o)\left(\frac{P_f \gamma_m}{P_m}\right)^{\frac{2}{\alpha_o}} + \rho(\gamma_m, \alpha_o)},$$

$$\mathbb{P}_f(j) = \exp\left(-\pi R_f^{\frac{2\alpha_j}{\alpha_o}}\left(p\lambda_f + \frac{P_m^{\frac{2}{\alpha_o}}}{P_f} \lambda_m\right) C(\alpha_o)\gamma_f^{\frac{2}{\alpha_o}}\right) + O(\Delta), \quad (5.13)$$

where the term $O(\Delta) \to 0$ *as* $\Delta \to 0$ *for all choices of parameters.*

Proof. See Appendix 5.8.2.   □

Since Δ is infinitesimal, the $O(\Delta)$ term is ignored in subsequent calculations. The above result for shared subchannel success probabilities can be easily extended to multitier HCNs, resulting in the following corollary.

Corollary 5.1 *Suppose a subchannel* j *is shared by* n *tiers of closed access BSs* $\{\Omega_i\}_{i=1}^{n}$, *where* $\Omega_i \sim PPP(\lambda_i)$ *are homogeneous and each BS in tier* i *has constant* transmit power P_i. Assuming each tier's UEs form a stationary point process, we have the following.

- If tier i employs the closest association scheme and has outdoor transmission, then the success probability for a tier i UE is

$$\mathbb{P}_{i,c}(j) = \frac{1}{1 + C(\alpha) \sum_{j \neq i} \frac{\lambda_j}{\lambda_i} \left(\frac{P_j \gamma_i}{P_i} \right)^{2/\alpha} + \rho(\gamma_i, \alpha)}. \tag{5.14}$$

- If in tier i each UE lies in an infinitesimally thin ring of radius R_i centered at its designated BS, then the success probability is

$$\mathbb{P}_{i,c}(j) = \exp\left[-\pi C(\alpha_o) R_i^2 \left(\sum_{j=1}^n \left(\frac{P_j}{P_i} \right)^{2/\alpha} \lambda_j \right) \gamma_i^{2/\alpha} \right] + O(\Delta), \tag{5.15}$$

where α is the universal path loss exponent and γ_i is the minimum success probability threshold in tier i.

Proof. The proof is similar to Appendix 5.8.2.   □

5.4.2   Success probability with open access femtocells

Suppose that FAPs operate using the open access policy described in Section 5.3.3. First, note that femtocell UEs' success probabilities are not affected by the open access in all subchannels. Moreover, macrocell UEs in $j' \in \mathcal{J}_m \backslash \mathcal{J}_f$ cannot access FAPs and thus $\mathbb{P}_{m,o}(j', \kappa) = \mathbb{P}_{m,c}(j')$. However, for $j \in \mathcal{J}_m \cap \mathcal{J}_f$, we have the following result.

Lemma 5.2 *In $j \in \mathcal{J}_m \cap \mathcal{J}_f$, we have $\mathbb{P}(F \to u) = \frac{p\lambda_f \kappa^2}{\lambda_m + p\lambda_f \kappa^2} = 1 - \mathbb{P}(M \to u)$, and the macrocell success probability is given by*

$$\mathbb{P}_{m,o}(j, \kappa) = \frac{\kappa^2 p\lambda_f}{\lambda_m \left(1 + \rho\left(\frac{\gamma_m \kappa^{\alpha_o} P_m}{P_f}, \alpha_o \right) \right) + \kappa^2 p\lambda_f (1 + \rho(\gamma_m, \alpha_o))}$$

$$+ \frac{\lambda_m}{\lambda_m (1 + \rho(\gamma_m, \alpha_o)) + \kappa^2 p\lambda_f \left(1 + \rho\left(\frac{\gamma_m P_f}{\kappa^{\alpha_o} P_m}, \alpha_o \right) \right)}. \tag{5.16}$$

Proof. See Appendix 5.8.3.   □

Remark: For $j \in \mathcal{J}_m \cap \mathcal{J}_f$, setting $\kappa = 0$, we have $\lim_{\kappa \to 0} \mathbb{P}_{m,o}(j, \kappa) = \mathbb{P}_{m,c}(j)$, which amounts to prohibiting u from communicating to F. More precisely, it means that $\mathbb{P}(M \to u) = 1$ and F can be arbitrarily close to u with non-zero probability. Therefore, we are back to (5.13) in Lemma 5.1, where $M \to u$ is interfered with by $\Phi_j \cup \Theta \backslash M$, with $(\Theta \backslash M) \cap D(u; V) = \emptyset$, and $\Phi_j \sim \mathrm{PPP}(p\lambda_f)$ spread throughout $\mathbb{R}^2$, with $D(x; r)$ denoting the disk centered at x with radius r.

Mathematically, the limit is justified as follows. As $\kappa \to 0$, the first term in (5.16) goes to zero, and in the second term, we have

$$\lim_{\kappa \to 0} \kappa^2 \left[1 + \rho \left(\frac{\gamma_m P_f}{\kappa^{a_o} P_m}, a_o \right) \right] = \left(\frac{\gamma_m P_f}{P_m} \right)^{\frac{2}{a_o}} \int_0^\infty \frac{1}{1 + u^{\frac{a_o}{2}}} du$$

$$= \left(\frac{\gamma_m P_f}{P_m} \right)^{\frac{2}{a_o}} C(a_o). \tag{5.17}$$

Thus, $\lim_{\kappa \to 0} \mathbb{P}_{m,o}(j, \kappa) = \mathbb{P}_{m,c}(j)$.

Remark: Although for $j \in \mathcal{J}_m \cap \mathcal{J}_f$, setting $\kappa > 1$ is not practical in general and is outside our range of interest, there is still physical significance on setting $\kappa \to \infty$:

$$\lim_{\kappa \to \infty} \mathbb{P}_{m,o}(j, \kappa) = \frac{p\lambda_f}{p\lambda_f [1 + \rho(\gamma_m, a_o)] + \lambda_m C(a_o) \left(\frac{P_m \gamma_m}{P_f} \right)^{\frac{2}{a_o}}}. \tag{5.18}$$

Indeed, setting $\kappa \to \infty$ means that a macrocell UE communicates with F with unit probability and is prohibited communicating with M. Then, the limit follows from switching the roles of MBSs and FAPs in (5.13). For $j \in \mathcal{J}_m \cap \mathcal{J}_f$, we note that $\mathbb{P}_{m,o}\left(j, (P_f/P_m)^{1/a_o}\right) = \mathbb{P}_m(j')$, where $j' \in \mathcal{J}_m \backslash \mathcal{J}_f$. This implies that by setting $\hat{\kappa} = (P_f/P_m)^{1/a_o}$ the cross-tier interference from FAPs to macrocell UEs is nullified. In fact, numerical evidence further verifies the observation that $\hat{\kappa}$ maximizes $\mathbb{P}_{m,o}(j, \kappa)$. In subsequent sections, we consider $\kappa = \hat{\kappa}$. Finally, we have the following generalization on the open access results to a multi-tier HCN.

Corollary 5.2 *Given that a subchannel j is shared by n sets of BSs $\{\Omega_i\}_{i=1}^n$, where $\Omega_i \sim \mathrm{PPP}(\lambda_i)$ are homogeneous and each BS in tier i has constant transmit power P_i, and a UE u can access all these tiers (open access). Consider the following policy: for $B_1, \ldots B_n$ being the geographically closest BSs to u in tiers $1, \ldots, n$ respectively, writing $V_i = \|u - B_i\|_2$, $B_i \to u$ iff $\kappa_i V_i = \min_{1 \le j \le n} \{\kappa_j V_j\}$, where $\kappa_1, \ldots, \kappa_n \in \mathbb{R}^+$ are constants. Then the UE has the following success probability:*

$$\sum_{i=1}^n \frac{\lambda_i}{\sum_{j=1}^n \lambda_j \left(\frac{\kappa_i}{\kappa_j} \right)^2 \left(1 + \rho \left(\gamma \frac{P_j}{P_i} \left(\frac{\kappa_j}{\kappa_i} \right)^\alpha, \alpha \right) \right)}, \tag{5.19}$$

where α is the universal path loss exponent and γ is the UE's threshold. Moreover, by putting $\kappa_i = P_i^{-1/\alpha}$, the success probability becomes $1/(1 + \rho(\gamma, \alpha))$, which is significantly larger than (5.18), and the effect of cross-tier interference is eliminated.

Proof. The proof is similar to that of Lemma 5.2.   □

5.5   Two-tier downlink throughput optimization

In this section, we quantify and optimize the network-wide downlink throughput of different subchannel allocations and FAP access control under QoS guarantees, which are defined in terms of success probability and per-tier throughput constraints.

5.5.1   Downlink throughput analysis

First, we define the UE/link throughput in a subchannel j as $\mathbb{P}(\text{success in } j) \log(1 + \gamma)$, where $\mathbb{P}(\text{success in } j) = \mathbb{P}(\text{SIR} > \gamma)$ is the success probability of the transmission in j and γ is a prescribed threshold. We assume that both macrocell and femtocell tiers employ *time division multiple access* (TDMA) and provide each UE (macro or femto) with unit time interval to communicate (no intracell interference). Under closed access FAPs, the downlink macrocell and femtocell sum user throughputs per subchannel are, respectively, given by

$$\mathcal{T}_{\text{m,c}}(j) = \mu_{\text{m}} \mathbb{P}_{\text{m,c}}(j) \log(1 + \gamma_{\text{m}}), \tag{5.20}$$

$$\mathcal{T}_{\text{f}}(j) = p\bar{c}_{\text{f}} \lambda_{\text{f}} \mathbb{P}_{\text{f}}(j) \log(1 + \gamma_{\text{f}}). \tag{5.21}$$

On the other hand, we define $\mathcal{T}_{\text{m,o}}(j, \hat{\kappa}) = \mu_{\text{m}} \mathbb{P}_{\text{m,o}}(j, \hat{\kappa}) \log(1 + \gamma_{\text{m}})$ as the macrocell user throughput per subchannel for the case of open access FAP. The aggregate throughputs under closed and open access FAP are denoted as $\mathcal{T}_{\text{c}}$ and $\mathcal{T}_{\text{o}}$, respectively, and are given by

$$\mathcal{T}_{\text{c}} = \sum_{j \in \mathcal{J}} (\mathcal{T}_{\text{m,c}}(j) + \mathcal{T}_{\text{f}}(j)), \tag{5.22}$$

$$\mathcal{T}_{\text{o}} = \sum_{j \in \mathcal{J}} (\mathcal{T}_{\text{m,o}}(j, \hat{\kappa}) + \mathcal{T}_{\text{f}}(j)). \tag{5.23}$$

Since each tier utilizes a certain number of shared and unshared subchannels, we define the net success probability to be the weighted average of the success probabilities. In the case of closed access FAP, we have

$$\mathbb{P}_{\text{m,c}}(\text{net}) = \frac{1}{|\mathcal{J}_{\text{m}}|} \sum_{j \in \mathcal{J}_{\text{m}}} \mathbb{P}_{\text{m,c}}(j), \tag{5.24}$$

$$\mathbb{P}_{\text{f}}(\text{net}) = \frac{1}{|\mathcal{J}_{\text{f}}|} \sum_{j \in \mathcal{J}_{\text{f}}} \mathbb{P}_{\text{f}}(j). \tag{5.25}$$

For the case of open access FAP, $\mathbb{P}_{\text{m,o}}(\text{net})$ is defined in a similar manner as in (5.24), except that $\mathbb{P}_{\text{m,c}}(j)$ is replaced with $\mathbb{P}_{\text{m,o}}(j, \hat{\kappa})$. Note that $\mathbb{P}_{\text{m,o}}(\text{net}) = \mathbb{P}_{\text{m,c}}(j)$ with $j \in \mathcal{J}_{\text{m}} \backslash \mathcal{J}_{\text{f}}$. By denoting $j \in \mathcal{J}_{\text{m}} \backslash \mathcal{J}_{\text{f}}$, $j' \in \mathcal{J}_{\text{f}} \backslash \mathcal{J}_{\text{m}}$, and $j'' \in \mathcal{J}_{\text{m}} \cap \mathcal{J}_{\text{f}}$, the aggregate network throughput for closed access FAP, $\mathcal{T}_{\text{c}}$, is shown in (5.26). Note that the aggregate network throughput for open access FAP, $\mathcal{T}_{\text{o}}$, is defined in a similar manner as in (5.26), except that $\mathbb{P}_{m,c}(net)$ is replaced with $\mathbb{P}_{m,o}(net)$.

$$
\begin{aligned}
\mathcal{T}_c(p, \mathcal{J}_m, \mathcal{J}_f, \mu_m, \lambda_f, \bar{c}_f, \gamma_m, \gamma_f) &= (|\mathcal{J}_m \backslash \mathcal{J}_f|) \mu_m \mathbb{P}_{m,c}(j) \log(1 + \gamma_m) \\
&+ p\bar{c}_f \lambda_f \mathbb{P}_f(j'') \log(1 + \gamma_f)) \\
&+ (|\mathcal{J}_f \backslash \mathcal{J}_m|) p\bar{c}_f \lambda_f \mathbb{P}_f(j') \log(1 + \gamma_f) \\
&+ |\mathcal{J}_f \cap \mathcal{J}_m| (\mu_m \mathbb{P}_{m,c}(j'') \log(1 + \gamma_m) \\
&+ p\bar{c}_f \lambda_f \mathbb{P}_f(j'') \log(1 + \gamma_f)) \\
&= |\mathcal{J}_m| \mu_m \mathbb{P}_{m,c}(net) \log(1 + \gamma_m) \\
&+ p|\mathcal{J}_f| \bar{c}_f \lambda_f \mathbb{P}_f(net) \log(1 + \gamma_f). \quad (5.26)
\end{aligned}
$$

5.5.2 Network throughput optimization

In the following, we provide flexible spectrum allocation schemes with closed and open access FAPs as a means to maximize the overall network throughput subject to success probabilities and per-tier throughput constraints. In the case of no restriction on $(p, \mathcal{J}_m, \mathcal{J}_f)$, the above optimization is analytically intractable. Nevertheless, closed-form results are obtained if we restrict $(p, \mathcal{J}_m, \mathcal{J}_f)$ to be either joint or disjoint allocation. For the case of closed access FAP, we define the following terminology.

- $(\tilde{p}_c^J, |\tilde{\mathcal{J}}_{m,c}^J|, |\tilde{\mathcal{J}}_{f,c}^J|)$ is the optimal joint allocation, i.e., the constrained network throughput maximizing allocation with $\tilde{\mathcal{J}}_{m,c}^J = \tilde{\mathcal{J}}_{f,c}^J = \mathcal{J}$.
- $(\tilde{p}_c^D, |\tilde{\mathcal{J}}_{m,c}^D|, |\tilde{\mathcal{J}}_{f,c}^D|)$ is the optimal disjoint allocation, i.e., the constrained network throughput maximizing allocation with $\tilde{\mathcal{J}}_{m,c}^D \cap \tilde{\mathcal{J}}_{f,c}^D = \emptyset$, $\tilde{\mathcal{J}}_{m,c}^D \cup \tilde{\mathcal{J}}_{f,c}^D = \mathcal{J}$.
- $(\tilde{p}_c^G, |\tilde{\mathcal{J}}_{m,c}^G|, |\tilde{\mathcal{J}}_{f,c}^G|)$ is the optimal general allocation, i.e., the constrained network throughput maximizing allocation *with no restriction on* $\tilde{\mathcal{J}}_{m,c}^G, \tilde{\mathcal{J}}_{f,c}^G$.

For the case of open access FAP, we define $(\tilde{p}_o^J, |\tilde{\mathcal{J}}_{m,o}^J|, |\tilde{\mathcal{J}}_{f,o}^J|)$, $(\tilde{p}_o^D, |\tilde{\mathcal{J}}_{m,o}^D|, |\tilde{\mathcal{J}}_{f,o}^D|)$ and $(\tilde{p}_o^G, |\tilde{\mathcal{J}}_{m,o}^G|, |\tilde{\mathcal{J}}_{f,o}^G|)$ in a similar manner. The constraints on the net success probabilities and per-tier throughputs are given by

$$
\begin{cases}
\mathbb{P}_m(net) \geq 1 - \epsilon_m, \\
\mathbb{P}_f(net) \geq 1 - \epsilon_f,
\end{cases} \quad (5.27)
$$

$$
\begin{cases}
\sum_{j \in \mathcal{J}_m} \mathcal{T}_m(j) \leq \theta_m \mathcal{T}, \\
\sum_{j \in \mathcal{J}_f} \mathcal{T}_f(j) \leq \theta_f \mathcal{T},
\end{cases} \quad (5.28)
$$

where $\epsilon_m, \epsilon_f, \theta_m, \theta_f \in [0, 1]$ such that $\theta_m + \theta_f \geq 1$. For notational convenience, we denote $\mathbb{P}_m(net)$ to represent $\mathbb{P}_{m,c}(net)$ or $\mathbb{P}_{m,o}(net)$, and $\mathcal{T}_m, \mathcal{T}$ to represent $\mathcal{T}_{m,c}, \mathcal{T}_c$ or $\mathcal{T}_{m,o}, \mathcal{T}_o$, respectively, depending on the access control of the FAP. The constraints above are introduced in our network throughput maximization problem so as to enforce minimum QoS in terms of success probabilities and per-tier throughput.

5.5.3   Optimal joint allocation with closed access femtocells

For general values of λ_m and λ_f, the optimization is intractable and one may resort to numerical optimization techniques. In order to gain some insights, we focus on sparse networks with very small λ_f and λ_m, where $\tilde{p}_c^J$ can be derived. Indeed, as we show in the numerical section, the joint allocation is preferred in sparse networks when femtocells operate in closed access.

Lemma 5.3 *In a closed access two-tier network when λ_f and λ_m are infinitesimally small, if $l_2 \leq u_1$, then $\tilde{p}_c^J = \underset{p \in \{l_2, \min\{u_1, u_2\}\}}{\arg\max} \; \mathcal{T}_c(p)$. Otherwise, the optimization problem for joint allocation is infeasible. We have*

$$u_1 = \min\left\{1, \frac{1}{a}\left(\frac{1}{1-\epsilon_m} - 1 - \rho\right)\right\},$$

$$l_2 = \beta\left(\frac{1-\theta_m}{\theta_m}\right),$$

$$u_2 = \beta\left(\frac{\theta_f}{1-\theta_f}\right),$$

where $a = \frac{\lambda_f}{\lambda_m} C(\alpha_o)\left(\frac{P_f \gamma_m}{P_m}\right)^{\frac{2}{\alpha_o}}$, $\rho = \rho(\gamma_m, \alpha_o)$ and $\beta(x) = \sqrt{\frac{1}{a}\left(x \frac{\mu_m \log(1+\gamma_m)}{\lambda_f \bar{c}_f \log(1+\gamma_f)} + \frac{(1+\rho)^2}{4a}\right)} - \frac{1+\rho}{2a}.$

Proof. Since λ_m and λ_f are infinitesimally small, by only considering the first-order terms in λ_f and λ_m, we see that, for a subchannel set $\mathcal{S}$, $\lambda_f \mathbb{P}_f(\mathcal{S}) = \lambda_f(1 - o(1))$ and

$$\mathcal{T}_c = |\mathcal{J}|\left(\frac{\mu_m \log(1+\gamma_m)}{1 + ap + \rho} + p\lambda_f \bar{c}_f \log(1+\gamma_f)\right) - o(\lambda_f). \tag{5.29}$$

Since λ_f and λ_m are infinitesimally small, we may consider that the femto success probability constraint is satisfied for all p. We also neglect the $o(\lambda_f)$ term. Now, optimizing the total throughput $\mathcal{T}_c$ for p with the macro success probability constraint (5.27) is the same as optimizing $\mathcal{T}_c$ in (5.29) for $p \in [0, u_1]$. In addition, optimizing with the throughput constraint (5.28) is the same as optimizing $\mathcal{T}_c$ for $p \in [l_2, u_2]$, where u_1, l_2 and u_2 are defined in Lemma 5.3. If $l_2 > u_1$, then there is no feasible solution. Otherwise, since $\mathcal{T}_c$ in (5.29) is convex in p with respect to the domain $[l_2, \min\{u_1, u_2\}]$, we know that the maximum of $\mathcal{T}$ in the interval occurs at the extreme points. Therefore, $\tilde{p}_c^J = \underset{p \in \{l_2, \min\{u_1, u_2\}\}}{\arg\max} \; \mathcal{T}_c(p).$ □

5.5.4   Optimal disjoint allocation with closed access femtocells

The optimal disjoint allocation in a two-tier network was derived in [19] for certain specific cases. Here, we derive mathematical expressions for the optimal disjoint allocation and the optimal network throughput for the general case.

Lemma 5.4 *With closed access FAP, the optimal disjoint allocation* $(\tilde{p}_c^D, \tilde{\mathcal{J}}_{m,c}^D, \tilde{\mathcal{J}}_{m,c}^D)$ *is given by*

$$
\tilde{p}_c^D =
\begin{cases}
\frac{1}{s} \min\left\{1, \log\left(\frac{1}{1-\epsilon_f}\right)\right\}, & \text{when } s \geq 1, \\
\min\left\{1, \frac{1}{s} \log\left(\frac{1}{1-\epsilon_f}\right)\right\}, & \text{when } s < 1,
\end{cases}
\tag{5.30}
$$

where $s = \lambda_f \pi R_f^{\frac{2a_j}{a_o}} C(a_o) \gamma_f^{\frac{2}{a_o}}$. *If* $\tilde{T}_{f,c}^D(j) > T_{m,c}^D(j')$, *then for* $\tilde{\mathcal{J}}_{m,c}^D, \tilde{\mathcal{J}}_{f,c}^D$, *we have*

$$
|\tilde{\mathcal{J}}_{f,c}^D| = \left\lfloor \frac{|\mathcal{J}|}{1 + \left(\frac{1}{\theta_f} - 1\right) \frac{\tilde{T}_{f,c}^D(j)}{T_{m,c}^D(j')}} \right\rfloor, \qquad \tilde{\mathcal{J}}_{m,c}^D = |\mathcal{J}| - \tilde{\mathcal{J}}_{f,c}^D.
$$

$$
\tag{5.31}
$$

Otherwise,

$$
|\tilde{\mathcal{J}}_{m,c}^D| = \left\lfloor \frac{|\mathcal{J}|}{1 + \left(\frac{1}{\theta_m} - 1\right) \frac{T_m^D(j')}{\tilde{T}_f^D(j)}} \right\rfloor, \qquad |\tilde{\mathcal{J}}_{m,c}^D| = |\mathcal{J}| - |\tilde{\mathcal{J}}_{m,c}^D|
\tag{5.32}
$$

where $T_{m,c}^D(j') = \frac{\mu_m \log(1+\gamma_m)}{1+\rho(\gamma_m, a_o)}$ *is the per-subchannel macro throughput* $(j' \in \mathcal{J}_m)$, *and* $\tilde{T}_{f,c}^D(j)$ *is the optimized per-subchannel femto throughput* $(j \in \mathcal{J}_f)$:

$$
\tilde{T}_{f,c}^D(j) =
\begin{cases}
(se)^{-1} \bar{c}_f \lambda_f \log(1+\gamma_f) & \text{for } 1-\epsilon_f \leq e^{-1}, s \geq 1 \\
s^{-1} \bar{c}_f \lambda_f (1-\epsilon_f) \log\left(\frac{1+\gamma_f}{1-\epsilon_f}\right) & \text{for } 1-\epsilon_f > e^{-1}, s \geq 1 \\
e^{-s} \lambda_f \bar{c}_f \log(1+\gamma_f) & \text{for } 1-\epsilon_f \leq e^{-s}, s < 1 \\
s^{-1} \bar{c}_f \lambda_f (1-\epsilon_f) \log\left(\frac{1+\gamma_f}{1-\epsilon_f}\right) & \text{for } 1-\epsilon_f > e^{-s}, s < 1.
\end{cases}
\tag{5.33}
$$

Proof. See Appendix 5.8.4. □

Furthermore, we have the following result for the optimal throughput.

Corollary 5.3 *When* $\mathcal{J}$ *is large, the optimal network throughput for the optimal disjoint allocation with closed access FAP is approximately given by*

$$
\tilde{T}_c^D = \frac{|\mathcal{J}|}{\frac{\theta_f}{\tilde{T}_{f,c}^D(j)} + \frac{1-\theta_f}{T_{m,c}^D(j')}}, \qquad \frac{|\mathcal{J}|}{\frac{\theta_m}{T_{m,c}^D(j')} + \frac{1-\theta_m}{\tilde{T}_{f,c}^D(j)}},
\tag{5.34}
$$

for $\tilde{T}_{f,c}^D(j) > T_{m,c}^D(j')$ *and* $\tilde{T}_{f,c}^D(j) \leq T_{m,c}^D(j')$, *respectively.*

Proof. This is proved by substituting (5.31) (respectively (5.32)) and (5.33) in (5.50) when $\tilde{T}_{f,c}^D(j) > T_{m,c}^D(j')$ (respectively $\tilde{T}_{f,c}^D(j) \leq T_{m,c}^D(j')$) and after some manipulations. □

Remark: We note that (5.31) (resp. (5.32)) could be zero if $\tilde{T}_{f,c}^D(j) \gg T_{m,c}^D(j')$ (resp. $\tilde{T}_{f,c}^D(j) \ll T_{m,c}^D(j')$) and $|\mathcal{J}|$ is not sufficiently large. This reflects a natural difficulty in allocating scarce resources to an extremely asymmetric two-tier network. On the other side, we see that $|\mathcal{J}|$ does not appear until the final step, and in fact the ratio

$\tilde{\mathcal{T}}_{f,c}^{D}(j)/\mathcal{T}_{m,c}^{D}(j')$ guides us on how large $|\mathcal{J}|$ should be; it should be large enough such that $|\mathcal{J}_m|$ and $|\mathcal{J}_f|$ are not zero in the above cases. Furthermore, to validate the assumption $p \in (0, 1]$, we should ensure that $|\mathcal{J}|$, hence $|\mathcal{J}_{m,c}^{D}|$, is large.

The optimal disjoint allocation and the optimal femto per-subchannel throughput are independent of λ_f when $\lambda_f \geq (\pi R_f^{\frac{2a_i}{a_o}} C(a_o)\gamma_f^{\frac{2}{a_o}})^{-1}$. This is because the optimal allocation and the throughput are only dependent on the optimal effective femto density $p_c^D \lambda_f$, and p_c^D is inversely proportional to λ_f when λ_f exceeds that threshold (cf. Remark 2 'Boundedness of the ASE' in [19]). More precisely, when the FAP deployment is very dense, the number of subchannels assigned to each FAP is reduced in order to minimize the likelihood of nearby FAPs using the same subchannels. Therefore, the increase in $\tilde{\mathcal{T}}_{f,c}^{D}(j)$ by increasing λ_f is exactly nullified by the decrease in the number of subchannels that each FAP is allocated to once λ_f is sufficiently large. Similarly to [19, 33], our result refutes the intuitive idea of increasing overall network throughput by simply employing more FAPs, even if we employ the optimized disjoint allocation to avoid cross-tier interference. Moreover, we remark that the stabilization of $\tilde{\mathcal{T}}_{f,c}^{D}(j)$ is not caused by our constraints in (5.27) and (5.28). Indeed, the constraints could be removed by putting $\epsilon_f = \theta_m = \theta_f = 1$. When $\lambda_f \geq (\pi R_f^{\frac{2a_i}{a_o}} C(a_o)\gamma_f^{\frac{2}{a_o}})^{-1}$, the unconstrained optimal $\tilde{\mathcal{T}}_{f,c}^{D}(j)$ is $\frac{\bar{c}_f \lambda_f \log(1+\gamma_f)}{se} = \frac{\bar{c}_f \log(1+\gamma_f)}{\pi R_f^{2a_i/a_o} C(a_o)\gamma_f^{2/a_o} e}$, which is still independent of λ_f.

5.5.5   Optimal joint allocation with open access femtocells

We determine the optimal joint allocation $\tilde{p}_o^J$, with open access FAPs setting $\kappa = \hat{\kappa} = (P_f/P_m)^{1/a_o}$ as stated before. Therefore, the total throughput is given by

$$\mathcal{T}_o = \frac{|\mathcal{J}_m|\mu_m \log(1+\gamma_m)}{1 + \rho(\gamma_m, a_o)}$$
$$+ p\lambda_f \bar{c}_f e^{-ps}(|\mathcal{J}_f \cap \mathcal{J}_m| e^{-d} + |\mathcal{J}_f \setminus \mathcal{J}_m|) \log(1+\gamma_f), \tag{5.35}$$

where $d = \lambda_m \pi R_f^{\frac{2a_i}{a_o}} \frac{P_m}{P_f}^{\frac{2}{a_o}} C(a_o)\gamma_f^{\frac{2}{a_o}}$ and $s = \lambda_f \pi R_f^{\frac{2a_i}{a_o}} C(a_o)\gamma_f^{\frac{2}{a_o}}$. By restricting to the joint allocation, i.e., $\mathcal{J} = \mathcal{J}_m = \mathcal{J}_f$, we have the following result on $\tilde{p}_o^J$.

Lemma 5.5 *For joint allocation with open access FAPs, the optimization for $\tilde{p}_o^J$ is infeasible if $u_1 < 0$. Otherwise, for $u_1 \geq 0$, if $s \leq 1$, we distinguish the following cases.*

- *If $F^{-1}(l_2) \leq \min\{1, u_1\}$, then $\tilde{p}_o^J = \min\{1, u_1, F^{-1}(u_2)\}$.*
- *If $F^{-1}(l_2) > \min\{1, u_1\}$, then the optimization is infeasible.*

If $s > 1$, we have

- *If $(se)^{-1} < l_2$, then the problem is infeasible.*
- *If $l_2 \leq (se)^{-1} < u_2$, writing $F^{-1}([l_2, (se)^{-1}]) = [a, b]$, then:*
  - *If $u_1 < a$, then the problem is infeasible.*
  - *If $u_1 \geq a$, then $\tilde{p}_o^J = \min\{u_1, s^{-1}\}$.*
- *If $u_2 \leq (se)^{-1}$, writing $[0, 1] \cap F^{-1}([l_2, u_2]) = [a_1, b_1] \cup [a_2, b_2]$, then:*
  - *If $u_1 < a_1$, then the problem is infeasible;*

- If $a_1 \leq u_1 < b_1$, then $\tilde{p}_o^J = u_1$;
- If $b_1 \leq u_1$, then $\tilde{p}_o^J = b_1$;

where

$$u_1 = \frac{1}{s}\left(\log\left(\frac{1}{1-\epsilon_f}\right) - d\right),$$

$$l_2 = \frac{1-\theta_m}{\theta_m}\eta,$$

$$u_2 = \frac{\theta_f}{1-\theta_f}\eta, \tag{5.36}$$

$F(p) = pe^{-sp}$ *for* $0 \leq p \leq 1$, *and*

$$\eta = \frac{\mu_m \log(1+\gamma_m)}{\lambda_f \bar{c}e^{-d}\log(1+\gamma_f)(1+\rho(\gamma_m,\alpha_o))}.$$

Proof. See Appendix 5.8.5.   □

5.5.6   Optimal disjoint allocation with open access femtocells

When disjoint allocation is used, there is no shared subchannel, and hence a macrocell UE cannot hand over its link from an MBS to an FAP. In this case, we always have $(\tilde{p}_o^D, |\tilde{\mathcal{J}}_{m,o}^D|, |\tilde{\mathcal{J}}_{f,o}^D|) = (\tilde{p}_c^D, |\tilde{\mathcal{J}}_{m,c}^D|, |\tilde{\mathcal{J}}_{f,c}^D|)$ and we write the optimal disjoint allocation as $(\tilde{p}^D, |\tilde{\mathcal{J}}_m^D|, |\tilde{\mathcal{J}}_f^D|)$ subsequently.

5.6   Numerical results

In the simulation results, the default values given in Table 5.1 are used unless otherwise stated. For Figs. 5.5 and 5.6, we simulated a two-tier network in a 40 km × 40 km square area with the reference receiver at the origin. Figs. 5.5 and 5.6 show that our analytical results (solid curves) closely match the corresponding simulated results (dashed curves), hence validating our analysis in Section 5.4. In Fig. 5.5 the macrocell and femtocell success probabilities are compared for varying p and $\lambda_f = 80\lambda_m = 8 \times 10^{-5}$. In Fig. 5.5, we see that for $j' \in \mathcal{J}_m\backslash\mathcal{J}_f$, $\mathbb{P}_m(j')$ is constant since it is independent of the spectrum allocation. By comparing the curves for unshared and shared subchannels, it is observed that cross-tier interference affects the macrocell tier more severely than the femtocell tier since the deployment of FAPs is significantly denser than that of MBSs. Furthermore, for $j \in \mathcal{J}_m \cap \mathcal{J}_f$, $\mathbb{P}_{m,c}(j)$ is unacceptably low for large p, implying that techniques such as open access [14] and interference cancelation [28] may be required for efficient spectrum sharing.

Fig. 5.6 shows how $\mathbb{P}_{m,o}(j, \kappa)$, where $j \in \mathcal{J}_m \cap \mathcal{J}_f$, varies with κ for different $p\lambda_m$. Interestingly, all curves attain a maximum at $\hat{\kappa} = (P_f/P_m)^{1/\alpha_o} = 0.316$, indicating that the success probabilities are maximized when the effect of cross-tier interference is eliminated. Note that the presence of co-channel FAPs has two contrasting effects on a

Table 5.1 Notations and parameters (© 2012 IEEE).

Symbol	Meaning	Default value
$\Theta\ (\Theta_j)$	Point process of MBSs (using subchannel j)	–
$\Phi\ (\Phi_j)$	Point process of FAPs (using subchannel j)	–
λ_m	Intensity of MBSs	$10^{-6}\,\mathrm{m}^{-2}$
λ_f	Intensity of FAPs	–
μ_m	Intensity of macrocell UEs	$1.5\times 10^{-5}\,\mathrm{m}^{-2}$
$\bar{c}_f$	Average number of femtocell UEs in an FAP cluster	4
R_f	Femtocell radius	40 m
Δ	Femtocell ring thickness	0.001 m
α_o	Outdoor path loss exponent	4
α_i	Indoor path loss exponent	3.7
P_m	Transmit power of MBS	43 dBm
P_f	Transmit power of FAP	23 dBm
$\mathcal{J}$	Set of all subchannels	$|\mathcal{J}| = 1000$
$\mathcal{J}_m$	Subchannels available to macro tiers	–
$\mathcal{J}_f$	Subchannels available to femto tiers	–
$p = k/|\mathcal{J}_f|$	Probability of an FAP transmitting in a subchannel	–
γ_m	QoS threshold of macrocell UE	4.8 dB
γ_f	QoS threshold of femtocell UE	4 dB
ϵ_m	Macro outage probability constraint	0.8
ϵ_f	Femto outage probability constraint	0.3
θ_m	Macro tier throughput constraint	0.75
θ_f	Femto tier throughput constraint	0.75
$\cdot^G,\ \cdot^D,\ \cdot^J$	Superscripts for general, disjoint and joint allocations	–

macrocell UE. On one hand, it causes cross-tier interference to the macrocell UEs. On the other hand, the availability of nearby FAPs allows the UE to hand over its connection to an FAP, improving its link reliability. For small κ, i.e., $\kappa \ll \hat{\kappa}$, we observe that $\mathbb{P}_{m,o}(j, \kappa)$ decreases as $p\lambda_f$ increases. This is due to the fact that for small κ, the handover from M to F is not favored since femtocell interference outweighs the connection availability for macrocell UEs. For large κ, i.e., $\kappa > \hat{\kappa}$, communication handover from M to F is favored, such that femtocells' connection availability for macrocell UEs outweighs the interference created, resulting in success probability increasing with $p\lambda_f$.

We compare next the performance of the optimal disjoint and joint allocations with the optimal general allocation for the cases of femtocells employing closed and open access. We simulate the optimal disjoint, joint and general allocations based on the analytical values of success probabilities. Using these values, we perform exhaustive search for the optimal disjoint, joint or general allocation. For the cases of disjoint and joint allocations, the analytical results coincide with the simulated ones. Fig. 5.7 illustrates how the optimal throughput $\tilde{\mathcal{T}}$ increases with increasing λ_f for fixed λ_m. While we know that $\tilde{\mathcal{T}}^D, \tilde{\mathcal{T}}_c^J \le \tilde{\mathcal{T}}_c^G$ and $\tilde{\mathcal{T}}^D, \tilde{\mathcal{T}}_o^J \le \tilde{\mathcal{T}}_o^G$, we would like to investigate when the optimal joint and disjoint allocations approximate the optimal general allocation. First, with open access FAPs, we observe that $(\tilde{p}^D, |\tilde{\mathcal{J}}_m^D|, |\tilde{\mathcal{J}}_f^D|)$ should not be used since $\tilde{\mathcal{T}}^D$ is always lower than $\tilde{\mathcal{T}}_o^G$ and $\tilde{\mathcal{T}}_o^J$. It is also observed that $\tilde{\mathcal{T}}_o^J \approx \tilde{\mathcal{T}}_o^G$ for all λ_f, reflecting that

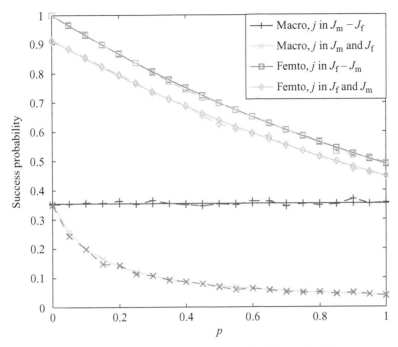

Figure 5.5 Success probabilities in different tiers and subchannels with respect to p. Solid lines are analytical results; dashed lines are numerical results (© 2012 IEEE).

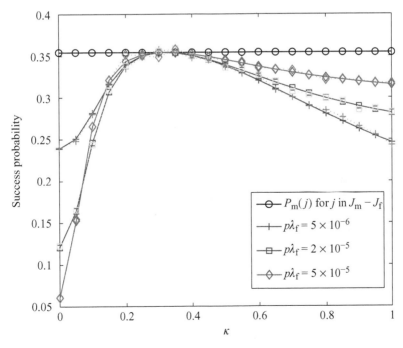

Figure 5.6 Macrocell success probabilities with respect to κ for different effective FAP density $p\lambda_f$. Solid lines are analytical results; dashed lines are numerical results (© 2012 IEEE).

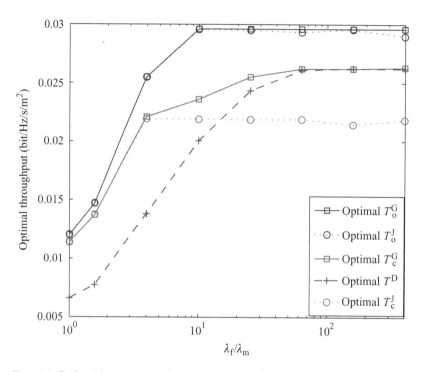

Figure 5.7 Optimal throughput with respect to macro–femto density ratio (© 2012 IEEE).

the optimal joint allocation is preferable since it is nearly optimal and more analytically tractable than the optimal general allocation. With closed access FAPs, we have that $\tilde{\mathcal{T}}_c^J \approx \tilde{\mathcal{T}}_c^G$ for small λ_f and $\tilde{\mathcal{T}}^D \approx \tilde{\mathcal{T}}_c^G$ for large λ_f. This confirms with our intuition that for sparse closed access FAP deployment, $(\tilde{p}_c^J, |\mathcal{J}|, |\mathcal{J}|)$ should be employed since the weak cross-tier interference incurred is overweighed by the efficient spectral reuse, and that in dense closed access FAP settings, $(\tilde{p}^D, |\tilde{\mathcal{J}}_m^D|, |\tilde{\mathcal{J}}_f^D|)$ is to be employed since cross-tier interference needs to be avoided. Finally, we make two macroscopic observations on the optimal throughput: first, we see that $\tilde{\mathcal{T}}_o^G$ and $\tilde{\mathcal{T}}_o^J$ are always greater than $\tilde{\mathcal{T}}_c^G, \tilde{\mathcal{T}}^D$, and $\tilde{\mathcal{T}}_c^J$; second, for all optimal allocations, the corresponding optimized throughput $\tilde{\mathcal{T}}$ stabilizes when λ_f becomes large. This reflects that the network throughput does not keep on increasing with λ_f even if optimal spectrum allocation is employed.

Figs. 5.8 and 5.9 illustrate how the optimal subchannel allocations vary with λ_f for fixed λ_m as a means to investigate how the optimal joint, disjoint and general allocations behave under different access control. With open access FAPs, we see that for the optimal general allocation, $|\tilde{\mathcal{J}}_{m,o}^G| = 1000$ for all λ_f, implying that a macrocell has to occupy all subchannels since sharing subchannels does not decrease $\mathbb{P}_{m,o}(\text{net})$ provided κ is chosen optimally. Nevertheless, sharing subchannels does decrease $\mathbb{P}_f(\text{net})$. Thus, in Figs. 5.8 and 5.9, we observe that both $|\tilde{\mathcal{J}}_{f,o}^G|$ and $\tilde{p}_o^G$ decrease as λ_f increases in order to mitigate intratier femtocell interference.

With closed access FAPs, in Fig. 5.8 we observe that $|\tilde{\mathcal{J}}_{m,c}^G| \approx 1000$ for all λ_f but $|\tilde{\mathcal{J}}_{f,c}^G|$ decreases from 1000 to 140 as λ_f increases. Moreover, in Fig. 5.9 we see that $\tilde{p}_c^G$ decreases

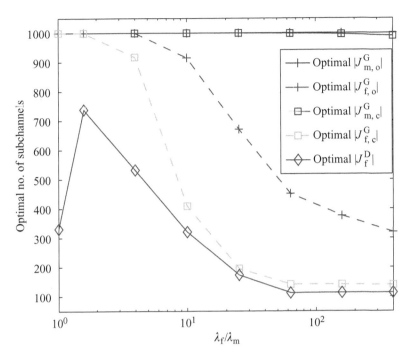

Figure 5.8 Optimal number of subchannels allocated with respect to macro–femto density ratio (© 2012 IEEE).

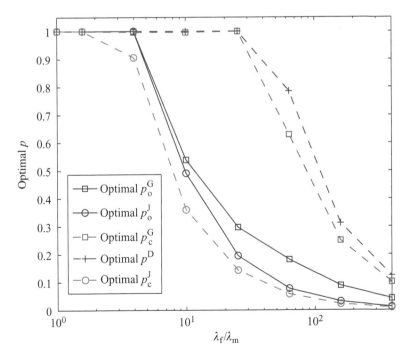

Figure 5.9 Optimal p with respect to macro–femto density ratio (© 2012 IEEE).

from 1 to 0.1 as λ_f increases. This means that as λ_f increases, the number of subchannels assigned to femtocells as well as the proportion of transmitting FAPs in a subchannel are significantly decreased under optimized general allocation. Due to this significant mitigation in femtocell interference, the macrocell should use nearly all subchannels for higher spectral reuse at the cost of moderate cross-tier interference. Furthermore, for the optimal disjoint allocation $(\tilde{p}^{\mathsf{D}}, |\tilde{\mathcal{J}}_{\mathrm{m}}^{\mathsf{D}}|, |\tilde{\mathcal{J}}_{\mathrm{f}}^{\mathsf{D}}|)$, we observe that in Fig. 5.8 for small λ_f there is a discontinuity. This is due to the transition of tightness in constraint from the lower bound in (5.28) to its upper bound. Moreover, in Fig. 5.9, we observe that $\tilde{p}^{\mathsf{D}}$ decreases as λ_f increases, indicating that the density of co-subchannel FAPs should be upper bounded in order to restrain cross-tier interference among femtocells.

5.7   Conclusion and future direction

In this chapter, we have explored a tractable two-tier stochastic network model and derived analytically the success probabilities in each tier under both femtocell open and closed access operation. The proposed framework also allowed us to generalize the results to multi-tier HCNs, as well as to obtain useful insights on spectrum allocation and reuse, namely the optimized joint and disjoint subchannel allocations. From extensive numerical results, we observed that the optimized joint and disjoint allocations provided the highest throughput among all subchannel allocations in sparse and dense closed access femtocell deployments, respectively. Furthermore, a simple location-based open access policy was proposed, showing that the optimized joint allocation with open access FAPs may obtain the highest achievable throughput for all femtocell density regimes.

There are many revenues for further investigation. Although the deployment of femtocells is seen to be a promising way of catering for the ever-increasing demand for higher data-rate wireless services, energy efficiency has become an important system operating parameter, which has to be taken into account for the long-term economic value of cellular networks. Thus an important question for future study is to understand the tradeoff in terms of energy efficiency and spatial reuse in HCNs. Another possible extension is to explore how the concept of cognitive radio can be applied more effectively in HCNs in order to further improve network capacity, coverage and reliability. Furthermore, this chapter only considered the downlink and the investigation of the proposed two-tier network model for the uplink case may be interesting.

5.8   Appendix

5.8.1   Derivation of $f_{\mathcal{R}}(r)$

Recall each femtocell UE is uniformly distributed in the infinitesimally thin ring with inner and outer radii R_f and $R_f + \Delta$, respectively. Denoting $\mathcal{R}$ as the distance between an FAP and its UE, for $R_f \leq r \leq R_f + \Delta$, we have

$$\mathbb{P}(\mathcal{R} \leq r) = \frac{\pi(r^2 - R_f^2)}{\pi((R_f + \Delta)^2 - R_f^2)} 1\ (R_f \leq r \leq R_f + \Delta). \tag{5.37}$$

We can obtain its PDF $f_{\mathcal{R}}(r)$ by differentiating (5.37) with respect to r within the range of support.

5.8.2   Proof of Lemma 5.1

For $j \in \mathcal{J}_m \backslash \mathcal{J}_f$, only MBSs transmit via subchannel j and $\mathbb{P}_{m,c}(j)$ can be found in [26]. Since no FAP is transmitting in subchannel j, $\mathbb{P}_f(j) = 0$. For $j \in \mathcal{J}_f \backslash \mathcal{J}_m$, only FAPs transmit via subchannel j so $\mathbb{P}_{m,c}(j) = 0$. For an FAP x, the set of interferers in subchannel j is $\Phi_j \backslash \{x\}$, where $\Phi_j \sim \text{PPP}(p\lambda_f)$ is the set of FAPs using subchannel j. From [31] and using $f_{\mathcal{R}}(r)$, the femtocell success probability is given by

$$
\mathbb{P}_f(j) \overset{(a)}{=} \int_{R_f}^{R_f + \Delta} \mathbb{P}\left(\frac{h_x g_{\alpha_i}(r)}{\sum_{y \in \Phi_j} h_y g_{\alpha_o}(y - u)} > \gamma_f \right) f_{\mathcal{R}}(r) dr
$$

$$
\overset{(b)}{=} \mathbb{P}\left(\frac{h_x g_{\alpha_i}(R_f)}{\sum_{y \in \Phi_j} h_y g_{\alpha_o}(y - u)} > \gamma_f \right) + O(\Delta)
$$

$$
\overset{(c)}{=} \exp\left(-p\lambda_f \pi R_f^{\frac{2\alpha_i}{\alpha_o}} C(\alpha_o) \gamma_f^{2/\alpha_o} \right) + O(\Delta), \tag{5.38}
$$

where (a) follows from the Slivnyak theorem [32], (b) follows from the fundamental theorem of calculus and (c) follows from [31], where $C(\alpha_o) = \frac{(2\pi)/\alpha_o}{\sin(2\pi/\alpha_o)}$. Note that the term $O(\Delta) \to 0$ as $\Delta \to 0$ for all choices of other parameters.

Next, for $j \in \mathcal{J}_m \cap \mathcal{J}_f$, the BSs in the sets Φ_j and Θ are transmitting via subchannel j, so for an MBS x transmitting to a macrocell UE u, the set of interferers is $\Phi_j \cup \Theta \backslash \{x\}$. We let $I_m^{(m)} = \sum_{y \in \Theta} P_m h_y g_{\alpha_o}(y - u)$ and $I_f^{(m)} = \sum_{y \in \Phi_j} P_f h_y g_{\alpha_o}(y - u)$, where the superscript denotes the type of transmission and the subscript denotes the type of interferers. By defining $V = \|x - u\|$ with PDF $f_V(r) = 2\pi \lambda_m r \exp(-\lambda_m \pi r^2)$, the macrocell success probability is given by

$$
\mathbb{P}_{m,c}(j) \overset{(a)}{=} \int_0^\infty \mathbb{P}\left(\frac{P_m h_x g_{\alpha_o}(x - u)}{I_m^{(m)} + I_f^{(m)}} > \gamma_m | \|x - u\|_2 \in dr \right) f_V(r) dr
$$

$$
\overset{(b)}{=} \int_0^\infty \mathbb{E}\left\{ \exp\left(-\frac{\gamma_m r^{\alpha_o}}{P_m}(I_m^{(m)} + I_f^{(m)}) \right) \right\} f_V(r) dr
$$

$$
\overset{(c)}{=} \int_0^\infty \exp\left(-\pi r^2 \lambda_m \rho(\gamma_m, \alpha_o) \right)
$$

$$
\times \exp\left(-p\lambda_f \pi r^2 C(\alpha_o) \left(\frac{P_f \gamma_m}{P_m} \right)^{2/\alpha_o} \right) f_V(r) dr
$$

$$
= \frac{1}{1 + p\frac{\lambda_f}{\lambda_m} C(\alpha_o) \left(\frac{P_f \gamma_m}{P_m} \right)^{\frac{2}{\alpha_o}} + \rho(\gamma_m, \alpha_o)}, \tag{5.39}
$$

where (a) follows from the Slivnyak theorem [32] (b) by the Laplace transform of $h_x \sim \exp(1)$ and (c) by the independence of $I_m^{(m)}$ and $I_f^{(m)}$ and the Laplace transforms of $I_m^{(m)}$ and $I_f^{(m)}$ [26, 31].

For an FAP x communicating with femtocell UE u, the set of interferers is $\Theta \cup \Phi_j \setminus \{x\}$. We let $I_m^{(f)} = \sum_{y \in \Theta} P_m h_y g_{a_o}(y - u)$ and $I_f^{(f)} = \sum_{y \in \Phi_j \setminus \{x\}} P_f h_y g_{a_o}(y - u)$; the femtocell success probability is given by

$$
\mathbb{P}_f(j) \overset{(a)}{=} \int_{R_f}^{R_f + \Delta} \mathbb{P}\left(\frac{P_f h_x g_{a_i}(r)}{I_m^{(f)} + I_f^{(f)}} > \gamma_f \right) f_{\mathcal{R}}(r) dr
$$

$$
\overset{(b)}{=} \int_{R_f}^{R_f + \Delta} \mathbb{E}\left\{ \exp\left(-\frac{\gamma_f r^{\alpha_i}}{P_f} I_m^{(f)} \right) \right\} \mathbb{E}\left\{ \exp\left(-\frac{\gamma_f r^{\alpha_i}}{P_f} I_f^{(f)} \right) \right\} f_{\mathcal{R}}(r) dr
$$

$$
\overset{(c)}{=} \int_{R_f}^{R_f + \Delta} \exp\left(-\pi r^{\frac{2\alpha_i}{\alpha_o}} \left(p\lambda_f + \left(\frac{P_m}{P_f}\right)^{\frac{2}{\alpha_o}} \lambda_m \right) C(\alpha_o) \gamma_f^{\frac{2}{\alpha_o}} \right) f_{\mathcal{R}}(r) dr
$$

$$
\overset{(d)}{=} \exp\left(-\pi R_f^{\frac{2\alpha_i}{\alpha_o}} \left(p\lambda_f + \left(\frac{P_m}{P_f}\right)^{\frac{2}{\alpha_o}} \lambda_m \right) C(\alpha_o) \gamma_f^{\frac{2}{\alpha_o}} \right) + O(\Delta), \quad (5.40)
$$

where (a) follows from the Slivnyak theorem [32], (b) follows from the Laplace transform of $h_x \sim \exp(1)$ and the independence between I_m and I_f, (c) follows from letting $s = r^{\alpha_i} \gamma_f / P_f$ in the Laplace transforms of $I_m^{(f)}$, and $I_f^{(f)}$ [31] and (d) follows from the fundamental theorem of calculus.

5.8.3   Proof of Lemma 5.2

First, we observe that, since Θ, Φ and $\tilde{\Theta}$ are stationary processes, the distributions of the distance $V \sim \|M - u\|$ and $W \sim \|F - u\|$ will remain the same for all macrocell and femtocell Tx–Rx pairs, respectively. The PDFs of V and W can be written as $f_V(r) = 2\pi \lambda_m r \exp(-\lambda_m \pi r^2)$ and $f_W(r) = 2\pi p\lambda_f r \exp(-p\lambda_f \pi r^2)$, respectively. Now, for $w \geq 0$ we have

$$
\mathbb{P}(w \leq W \leq \kappa V) = \int_w^\infty 2\pi r p\lambda_f \exp\left(-\pi r^2 p\lambda_f\right) \exp\left(-\pi r^2 \frac{\lambda_m}{\kappa^2}\right) dr
$$

$$
= \frac{p\lambda_f \kappa^2}{\lambda_m + p\lambda_f \kappa^2} \exp\left(-\pi w^2 \left(\frac{\lambda_m}{\kappa^2} + p\lambda_f\right)\right).
$$

Thus, by setting $w = 0$, we obtain $\mathbb{P}(F \to u) = \mathbb{P}(W \leq \kappa V) = \frac{p\lambda_f \kappa^2}{\lambda_m + p\lambda_f \kappa^2}$ and the PDF of W conditioned on the event that $W \leq \kappa V$ is given by

$$
f_{W|W \leq \kappa V}(w|W \leq \kappa V) = 2\pi w \left(\frac{\lambda_m}{\kappa^2} + p\lambda_f\right) \exp\left(-\pi w^2 \left(\frac{\lambda_m}{\kappa^2} + p\lambda_f\right)\right). \quad (5.41)
$$

Similarly, we can derive

$$
\mathbb{P}(\kappa v \leq \kappa V < W) = \frac{\lambda_m}{\lambda_m + p\lambda_f \kappa^2} \exp\left(-(\lambda_m + p\lambda_f \kappa^2)\pi v^2\right).
$$

By setting $v = 0$, we obtain $\mathbb{P}(\mathsf{M} \to u) = \mathbb{P}(W > \kappa V) = \frac{\lambda_\mathrm{m}}{\lambda_\mathrm{m} + p \lambda_\mathrm{f} \kappa^2}$ and the PDF of V conditioned on the event that $W > \kappa V$ is given by

$$f_{V|W > \kappa V}(v | W > \kappa V) = 2\pi v \left(\lambda_\mathrm{m} + p \lambda_\mathrm{f} \kappa^2\right) \exp\left(-\pi v^2 \left(\lambda_\mathrm{m} + p \lambda_\mathrm{f} \kappa^2\right)\right). \quad (5.42)$$

Now, we compute the success probability conditioned on F being designated to u, i.e., $W \le \kappa V$. This implies that there is no FAP in the disk $D(u; W)$ and no MBS in the disk $D(u; W/\kappa)$, where $D(x; r)$ denotes the disk centered at x with radius r. Letting $\mathcal{I}_\mathrm{m}^{(\mathrm{f})} = \sum_{x \in \Theta} P_\mathrm{m} h_x g_{a_\mathrm{o}}(x - u)$ and $\mathcal{I}_\mathrm{f}^{(\mathrm{f})} = \sum_{y \in \Phi_j} P_\mathrm{f} h_y g_{a_\mathrm{o}}(y - u)$, the conditional success probability is given by

$$\mathbb{P}(\mathrm{success} | \mathsf{F} \to u) = \int_0^\infty \mathbb{E}\left\{\exp\left(-\frac{r^{a_\mathrm{o}} \gamma_\mathrm{m}}{P_\mathrm{f}} \mathcal{I}_\mathrm{m}^{(\mathrm{f})}\right)\right\}$$

$$\times \mathbb{E}\left\{\exp\left(-\frac{r^{a_\mathrm{o}} \gamma_\mathrm{m}}{P_\mathrm{f}} \hat{\mathcal{I}}_\mathrm{f}^{(\mathrm{f})}\right)\right\} f_{W | W \le \kappa V}(r | W \le \kappa V) \mathrm{d}r. \quad (5.43)$$

Conditioning on $W \in [r, r + \mathrm{d}r)$, we have the Laplace transform for $\mathcal{I}_\mathrm{m}^{(\mathrm{f})}$ as follows:

$$\mathbb{E}\left\{\exp\left(-s\mathcal{I}_\mathrm{m}^{(\mathrm{f})}\right)\right\} \overset{(\mathrm{a})}{=} \mathbb{E}\left\{\prod_{x \in \Theta} \frac{1}{1 + s P_\mathrm{m} \|x - u\|^{-a_\mathrm{o}}}\right\}$$

$$\overset{(\mathrm{b})}{=} \exp\left(-2\pi \lambda_\mathrm{m} \int_{v = r/\kappa}^\infty 1 - \frac{1}{1 + s P_\mathrm{m} v^{-a_\mathrm{o}}} v \mathrm{d}v\right)$$

$$= \exp\left(-\pi \lambda_\mathrm{m} \frac{r^2}{\kappa^2} \rho\left(\frac{s \kappa^{a_\mathrm{o}} P_\mathrm{m}}{r^{a_\mathrm{o}}}, a_\mathrm{o}\right)\right) \quad (5.44)$$

where (a) follows from the Laplace transform of $h_x \sim \exp(1)$ and (b) follows from Campbell's theorem [32] and the fact that $\Theta(D(u, r/\kappa)) = \emptyset$. Note that $\rho(\gamma, a_\mathrm{o}) = \int_{\gamma^{\frac{-2}{a_\mathrm{o}}}}^\infty \gamma^{\frac{2}{a_\mathrm{o}}} / (1 + u^{\frac{a_\mathrm{o}}{2}}) \mathrm{d}u$.

Since $\Phi_j(D(u, r)) = \emptyset$, the Laplace transform for $\mathcal{I}_\mathrm{f}^{(\mathrm{f})}$ is given by

$$\mathbb{E}\left\{\exp\left(-s\mathcal{I}_\mathrm{f}^{(\mathrm{f})}\right)\right\} = \exp\left(-\pi r^2 p \lambda_\mathrm{f} \rho\left(\frac{s P_\mathrm{f}}{r^{a_\mathrm{o}}}, a_\mathrm{o}\right)\right). \quad (5.45)$$

Substituting (5.44) and (5.45) into (5.43) with $s = \frac{\gamma_\mathrm{m} r^{a_\mathrm{o}}}{P_\mathrm{f}}$ and using (5.41), we obtain

$$\mathbb{P}(\mathrm{success} | \mathsf{F} \to u) = \frac{\lambda_\mathrm{m} + \kappa^2 p \lambda_\mathrm{f}}{\lambda_\mathrm{m}\left(1 + \rho\left(\frac{\gamma_\mathrm{m} \kappa^{a_\mathrm{o}} P_\mathrm{m}}{P_\mathrm{f}}\right)\right) + \kappa^2 p \lambda_\mathrm{f}(1 + \rho(\gamma_\mathrm{m}, a_\mathrm{o}))}. \quad (5.46)$$

Following the steps from (5.43) to (5.46), we derive the Laplace transforms for $\mathcal{I}_\mathrm{m}^{(\mathrm{m})}$ and $\mathcal{I}_\mathrm{f}^{(\mathrm{m})}$ as follows:

$$\mathbb{E}\left\{\exp(-s\mathcal{I}_\mathrm{m}^{(\mathrm{m})})\right\} = \exp\left(-\pi r^2 \lambda_\mathrm{m} \rho\left(\frac{s P_\mathrm{m}}{r^{a_\mathrm{o}}}, a_\mathrm{o}\right)\right), \quad (5.47)$$

$$\mathbb{E}\left\{\exp(-s\mathcal{I}_\mathrm{f}^{(\mathrm{m})})\right\} = \exp\left(-\pi (\kappa r)^2 p \lambda_\mathrm{f} \rho\left(\frac{s P_\mathrm{f}}{(\kappa r)^{a_\mathrm{o}}}, a_\mathrm{o}\right)\right). \quad (5.48)$$

The success probability conditioned on M being designated to u can be written as

$$\mathbb{P}(\text{success}|M \to u) = \frac{\lambda_m + \kappa^2 p \lambda_f}{\lambda_m(1 + \rho(\gamma_m, a_o)) + \kappa^2 p \lambda_f \left(1 + \rho\left(\frac{\gamma_m P_f}{\kappa^{a_o} P_m}, a_o\right)\right)}. \quad (5.49)$$

Using (5.46) and (5.49), we obtain our result from $\mathbb{P}_{m,o}(j, \kappa) = \mathbb{P}(\text{success}|F \to u)\mathbb{P}(F \to u) + \mathbb{P}(\text{success}|M \to u)\mathbb{P}(M \to u)$.

5.8.4   Proof of Lemma 5.4

As noted in Section 5.5.6, we omit the subscript for open/closed access FAP since it makes no difference under disjoint allocation. The throughput for disjoint allocation $(p^D, |\mathcal{J}_m^D|, |\mathcal{J}_f^D|)$ is given by

$$\begin{aligned}
&\mathcal{T}^D\left(p^D, |\mathcal{J}_m^D|, |\mathcal{J}_f^D|, \mu_m, \lambda_f \bar{c}_f, \gamma_m, \gamma_f\right) \\
&= |\mathcal{J}_m^D|\mu_m \mathbb{P}_m(\mathcal{S}) \log(1 + \gamma_m) + p^D |\mathcal{J}_f^D|\lambda_f \bar{c}_f \mathbb{P}_f(\mathcal{S}) \log(1 + \gamma_f) \\
&= \frac{|\mathcal{J}_m^D|\mu_m \log(1 + \gamma_m)}{1 + \rho(\gamma_m, a_o)} + p^D |\mathcal{J}_f^D|\lambda_f \bar{c}_f e^{-p^D s} \log(1 + \gamma_f), \quad (5.50)
\end{aligned}$$

where $s = \lambda_f \pi R_f^{\frac{2a_i}{a_o}} C(a_i)\gamma_f^{\frac{2}{a_i}}$. In the following, we write $j \in \mathcal{J}_f^D$ and $j' \in \mathcal{J}_m^D$. The only success probability constraint is $\mathbb{P}_f(j) \geq 1 - \epsilon_f$, while the per-tier throughput constraints are the same as in (5.28). There is no constraint for $\mathbb{P}_m(j')$ since it is independent of the subchannel allocation. We could optimize $(\tilde{p}^D, |\tilde{\mathcal{J}}_m^D|, |\tilde{\mathcal{J}}_f^D|)$ with respect to p^D first, since p^D is assumed to be independent of $\mathcal{J}_m^D$ and $\mathcal{J}_f^D$. Accordingly, we have $\frac{\partial \mathcal{T}^D}{\partial p^D} = |\mathcal{J}_f^D|\lambda_f \bar{c}_f \log(1 + \gamma_f)(1 - p^D s)e^{-p^D s}$.

If $s < 1$, $\mathcal{T}^D$ is an increasing function of $p^D \in [0, 1]$. If $s \geq 1$, $\mathcal{T}^D$ is increasing in the range $p^D \in [0, s^{-1}]$, decreasing in the range $p^D \in [s^{-1}, 1]$ and has a maximum at $p^D = s^{-1}$. On the other hand, from (5.27), we have $e^{-sp^D} \geq 1 - \epsilon_f$, which implies $p^D \leq \frac{1}{s} \log \frac{1}{1-\epsilon_f}$. Taking the constraint into account, we arrive at (5.30) for $\tilde{p}^D$ and the corresponding optimized $\tilde{\mathbb{P}}_f(j) = e^{-\tilde{p}^D s}$ is given by

$$\tilde{\mathbb{P}}_f(j) = \begin{cases} \max\{1 - \epsilon_f, e^{-1}\} & \text{when } s \geq 1, \\ \max\{1 - \epsilon_f, e^{-s}\} & \text{when } s < 1. \end{cases} \quad (5.51)$$

The per-subchannel femto throughput is $\mathcal{T}_f^D(j) = p^D \lambda_f \mathbb{P}_f(j) \log(1 + \gamma_f)$. By substituting the optimal $\tilde{p}^D$ and $\tilde{\mathbb{P}}_f(\mathcal{S})$ into the previous expression, we have the optimized $\mathcal{T}_f^D(j)$, i.e. $\tilde{\mathcal{T}}_f^D(j)$, in (5.33). Next, with $\tilde{p}^D$, we optimize $\mathcal{T}^D$ for $(|\mathcal{J}_m^D|, |\mathcal{J}_f^D|)$, where $|\mathcal{J}_m^D| + |\mathcal{J}_f^D| = |\mathcal{J}|$, under throughput constraints in (5.28). This optimization is easy to solve since (5.50) is linear in $|\mathcal{J}_m^D|$ and $|\mathcal{J}_f^D|$ (cf. Theorem 1 in [19]). Now, the per-tier throughput for the macro tier is $\mathcal{T}_m^D(j') = \frac{\mu_m \log(1+\gamma_m)}{1 + \rho(\gamma_m, a_o)}$. If $\tilde{\mathcal{T}}_f^D(j) > \mathcal{T}_m(j')$, then we should allocate as many subchannels to the femto tier as possible in order to maximize the throughput. This means that the constraint $\tilde{\mathcal{T}}_f^D \leq \theta_f \mathcal{T}^D$ is tight and we have

$$|\tilde{\mathcal{J}}_f^D| = \max_{1 \leq |\tilde{\mathcal{J}}_f^D| \leq |\mathcal{J}|} \left\{ |\tilde{\mathcal{J}}_f^D| : \frac{|\tilde{\mathcal{J}}_f^D| \tilde{\mathcal{T}}_f^D(j)}{|\tilde{\mathcal{J}}_f^D| \tilde{\mathcal{T}}_f^D(j) + |\tilde{\mathcal{J}}_m^D| \tilde{\mathcal{T}}_m^D(j')} \leq \theta_f \right\}$$

$$\Leftrightarrow |\tilde{\mathcal{J}}_f^D| = \left\lfloor \frac{|\mathcal{J}|}{1 + \left(\frac{1}{\theta_f} - 1 \right) \frac{\tilde{\mathcal{T}}_f^D(j)}{\tilde{\mathcal{T}}_m^D(j')}} \right\rfloor, \tag{5.52}$$

and $|\tilde{\mathcal{J}}_m^D| = |\mathcal{J}| - |\tilde{\mathcal{J}}_f^D|$. Similarly, if $\mathcal{T}_m^D(j') > \tilde{\mathcal{T}}_f^D(j)$, then we should allocate as many subchannels to the macro tier as possible, such that the constraint for macro $\mathcal{T}_m^D \leq \theta_m \mathcal{T}$ is tight and we arrive at (5.32) by similar calculations.

5.8.5 Proof of Lemma 5.5

For the joint allocation with open access FAPs, we have

$$\mathcal{T}_o = |\mathcal{J}| \left(\frac{\mu_m \log(1 + \gamma_m)}{1 + \rho(\gamma_m, \alpha_o)} + F(p) \lambda_f \bar{c}_f e^{-d} \log(1 + \gamma_f) \right) \tag{5.53}$$

with $F(p) = p e^{-ps}$. The macrocell success probability constraint is irrelevant since it is independent of spectrum allocation, and the femtocell success probability constraint (5.27) and throughput constraints (5.28) translate to $0 < p \leq u_1$ and $l_2 \leq F(p) \leq u_2$, respectively. Therefore, if $u_1 \leq 0$, then the constraint on femtocell success probability is violated (the probability is too low) and the problem becomes infeasible. On the other hand, if $u_1 > 0$, we have the following two cases for s.

- Case 1: $s \leq 1$
  $F(p)$ is an increasing function of p for $p \in [0, 1]$, hence invertible. Therefore, the feasible region for the problem is $I_1 = [0, 1] \cap (0, u_1] \cap [F^{-1}(l_2), F^{-1}(u_2)]$. If $I_1 = \emptyset$, then the problem is infeasible. Otherwise, $\tilde{p}_o^J$ is the largest point in I_1, which is the derived expression.
- Case 2: $s > 1$
  $F(p)$ is not monotonic in $[0, 1]$; it attains maximum at $p = s^{-1}$ with maximal value $F(s^{-1}) = (se)^{-1}$, and it is increasing in the range $[0, s^{-1}]$ and decreasing in $[s^{-1}, 1]$. We have the following cases for the value of $(se)^{-1}$:
  - If $(se)^{-1} < l_2$, then the femto throughput is too low, such that (5.28) is violated and the problem is infeasible.
  - If $l_2 \leq (se)^{-1} < u_2$, then the feasible region is $I_2 = [0, u_1] \cap [a, b]$, where $[a, b] = F^{-1}([l_2, (se)^{-1}))$, which corresponds to the throughput constraint. If $I_2 = \emptyset$, the problem is infeasible. Otherwise, since F maximizes at s^{-1}, the optimal p is $\tilde{p}_o^J = \min\{u_1, s^{-1}\}$.
  - If $u_2 \leq (se)^{-1}$, by the fact stated about F, we see that the feasible region for throughput constraint $F^{-1}(l_2, u_2) \cap [0, 1]$ consists of two disjoint intervals $[a_1, b_1] \cup [a_2, b_2]$, where $a_1 \leq b_1 \leq a_2 \leq b_2$. Therefore, the feasible region for the problem is

$I_3 = [0, u_1] \cap ([a_1, b_1] \cup [a_2, b_2])$. Thus, if $u_1 < a_1$, then $I_3 = \emptyset$, and the problem is infeasible. Otherwise, if $a_1 < u_1 < b_1$, since $u_1 < b_1 \leq s^{-1}$, F is increasing in $[0, u_1]$, thus the optimal p is u_1. Finally, if $b_1 \leq u_1$, since F attains a maximum on b_1 and a_2 on the domain I_3, we can take $\tilde{p}_o^J = b_1$.

Copyright notice

References

[1] H. Claussen, L. T. W. Ho and L. G. Samuel, An overview of the femtocell concept. *Bell Labs Technical Journal*, **13**:1 (2008), 221–246.

[2] V. Chandrasekhar, J. G. Andrews and A. Gatherer, Femtocell networks: a survey. *IEEE Communication Magazine*, **46**:9 (2008), 59–67.

[3] D. López-Pérez, İ. Güvenc, G. de la Roche, M. Kountouris, T. Q. S. Quek and J. Zhang, Enhanced intercell interference coordination challenges in heterogeneous networks. *IEEE Communication Magazine*, **18**:3 (2011), 22–30.

[4] V. Chandrasekhar, J. G. Andrews, T. Muharemovic, Z. Shen and A. Gatherer, Power control in two-tier femtocell networks. *IEEE Transaction in Wireless Communications*, **8**:8 (2009), 4316–4328.

[5] C. W. Tan, S. Friedland and S. H. Low, Spectrum management in multiuser cognitive wireless networks: optimality and algorithm. *IEEE Journal on Selected Areas in Communications*, **29**:2 (2011), 421–430.

[6] D. T. Ngo, L. B. Le and T. Le-Ngoc, Distributed Pareto-optimal power control in femtocell networks. In *Proceedings of the IEEE International Symposium on Personal, Indoor and Mobile Radio Communications (PIMRC)*, Toronto (2011), pp. 222–226.

[7] H.-S. Jo, C. Mun, J. Moon and J.-G. Yook, Interference mitigation using uplink power control for two-tier femtocell networks. *IEEE Transactions in Wireless Communications*, **8**:10 (2009), 4906–4910.

[8] Y. Jeong, T. Q. S. Quek and H. Shin, Beamforming optimization for multiuser two-tier networks. *Journal of Communications and Networks*, **13**:4 (2011), 327–338.

[9] S. Akoum, M. Kountouris and R. W. Heath Jr., On imperfect CSI for the downlink of a two-tier network. In *Proceedings of the IEEE International Symposium on Information Theory (ISIT)*, Saint-Petersburg, Russia (2011), pp. 553–557.

[10] Y. Jeong, H. Shin and M. Z. Win, Interference rejection combining in two-tier femtocell networks. In *Proceedings of the IEEE International Symposium on Personal, Indoor and Mobile Radio Communications (PIMRC)*, Toronto (2011), pp. 137–141.

[11] I. Güvenc, M.-R. Jeong, F. Watanabe and H. Inamura, A hybrid frequency assignment for femtocells and coverage area analysis for co-channel operation. *IEEE Communications Letters*, **12**:12 (2008), 880–882.

[12] H.-S. Jo, P. Xia and J. G. Andrews, Open, closed, and shared access femtocells in the downlink. *CoRR*, abs/1009.3522 (2010).

[13] R. Madan, J. Borran, A. Sampath, N. Bhushan, A. Khandekar and T. Ji, Cell association and interference coordination in heterogeneous LTE-A cellular networks. *IEEE Journal on Selected Areas in Communications*, **28**:9 (2010), 1479–1489.

[14] P. Xia, V. Chandrasekhar and J. G. Andrews, Open vs. closed access femtocells in the uplink. *IEEE Transactions in Wireless Communications*, **9**:12 (2010), 3798–3809.

[15] C.-H. Ko and H.-Y. Wei, On-demand resource-sharing mechanism design in two-tier OFDMA femtocell networks. *IEEE Transactions on Vehicular Technology*, **60**:3 (2011), 1059–1071.

[16] S.-M. Cheng, S.-Y. Lien, F.-S. Chu and K.-C. Chen, On exploiting cognitive radio to mitigate interference in macro/femto heterogeneous networks. *IEEE Wireless Communications Magazine*, **18**:3 (2011), 40–47.

[17] A. Adhikary, V. Ntranos and G. Caire, Cognitive femtocells: breaking the spatial reuse barrier of celluar systems. In *Proceedings of the IEEE Information Theory and Applications Workshop (ITA)*, San Diego, CA (2011), pp. 1–10.

[18] S. Akoum, M. Zwingelstein-Colin, R. W. Heath Jr and M. Debbah, Cognitive cooperation for the downlink of frequency reuse small cells. *EURASIP Journal on Advances in Signal Processing* (2011).

[19] V. Chandrasekhar and J. G. Andrews, Spectrum allocation in tiered cellular networks. *IEEE Transactions in Communications*, **57**:10 (2009), 3059–3068.

[20] J.-H. Yun and K. G. Shin, Adaptive interference management of OFDMA femtocells for co-channel deployment. *IEEE Journal on Selected Areas in Communications*, **29**:6 (2011), 1225–1241.

[21] F. Baccelli and B. Blaszczyszyn, Stochastic geometry and wireless networks. *Foundations and Trends in Networking*, **3**:3/4 (2009), 249–449.

[22] M. Win, P. Pinto and L. Shepp, A mathematical theory of network interference and its applications. In *Proceedings of the IEEE*, **97**:2 (2009), 205–230.

[23] M. Haenggi, J. G. Andrews, F. Baccelli, O. Dousse and M. Franceschetti, Stochastic geometry and random graphs for the analysis and design of wireless networks. *IEEE Journal on Selected Areas in Communications*, **27**:7 (2009), 1029–1046.

[24] A. Rabbachin, T. Q. S. Quek, P. Pinto, I. Oppermann and M. Z. Win, Non-coherent UWB communications in the presence of multiple narrowband interferers. *IEEE Transactions in Wireless Communications*, **9**:11 (2010), 3365–3379.

[25] A. Rabbachin, T. Q. S. Quek, H. Shin and M. Z. Win, Cognitive network interference. *IEEE Journal on Selected Areas in Communications*, **29**:2 (2011), 480–493.

[26] J. G. Andrews, F. Baccelli and R. K. Ganti, A tractable approach to coverage and rate in cellular networks. *IEEE Transactions in Communications*, **59**:11 (2011), 3122–3134.

[27] V. Chandrasekhar and J. G. Andrews, Uplink capacity and interference avoidance for two-tier femtocell networks. *IEEE Transactions in Wireless Communications*, **8**:7 (2009), 3498–3509.

[28] K. Huang, V. K. N. Lau and Y. Chen, Spectrum sharing between cellular and mobile ad hoc networks: transmission–capacity trade-off. *IEEE Journal on Selected Areas in Communications*, **27**:7 (2009), 1256–1267.

[29] W. C. Cheung, T. Q. S. Quek and M. Kountouris, Stochastic analysis of two-tier networks: effect of spectrum allocation. In *Proceedings of the IEEE International Conference on Acoustics, Speech, and Signal Processing (ICASSP)*, Prague (2011), pp. 2964–2967.

[30] M. Andrews, V. Capdevielle, A. Feki and P. Gupta, Autonomous spectrum sharing for unstructured cellular networks with femtocells. *Bell Labs Technical Journal*, **15**:3 (2010), 85–97.

[31] F. Baccelli, B. Blaszczyszyn and P. Mühlethaler, Stochastic analysis of spatial and opportunistic ALOHA. *IEEE Journal on Selected Areas in Communications*, **27**:7 (2009), 1105–1119.

[32] D. Stoyan, W. S. Kendall and J. Mecke, *Stochastic Geometry and its Applications*, 2nd edn (Wiley, 1996).

[33] M. Haenggi and R. K. Ganti, Interference in large wireless networks. *Foundations and Trends in Networking*, **3**:2 (2009), 127–248.

[34] R. K. Ganti and M. Haenggi, Interference and outage in clustered wireless ad hoc networks. *IEEE Transactions on Information Theory*, **55**:9 (2009), 4067–4086.

[35] M. Haenggi, J. G. Andrews, F. Baccelli, O. Dousse and M. Franceschetti, Stochastic geometry and random graphs for the analysis and design of wireless networks. *IEEE Journal on Selected Areas in Communications*, **27**:7 (2009), 1029–1046.

[36] M. Z. Win, P. C. Pinto and L. A. Shepp, A mathematical theory of network interference and its applications. *Proceedings of the IEEE*, **97**:2 (2009), 205–230.

[37] J. Kingman, *Poisson Processes* (Oxford University Press, 1993).

[38] D. J. Daley and D. Vere-Jones, *An Introduction to the Theory of Point Processes*, 2nd edn (Springer, 2003), vol. 1.

[39] F. Baccelli, B. Blaszczyszyn and P. Muhlethaler, An ALOHA protocol for multihop mobile wireless networks. *IEEE Transactions on Information Theory*, **8**:6 (2006), 569–586.

[40] S. Chandrasekhar, Stochastic problems in physics and astronomy. *Review of Modern Physics*, **15**:1 (1943), 1–89.

[41] M. Y. Vardi, L. Shepp and L. Kaufman, A statistical model for positron emission tomography. *Journal of the American Statistical Association*, **80**:389 (1985), 8–20.

[42] D. L. Snyder, Filtering and detection for doubly stochastic Poisson processes. *IEEE Transactions on Information Theory*, **18**:1 (1972), 91–102.

[43] J. R. Pierce, Optical channels: practical limits with photon-counting. *IEEE Transactions on Information Theory*, **26**:12 (1978), 1819–1821.

[44] J. R. Pierce, E. C. Posner and E. R. Rodemich, The capacity of the photon counting channel. *IEEE Transactions on Information Theory*, **27**:1 (1981), 61–77.

[45] J. L. Massey, Capacity cutoff rate, and coding for direct detection optical channel. *IEEE Transactions in Communications*, **29**:11 (1981), 1615–1621.

[46] P. Pinto, A. Giorgetti, M. Z. Win and M. Chiani, A stochastic geometry approach to coexistence in heterogeneous wireless networks. *IEEE Journal on Selected Areas in Communications*, **27**:7 (2009), 1268–1282.

[47] P. Pinto and M. Z. Win, Communication in a Poisson field of interferers – Part I: Interference distribution and error probability. *IEEE Transactions in Wireless Communications*, **9**:7 (2010), 2176–2186.

[48] P. Pinto and M. Z. Win, Communication in a Poisson field of interferers – Part II: Channel capacity and interference spectrum. *IEEE Transactions in Wireless Communications*, **9**:7 (2010), 2187–2195.

[49] A. Rabbachin, T. Q. S. Quek, P. Pinto, I. Oppermann and M. Z. Win, Non-coherent UWB communications in the presence of multiple narrowband interferers. *IEEE Transactions in Wireless Communications*, **9**:11 (2010).

[50] E. Sousa, Performance of a spread spectrum packet radio network link in a Poisson field of interferers. *IEEE Transactions on Information Theory*, **38**:6 (1992), 1743–1754.

[51] J. Ilow, D. Hatzinakos and A. Venetsanopoulos, Performance of FH SS radio networks with interference modeled as a mixture of Gaussian and alpha-stable noise. *IEEE Transactions in Communications*, **46**:4 (1998), 509–520.

[52] F. Baccelli, B. Błaszczyszyn and F. Tournois, Spatial averages of coverage characteristics in large CDMA networks. *Wireless Networks*, **8**:6 (2002), 569–586.

[53] X. Yang and A. P. Petropulu, Co-channel interference modeling and analysis in a Poisson field of interferers in wireless communications. *IEEE Transactions in Signal Processing*, **51**:1 (2003), 64–76.

[54] J. Orriss and S. K. Barton, Probability distributions for the number of radio transceivers which can communicate with one another. *IEEE Transactions in Communications*, **51**:4 (2003), 676–681.

[55] S. P. Weber, X. Yang, J. G. Andrews and G. de Veciana, Transmission capacity of wireless ad hoc networks with outage constraints. *IEEE Transactions Information Theory*, **51**:12 (2005), 4091–4102.

[56] H. Q. Nguyen, F. Baccelli and D. Kofman, A stochastic geometry analysis of dense IEEE 802.11 networks. In *Proceedings of the IEEE Conference on Computer Communications*, Anchorage, AK (2007), pp. 1199–1207.

[57] O. Dousse, F. Baccelli and P. Thiran, Impact of interferences on connectivity in ad hoc networks. *IEEE/ACM Transactions on Networking*, **13**:2 (2005), 425–436.

[58] R. Niu and P. K. Varshney, Performance analysis of distributed detection in a random sensor field. *IEEE Transactions on Signal Processing*, **56**:1 (2008), 339–349.

[59] S. Srinivasa and M. Haenggi, Distance distributions in finite uniformly random networks: theory and applications. *IEEE Transactions on Vehicular Technology*, **59**:2 (2010), 940–949.

[60] O. Dousse, M. Franceschetti and P. Thiran, On the throughput scaling of wireless relay networks. *IEEE Transactions on Information Theory*, **52**:6 (2006), 2756–2761.

[61] L. Song and D. Hatzinakos, Cooperative transmission in Poisson distributed wireless sensor networks: protocol and outage probability. *IEEE Transactions in Wireless Communications*, **5**:10 (2006), 2834–2843.

[62] A. Ghasemi and E. S. Sousa, Interference aggregation in spectrum-sensing cognitive wireless networks. *IEEE Journal of Selected Topics in Signal Processing*, **2**:1 (2008), 41–56.

[63] R. Menon, R. M. Buehrer and J. H. Reed, On the impact of dynamic spectrum sharing techniques on legacy radio systems. *IEEE Transactions in Wireless Communications*, **7**:11 (2008), 4198–4207.

[64] W. Ren, Q. Zhao and A. Swami, Power control in spectrum overlay networks: how to cross a multi-lane highway. *IEEE Journal on Selected Areas in Communications*, **27**:7 (2009), 1283–1296.

[65] A. Rabbachin, T. Q. S. Quek, H. Shin and M. Z. Win, Cognitive network interference. *IEEE Journal on Selected Areas in Communications*, **29**:2 (2011), 480–493.

[66] H. Inaltekin, M. Chiang, H. V. Poor and S. B. Wicker, The behavior of unbounded path-loss models and the effect of singularity on computed network characteristics. *IEEE Journal on Selected Areas in Communications*, **27**:7 (2009), 1078–1092.

[67] R. K. Ganti and M. Haenggi, Interference and outage in clustered wireless ad hoc networks. *IEEE Transactions on Information Theory*, **55**:9 (2009), 4067–4086.

[68] K. Gulati, B. L. Evans, J. G. Andrews and K. R. Tinsley, Statistics of co-channel interference in a field of Poisson and Poisson–Poisson clustered interferers. *IEEE Transactions on Signal Processing*, **58**:11 (2010).

[69] V. Chandrasekhar and J. G. Andrews, Spectrum allocation in tiered cellular networks. *IEEE Transactions in Communications*, **57**:10 (2009), 3059–3068.

[70] V. Chandrasekhar, M. Kountouris and J. G. Andrews, Coverage in multi-antenna two-tier networks. *IEEE Transactions in Wireless Communications*, **8**:10 (2009), 5314–5327.

[71] V. Chandrasekhar and J. G. Andrews, Uplink capacity and interference avoidance for two-tier femtocell networks. *IEEE Transactions in Wireless Communications*, **8**:7 (2009), 3498–3509.

[72] Y. Kim, S. Lee and D. Hong, Performance analysis of two-tier femtocell networks with outage constraints. *IEEE Transactions in Wireless Communications*, **9**:9 (2010), 2695–2700.

[73] S.-M. Cheng, W. C. Ao and K.-C. Chen, Downlink capacity of two-tier cognitive femto networks. In *Proceedings of the IEEE International Symposium on Personal, Indoor and Mobile Radio Communications (PIMRC)*, Istanbul (2010), pp. 1301–1306.

6 Self-organization

Fredrik Gunnarsson

6.1 Introduction

Heterogeneous cellular network (HCN) deployments may imply an order of magnitude more network nodes than conventional homogenous macrocell deployments. Therefore, it is important that the integration and operation of these new nodes require minimal manual efforts from operators. *Self-organizing network* (SON) features can be seen as essential enablers to facilitate service as well as network deployment and management. The main objectives of SONs are to reduce the deployment costs, simplify network management (managing a plethora of *radio access technologies* (RATs) without significantly increasing operational expenses) and enhance network performance.

Within the *3rd Generation Partnership Project* (3GPP) *Long Term Evolution* (LTE), SON was among the early system requirements, and SON features were already included in the first 3GPP LTE release, i.e., Release 8 [1]. SON work items in 3GPP [2, 3] have been inspired by the SON studies and the set of requirements defined by the operators' alliance *Next Generation Mobile Networks* (NGMN) [4]. This chapter addresses HCN aspects of SON, although these automation features are applicable to other types of network deployment as well. The main focus is on LTE, but *Universal Mobile Telecommunication System* (UMTS) and multiple RATs will also be considered where applicable. More general discussions about SON can be found in [5–9], and a discussion with special focus on femtocells can be found in [10].

SON operations are supported by the *operation, administration, and maintenance* (OAM) architecture, which is presented in Section 6.2, and are commonly divided into four key components/phases: planning, self-configuration, self-optimization and self-healing, as illustrated in Fig. 6.1.

Planning and self-configuration concern tasks that are carried out to introduce new or replan existing site installations, and automate installation and configuration procedures. Planning and self-configuration are addressed in Section 6.3. While in operation, self-optimization tunes parameters that dictate algorithmic behaviors based on empirical network observations. Section 6.4 analyzes self-optimization and provides relevant insights about SON for HCNs. Self-healing concerns tasks that are carried out to detect, and if possible to compensate for, failures and disruptive events (e.g., malfunctioning equipment). Self-healing is addressed in Section 6.5. Section 6.6 discusses some performance monitoring aspects, and Section 6.7 provides a summary of the chapter.

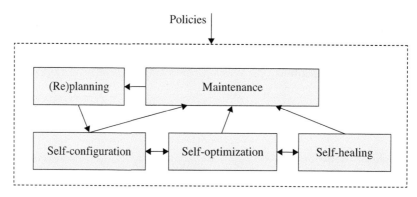

Figure 6.1 Key components of self-organizing networks.

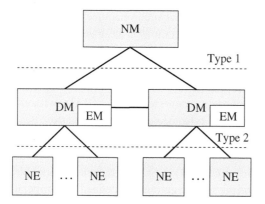

Figure 6.2 Management architecture with NM dictating the policies to, and receiving key performance indicators from, functions in the DM/EM and the NEs. The two interfaces from top to bottom are referred to as Type 1 (Itf-N) and Type 2 (e.g., Itf-S or TR-069), respectively.

6.2   Management architecture

Network operations are maintained and supervised through a management architecture, as illustrated in Fig. 6.2. The operator interacts with the network at a high level through the *network management* (NM) or *network management system* (NMS), which in turn interacts with the *domain manager* (DM)/*element manager* (EM) through the standardized Type 1 interface, i.e., Interface-North (Itf-N). The DM/EM manages individual *network elements* (NEs), typically through the vendor specific Type 2 interface, sometimes referred to as Interface-South (Itf-S). One exception is the *Home evolved NodeB* (HeNB), which can be managed through the standardized interface TR-069 or vendor specific *HNB/HeNB management system* (HMS) [11–13].

Basically, *policies* in terms of reference values and targets, weights and configuration parameters are fed from the NM to the DM/EM, and from the DM/EM to the NEs. Similarly, *performance information* is fed from the NEs to the DM/EM and onwards to the NM. The performance information is typically defined as counters of events. For example, a *user equipment* (UE) will complete several steps on the way to be fully connected to the network, and each step is monitored via counters, such as the number of access attempts and the number of successful attempts. These counters may be reported and afterwards reset regularly. The performance information can be refined, for example, through averaging. Refined performance indicators are often referred to as *key performance indicators* (KPIs).

Moreover, automatic (re)configuration in the NE may result in a configuration notification fed upwards as part of the *configuration management* (CM). Another example is *fault management* (FM) with alarms fed upwards, where several alarms may be associated with one another and represented by one indicating the root cause.

As described in [13], SON functions can be classified into different types, according to how they are mapped into the network architecture.

- **NM-centralized SON** operates to meet centralized policies defined in the NM, reconfiguring NE parameters based on network information fed back from the NEs. Feedback (from NEs to NM) and instructions (from NM to NEs) are transmitted through the DM over the Itf-N interface. NM-centralized SON takes benefit from its ability to consider long-term data, as well as multiple cells, RATs and UE vendors. It is based on the performance indicators and policies defined in the OAM specifications in 3GPP. Its main drawback is the delay in decision making due to data gathering, processing and distribution, preventing prompt reactions to crucial events.
- **DM-centralized SON** reconfigures NE parameters based on the information obtained through network monitoring over the Itf-S interface. The NM receives policies and provides performance information over the Itf-N interface. Its main benefits and drawbacks are similar to NM-centralized SON, except that DM-centralized SON is typically vendor specific. As such it is based on vendor specific performance information, and may use vendor specific configurations, policies, targets, weights etc.
- **Distributed SON** is implemented in the NEs (typically *evolved NodeBs* (eNBs) in the case of *Evolved UTRAN* (E-UTRAN), and *Radio Network Controller* (RNC) in the case of *UMTS Terrestrial Radio Access Network* (UTRAN)), and receives policies from and provides KPIs to the NM through the DM over the Itf-N/S interfaces. In this chapter, there is an ambition to outline some central parts of these interfaces in relation to SON. It is able to react fast to reports from served UEs and information obtained from other NEs via X2 [14] and S1 [15] interfaces, respectively, in the case of E-UTRAN, and Iub [16] and Iur [17] interfaces, respectively, in the case of UTRAN. Moreover, information can be shared between RATs via the *RAN information management* (RIM) transfer mechanisms [15, 18].
- **Hybrid SON** is essentially a combination of both centralized and distributed SON functional components.

6.3   Self-configuration

Traditionally, deployment of a new *base station* (BS) meant extensive manual work from operators' staff, e.g., planning the new site location, installing the BS and configuring the network equipment. Automatic mechanisms that reduce such manual work make planning, installation and configuration less time consuming, less personnel intensive and less prone to errors. Both the 3GPP and the NGMN SON requirements include self-configuration as a key feature of future wireless networks. Self-configuration can be divided into a planning phase and an installation phase.

6.3.1   Planning

Some planning efforts are still necessary. The planning phase identifies site location candidates, where new BSs should be deployed. If the deployment of new BSs is for an area where cellular networks are already provided, with either a different or the same RAT, then measurements from the existing network can provide valuable information for BS configuration, which will be discussed in Section 6.6. The site survey and network expansion activity will give a set of candidate sites where BSs may be deployed.

BSs may be associated with an initial set of site-specific parameters. This set of parameters may be configured through the 3GPP *automatic radio configuration data handling function* (ARCF) [19], and may include cell identities [20–23], pre-configured neighboring relations [24], antenna configurations, transmit power levels, operational carrier etc. Note that some information may be network specific and valid across the entire network, such as the *Dynamic Host Configuration Protocol* (DHCP) server information that provides *Internet Protocol* (IP) addresses and further connectivity information. Other information may be BS specific, and possibly subject to self-optimization. Further details will be discussed in Section 6.4.

An important aspect that self-configuration deals with is cell identity selection. Cells broadcast both a globally unique cell identity and a waveform associated with a non-unique cell identity. Each E-UTRAN cell broadcasts a *physical layer cell identity* (PCI), and as part of the system information a *public land mobile network* (PLMN) identity (or several of them in the PLMN identity list) and a cell identity. The cell identity is a unique 28-bit cell identity representing the cell within the PLMN context. The PCI is mapped to synchronization and reference signals that UEs use for cell search, cell identification, measurements, phase reference etc. The number of available PCIs is limited per frequency carrier. For example, there are 504 PCIs defined in LTE. This means that PCIs may need to be reused in dense network deployments. In *Radio Resource Control* (RRC) [25], the combination of PLMN (the first PLMN in case of a list of PLMNs) and the 28-bit cell identity is referred to as the *Evolved Cell Global Identifier* (ECGI). S1AP [15] specifies the cell identities used among eNBs and the core network, while X2AP [14] specifies the cell identities used among eNBs. In the former case, the core network only identifies eNBs (not cells within an eNB), since it is irrelevant which cell within an eNB is serving a UE in that context. In the latter case, the 28-bit cell identity is divided into two parts: a 20-bit identity referring to the eNB and an eight-bit identity referring to the cell within the eNB. HeNBs are identified by 28 bits and consequently can only serve one cell each. Fig. 6.3 illustrates the relation between eNB and cell identities.

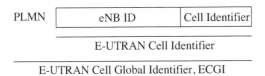

Figure 6.3 Relation and different eNB and cell identities in LTE.

The planning phase either selects a PCI in a centralized manner, or determines a PCI list from which the cell may select a PCI in a distributed manner [13], which will be discussed in Section 6.4.2. Preferably, the PCIs should be locally unique, which means that *low-power nodes* (LPNs) covered by one macrocell should have distinct PCIs. This should be possible with a reasonably large number of LPNs. Indeed, the worst case scenario analysis using 3GPP scenarios and with up to 75 LPNs per macrocell [26] has shown that the required number of PCIs to guarantee the tractability of the assignment problem is well below the number of available PCIs. However, a *closed subscriber group* (CSG), together with possibly other LPNs, may be configured with PCIs from predefined ranges, which can make the PCI assignment problem intractable.

The planning phase concerns other RATs as well. For example, in UTRAN the cell is uniquely identified by its *UTRAN Cell Identity*, which consist of a cell identity that is unique within the *Radio Network Subsystem* (RNS) and an RNC identity. In the physical layer, the cell is identified by its *primary scrambling code* (PSC). Moreover, since the RNC is responsible for mobility in UTRAN, the planned neighbor relations are configured in the RNC.

6.3.2   Installation

The self-configuration installation procedure [4, 19] includes the following phases.

- Establishment of basic connectivity, such as providing an eNB IP address, gateway information and OAM connectivity information.
- Connection to the OAM system for downloading required software together with transport and radio configuration data, as prepared by ARCF in the planning step; see Section 6.3.1. This includes a binding step, where the cell equipment is associated with the site and the prepared configuration data. This binding step can be fully automatic, or involve actions taken at the site by operators' staff. More details about this process are given in [27].
- Installation of software, establishment of S1 links and X2 links (if there are planned neighbor relations). After self-testing, the BS is operational and ready to take network traffic.

The installation procedure is typically vendor specific. However, in case of *relay node* (RN) (see Section 10.2) deployments, parts of the installation need to be standardized to address certain issues related to the following cases: (1) when not all BSs can act as *donor eNBs* (DeNBs) to support RNs, and (2) when RNs from multiple vendors need to

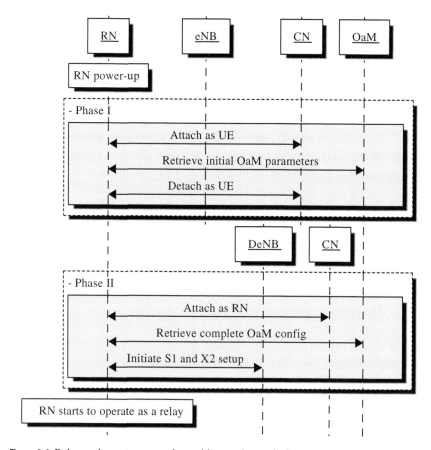

Figure 6.4 Relay node startup procedure with two phases: i) the RN connects as a UE, ii) the RN connects with an RN indication.

find their respective OAM system. The RN installation procedure [2], as illustrated in Fig. 6.4, can be divided into two phases. In the first phase, the RN attaches to the eNB as a UE, and uses the provided connectivity to access the OAM system and download software and configuration information, including a list of eNBs supporting RNs. Then, in the second phase, the RN disconnects and attaches as an RN to a DeNB, by including an RN indication to the eNB during the connection establishment. This will trigger the self-configuration of the RN for operation, including dedicated establishment of the backhaul links (possibly with resource time division to handle RNs with poor isolation between transmit and receive antennas).

One target of the RN self-configuration is to assign ECGIs to the cells served by the RN. The ECGI is derived from the eNB *Identifier* (ID) of the DeNB and an eight-bit cell identity specified for each RN cell, which is unique in the DeNB. Thus, in the second phase, the RN provides the eNB ID of its serving DeNB to its OAM system, and in return receives the ECGIs to be assigned to each of its served cells. The uniqueness of the cell identities under the DeNB must be ensured by the OAM system, e.g., by NM assigning a cell identity range for each RN vendor. From the perspective of the core network and other eNBs, this ECGI assignment procedure ensures that RN cells are perceived as any cells of the DeNB.

Another target of the RN self-configuration is to discover its vendor-specific OAM system. A dedicated procedure is based on *domain name system* (DNS) lookup tables [28, 29]. In this procedure, the RN benefits from its embedded UE components, since there is a *subscriber identity module* (SIM) within the RN. In order to derive a fully qualified domain name, and from the SIM, the RN obtains the operator identity (i.e., *mobile network code* (MNC) and *mobile country code* (MCC)), as well as the RN vendor code (i.e., *International Mobile Equipment Identity Type Allocation Code* (IMEI-TAC)). Then, this domain name is sent to a DNS for lookup, and the OAM system *transport network layer* (TNL) address is provided in return. For example, an operator with operator identity MCC = 234 and MNC = 15, together with an RN vendor code 12345678, would result in the following domain name:

```
12345678<IMEI-TAC>.eutran-rn.oam.015<MNC>.234<MCC>.3gppnetwork.org,
```

where a zero is added since MNC is encoded with less than three digits.

The RNs may also be supported by multiple DM/EM systems. For example, if the network is large and with many RNs, then one single DM/EM system cannot manage all the RNs in the entire network. Therefore, the RNs need to be directed to the DM/EM that is assigned to serve RNs at its particular network region. In this case, the *tracking area code* (TAC)/indexTracking area of the DeNB to which the RN is connected is used as a key to associate the RN to the correct DM/EM region, so as to enable regional DM/EM systems. This procedure is currently specified only for RNs [28, 29], but it can naturally be extended to other types of HCN node. In some cases, the operator identity may still be derived through an SIM (if there is any UE component embedded in the BS), while the operator identity information (together with vendor specific information) can be obtained by other means. For example, it can be preconfigured in the BS by the installation engineer. Information such as the BS type could also be included in the fully qualified domain name to enable different DM/EM for different node types.

In the installation phase, LPNs may also provide information to the OAM system. This information may include the BS type, estimated LPN position (e.g., based on a *global navigation satellite system* (GNSS)), details about neighboring BSs (such as used PCIs), observed transmit powers etc. These parameters can be measured or detected by LPNs through its *network monitoring mode*, as illustrated in Fig. 6.5. Typically, LPNs use a UE functionality, which is mandatory for RNs but also common to other types of LPN, to obtain information from the environment. This information can be used to automatically assign an initial BS configuration in the binding step, and automatically derive BS parameters at the installation phase.

6.4 Self-optimization

Self-optimization requirements have been defined by operators through the 3GPP and/or the NGMN. This section addresses a selected set of SON use cases based on such

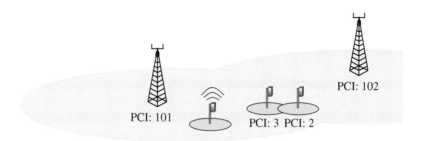

Figure 6.5 Network monitoring at LPNs is an efficient means to consider the local environment in the initial LPN configuration.

requirements. Our target is to describe SON requirements and key behaviors, and present current solutions to exemplify SON use cases. E-UTRAN features, as well as some UTRAN features, are considered. Centralized SON features are essentially applicable to any RAT and multiple RATs. Distributed SON features are implemented in the eNB in E-UTRAN and rely on the S1 and X2 interfaces for inter-node aspects. In UTRAN, inter-node aspects are facilitated by the RNC, which is a key SON enabler, while some distributed SON features can be considered for the NodeB.

6.4.1   Automatic neighbor relation

Traditionally, one of the main operator configuration/optimization efforts has been related to the establishment of neighboring relations among cells served by BSs. These neighboring relations can be derived in the planning step, as described in Section 6.3.1, and take up a significant part of operations expenditures. Therefore, the automation of neighboring relations management is a key SON feature, especially in HCNs, where the number of nodes and potential neighboring relations increase dramatically.

The *Automatic Neighbor Relation* (ANR) SON function can be seen as automatic management of the *neighbor relation table* (NRT) in the *radio access network* (RAN), located in eNBs in E-UTRAN and in the RNC in UTRAN. The NRT includes neighboring relation information for each served cell, as well as information about their neighboring cells. A neighboring cell relation is a cell-to-cell relation, while an X2 link is between two eNBs, and an Iur interface is between two RNCs. Moreover, neighboring cell relations are unidirectional, while X2 and Iur interfaces are bidirectional. A neighboring cell may operate at a different carrier frequency and/or with a different RAT.

The neighboring cell relation information includes the following.

- Neighboring cell RAT and *cell identities* (which are PCI and ECGI in case of E-UTRAN, or PSC and the UTRAN cell identity in case of UTRAN).
- Attribute values:
  - *No Remove*. If set, the neighboring cell relation cannot be removed by ANR in RAN.
  - *No Handover* (HO). If set, the neighboring cell relation cannot be used for HOs.
  - *No X2* (for intra-LTE relations only). If set, the eNB is prevented from using an X2 interface to initiate procedures for the eNB to serve the neighboring cell.

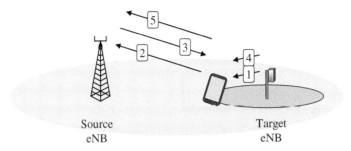

Figure 6.6 The UE ANR in E-UTRAN is a UE reporting procedure divided into five steps, where the serving eNB can request a UE to acquire and report relevant neighboring cell system information such as ECGI, TAC and CSG information in E-UTRAN.

Statistics are gathered for each neighboring cell relation, and are reported to the OAM system [30]. The OAM system is notified every time ANR updates the NRT, but can also modify cells' NRTs by itself, e.g., changing attributes of individual entries [31]. Rarely used or incorrectly configured neighboring relations can be candidates for automatic removal by ANR, if allowed by the corresponding attribute. Removal criteria can be based on, e.g., the time since last update, or the number of usages over a fairly long time interval.

E-UTRAN

The LTE ANR SON function [2, 31–33] is located in the eNB. It supports management of neighboring cell relations from E-UTRAN to E-UTRAN, UTRAN, *GSM EDGE Radio Access Network* (GERAN) and CDMA2000 cells [31]. Based on the UE ANR feature, an eNB can request a UE to decode and report neighboring cell information through its broadcast system information. Fig. 6.6 illustrates the UE ANR feature, which can be divided into five steps.

1. A UE detects the physical layer cell identity (e.g., PCI in E-UTRAN and PSC in UTRAN) of a candidate cell.
2. If measurements of the candidate cell meet a configured reporting criterion, then a measurement report is sent from the UE to its serving eNB.
3. The eNB has reasons to learn more about the candidate cell (e.g., the cell identity has never been reported before), and requests the UE to acquire the system information from the candidate cell and report relevant parts such as the unique cell identity, CSG indications, CSG identity, TAC etc. In order to facilitate the acquisition, the serving eNB may configure a measurement gap to inform the UE that no data will be transmitted during the measurement period. Alternatively, it may allow the UE to use autonomous gaps, in which the UE can neglect its scheduled data when necessary to enable the acquisition of candidate cell system information. As for any measurement and report configurations, the serving eNB uses the RRC RRCConnectionReconfiguration message [25] for this purpose, indicating that the UE shall report CGI for the candidate cell.

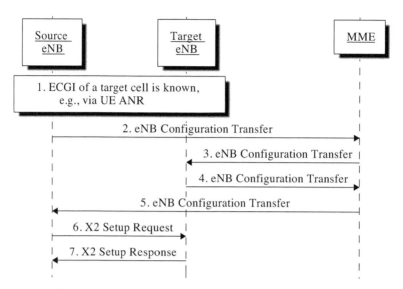

Figure 6.7 Target eNB transport network layer address recovery.

4. The UE acquires the relevant candidate cell system information.
5. The UE reports the relevant candidate cell system information to the serving eNB.

UE ANR support is needed in order to complete the ANR procedure. This is indicated by the UE via feature group indicators [25]. There are separate indicators for intrafrequency E-UTRAN, interfrequency E-UTRAN and inter-RAT neighbors.

The UE ANR procedure is completed when the serving eNB has sufficient information to initiate an HO to the candidate cell, and the NRT can be updated with a new neighboring relation between the serving cell and the candidate cell. The NRT attributes are set according to default parameters provided by the OAM system. Field experiences with ANR for intrafrequency E-UTRAN neighboring cells indicate that it is possible to use the UE ANR for both uniquely identifying candidate cells and successfully completing a subsequent HO [34]. This means that HO relations are established when needed, without jeopardizing the success rate of HOs.

In theory, it is always possible to use the UE ANR procedure for every HO to validate candidate cells. UE ANR is also the proposed procedure for CSG cell detection [2], where the UE reports the CSG ID of the target cell together with the ECGI to its serving cell. Moreover, UEs are provided with a white list of corresponding allowed CSG IDs from higher signaling layers, which they are authorized to use. This information is used by the UE to indicate its CSG membership status to the serving cell. Further details are given in Chapter 4.

The ECGI reported by the UE ANR procedure can be used for intra-LTE S1 HOs. Moreover, it can be used to retrieve the TNL address of the eNB serving a neighboring cell, which in turn can be used to establish an X2 interface between the two BSs, as illustrated in Fig. 6.7.

In more detail, the process for establishing an X2 link is as follows.

1. The source eNB is aware of the ID of a target eNB, e.g., through the UE ANR procedure.
2. The source eNB queries the *Mobility Management Entity* (MME) with an S1AP eNB Configuration Transfer message [15], including SON Information with a TNL address request as well as the source and target eNB IDs.
3. The MME relays the SON Information in an S1AP MME Configuration Transfer message to the eNB identified by the target eNB ID.
4. The target eNB answers the MME with an S1AP eNB Configuration Transfer message, including SON Information with its TNL address as well as the source and target eNB IDs.
5. The MME relays the SON Information in an S1AP MME Configuration Transfer message to the eNB identified by the source eNB ID.
6. The source eNB initiates X2 connectivity, and when completed sends an X2AP X2 Setup Request message [14] over the established X2 link to the target eNB, including information about its served cells.
7. The target eNB responds with analogous information through an X2AP X2 Setup Response message.

The information about the served cells obtained over the X2 interface is comprehensive, and includes, e.g.:

- cell identities such as PCI and ECGI, CSG indication and CSG ID if applicable, tracking area information and all supported PLMNs;
- neighboring cell information including PCI and ECGI;
- downlink and uplink carrier frequencies and bandwidths (in the case of *frequency division duplexing* (FDD)), and downlink/uplink time separation (in the case of *time division duplexing* (TDD));
- antenna configuration such as the number of antenna ports;
- random access procedure configuration.

The information about the served cells may change over time, and can be updated when it changes through an X2AP eNB Configuration Update message. An eNB may use the served cell information from neighboring eNBs to identify candidate cells through the PCIs reported by the UEs. This can be particularly useful in case of LPNs deployed under an umbrella macrocell. The eNB has local knowledge about neighboring cells, and LPNs may identify most of its neighboring PCIs reported by its UEs using the eNB served cell information.

This is also applicable in case of RNs. Since X2 between RN and DeNB is established before the RN is operational, the RN can be preconfigured with the PCI-ECGI mappings and neighbor relations from the serving DeNB. However, there may still be neighboring cells to the RN that are not known by its DeNB. One example is the deployment in Fig. 6.8, where an RN is deployed in a coverage hole between two macrocell eNBs. The signaling is analogous to that of Fig. 6.6, but the reported ECGI contains an eNB ID that is associated to the target DeNB.

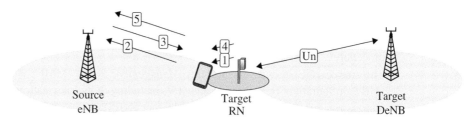

Figure 6.8 UE ANR procedure in case the candidate cell is served by an RN.

Similar to Fig. 6.7, the source eNB may initiate TNL address recovery to the target eNB. However, the target eNB ID is associated with the DeNB of the RN. This will lead to the establishment of an X2 link between the source eNB and the target DeNB. Note that the RN is hidden under its DeNB and is invisible to the core network and other eNBs.

UTRAN

The neighboring cell relations are used in a different way in UTRAN. The UE is informed by its serving RNC about neighboring cell relations in the lists: 32 intrafrequency UTRAN neighbors including the serving cell(s), 32 interfrequency UTRAN neighbors and 32 inter-RAT neighbors. When reporting, the UE indicates the concerned cell by its list index. It is also possible to report unlisted cells, denoted as *detected cells* by their cell identities [35], but the reporting requirements are much less strict [36] – a detected cell shall be detected and reported within 30 seconds.

The UTRAN ANR feature [37] is a bit different from the feature in LTE. Its OAM requirements [31] are essentially the same, but the UE is configured with logging criteria and the identified cells are included in a log that is reported in bulk. The logging criteria are similar to HO reporting criteria. The UE indicates that a UTRAN ANR log is available when either a configurable number of items have been gathered or a configurable timer has expired. The mechanism to retrieve the log from the UE is detailed in [35].

The RNC and/or the OAM system can also gather information from detected cell reports, which indicate missing neighbor relations. If the detected cell can be uniquely identified within the RNS, then a corresponding neighbor relation can be established.

6.4.2   Automatic cell identity management

Non-unique PCIs are assigned to the cells as part of the self-configuration phase. Ideally, these non-unique PCIs should be locally unique to let candidate cells be identifiable only by their PCIs. However, it is difficult to predict potential neighboring cell relations before deployment, which means that local uniqueness may be violated when BSs become operational. Fig. 6.9 illustrates two typical violations.

- *PCI confusion*, where one cell has more than one neighboring cell with the same PCI. This means that the reported PCI does not uniquely identify candidate cells. In Fig. 6.9, the macrocell experiences PCI confusion for both PCI 3 and PCI 4.

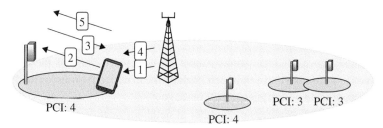

PCI: 4   PCI: 4   PCI: 3   PCI: 3

Figure 6.9 PCI confusions and collisions. Confusions can be detected via the UE ANR procedure, while collisions are difficult to detect.

- *PCI collision*, where overlapping cells are assigned the same PCI, which means that a UE served by one of the cells will neither discover nor report the other overlapping cell. In Fig. 6.9, the two picocells with PCI 3 experience PCI collision.

E-UTRAN

PCI confusions can be resolved via the UE ANR procedure as illustrated in Fig. 6.9, where a UE served by the leftmost picocell discovers the macrocell and requests the UE to report the macrocell ECGI. This information is used to set up X2 between the leftmost pico eNB and the macro eNB. When X2 is established, the macro eNB will be aware from the served cell information that the pico eNB is serving a cell with the same PCI as one of its existing neighboring relations. An alternative is that the cell subject to the PCI confusion (the macrocell in the example) suspects PCI confusion due to frequent HO failures to the picocell and therefore orders UEs to also report the ECGI of the cell with the suspected PCI. In the case of confusion, different ECGIs will be reported at different times. Thereby, the PCI confusion is detected [32, 33].

According to [2, 13], the OAM system is notified about the detected PCI confusion, and can initiate a centralized PCI reassignment mechanism, which proposes a new PCI to the picocell based on the neighboring relation information in the OAM system. Alternatively, the OAM system may provide the pico eNB with a set of available PCIs to select from, and authorize the pico eNB to distributedly select an alternative PCI. The pico eNB uses the served cell information provided by eNBs serving its neighboring cells to solve the detected PCI confusion. Then, the pico eNB removes from the set of available PCIs the PCIs that have been

- reported by an embedded receiver in the LPN (e.g., network monitoring mode),
- reported by served UEs,
- assigned to neighboring cells and their neighboring cells (i.e., neighbors of neighbors), as signaled over X2AP from neighboring eNBs.

As a result, the set of remaining PCIs is locally confusion free and thus safe to use. After the selected confusion-free set has been adopted, the pico eNB informs its neighbors about the change via an X2AP eNB Configuration Update message. This PCI reassignment means that none of the neighboring cells of the pico eNB (including the macrocell) will experience PCI confusion after the PCI reassignment [2, 33] provided that it is possible for the eNB to determine a confusion-free PCI. If not, the eNB notifies the OAM system [13].

As stated above, PCI collisions cannot be detected by any of the colliding cells. However, it is possible that a third cell (usually the macrocell) observes the PCI collision as a PCI confusion and triggers a PCI reassignment.

In HCNs, with the increasing number of neighbor relations in the network, the PCI assignment problem with the objective of locally unique PCI assignments might be intractable. However, most LPNs only have macrocells as neighbors and do not become neighbors to other LPNs. Such few to many neighbor relations significantly help the assignment tractability, and locally unique assignments should be available even in worst case deployments with up to 75 LPNs per macrocell [26, 38].

UTRAN

PSC confusion can be detected in a similar manner via the UTRAN ANR mechanism. The compiled logs from different UEs will reflect more than one cell with a specific PSC but with different UTRAN Cell Identities. If the PSC in confusion is associated to two cells controlled by the same RNC, then the RNC can determine the PSC confusion based on HO procedure information. UEs from the two picocells in confusion will make HOs to the macrocell, or at least will report the macrocell as a detected cell. This will enable the RNC to detect the PSC confusion.

6.4.3   Random access optimization

The random access procedure is for UEs to establish uplink time synchronization with the network and to notify their presence to the network. In LTE, the random access procedure [39] is performed for any of the following five events:

(1) initial access of an idle UE;
(2) re-establishment after *radio link failure* (RLF);
(3) HO to a different cell;
(4) downlink data transmission to a UE, which is out of time-synchronization;
(5) uplink data transmission from an out-of-synch UE.

At events 3, 4 and 5, the serving BS can reserve resources and avoid ambiguities to facilitate the random access procedure (i.e., non-contention based procedure). However, in the general case, the possibility of a collision between different UEs' access attempts needs to be handled through a contention-based procedure. The UE is made aware, either via broadcast or dedicated signaling, of the random access procedure, which is divided into four steps as illustrated in Fig. 6.10.

In the first step, the UE selects a *random access preamble* and a *resource unit*, with is either selected randomly or assigned by the serving BS. There are two main requirements on the sequence comprising the random access preamble: good auto-correlation properties to allow precise arrival time estimation and low cross-correlation properties with other preambles to suppress interference from other UEs. A sequence

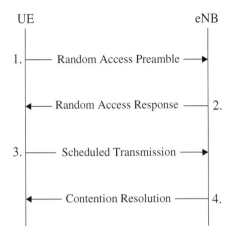

Figure 6.10 The four steps of the random access procedure in LTE.

that has ideal such properties is the Zadoff–Chu sequence (also known as the root sequence) [40, 41]. Multiple orthogonal preamble sequences can be derived from one Zadoff–Chu sequence by cyclically shifting the sequence. Each cell is assigned 64 preambles [40]. For small cells with radius of up to 1.5 km, all 64 preambles can be derived from a single root sequence, which are therefore mutually orthogonal. In larger cells, not all preambles can be derived from a single root sequence, and multiple root sequences must be allocated to a cell. Although preambles derived from different root sequences are not orthogonal to one another, the cross-correlation is low.

Root sequence assignments can be based on neighboring cell relation information, since it is advisable to assign different root sequences to neighboring cells to avoid random access attempts being overheard and interpreted as true access attempts by neighboring cells. Root sequence assignments can also be based on historical and current uplink data, as well as random access traffic observations in the BS, which can be differentiated into the different random access event types listed above.

When a UE initiates a random access procedure, it should retrieve the *transmission format* (e.g., single or repeated preamble, short or long cyclic prefix) and the available random access resource unit configurations. The *preamble transmit power* is determined by estimating the downlink path loss PL from the downlink reference signal (or pilot signal) and using parameters such as the desired received power (P_{0_RACH}), the power ramping step (Δ_{RACH}) considered for each retransmission attempt and the format-based offset ($\Delta_{Preamble}$). The latter is equal to 0 dB for formats with a single preamble and -3 dB for formats with two preambles. UEs also monitor the number of preamble transmission attempts (m). Then, the preamble transmit power is set as follows:

$$P_{RACH} = \min(P_{max}, P_{0\_RACH} - PL + (m - 1)\Delta_{RACH} + \Delta_{Preamble}), \qquad (6.1)$$

and the selected preamble is sent with the determined format and transmit power in the selected resource unit.

In the second step, the UE receives a random access response. In case of a non-contention-based procedure the procedure is completed, while in case of a contention-based access the third step (i.e., the UE sends a UE-specific identity) and the fourth step (i.e., the network echoes the identity) are needed to resolve ambiguities. For further details, see [2, 42]. If the UE fails to complete the procedure, the network may instruct the UE to consider a random *backoff time* to even out access congestion peaks.

In LTE, the following *random access channel* (RACH) parameters are considered as candidates for self-optimization [2, 39].

- Root sequence assignment to each cell, from which random access preambles are generated.
- Resource allocation between random access and uplink data.
- Preamble split between preambles used for non-contention- and contention-based access. The former preambles are reserved by the network for specific UEs, while the latter preambles form a preamble set from which UEs can randomly select their preambles.
- Backoff parameter value, which dictates the backoff time randomization.
- Transmit power control parameters in (6.1).

The RACH optimization target [43] can be set in terms of the following.

- *Access probability*. This is the probability of a UE having completed access after a certain number of random access attempts. The target can be set with respect to the access probability AP_m at attempt m, which is the probability that the UE has access after attempt m. For example, $AP_1 = 0.5$ and $AP_3 = 0.99$.
- *Access delay (AD) probability*. This is the probability distribution of the access delay experienced by accessing UEs. The access delay is defined as the time duration for a random access procedure to complete once it is initiated by a UE. The target is then formulated as *cumulative distribution function* (CDF) requirements in percentiles, $CDF_{AD}(AD_{tgt}) = P(AD \leq AD_{tgt})$. For example, a target $CDF_{AD}(60\,ms) = 0.90$ means that 90% of the access delays need to be 60 ms or less in order to meet the target.

In order to assist access probability and delay estimation [44], the UE can be instructed to provide a *RACH report*, which includes the number of preamble transmission attempts needed before the random access procedure successfully completes, and whether or not the procedure has been subject to a contention resolution failure. The RACH report can be provided to the serving eNB in a *UE information response* when triggered by a *UE information request* [25] after a completed random access procedure. Release 11 also adds a more detailed report for connection establishment failures related to the random access procedure. This includes the number of preambles sent, whether contention resolution was successful, and whether the UE used maximum power during random access transmissions.

The random access preamble assignment problem is essentially a classical reuse problem, except that different root sequences have slightly different properties. In the case of LPNs, the service area is typically so small that one root sequence is sufficient to generate all preambles. Furthermore, if a range of PCIs is reserved for LPNs, then each reserved PCI can be associated with a root sequence to reduce the two assignment problems into one. This is always the case in UTRAN, since the random access preambles are derived based on the physical cell identity. Another implication of the LPN small service area is that transmission formats with one preamble and a short cyclic prefix should be sufficient.

Another example of RACH self-optimization is the sharing of resources between random access and uplink data, as well as between non-contention-based and contention-based random access [45, 46], so that resource allocations are automatically adjusted to timely meet UE demands.

Yet another example of RACH self-optimization is power control parameter tuning [44, 47] to compensate for uplink interference variations, while maintaining an acceptable access probability or access delay. Automatic power control parameter adjustments are not supported in UTRAN, since a comparable RACH report is not available from the UEs.

6.4.4 Mobility robustness optimization

Mobility is a key feature in terrestrial cellular networks, and its robustness is crucial. As a result, *mobility robustness optimization* (MRO) is a key SON feature. MRO requirements [43] for intra-LTE mobility are usually specified in terms of acceptable *handover failure* (HOF) rates. Moreover, MRO should always consider the minimization of the number of *unnecessary HOs*, provided that the HOF rate targets configured by the operator are satisfied. The LTE MRO function can be located either in the DM/EM or in the eNB. The corresponding requirements can also be formulated for HOs within any RAT or between any two RATs.

E-UTRAN
In LTE, HOs are UE assisted, which means that the UE is configured by its serving BS to send a *measurement report* (MR) when a reporting criterion is met. One typical criterion for intra-frequency measurements may be based on the Event A3 condition [25], which can be slightly simplified as

$$\gamma_i^{(dB)} + \lambda_i^{(dB)} > \gamma_s^{(dB)} + \lambda_s^{(dB)} + \Delta, \tag{6.2}$$

where the subscripts s and i indicate the serving cell and the candidate cell, respectively, Δ is the HO margin in dB, γ is the reported triggering quantity, which can be either *reference signal received power* (RSRP) or *reference signal received quality* (RSRQ), and λ is the cell-specific offset. The MR is sent when the criterion has been fulfilled during a configurable *time-to-trigger* (TTT).

When the UE sends an MR, the source eNB may initiate the HO procedure to the target eNB serving the reported cell. If the target eNB is able to accept the UE, it may specify random access information to the source eNB. At this stage, the target cell is considered to be *prepared* for HO. The source eNB forwards the random access information to the UE via an HO command. A successful random access enables the UE to send an HO complete message to the target eNB. The basic HO steps are illustrated in Fig. 6.11, and more details are provided in Chapter 9.

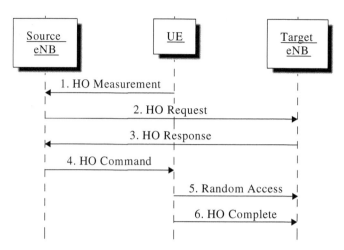

Figure 6.11 An example of the LTE HO procedure with X2 support in brief.

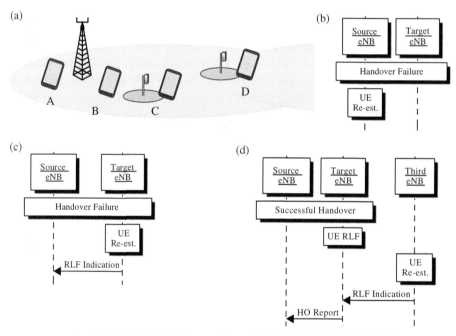

Figure 6.12 HOF categories and associated mechanisms to identify the HOF cause: (b) too early HO, (c) too late HO, and (d) HO to the wrong cell.

MRO is mainly about avoiding HOFs by adjusting HO parameters, including the MR triggering criteria. As shown in Fig. 6.12, it is instructive to divide HOFs into the following categories [25].

- **Too early HO**. The RLF occurs while the UE is served by the source cell before the HO is initiated or during HO, and the UE attempts to re-establish the connection in the source cell via an RRC Connection Re-establishment message. One example is UE B, which fails handing over to the small cell and re-establishes connection in the macrocell.

- **Too late HO**. The failure occurs while the UE is served by the source cell before the HO is initiated or during HO, and the UE attempts to re-establish the connection in the target cell or in a cell different from the source cell. A typical example is UE C, which is subject to an RLF to its serving LPN before the HO to the macrocell is completed.
- **HO to wrong cell**. The failure occurs after a successful HO from a source cell to a target cell, and the UE attempts to re-establish the connection in a third cell. One example is UE D, which successfully completes the HO to a small cell, but shortly after is subjected to an RLF, and re-establishes connection in a different macrocell other than the original macrocell.

These failure cases are identified as critical, and therefore dedicated mechanisms are designed to enable the determination of HOF root cause. Figs. 6.12(b)–(d) provide signaling charts that describe these mechanisms.

One key component in HOF analysis is the information [25] provided by the UE when it re-establishes the connection with the RAN after a failure. This includes information about its previous serving cell and the UE identity while in that cell (i.e., the *cell radio network temporary identifier* (C-RNTI)). In addition, the UE may indicate in any of the RRC connection management messages that it has RLF information available. The RLF information can be retrieved by the serving BS via a *UE information request*. In the RLF report in the *UE information response*, the UE may include cell identities of its former serving cell as well as the re-establishment cell. Moreover, the UE may include information about radio signal strength and quality measurements of the serving and neighboring cells (same or different frequency carrier and RAT) taken right before the RLF [25, 48], and location information on an availability basis. Release 11 also adds that the UE can report the time elapsed between the failure time and the reporting time. Moreover, the UE can report the failure cause as either insufficient downlink signal strength, uplink link problems, or random access problems. It can also report radio link failures when being handed over from a UTRAN cell to an E-UTRAN cell, possibly re-establishing in a different UTRAN cell. Similar to the RLF, information can also be reported if a previous RRC connection establishment procedure has failed.

Fig. 6.12(b) illustrates the detection of too early HOs, where the HO procedure from a cell served by a source eNB to a cell served by a target eNB has failed, and the UE re-establishes connection in the source cell. This situation can be detected via the re-establishment information from the UE as described above. Since the UE provides information about the previous connection and the RRC UE context is stored in the source cell, the UE can be associated with its previous RRC UE context. A too early HO could have been prevented if the HO criterion to the target cell were configured differently: for example, a higher HO margin or neighboring cell-specific offset λ_i in (6.2), a longer TTT, or a combination of those two.

The too early HO can also be between RATs, but then the UE may restablish the connection in the target RAT after the failure. In order to inform the source RAT about the failed inter-RAT HO, Rel. 11 SON enhancements introduces the possibility to send a *failure event report* via RIM SON transfer including also the RLF report [2, 15].

The detection of too late HOs is similar to the detection of too early HO. As shown in Fig. 6.12(c), the HO procedure from a cell served by a source eNB to a cell served by a target eNB has failed, and the UE re-establishes connection in the target cell. if the target cell has been prepared, the UE information indicating the source cell identity can be used to associate the UE with its RRC UE content in the target cell. The target eNB informs the source eNB about the RLF via an *RLF Indication* over X2 [14], and the target eNB may include the information in the RLF report from the UE. A too late HO could also be caused by a coverage hole at the cell border between two cells. A coverage hole can be identified through the radio measurements taken by the UE prior to the failure [48], where the UE reports insufficient radio signal quality with respect to all measured cells. A too late HO could have been prevented if the HO criterion to the target cell were configured differently: for example, a lower HO margin or neighboring cell-specific offset λ_i in (6.2), a shorter TTT, or a combination of those two.

HO to wrong cell is slightly more complicated to disclose. As shown in Fig. 6.12(d), the UE has successfully completed the HO from a source eNB, e.g., a macro eNB, to a target eNB, e.g., an LPN. However, right after the completed HO, the UE is subjected to an RLF, and the UE re-establishes connection in a cell served by a third eNB. The UE provides the re-establishment information and possibly the RLF report to the third eNB, which triggers an RLF indication over X2 to the target eNB. The target cell then can provide the source eNB with information about the failure, subsequent to the completed HO, via an *HO report* over X2 [14]. An HO to the wrong cell could be prevented if the first target cell-specific offset λ_i in (6.2) were higher, or the third cell-specific offset were lower, to favor HOs towards the third cell rather than the first target cell. This is also related to the UE velocity, especially in the case when the target eNB is an LPN, since high-velocity UEs may quickly pass through small coverage areas of LPNs without benefitting from their service.

There is also an ambition to avoid unnecessary HOs, such as the following.

- *HO oscillations*, where the serving cell changes back and forth between cell pairs. These can be detected by observing time-stamped entires of the *last visited cell information* in the *UE history information* that follows the UE from one serving cell to another [49].
- *HO with short stay*, where a UE only briefly stays at a cell after HO before moving on to the next cell. Short-stay HOs can be detected from the *UE history information*. Information about detected short-stay events can be sent to the source cell.
- *Unnecessary inter-RAT HO*, where an HO is triggered to avoid poor E-UTRAN coverage, but the E-UTRAN coverage remains acceptable after the HO. It can be detected in the target RAT (e.g., UTRAN or GERAN) by configuring inter-RAT measurements of the source E-UTRAN cell. If the coverage of the source E-UTRAN cell is evaluated to be acceptable, then the inter-RAT HO is considered unnecessary [2]. Information about the unnecessary HO is signaled as an HO report via RIM, using the *SON transfer* mechanism [15].

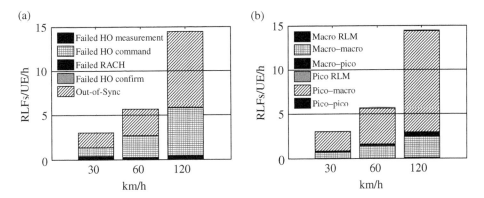

Figure 6.13 Handover failure issues.

It is not reasonable to adjust the HO criteria based on isolated events. Instead, HO failure statistics may be gathered over a time window, and then processed and analyzed in the eNB to determine whether HO parameter changes should be considered.

The HO procedure is complicated to adjust if it is based on quite many parameters. One way to reduce complexity is to introduce dependencies between parameters, e.g., between TTT and HO margin, which are key parameters in the MR triggering [50–52]. An alternative is to let some parameters remain fixed, while adjusting others. In [53], the authors show robustness improvements in realistic mobility scenarios when using the latter approach.

The mix of macrocell BSs and LPNs implies some new challenges. More specifically, the HO failure causes can be quite different compared to the macrocell-only scenario [54]. Fig. 6.13[1] provides failure statistics for various situations and signaling steps in HCNs. We can see that handouts from LPN to macrocells are relatively error prone. The same observation is made in [55], where different HO parameters are considered for different cell transitions (e.g., macro–macro, macro–LPN, LPN–LPN) to avoid too late HOs. Similar results can also be found in [56], where the focus is on HO behavior when UEs drive through LPN coverage areas causing HOs to wrong cells.

SON enhancements in Rel. 11 include the optional possibility to associate HOs with a mobility identity. Subsequent RLF indications and HO reports refer to the mobility identity if used, which means that different UE groups, for example classified by UE velocity, capability, services, etc., can be configured with the same HO parameters and mobility identity, and HO issues can then be associated with the specific UE group to enable MRO separately for each group. The same can also be achieved by associating the C-RNTIs with a specific group at the source cell.

Mobility into CSG cells is handled differently from mobility to non-CSG cells, as described in Chapter 4. It is thus associated with some specific failure cases involving, e.g., the UE capability to acquire the CSG system information and to accurately assess its proximity to CSG cells to which it is authorized.

---

[1] The author acknowledges Stefan Wager for providing the HCN HO simulation evaluations.

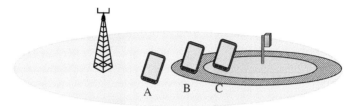

Figure 6.14 Cell range expansion of LPNs is a form of mobility load balancing.

Since MRO formulated in 3GPP is intended to be driven by failures related to mobility, it is important that the considered failures are actually due to mobility [57]. Other reasons for failure may include admission reject in the target eNB, load balancing from the source eNB to the target eNB, actions to prepare a cell for node maintenance, etc. Such failures should not be confused with the failures due to mobility.

UTRAN

There are no requirements specified for UTRAN MRO to date. However, it can be noted that, for intra-RNS mobility, the serving RNC is capable of detecting too early/late HO and HO to wrong cell, where similar approaches to those in E-UTRAN can be considered. Moreover, mobility in UTRAN is also a bit different from that in E-UTRAN, since some UTRAN physical channels such as dedicated channels are subjected to soft/softer HOs, while others are only provided by one cell. This will impact some MRO aspects and requirements.

6.4.5   Mobility load balancing

The load balancing mechanism in 3GPP is essentially formulated as a control problem with requirements to avoid cell overload [43]. The objective of load balancing is to manage uneven traffic distributions, while minimizing the number of needed HOs and redirections. The targets can be specified in terms of typical cell overload indications, connection establishment failure rate, radio access bearer setup failure rate, etc. The same targets as specified for MRO can also be considered to avoid jeopardizing mobility robustness. The mobility load balancing function is located in the eNB.

One concern in HCNs is that LPNs may attract too little traffic, so that macrocell offloading strategies are needed. One such technique is *cell range expansion*, where a *range expansion bias* (REB) is considered between macrocells and LPNs for some or all UEs, as illustrated in Fig. 6.14. The adjustment of the REB can be seen as mobility load balancing. Such adjustments need to consider the capabilities of UEs, which are different for different 3GPP releases. An LTE Rel. 8/9 UE shall support detection of a non-serving cell that is 6 dB weaker than the serving cell, while an LTE Rel. 10 UE with support for time domain *intercell interference coordination* (ICIC) shall detect a non-serving cell that is 7.5 dB weaker than the serving cell. A UE capable of interference cancellation shall be able to detect up to 9 dB weaker non-serving cells; see Chapter 7 and [58].

Load balancing algorithms also need some load information from neighboring cells. In 3GPP, eNBs can share *resource status* information through X2 interfaces [14]. Resource status information includes the following.

- **Hardware load indication**, which is a crude measure of the hardware resource utilization, quantized into low, mid, high, or overload.
- **Transport network layer load indication**, which describes the transport load situation, quantized into low, mid, high, or overload.
- **Radio resource status**, which indicates the percentage of scheduled resources with respect to the available resources, and is divided into guaranteed and non-guaranteed bit rate usage.

Some examples of load balancing algorithms can be found in [59, 60], which consider sharing based on radio resource utilization. An alternative is to share throughput information among nodes, and use throughput statistics for load balancing [61]. The load balancing actions can be considered for all UEs, or only for a selected set of them, e.g., only for UEs with specific services, resource requirements, etc. (see also Chapter 9).

There are no requirements specified for intra-UTRAN load balancing to date. Similar to the UTRAN MRO case, the controlling RNC is capable of observing the load situation in intra-RNC cells, and considers adequate countermeasures to avoid cell overload. The cell load can be monitored over Iub [16], via cell-specific common measurement reporting from the NodeB. The cell-specific common MR includes transmission power measurements, both total transmission power and transmission power excluding channels used for *High Speed Downlink Packet Access* (HSDPA) and the Enhanced Uplink, and received power levels, both total received power and received power associated with the Enhanced Uplink.

For inter-RAT and interfrequency load balancing, the situation can be alleviated by considering idle mode or connected mode actions. In the former, only the cell selection procedure of idle UEs is affected, and once connected the UEs stay with the selected RAT/frequency. In the latter, connected UEs can also be subjected to load sharing actions. Inter-RAT load information can be shared transparently over the *core network* (CN) via RIM and SON transfer signaling [15]. It enables transfer of load information on demand, or via event-triggering when the cell load has changed more than configured, between E-UTRAN, UTRAN, and GERAN. The load information is relative to a maximum load. It is possible to configure a cell-specific capacity and indicate overload.

6.4.6 Transmission power tuning

The transmission power levels of pilots, synchronization signals and broadcast information relate to the cell coverage and impact the cell selection mechanisms. Moreover, the transmission power levels of downlink data and control symbols are related to the reception performance. Section 6.4.5 describes load balancing mechanisms, where UEs are assigned to cells not corresponding to the most favorable downlink radio links. An alternative is to adjust the pilot power level of the LPN and thereby increase or decrease its coverage, as illustrated in Fig. 6.15. A typical example is a possibly CSG LPN

Figure 6.15 LPN pilot power tuning.

deployed indoors. It is desirable to limit its interactions with outdoor UEs, in particular if they are unauthorized to use the LPN for services.

One indication of such interactions is excessive HO events, together with possible admission rejects if the LPN is a CSG cell. A too-high pilot power leads to triggering of HO events outdoors, and a too-low pilot power corresponds to plenty of HO events indoors. The number of HO events can be minimized if the pilot power is matching the desired LPN coverage [62], as shown in Fig. 6.15. Similarly, if authorized accesses dominate the total number of access attempts, then the pilot power can be increased, while if the number of unauthorized accesses increases, then the pilot power should be decreased [63].

A more refined analysis can be based on the downlink performance of LPN UEs, characterized by radio channel and interference conditions reported by UEs served by the LPN. The pilot power can be adjusted to retract UEs with unfavorable radio conditions to surrounding cells, while attracting UEs with adequate conditions. One example is to base these transmission power adjustments on UE measurements such that the received power at an intended cell radius is equal between an external interferer and the cell itself [64]. Thereby, a reasonable cell border performance is obtained.

Transmission power adjustments can also be considered for macrocells. For example, when one carrier is used by the macrocell at a reduced power level to protect LPNs from downlink interference, the macrocell can use other carriers at full power. This is an attractive alternative to completely reserving one carrier for LPN usage. What constitutes an adequate power reduction of the protected carrier depends on the deployment. The LPNs can signal the interference information reported by their served UEs to the macro BS. Basically, the LPN indicates whether the observed interference power is acceptable or not, e.g., via the existing *relative narrowband transmit power* (RNTP) signaling in the Load information message over X2 with some modifications [14, 65].

A similar issue concerns time-domain ICIC in terms of *almost blank subframes* (ABSs) with reduced power; see Chapter 7. The reduced power level of ABSs can be self-optimized to ensure sufficient protection for LPNs during the protected subframes. An alternative is to blank data and control symbols during protected subframes, where the level of protection is controlled by adjusting the number of protected subframes.

The three strategies, to adjust the total carrier power, to adjust the power of protected subframes, and to consider blank data and control symbols during protected subframes, have been compared through simulations [66]. The conclusion is that power adjustments of either the entire carrier or the protected subframes provide additional benefits, given that the power adjustment is implemented properly.

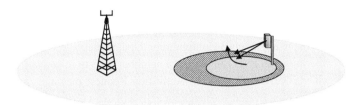

Figure 6.16 Coverage and capacity optimization based on adjustments of the antenna tilt.

6.4.7   Coverage and capacity optimization

While most self-optimization functions described above are considered for implementation in BSs, there are centralized or hybrid aspects of some of them. *Coverage and capacity optimization* (CCO) concerns aspects that involve multiple cells, and thus is more suitable for centralized and/or hybrid SON operations.

The 3GPP requirements on CCO focus on providing adequate coverage and capacity, as well as tradeoffs by reconfiguring antenna parameters and BS transmission powers [6, 43, 67]. The latter can also be addressed in a distributed fashion, as discussed in Section 6.4.6. Furthermore, the scope of CCO contains spectrum management as well. CCO can also benefit from network observations that will be discussed in Section 6.6.

LPN antenna tilt optimization has a comparable impact to transmit power tuning, which has been described in Section 6.4.6. Basically, LPN uptilt increases the LPN coverage (similar to increasing transmit power), and LPN downtilt decreases the LPN coverage (similar to decreasing transmit power), as illustrated in Fig. 6.16. The possibility of benefiting from LPN antenna tilting depends on the deployment, as well as the LPN capability. It is presently most common that LPNs have fixed antennas, which cannot be adjusted.

In general, antenna tilt optimization is typically considered at a slow time scale, while interference coordination and load balancing are considered at a faster time scale. It has been concluded that interference coordination and load balancing are complementary to each other [61]. Load balancing affects intercell interference, which needs to be managed either via distributed ICIC or via a centralized CCO mechanism, such as antenna tilting.

Another configuration option is the selection of operational carrier. For *Home NodeBs* (HNBs) and HeNBs, it is possible to let them select an operational carrier from a set of carriers proposed by the OAM system [68, 69]. Femtocells can take advantage of the femtocell listening mode to observe the interference situation on different carriers, and select the carrier with the least interference. For macro-, micro-, and picocells, the operational carrier is selected by the OAM system, and the carrier frequency is part of the ARCF. Note that carrier selection is not time critical, since changing a carrier is a complicated procedure that involves a cell restart. In HCNs, the OAM system can first determine which macrocell carrier(s) should be protected, and thereafter assign carriers to LPNs. If the macrocell and LPNs use the same carrier, they need to agree on certain interference protection via distributed inter-node signaling.

Some multi-cell mechanisms relating to coverage and capacity can be implemented in the BSs as well. One example is energy saving mechanisms, to which there is signaling support between base stations as well as RATs. If an eNB in LTE has switched off a particular cell in order to lower the energy consumption, it will notify neighboring eNBs via X2AP [14] using an eNB Configuration Update message with a Deactivation indication. Furthermore, an eNB can request a neighbor eNB to re-activate a previously switched-off cell via X2AP using a Cell Activation Request message. As a response when cells have been re-activated, the neighbor eNB will use a Cell Activation Response message. Rel. 11 has introduced some inter-RAT support via RIM [15], where it is possible to transfer cell activation/deactivation information between RATs via SON transfer messages using Energy Saving Indication. For E-UTRAN cells, it is also possible to request activation via RIM. For further discussions concerning energy efficiency, see Chapter 14.

6.5   Self-healing

When introducing a large number of LPNs, which are less complex and sophisticated than macrocell BSs, there is an obvious risk that the cost of fault management may grow proportionally with the number of LPNs. One objective of SON is to scale down the cost of fault management, without increasing the complexity significantly. SON features should preferably be implemented where they are most effective in terms of performance and operational costs. Self-healing functions can be implemented in the OAM system, the BSs, or both.

Self-healing aims to reduce the manual efforts for fault detection, compensation, and recovery, as well as hardware repair and replacement. Faults can be of very different kinds, e.g., associated with the transport network, BS hardware, software, antennas, and communication protocols such as OAM, S1, and X2.

The use cases of self-healing can be summarized as follows.

- **Cell outage detection**, which detects cells not carrying traffic or not responding to any request or communication.
- **Cell outage recovery**, which restarts cell modules, e.g., by replacing or repairing faulty BS hardware to mitigate the faults and errors.
- **Cell outage compensation**, which compensates cell outages, e.g., by involving neighboring cells when recovery actions have failed.
- **Return from cell compensation**, which resumes the state prior to cell outage with a smooth transition between the faulty and compensating cells.

In 3GPP, the self-healing procedure [70] is based on one or more *Triggering Conditions of Self-Healing* (TCoSHs), which are required to be monitored continuously by the self-healing function. When a TCoSH is met, an appropriate self-healing process is triggered. The self-healing process gathers necessary information such as measurements and results from dedicated tests. Based on the gathered information, the self-healing process analyses the situation and proposes recovery actions if needed. Before initiating any action, current configuration information is backed up and the OAM system is notified about the proposed recovery actions. Relevant information is also logged during recovery.

Examples of cell outage compensation include the reconfiguration of a set of neighboring cells to extend their coverage to alleviate outage situations. In the downlink, the coverage of neighboring cells is automatically extended if the faulty cell has discontinued its downlink transmission, and thereby no induced interference. This technique is further explored in [7, 71]. In the uplink, outage can be mitigated by reducing the uplink power of UEs in surrounding cells to limit the uplink interference in the compensating cells. The limited uplink interference means that the uplink coverage of those cells is extended.

LPNs may be deployed to improve coverage or capacity. In the latter case, coverage may have been provided even without the LPN, which means that an outage of an LPN mainly affects the capacity but not the coverage of the network. One indication of whether a cell is a coverage cell or a capacity cell is the *covered by* relation indication, which describes to what extent a cell is covered by a neighboring cell [72]. The establishment of the relation and to what extent a cell is covered by a neighboring cell can be based on the observed data that will be described in Section 6.6. The main objective of those relations is to support energy saving actions, such as disabling capacity cells when they are not required, but the relations can also be used for self-healing purposes.

6.6 Performance monitoring

Network performance monitoring aspects related to HCN mainly concern (1) increasing network, traffic and channel knowledge to support network expansion through extensive data, possibly geo-localized, and (2) decreasing measurements to prevent excessive OAM signaling to avoid performance data scaling linearly with the number of deployed LPNs. These two aspects are addressed in the following subsections.

6.6.1 Minimization of drive tests

Operators rely on planning tools to some extent. Traditionally, refined information about the actual radio network performance is obtained through drive tests. However, drive tests have the drawback of being costly and time consuming, and are typically limited to roads and not UEs' most common locations. One attractive alternative is to use UEs as probes. In 3GPP, this alternative is referred to as *minimization of drive tests* (MDT) [73], which was introduced in Rel. 10 with enhancements in Rel. 11.

MDT is based on the 3GPP trace functionality [74], and enables the operator to configure and initiate trace logging either towards a specific UE or a particular cell or area. The trace information is forwarded to the *Trace Collection Entity* (TCE). UE-specific activation is handled through the core network using a signaling-based procedure, while cell-based activation is triggered through the OAM system using a management-based procedure [15]. One advantage with UE-specific trace logging initiation is that the UE capability can be considered when selecting UEs for MDT. Moreover, there are two modes of operation, either *Logged MDT* or *Immediate MDT*, with the main difference being that the former is for idle UEs, while the latter is for connected UEs.

In the case of logged MDT, the UE is configured to measure and store radio signal strength and quality measurements, cell and area identities, and location information if available. This concerns E-UTRAN, UTRAN, GERAN, and CDMA2000 cells. When the UE connects to the network, it indicates that it has logged MDT available. The network can fetch the logged MDT information using the *UE information request* message, and receive the information in return through the *UE information response*. The UE maintains the MDT configuration when it connects to the network, as well as during mobility. When the UE returns to idle state, the logging resumes provided that the UE is still served by one of the configured cells or resides in one of the configured tracking areas. Similarly, the UE can log information about RLFs including radio signal strength and quality [48] in an RLF report, as discussed in Section 6.4.4.

Rel. 11 enhancements include acessibility enhancements [73], where the UE logs failed RRC connection establishments for LTE and UMTS. For each connection establishment failure, the UE stores time stamps, the global cell identity of the cell the UE failed to access, radio measurements, and location information if available. In addition, in the case of LTE, the UE stores information similar to the RACH report in Section 6.4.3 including the number of transmitted RA preambles and whether contention was detected, but also an indication whether the maximum transmission power was used. In the case of UMTS, the access information that can be provided is similar.

In the case of immediate MDT, the UE can be configured with measurements to be performed involving reporting triggers and criteria as utilized for *radio resource management* (RRM). In addition, there are measurements performed in the base station. E-UTRAN measurements include the following.

- **M1** RSRP and RSRQ measurements by the UE.
- **M2** *power headroom* (PH) measurements by the UE.
- **M3** received interference power measurements by the eNB.
- **M4** data volume measurements for downlink and uplink separately per *quality of service* (QoS) class by the eNB.
- **M5** scheduled IP throughput for downlink and uplink separately per *Radio Access Bearer* (RAB) and per UE by the eNB.

Measurements M1 and M2 are supported in Rel. 10, while Rel. 11 MDT enhancements introduce M3, M4, and M5. UTRAN measurements are similar. The PH is the power difference between the UE's maximum uplink transmit power and the uplink transmit power that the UE would have used if there were no uplink transmit power limit. This means that PH can be positive or negative, and the latter indicates that the UE is uplink transmit power limited.

MDT information can also be geo-localized, either via GNSS information provided by the UE if available, or based on various network-based localization mechanisms [73, 75], either by forwarding raw localization information to TCE, or by computing a location estimate in the eNB and forwarding that to the TCE. The Rel. 11 enhancements include that the eNB can request the UE to attempt to make GNSS information available, as well as including GNSS estimation accuracy information if available. Furthermore, the UE can be requested to report round-trip time information (RX-TX time difference information) to support network-based information.

The information from geo-localized active UEs can be used to generate spatial information about the radio network, which constitutes valuable information for analysis and network expansion plans [6]. One use case is to automatically identify coverage holes and traffic hotspots to support LPN rollout planning.

6.6.2 Heterogeneous cellular network monitoring

When HCNs are deployed, it is likely that many aspects concerning the variety of node types are literally heterogeneous. Thus, the interest in monitoring the nodes and the monitoring scope itself can be very different [76], depending on the time of the day and week, traffic volume, node type, node importance, etc.

The main objective of HCN monitoring is to avoid the data volume not scaling linearly with the number of LPNs at all times in all situations. Therefore, it is important to enable configuration of the management scope for each node. Different criteria can be adopted to disable the traditional monitoring scope, where all information is provided, and allow us to retrieve information only when requested from the OAM system. The information can be performance measurements, alarms, notifications, etc.

The monitoring scope can also depend on the node importance. Traditionally, all nodes are considered equally important from a monitoring perspective. This can be questioned considering the variety of node types in an HCN. The node importance may directly depend on the node type or the cell size, but there could be many other reasons.

The reduction of management data can be through data filtering or aggregation. For example, some alarms are not presented, and some measurements are not presented as often as they are made available in the node but instead once every hour, day, or upon request. In addition, some counters associated to a specific cell are normally handled separately per cell relation, which may lead to extensive amount of management data, particularly if the few cell relations are rarely used. The management data may be aggregated across all cell relations, and summarized in one counter per performance indicator for the entire cell instead of one per cell relation. Furthermore, data associated with each cell served by a BS may be aggregated to represent the entire BS.

6.7 Summary and conclusions

The SON paradigm is providing means to facilitate network and service management, while supporting radio network robustness and optimized coverage and capacity. SON is especially important to HCNs due to the increasing number of nodes that need to be managed. SON enables smooth planning and introduction of LPNs into an existing macrocell deployment. Continuous self-optimization and self-healing ensure that the LPNs are able to provide the intended capacity without requiring significant management and monitoring efforts. Smart nodes, either macro BSs or LPNs, with SON features controlled from intuitive system-wide policies, enable efficient overall OAM.

In order to completely support business objectives, SON needs to span all RATs. Since there is no single ideal SON function location, SON should be implemented where it is most efficient considering both high and low level requirements and aspects. Different deployment options have been described in relation to the 3GPP management architecture. SON functionalities have been divided into self-configuration, self-optimization, and self-healing. The associated 3GPP SON feature requirements and signaling support have been described together with 3GPP specification references. Moreover, key SON features, such as ANR, identity management, MRO, *mobility load balancing* (MLB), and CCO, have been described in more detail with references to the literature.

References

[1] 3GPP, *Evolved Universal Terrestrial Radio Access Network (E-UTRAN); Self-Configuring and Self-Optimizing Network (SON) Use Cases and Solutions* (TS 36.902).

[2] 3GPP, *Evolved Universal Terrestrial Radio Access (E-UTRA) and Evolved Universal Terrestrial Radio Access Network (E-UTRAN); Overall Description; Stage 2* (TS 36.300).

[3] 3GPP, *Telecommunication Management; Self-Organizing Networks (SON) Policy Network Resource Model (NRM) Integration Reference Point (IRP); Requirements* (TS 32.521).

[4] Next Generation Mobile Networks (NGMN), *Recommendation on SON and O&M Requirements* (2008).

[5] C. Prehofer and C. Bettstetter, Self-organization in communication networks: principles and design paradigms. *IEEE Communication Magazine*, **43**:7 (2005), 78–85.

[6] J. Ramiro and K. Hamied, *Self-Organizing Networks (SON): Self-Planning, Self-Optimization and Self-Healing for GSM, UMTS and LTE* (Wiley, 2011).

[7] S. Hämäläinen, H. Sanneck and C. Sartori, *LTE Self-Organising Networks (SON): Network Management Automation for Operational Efficiency* (Wiley, 2011).

[8] J. Belschner, P. Arnold, H. Eckhardt, E. Kuhn, E. Patouni, A. Kousaridas, N. Alonistioti, A. Saatsakis, K. Tsagkaris and P. Demestichas, Optimisation of radio access network operation introducing self-x functions: use cases, algorithms, expected efficiency gains. In *Proceedings of the IEEE Vehicular Technology Conference (VTC) Spring* (2009), pp. 1–5.

[9] J. van den Berg, R. Litjens, A. Eisenbltter, M. Amirijoo, O. Linnell, C. Blondia, T. Kuerner, N. Scully, J. Oszmianski and L. Schmelz, Self-organisation in future mobile communication networks. In *ICT – Mobile Summit* (2008).

[10] J. Zhang and G. De la Roche, *Femtocells: Technologies and Deployment* (Wiley, 2010).

[11] 3GPP, *Telecommunication Management; Architecture* (TS 32.102).

[12] 3GPP, *Telecommunication Management; Integration Reference Point (IRP) Concept and Definitions* (TS 32.150).

[13] 3GPP, *Telecommunication Management; Self-Organizing Networks (SON); Concepts and Requirements* (TS 32.500).

[14] 3GPP, *Evolved Universal Terrestrial Radio Access (E-UTRA); X2 Application Protocol (X2AP)* (TS 36.423).

[15] 3GPP, *Evolved Universal Terrestrial Radio Access Network (E-UTRAN); S1 Application Protocol (S1AP)* (TS 36.413).

[16] 3GPP, *UTRAN Iub Interface Node B Application Part (NBAP) Signaling* (TS 25.433).

[17] 3GPP, *UTRAN Iur Interface Radio Network Subsystem Application Part (RNSAP) Signaling* (TS 25.423).

[18] 3GPP, *UTRAN Iu Interface Radio Access Network Application Part (RANAP) Signaling* (TS 25.413).

[19] 3GPP, *Telecommunication Management; Self-configuration of Network Elements; Concepts and Requirements* (TS 32.501).

[20] U. Patel and B. Gohil, Cell identity assignment techniques in cellular network: a review. In *Proceedings of the IEEE International Conference on Computer Science and Information Technology (ICCSIT)* (2010), vol. 2, pp. 594–596.

[21] S. Kourtis, Code planning strategy for UMTS-FDD networks. In *Proceedings of the IEEE Vehicular Technology Conference (VTC) Spring* (2000), pp. 815–819.

[22] R. Joyce, T. Griparis, G. Conroy, B. Graves and I. Osborne, A novel code planning approach for a WCDMA network. In *Proceedings of the International Conference on 3G Mobile Communication Technologies* (2003), pp. 31–36.

[23] T. Bandh, G. Carle, H. Sanneck, L. Schmelz, R. Romeikat and B. Bauer, Optimized network configuration parameter assignment based on graph coloring. In *Proceedings of the IEEE Network Operations and Management Symposium (NOMS)* (2010), pp. 40–47.

[24] H. Olofsson, S. Magnusson and M. Almgren, A concept for dynamic neighbor cell list planning in a cellular system. In *Proceedings of the IEEE International Symposium on Personal, Indoor and Mobile Radio Communications* (1996), vol. 1, pp. 138–142.

[25] 3GPP, *Evolved Universal Terrestrial Radio Access (E-UTRA); Radio Resource Control (RRC); Protocol Specification* (TS 36.331).

[26] O. Teyeb, G. Mildh and A. Furuskär, Physical cell identity assignment in heterogeneous networks. In *Proceedings of the IEEE Vehicular Technology Conference (VTC) Fall* (2012).

[27] T. Bandh and H. Sanneck, Automatic site identification and hardware-to-site-mapping for base station self-configuration. In *Proceedings of the International Workshop on SON, IEEE Vehicular Technology Conference (VTC) Spring* (2011), pp. 1–5.

[28] 3GPP, *Domain Name System Procedures; Stage 3* (TS 29.303).

[29] 3GPP, *Numbering, Addressing and Identification* (TS 23.003).

[30] 3GPP, *Telecommunication Management; Performance Management (PM); Performance Measurements Evolved Universal Terrestrial Radio Access Network (E-UTRAN)* (TS 32.425).

[31] 3GPP, *Telecommunication Management; Automatic Neighbour Relation (ANR) Management; Concepts and Requirements* (TS 32.511).

[32] M. Amirijoo, P. Frenger, F. Gunnarsson, H. Kallin, J. Moe and K. Zetterberg, Neighbor cell relation list and measured cell identity management in LTE. In *Proceedings of the IEEE Network Operations and Management Symposium* (2008), pp. 152–159.

[33] M. Amirijoo, P. Frenger, F. Gunnarsson, H. Kallin, J. Moe and K. Zetterberg, Neighbor cell relation list and physical cell identity self-organization in LTE. In *Proceedings of the IEEE International Conference on Communications, Workshops* (2008), pp. 37–41.

[34] A. Dahlen, A. Johansson, F. Gunnarsson, J. Moe, T. Rimhagen and H. Kallin, Evaluations of LTE automatic neighbor relations. In *Proceedings of the International Workshop on SON, IEEE Vehicular Technology Conference (VTC) Spring* (2011), pp. 1–5.

[35] 3GPP, *UTRAN Radio Resource Control (RRC) Protocol Specification* (TS 25.331).

[36] 3GPP, *UTRAN Requirements for Support of Radio Resource Management (FDD)* (TS 25.133).

[37] 3GPP, *Automatic Neighbor Relation (ANR) for UTRAN; Stage 2* (TS 25.484).

[38] J. Lim and D. Hong, Management of neighbor cell lists and physical cell identifiers in self-organizing heterogeneous networks. *Journal of Communications and Networks*, **13**:4 (2011), 367–376.

[39] 3GPP, *Evolved Universal Terrestrial Radio Access (E-UTRA); Medium Access Control (MAC) Protocol Specification* (TS 36.321).

[40] 3GPP, *Evolved Universal Terrestrial Radio Access (E-UTRA); Physical Channels and Modulation* (TS 36.211).

[41] B. Popovic, Generalized chirp-like polyphase sequences with optimum correlation properties. *IEEE Transactions on Information Theory*, **38**:4 (1992), 1406–1409.

[42] E. Dahlman, S. Parkvall and J. Sköld, *4G LTE/LTE-Advanced for Mobile Broadband* (Academic, 2011).

[43] 3GPP, *Telecommunication Management; Self-Organizing Networks (SON) Policy Network Resource Model (NRM) Integration Reference Point (IRP); Information Service (IS)* (TS 32.522).

[44] M. Amirijoo, P. Frenger, F. Gunnarsson, J. Moe and K. Zetterberg, On self-optimization of the random access procedure in 3G long term evolution. In *Proceedings of the IEEE Integrated Network Management (IM) Workshops* (2009), pp. 177–184.

[45] S. Choi, W. Lee, D. Kim, K.-J. Park, S. Choi and K.-Y. Han, Automatic configuration of random access channel parameters in LTE systems. In *Proceedings of the IFIP Wireless Days* (2011), pp. 1–6.

[46] O. Yilmaz, J. Hamalainen and S. Hamalainen, Self-optimization of random access channel in 3GPP LTE. In *Proceedings of the International Wireless Communications and Mobile Computing Conference* (2011), pp. 1397–1401.

[47] M. Amirijoo, P. Frenger, F. Gunnarsson, J. Moe and K. Zetterberg, Towards random access channel self-tuning in LTE. In *Proceedings of the IEEE Vehicular Technology Conference (VTC) Spring* (2009), pp. 1–5.

[48] J. Puttonen, J. Turkka, O. Alanen and J. Kurjenniemi, Coverage optimization for minimization of drive tests in LTE with extended RLF reporting. In *Proceedings of the IEEE International Symposium on Personal Indoor and Mobile Radio Communications* (2010), pp. 1764–1768.

[49] P. Bergman, J. Moe and F. Gunnarsson, Self-optimizing handover oscillation mitigation algorithms and field evaluations. In *Proceedings of the International Workshop on SON, IEEE International Symposium on Wireless Communication Systems (ISWCS)* (2012), pp. 1–5.

[50] T. Jansen, I. Balan, J. Turk, I. Moerman and T. Kuerner, Handover parameter optimization in LTE self-organizing networks. In *Proceedings of the IEEE Vehicular Technology Conference (VTC) Fall* (2010), pp. 1–5.

[51] T. Jansen, I. Balan, S. Stefanski, I. Moerman and T. Kurner, Weighted performance based handover parameter optimization in LTE. In *Proceedings of the International Workshop on SON, IEEE Vehicular Technology Conference (VTC) Spring* (2011), pp. 1–5.

[52] I. Balan, T. Jansen, B. Sas, I. Moerman and T. Kuerner, Enhanced weighted performance based handover optimization in LTE. In *Proceedings of the Future Network Mobile Summit* (2011), pp. 1–8.

[53] J. Rodriguez, I. de la Bandera, P. Munoz and R. Barco, Load balancing in a realistic urban scenario for LTE networks. In *Proceedings of the International Workshop on SON, IEEE Vehicular Technology Conference (VTC) Spring* (2011), pp. 1–5.

[54] ST-Ericsson, *RRC Re-Establishment in Hetnet HO Failure R2-122533* (Prague, 2012).

[55] H.-D. Bae, B. Ryu and N.-H. Park, Analysis of handover failures in LTE femtocell systems. In *Proceeding of the Australasian Telecommunication Networks and Applications Conference* (2011), pp. 1–5.

[56] D. Rose, T. Jansen and T. Kurner, Modeling of femto cells – simulation of interference and handovers in LTE networks. In *Proceedings of the IEEE Vehicular Technology Conference (VTC) Spring* (2011), pp. 1–5.

[57] ST-Ericsson, *Intra LTE Mobility and Non-Mobility Related Failures (R3-120725)* (Cabo San Lucas, Mexico, 2012).

[58] 3GPP, *Evolved Universal Terrestrial Radio Access (E-UTRA); Requirements for Support of Radio Resource Management* (TS 36.133).

[59] Q.-T. Nguyen-Vuong, N. Agoulmine and Y. Ghamri-Doudane, Novel approach for load balancing in heterogeneous wireless packet networks. In *Proceedings of the IEEE Network Operations and Management Symposium* (2008), pp. 26–31.

[60] A. Lobinger, S. Stefanski, T. Jansen and I. Balan, Load balancing in downlink LTE self-optimizing networks. In *Proceedings of the IEEE Vehicular Technology Conference (VTC) Spring* (2010), pp. 1–5.

[61] S. Klein, I. Karla and E. Kuehn, Potential of INTRA-LTE, intra-frequency load balancing. In *Proceedings of the IEEE Vehicular Technology Conference* (2011), pp. 1–5.

[62] H. Claussen, L. Ho and L. Samuel, Self-optimization of coverage for femtocell deployments. In *Proceedings of the Wireless Telecommunications Symposium* (2008), pp. 278–285.

[63] S. Jang, Y. Lee, J. Lim and D. Hong, Self-optimization of single femto-cell coverage using handover events in LTE systems. In *Proceedings of the Asia-Pacific Conference on Communications* (2011), pp. 28–32.

[64] H. Claussen, L. T. W. Ho and L. G. Samuel, An overview of the femtocell concept. *Bell Labs Technical Journal*, **13**:1 (2008), 221–245.

[65] Qualcomm (rapporteur), *DL Interference Solutions for Carrier Based Hetnet (Email Discussion Summary) R3-120483* (Cabo San Lucas, Mexico, 2011).

[66] R. Combes, Z. Altman, M. Haddad and E. Altman, Self-optimizing strategies for interference coordination in OFDMA networks. In *Proceedings of the International Conference on Communications (ICC) Workshop* (2011), pp. 1–5.

[67] Ericsson, *Enhanced NM Centralized Coverage and Capacity Optimization S5-121246* (Sarajevo, Bosnia and Herzegovina, 2012).

[68] F. Bernardo, R. Agusti, J. Cordero and C. Crespo, Self-optimization of spectrum assignment and transmission power in OFDMA femtocells. In *Proceedings of the Advanced International Conference on Telecommunications* (2010), pp. 404–409.

[69] X. Chu, Y. Wu, D. Lopez-Perez and X. Tao, On providing downlink services in collocated spectrum-sharing macro and femto networks, *IEEE Transactions on Wireless Communications*, **10**:12 (2011), 4306–4315.

[70] 3GPP, *Telecommunication Management; Self-Organizing Networks (SON); Self-healing Concepts and Requirements* (TS 32.541).

[71] M. Amirijoo, L. Jorguseski, R. Litjens and L. Schmelz, Cell outage compensation in LTE networks: algorithms and performance assessment. In *Proceedings of the International Workshop on SON, IEEE Vehicular Technology Conference (VTC) Spring* (2011), pp. 1–5.

[72] 3GPP, *Telecommunication Management; Evolved Universal Terrestrial Radio Access Network (E-UTRAN) Network Resource Model (NRM) Integration Reference Point (IRP); Information Service (IS)* (TS 32.762).

[73] 3GPP, *Radio Measurement Collection for Minimization of Drive Tests (MDT); Overall Description; Stage 2* (TS 37.320).

[74] 3GPP, *Telecommunication Management; Subscriber and Equipment Trace; Trace Control and Configuration Management* (TS 32.422).

[75] A. Kangas, I. Siomina and T. Wigren, Positioning in LTE. In *Handbook of Position Location: Theory, Practice, and Advances*, ed. S. A. Zekavat and R. M. Buehrer (Wiley, 2011), pp. 1081–1127. http://dx.doi.org/10.1002/9781118104750.ch32

[76] 3GPP, *Telecommunication Management; Study on Management of Evolved Universal Terrestrial Radio Access Network (E-UTRAN) and Evolved Packet Core (EPC)* (TS 32.816).

7 Dynamic interference management

Ismail Güvenç, Fredrik Gunnarsson and David López-Pérez

As discussed in Chapter 1, *heterogeneous cellular networks* (HCNs) with *low-power nodes* (LPNs) are important for improving the capacity and coverage of next generation broadband wireless communication systems. However, interference problems in HCNs pose an important challenge, and thus efficient interference management techniques are required to fully benefit from their deployments. The main contribution of this chapter is to review interference problems and interference management techniques for HCNs. A general notion of *macrocell base stations* (MBSs) and LPNs is adopted, but the simulations are based on *Long Term Evolution* (LTE) scenarios with macro eNBs and pico eNBs or femto HeNBs. More specifically, cell-selection and interference coordination methods are discussed, including mechanisms recently proposed in the *3rd Generation Partnership Project* (3GPP) LTE, and their performances are evaluated through system-level simulations. In such simulations, LTE-specific notation is used, and *macrocell user equipment* (MUE) and *picocell user equipment* (PUE) denotes UEs served by macro eNBs and pico eNBs, respectively.

This chapter is organized as follows. First, Section 7.1 reviews the main reasons for excessive intercell interference in HCNs. In Section 7.2, due to its significance, *range expansion* (RE) for HCNs is treated in more detail. Some example simulation results that demonstrate the *downlink* (DL)/*uplink* (UL) coverage imbalance in heterogeneous deployments are also provided. Section 7.3 gives a high-level overview of *intercell interference coordination* (ICIC) methods that are applicable to HCNs, and the next three sections are dedicated to specific ICIC approaches: frequency-domain, power-based, and time-domain ICIC techniques are discussed in Section 7.4, Section 7.5, and Section 7.6, respectively. In Section 7.7, some example performance results are provided, and finally Section 7.8 concludes the chapter.

7.1 Excessive intercell interference

The massive deployment of low-power cells overlaying macrocells will create a large number of new cell boundaries, in which *user equipments* (UEs) will suffer from strong intercell interference, thus degrading the performance of the entire network. In addition to the large number of newly created cell boundaries, the interference problem in HCNs is especially challenging due to UEs not always being served by the most favorable *base station* (BS), radio-propagation-wise, in DL or UL. This is due to the transmit

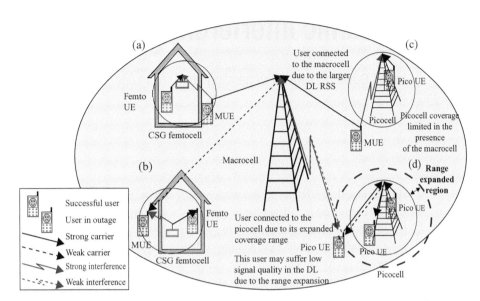

Figure 7.1 Dominant DL and UL cross-tier interference scenarios in HCNs: (a) macrocell UE jamming the UL of a femtocell; (b) femtocell jamming the DL of a macro user; (c) macrocell UE jamming the UL of a nearby picocell; (d) range-expanded picocell (mitigates UL interference from macrocell to picocells) (reproduced with permission from IEEE©).

power differences between the MBS and the LPN, and the implementation of RE and *closed subscriber group* (CSG) access for the LPNs. These scenarios are illustrated in Fig. 7.1, and discussed further in the following subsections.

7.1.1   Transmission power difference between nodes

A common strategy for serving cell selection in wireless networks is based on DL pilot *reference signal strength* (RSS) or *reference signal quality* (RSQ), evaluated at the UE and reported to the network. Network mechanisms (handover mechanisms) then usually have the objective that the most favorable cell in the DL shall serve the UE. This does not always mean that the UE is served by the most favorable cell in the UL, resulting in *UL/DL coverage imbalance*. For example, different BSs may have receiver chains with different sensitivity and low-noise amplifier configuration, and may be subject to different UL interference levels [1, 2].

However, the main reason for UL/DL coverage imbalance in HCNs is the transmission power difference between MBSs and LPNs. This implies that the UE connected to an MBS may have more favorable path gain to/from a non-serving LPN than to/from the serving MBS, since the selection is evaluated based on received pilot RSS or RSQ, and not on path loss.

Moreover, due to the DL-based cell selection procedure, UEs connected to an MBS may cause severe UL interference to LPNs in their vicinity due to lower path loss. Fig. 7.1(c) illustrates an MUE, served by a macrocell, which induces significant UL interference to a nearby picocell. Note also that, due to lower path loss to the picocell, this MUE might use much less UL transmit power if it would be connected to the picocell.

7.1.2 Low-power node range expansion

One drawback in HCN deployments with serving cell selections based on DL pilot RSS is that a UE served by an MBS may have more favorable UL to a non-serving LPN. As described in the previous section, such aggressor UEs may induce critical UL interference to the LPNs that become victim base stations. Another drawback is that LPNs may attract very little traffic and thereby macrocell traffic may not be offloaded as desired. In order to address these issues, an artificially biased serving cell selection can be considered, where the DL pilot is evaluated considering a bias in favor of the LPN to increase its DL coverage footprint (Fig. 7.1(d)). This is sometimes referred to as RE or as *cell range expansion* (CRE), the bias as *range expansion bias* (REB) or *cell selection offset* (CSO), and the coverage extension region is referred to as the *RE region*.

RE can thus be used to make the serving cell selection more UL relevant, and the UEs in the RE region have the most favorable uplink to their serving LPN. In contrast, these UEs have a more favorable DL from the non-serving aggressor MBS, thus creating DL interference issues for those victim UEs (Fig. 7.1(d)). Due to its significance, RE and its impact on HCNs will be discussed in more detail in the following section.

7.1.3 Closed subscriber group access

The fact that some cells may operate in CSG mode (e.g., restricted access cells) implies that nonsubscribers cannot always connect to the BS with the most favorable DL pilot RSS. This leads to significant cross-tier interference problems [3, 4]. Fig. 7.1 depicts a challenging scenario for ICIC, in which different non-subscribers walk nearby houses hosting a CSG *femtocell access point* (FAP). In the UL, non-subscriber (a) is an aggressor UE and uses a high transmit power to compensate for the path losses to its distant serving macrocell, thus inducing interference to the UL of the nearby CSG cell, which becomes a victim base station (Fig. 7.1(a)). In the DL, a CSG FAP can be an aggressor base station that interferes with the DL reception of non-subscriber (b) also connected to a distant serving macrocell, thus becoming a victim UE (Fig. 7.1(b)).

7.2 Range expansion

In the 3GPP, RE has been recently investigated for increasing the DL coverage footprint of LPN by adding a positive cell individual offset to the DL pilot RSS of the LPNs during the serving cell-selection procedure [5–11]. With a larger REB, more UEs are offloaded from the MBS to LPNs, at the cost of increased co-channel DL interference for range-expanded UEs. In these standard contributions, while RE is shown to degrade the throughput of the overall network, it is also shown to improve the sum capacity of MUEs due to offloading.

The idea of a biased serving cell selection is not new. It has been discussed for a long time as a means to share and balance the load between adjacent cells. For example, [12] discusses a handover criterion for *Global System for Mobile Communications* (GSM) networks with cell individual offsets that in turn are functions of the respective cell load. A similar approach is presented in [13], where the handover margin between two adjacent cells is adjusted based on cell loads.

In addition, due to the interest from standard organizations for investigating the merits of RE, there are also several recent RE-related works available in the academic literature. In [14], closed form analytical expressions of outage probability with RE in HCNs have been provided, which verifies that RE, without ICIC, degrades the outage probability of the overall network. In [15], joint cell-selection and scheduling for picocells and femtocells have been discussed for the DL. However, UL aspects of RE have not been covered. The DL/UL imbalance problem and tradeoffs for cell selection have been presented and investigated in [16–18]. Moreover, in [19–23], DL performance with different REB values has also been studied for picocell deployments, and resource partitioning is proposed to combat the detrimental effects of increased co-channel DL interference for range-expanded PUEs.

7.2.1   Definition of range expansion

The serving cell reselection mechanism (handover) is typically supported by event-triggered measurement reporting from the UE to its serving cell. The triggering condition is based upon DL pilot measurements[1] $\gamma_i^{(dB)}$, possibly considering a configurable *cell individual offset* (CIO) $\lambda_i^{(dB)}$ for each cell i. If the triggering condition is configured to trigger a report when a candidate cell is evaluated as stronger than the serving cell considering the CIO, and the serving cell is changed accordingly (handover), then the serving cell (re)selection can be described as

$$\hat{i} = \arg \max_{i \in \mathcal{C}} \left\{ \gamma_i^{(dB)} + \lambda_i^{(dB)} \right\}, \tag{7.1}$$

where $\mathcal{C}$ is the set of cell indices that include the currently serving cell as well as the candidate cells.

With RE, a UE adds a positive REB to the DL pilot measurements from an LPN, such that

$$\begin{cases} \lambda_i^{(dB)} = 0, & \text{if } i \text{ is a macrocell,} \\ \lambda_i^{(dB)} > 0, & \text{if } i \text{ is a picocell.} \end{cases} \tag{7.2}$$

Moreover, the REB may be UE specific, for example reflecting the UE capabilities and the transmit power difference between the MBS and the LPN, as well as the benefits of offloading a particular UE. Typical values of REB $\lambda_i^{(dB)}$ may range from few dB to

---

[1] Such as *reference signal received power* (RSRP) and *reference signal received quality* (RSRQ) in LTE, and pilot RSS and *signal to interference plus noise ratio* (SINR) in more general terms.

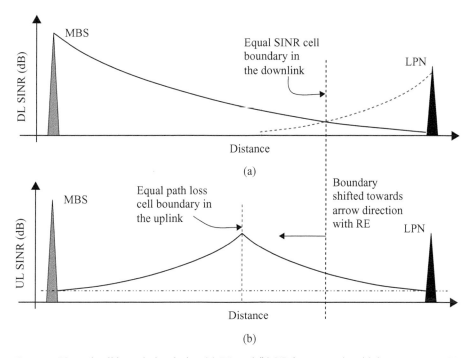

Figure 7.2 Natural cell boundaries during (a) DL and (b) UL in a scenario with heterogeneous DL transmission powers (reproduced with permission from IEEE©).

values of over 10 dB. Fig. 7.2 illustrates that RE, as in (7.1), increases the DL coverage footprint of LPNs, and thus offloads UEs from the MBSs to the LPNs.

In LTE, the report triggering condition for intrafrequency measurements may be based on the Event A3 condition [24], which is triggered when

$$\gamma_i^{(dB)} + \lambda_i^{(dB)} > \gamma_s^{(dB)} + \lambda_s^{(dB)} + \Delta, \tag{7.3}$$

where the subscript "s" indicates the serving cell, "i" indicates candidate cells and Δ is the handover margin in dB. The actual report is triggered when the criterion has been fulfilled during a configurable *time to trigger*. A zero handover margin and handover decisions directly based on the triggered A3 events will imply a cell reselection according to (7.1).

7.2.2 Downlink/uplink coverage imbalance

The natural cell boundaries between LPNs and MBSs are *different in DL and UL* as opposed to those between MBSs in a homogeneous network deployment. Fig. 7.2(a) shows that the DL SINRs observed from an MBS and an LPN are equivalent at a location that is closer to the LPN than to the MBS, which defines the natural DL cell boundary. In contrast, in the UL, since the UE transmits with the same transmit power to all cells, the location of the natural UL cell boundary is where the path losses to the MBS and the LPN are equivalent.

In practice, with traditional deployments, it is typically not feasible to associate a UE with different cells in the DL and the UL,[2] and the UE serving cell selection is based on DL measurements. Moreover, fractional UL power control in LTE is based on path loss compensation [25, 26]:

$$P = \min \left\{ P_{\mathrm{max}}, P_0 + 10 \log_{10} M + \alpha L_{\mathrm{DL}} \right\} \qquad (7.4)$$

where P_{max} is the maximum UE transmit power, P_0 is a UE-specific (optionally cell-specific) target received power, M is the number of *physical resource blocks* (PRBs) assigned to the UE, α is the cell-specific pathloss compensation factor (equal to 1 for full compensation), and L_{DL} the estimated DL path loss. In addition, there can be additional corrections based on decoding performance [25, 27].

Hence, the UL/DL imbalance may cause interference problems in the UL, since a UE that has a lower path loss with an LPN may be forced to connect (due to the large transmit power difference) to the umbrella MBS, where the path loss is higher. This means that the UE (aggressor UE) has to transmit with a high transmit power, which causes high UL interference to the nearby LPN (victim base station). In contrast, with RE, UEs in similar situations may be offloaded to the LPN, where the UL transmit powers of UEs (and hence, the interference level present in the network) will be reduced due to decreased path loss with the LPN. Therefore, another important merit of RE is that, due to UL power control, RE reduces the total UL interference observed in the network. The drawback is that these offloaded UEs become victim UEs and observe strong DL interference from the MBS as the aggressor base station.

7.2.3   Behavior of range expansion

In order to demonstrate the different tradeoffs associated with RE in the DL and the UL, system-level simulation results are presented in this section. Note that ICIC is not implemented for these particular simulations. Simulation parameters are mostly based on [28], and are specified by the 3GPP for evaluating picocell scenarios (see also Chapter 3).

A simulation scenario with seven neighboring eNBs is considered, where each eNB is further divided into three sectors. Two co-channel pico eNBs are positioned uniformly at the sector edge for all eNB sectors, and the system throughput for the center eNB is calculated. Note that the eNB of interest observes interference from all co-channel macrocells and picocells. Fractional UL power control is applied at the UEs, where a UE's UL transmit power is given by (7.4).

Results in Fig. 7.3(a) show that the UL sum throughputs of both macrocells and picocells are improved with increasing REB values. The reason for the improvement is that, with increasing REB values, the picocell coverage area in Fig. 7.2(a) is shifted closer to its ideal UL cell boundary in Fig. 7.2(b). Since UEs get connected to closer nodes, they decrease their UL transmit powers, and thus cause lower UL interference to other nodes. Moreover, low-SINR UEs that were originally connected to the macrocell are

---

[2] However, in more coordinated deployments, this can be feasible. Systems with soft handover can also be seen as enabling this to some extent.

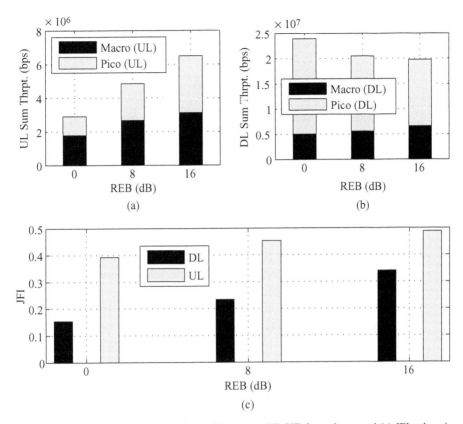

Figure 7.3 (a) Average DL UE throughput, (b) average UL UE throughput, and (c) JFI values in the DL/UL with different range expansion bias values (reproduced with permission from IEEE©).

now offloaded to the cell-edge picocells, which further improves the aggregate macrocell UL throughput.

In contrast, the DL sum throughputs in Fig. 7.3(b) decrease with increasing REB values. While RE improves the MUE DL throughput due to offloading of low-SINR MUEs to picocells, a dominant degradation in the aggregate PUE DL throughput is observed for large REB values. This is because range-expanded PUEs observe low SINRs; based on the cell-selection procedure in (7.1), the SINRs of range-expanded PUEs are always unfavorable.

Finally, *Jain's fairness index* (JFI) values, which jointly consider all MUEs and PUEs, are illustrated in Fig. 7.3(c) for different REB values, where the JFI is defined as

$$\mathrm{JFI}(R_1, R_2, \ldots, R_{N_{\mathrm{tot}}}) = \frac{\left(\sum R_j\right)^2}{N_{\mathrm{tot}} \sum R_j^2}, \tag{7.5}$$

with j denoting the UE index, R_j denoting the throughput of UE j, and N_{tot} denoting the total number of UEs in the system. Fig. 7.3(c) shows that JFI increases with REB values for both the UL and the DL, since larger numbers of UEs are able to benefit from the underutilized spectrum resources of *pico evolved NodeBs* (PeNBs). Moreover, due to implementation of fractional UL power control, throughputs are more fairly distributed among UEs in the UL, when compared with throughputs in the DL.

7.3   Intercell interference coordination

In *orthogonal frequency division multiple access* (OFDMA) systems, intracell interference is under strict control, and may even be avoided altogether. Exceptions to this include multi-user *multiple input multiple output* (MIMO), where several UEs share the same time–frequency resources. Therefore, the challenge is the intercell interference. As discussed in [29], intercell interference can be seen as a collision between *resource blocks* (RBs), and the collision probability is related to the negative impact from interference on the overall system capacity.

The intercell interference challenges summarized in Section 7.1 may significantly degrade the overall HCN performance, thus making the need for ICIC schemes obvious to guarantee its proper operation. These ICIC schemes need to address coordination of intercell interference in both control and data channels.

Information messages may be exchanged between BSs to facilitate coordination. For example, in 3GPP LTE, an *evolved NodeB* (eNB) may be connected to neighboring eNBs, through an X2 interface [30], where they can exchange load and resource utilization information messages.

Another source of information is UE reports about the perceived DL situation, as well as the observed UL challenges on the transmission side. For example, it can be very valuable if the UEs are able to sense, detect and report information to their serving cells, concerning potential interfering cells being present in their vicinity. In this way, the UE's serving cell in collaboration with the potential interfering cells may coordinate their resource allocations in terms of frequency, power, and time to mitigate intercell interference. However, it is also quite likely that UEs will have difficulties detecting LPNs transmitting at significantly lower power levels compared with MBS.

The intercell interference coordination schemes can be divided into the following categories [31, 32]:

- time-domain techniques,
- frequency-domain techniques,
- power-based techniques,
- antenna-based techniques.

While the three first approaches will be reviewed in more detail in the upcoming sections, antenna-based techniques will be tackled in more detail in Chapter 11.

7.4   Frequency-domain intercell interference coordination

Traditionally, *frequency domain* (FD) ICIC has been implemented as channel allocation mechanisms. For example, 2G systems such as GSM had co-channel reuse, which means that the same frequency channel is not used in every cell. This can be seen as static ICIC. It has also been identified that, in order to properly manage traffic variation, it is reasonable to dynamically assign frequency channels. The benefits of such dynamic management include reduced call drop and block rate, and improved system capacity [33–35].

7.4.1 Frequency-domain intercell interference coordination in LTE

In LTE, FD ICIC was discussed at an early stage, and X2 interface support for information exchange between eNBs is included in 3GPP specifications [29, 30] as the following information elements of the Load Information message.

- *Relative narrowband transmit power* (RNTP): The cell-specific RNTP indicates per RB whether the intended transmission power in the considered future time interval for non-reference symbols is below a threshold *RNTP_thres*. Both the RNTP indicator bitmap and the threshold can be communicated as part of X2 load information. By construction, the indicator can both be used to indicate frequency selective power allocations as well as scheduling strategies, since both will have an impact on the average transmit power per RB in the considered future time interval.
- *UL overload indicator* (OI): The cell-specific OI states per RB the measured average interference plus thermal noise power, quantized into high, medium, and low interference.
- *UL high-interference indicator* (HII): Using the HII indicator, a certain cell informs a specified neighboring cell about interference sensitivity of its scheduled UEs per RB quantized into high and low sensitivity.

Note that these signalings among different eNBs are used for information purposes only, and that the transmitting side cannot mandate any action related to the scheduling decisions on the receiving side.

In relation to the typical HCN interference scenarios, an MBS can announce via RNTP which parts of the frequency spectrum it intends to use more frequently. Picocells can exploit such information to schedule their range-expanded UEs at frequency resources that are used less frequently and thus subject to less interference. For the UL, the MBS may use the HII to announce the frequency resources where it intends to schedule its cell-edge UEs – UEs that may induce critical UL interference to LPNs. Moreover, using the OI, the LPN can indicate to the MBS in which parts of the frequency spectrum it receives significant UL interference. However, neither of these address the control channel performance.

7.4.2 Carrier-based intercell interference coordination

FD ICIC in LTE is typically associated with per RB signaling, and ICIC via scheduling decisions. However, another alternative for FD ICIC is plausible if multiple carriers are available. The operational carrier of a cell is configured via the *operation, administration, and maintenance* (OAM) system, and BSs may be capable of operating multiple cells with different carriers over the same service area.

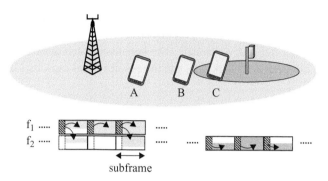

Figure 7.4 Carrier-based ICIC, where the LPN is assigned a protected carrier f_2, and the MBS aims at only scheduling UEs at the same carrier as the LPN when inter-cell interference is negligible.

One natural approach to multiple carrier operation is to move UEs from one carrier to another based on the load of each carrier in order to balance the load [36], or based on the observed radio signal strengths and quality of individual UEs in order to ensure coverage and quality of service [37]. Another option is available for UEs supporting carrier aggregation, where more than one carrier can be activated at the same time for a particular UE. One of the carriers is the primary carrier, while one or more other carriers are secondary carriers [38, 39]. Scheduling information to the UE can be transmitted on the same carrier as data is transmitted on, or can be transmitted on the primary carrier while data is transmitted on both primary and secondary carriers (cross-carrier scheduling).

The main interference challenge in HCNs is the strong interfering links to/from non-serving cells as discussed in Section 7.1. With multiple carriers, it is possible to separate the aggressors and victims to different frequency carriers, and thereby avoid interference. Fig. 7.4 illustrates a typical scenario with two carriers f_1 and f_2 assigned to an MBS, one carrier f_2 assigned to an LPN and three different UEs that are managed differently as discussed below. Moreover, the MBS may operate the carrier f_2 using other means of downlink interference reduction, such as power-based ICIC as in Section 7.5 or time-domain ICIC as in Section 7.6.

- UE A is not interfered in the downlink by and not interfering in the uplink to the LPN, and can therefore use either of, or both of, the two carriers. In order to avoid control channel interference, the MBS may use cross-carrier scheduling from carrier f_1 to blank the control channel on carrier f_2. In such a case, only the UEs capable of carrier aggregation can be scheduled by the MBS on f_2, as exemplified with subframes 1 and 3 in Fig. 7.4.
- UE B may induce UL interference to the LPN, and is therefore only scheduled on carrier f_1, e.g., the second subframe in Fig. 7.4. Alternatively, UE B may be served by the LPN via range expansion. Then, DL intercell interference may become a potential issue, but can be avoided since the LPN operates only on the protected carrier f_2, where the MBS may consider other interference reduction means as stated above.

- UE C is served by the LPN and is scheduled on the protected carrier f_2 in Fig. 7.4. If the LPN only allows access to CSG mobiles and UE C is not a member of the CSG, then it is in a similar situation to UE B, and MBS will schedule the mobile only on carrier f_1 as exemplified with subframe 2 in Fig. 7.4.

Note that the protected carrier is sometimes referred to as an *escape carrier*. Moreover, the benefits of such coordination are further explored in [40], and the different carrier-based ICIC mechanisms in relation to LTE are discussed in [41, 42].

7.4.3   Uplink interferer identification

When a BS receives a UL interference overload indication from a victim BS, it is not obvious which UE is being the aggressor UE and causing the overload. The typical situation could be an MBS receiving a UL interference overload indication from an LPN, but none of its served MUEs is able to detect the DL of the LPN due to its low transmission power.

Therefore, it is relevant to consider different methods to identify the unknown UL interferer. Some alternatives are discussed in 3GPP [42], and are summarized in the following.

- **OI from LPN to MBS and historical scheduling information in MBS:** The OI is frequency selective, and this information can be correlated to historical scheduling information in the MBS to identify the UL interferer. The accuracy depends on the interference averaging in the LPN to compile the OI, as well as how persistent the MBS scheduling is.
- **MUE and LPN location:** If an approximate location of the MUE is available and the MBS is aware of LPN locations within its serving area, then it can be possible to use such information to identify the UL interferer. Moreover, fingerprinting solutions, where the MUE reports radio and timing measurements in relation to detected BSs, could be used to obtain a proximity indication of MUEs with respect to the interfered LPN.
- **LPN tries to detect the MUE uplink:** If the MBS informs the interfered LPN about UL configurations of suspected MUEs, then the LPN can detect and identify the interferer, and report back the identification to the MBS. Upon detection, this gives a unique identification provided that the UL configuration is unique. Examples of LTE UL configurations include a random access configuration, a channel sounding configuration, a demodulation pilot etc.

7.5   Power-based intercell interference coordination

Power-based ICIC mainly aims at reducing, rather than avoiding, the intercell interference. This is essentially how *code division multiple access* (CDMA) systems traditionally handle both intracell and intercell interference, even though both time- and frequency-domain interference management are considered in evolved releases. In order to adjust the power adequately, it is important to have some kind of information about the victim receiver, and in this section this is addressed for the UL and the DL separately.

7.5.1   Uplink power-based intercell interference coordination

The UL/DL imbalance together with serving cell selection based on DL RSS evaluations may commonly lead to situations in which UEs with better UL to a nearby LPN connect to a far MBS. This may induce significant intercell interference to the LPN in the UL, as discussed in Section 7.1. One way of reducing the negative impact from this UL intercell interference problem is to make UEs served by LPNs transmit at higher powers. In this line, if the UL power control is aiming at a specific signal to interference plus noise ratio at the LPN receiver, then the interference can be compensated. A similar approach is to introduce artificial noise and/or attenuator at the LPN receiver [43], known as desensitization.

In LTE, the UL transmission power is controlled according to the fractional UL power control in (7.4). This means that the control mechanism strives to meet a desired target uplink received power P_0, which is cell specific, but UE-specific adjustments are possible. Moreover, some additional corrections consider the decoding performance, making fractional UL power control react to excessive intercell interference and gradually increase the UL transmit power. An alternative is to configure a higher cell-specific desired target uplink received power P_0, which means that the UE is using a more appropriate uplink power level already from the initiation of the connection. One such configuration is proposed in [44], where the cell-specific target UL received power P_0 is different for MBS and LPN according to

$$P_0^{\text{LPN}} = P_0^{\text{MBS}} + (P_{\text{MBS}} - P_{\text{LPN}}), \qquad (7.6)$$

where P_0^{LPN} and P_0^{MBS} are the cell-specific target UL received power P_0 for the LPN and the MBS, respectively, and P_{LPN} and P_{MBS} are the corresponding DL transmission powers, all powers given in dBm. This strategy is disclosed to lead to significant LPN cell-edge performance improvements, at the expense of some MBS cell-edge performance degradation.

One consequence of range expansion is that the remaining interfering macro UEs are further away from the LPN compared with operating without RE. This means that there is less need to compensate with uplink power control settings. Given a range expansion bias of REB dB, then the cell-specific target UL received power for LPN in (7.6) should be altered according to

$$P_0^{\text{LPN}} = P_0^{\text{MBS}} + (P_{\text{MBS}} - P_{\text{LPN}}) - \text{REB}. \qquad (7.7)$$

7.5.2   Downlink power-based intercell interference coordination

DL transmission power adjustments as a general intercell interference management mechanism is seen as inferior to other mechanisms in [45]. However, scenarios with co-channel CSG cells are different from those studied in [45], since victim UEs without access to the CSG cell, but within the CSG cell coverage area, receive excessive interference from the aggressing CSG cell. This holds true for CSG cells of any type, but the most commonly discussed example is CSG HeNBs. Therefore, CSG cells will be referred to as HeNBs in the following. These issues and some solution proposals are discussed in [40, 46], and are reviewed in the following.

While reducing the transmit power, the HeNB also reduces the total capacity of *home user equipments* (HUEs), but may significantly improve the performance of victim MUEs. Let P_{max} and P_{min} be the maximum and minimum HeNB transmit powers, respectively, P_M be the received transmit power from the strongest co-channel eNB at the HeNB, and α and β be two scalar DL power control variables. Then, the different DL power control approaches proposed at the 3GPP for HeNB can be listed as follows (all values are in dBm) [46].

1. *DL power control based on the strongest eNB received power at the HeNB:* The HeNB transmit power can be written as

$$P_{tx} = \max(\min(\alpha P_M + \beta, P_{max}), P_{min}) \,. \tag{7.8}$$

2. *DL power control based on path loss between the HeNB and the victim MUE:* The HeNB transmit power can be written as

$$P_{tx} = \mathrm{med}(P_M + P_{ofst}, P_{max}, P_{min}) \,, \tag{7.9}$$

where the transmit power offset P_{ofst} is defined by

$$P_{ofst} = \mathrm{med}(P_{ipl}, P_{ofst-max}, P_{ofst-min}) \,, \tag{7.10}$$

with P_{ipl} being a transmit power offset value that captures both the indoor path loss and the penetration loss between the HeNB and the nearest MUE, and $P_{ofst-max}$ and $P_{ofst-min}$ being the minimum and maximum values of P_{ofst}, respectively.

3. *DL power control based on an objective SINR of HUE:* In this approach, the received SINRs of HUEs should meet a target value, and thus the transmit power of its serving HeNB is reduced appropriately to achieve the HUE SINR target using the following expression:

$$P_{tx} = \max(P_{min}, \min(\widehat{PL} + P_{rec,HUE}, P_{max})) \,, \tag{7.11}$$

where

$$P_{rec,HUE} = 10 \log_{10}\left(10^{I/10} + 10^{N_0/10}\right) + \mathrm{SINR}_{tar} \,, \tag{7.12}$$

with I being the power of the interference detected by the served HUE, N_0 being the background noise power, SINR_{tar} being the target SINR of HUEs, and $\widehat{PL}$ being the path loss estimate between the HeNB and the HUE.

4. *DL power control based on an objective SINR of MUE:* In this approach, the received SINRs of MUEs should meet a target value, and thus the transmit power of the aggressing HeNB is reduced appropriately to achieve the MUE SINR target using the following expression:

$$P_{tx} = \max(\min(\alpha P_{SINR} + \beta, P_{max}), P_{min}) \,, \tag{7.13}$$

with P_{SINR} being the target SINR of HUEs, where only the nearest HeNB interference is considered.

While in the first and third schemes HeNBs are able to adjust their transmit power without any knowledge about victim MUEs or nearby eNBs/HeNBs, either through direct HeNB sensing or HUE channel quality indicators, respectively, in the second and fourth schemes HeNBs require some information from them. Therefore, the latter schemes may provide a tailored ICIC to victim MUE circumstances, which may result in enhanced MUE and HUE performance, at the expense of a more complex and expensive hardware at the HeNB.

7.6   Time-domain intercell interference coordination

The frequency-domain and power-based ICIC schemes may not be effective for the type of HCN interference scenarios presented in Fig. 7.1, especially if only one carrier is available. One example is DL control channel protection in systems where control and data symbols are time-multiplexed as they are in LTE. Therefore, it is attractive to also coordinate the interference in the time domain.

Some symbols should carry control information to allow the system to handle idle mobiles and support mobility procedures. These are mainly the common reference signal symbols (pilot), the primary and secondary synchronization symbols, and other symbols that carry system information (e.g., broadcast channel symbols). The rest of the symbols can be muted as discussed in Sections 7.6.1–7.6.2, or be transmitted with reduced power as in Section 7.6.3.

Physical downlink control channel (PDCCH) interference management has been identified as a key issue to be addressed by the time-domain ICIC in the 3GPP work item [47]. The DL control region occupies the first symbol(s) of the resource blocks in LTE [38]. Time-domain methods aim to suppress the interference to this control region and they can be classified into two main categories as follows [48].

- **Subframe alignment** where control regions are time-aligned between the eNBs and the LPNs. To protect the control channels of the MUEs, the LPN has to coordinate its transmission of control channels. More specifically, either no PDCCH is transmitted by the LPN in some coordinated subframes, or they are transmitted with lower power (lightly loaded PDCCH [48]). On the other hand, interference from reference, synchronization, and some system information symbols are still present at the MUE.
- **Orthogonal frequency division multiplexing (OFDM) symbol shift** where the victim control regions of the eNB are time-aligned with the LPN's data regions to avoid interference from synchronization and system information symbols of the LPN. To prevent the interference from the data region symbols of the LPN to the MUEs, the LPN either mutes the corresponding data symbols, or it applies a subframe blanking [48].

In Fig. 7.1, two DL intercell interference scenarios and corresponding time-domain ICIC methods are illustrated: a CSG case with a CSG femtocell in a macrocell, and an open access case with a range-expanded picocell in a macrocell. In the first case, the victim UEs are the UEs within the CSG femtocell coverage but without access authorization and therefore served by the macrocell. These MUEs are critically interfered

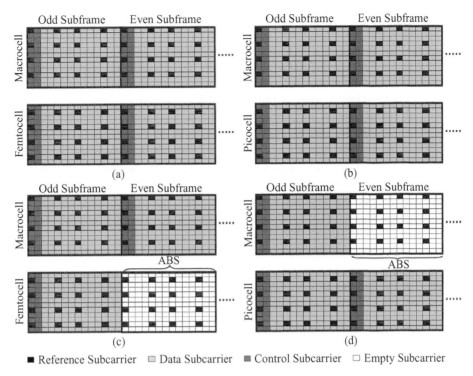

■ Reference Subcarrier ▢ Data Subcarrier ▣ Control Subcarrier ☐ Empty Subcarrier

Figure 7.5 Illustration of ABSs for time-domain ICIC for 3GPP heterogeneous networks.
(a) Macrocell and CSG femtocell subframes without any time-domain ICIC. (b) Macrocell and picocell subframes without any time-domain ICIC. (c) Macrocell and CSG femtocell subframes, with femtocell ICIC ABS during even subframes. (d) Macrocell and picocell subframes, with macrocell ICIC ABS during even subframes (reproduced with permission from IEEE©).

with by the CSG femtocells and need interference coordination from aggressor CSG femtocells. In the second case, range-expanded PUEs are victim UEs and need interference coordination from the aggressor macrocell. For simplicity, we assume that protected subframes are realized for even subframes.

When the subframes of macrocells and CSG femtocells are aligned in the time domain as in Fig. 7.5(a), their control and data regions overlap with each other. If the femtocells implement *almost blank subframe* (ABS) as in Fig. 7.5(c), then the macrocell can schedule MUEs critically interfered by femtocells to the subframes overlapping with these ABSs of femtocells.

In the case of range-expanded PUEs in picocells, such UEs may experience significant control and data region interference from the macrocell if the subframes are time-aligned as in Fig. 7.5(b). However, this interference problem may be alleviated through using ABSs at macrocells, and scheduling range-expanded PUEs within the subframes overlapping with the ABSs of macrocells (Fig. 7.5(d)).

In the second category of time-domain ICIC methods, as discussed earlier, the subframe boundary of the LPNs is shifted by a number of OFDM symbols with respect to the subframe boundary of the MBSs to prevent the overlap between their control regions [48] and thereby also to avoid interference from synch and system information symbols. Due to its larger impact in 3GPP standardization, the focus in the rest of this section is on the subframe alignment.

If the interference reduction could be even more coordinated between the nodes, then the reduction also could reflect the benefits more exactly. Such coordinated management is often referred to as *coordinated multi-point* (CoMP) reception and transmission [49]. In this chapter, such coordination is not considered.

Note also that the aggressor cells still need to transmit reference signals in the protected subframes, which occupy a limited portion of the whole subframe. These reference signals may still cause some severe interference problems in dominant interference settings [48]. Techniques to suppress such interference on the UE side are discussed in Section 7.6.3.

7.6.1   Almost blank subframes

This section describes the ABSs illustrated in Fig. 7.5 in further detail, which are used in the time-domain ICIC mechanism standardized in 3GPP Rel.10. Some signaling details are covered, between nodes, as well as between the mobile and its serving base station. An example is also provided.

In LTE, the *ABS patterns* are described by a bitmap, where '1' indicates an ABS, and '0' an ordinary subframe. Extensive inter-eNB signaling is possible in LTE via the X2 interface [30]. The *Load Information* message was described in Section 7.4.1, and it can also include the information element *ABS Information*. It contains either the ABS patterns, or an indication that ABS is inactive. A base station can request the ABS information from another base station using the Load Information message with the information element *Invoke Indication* set to *ABS Information*.

It is also possible for an aggressor cell to request the ABS status from a victim cell, to which the victim cell responds with a *Resource Status Update* message including the information element *ABS Status*. This attribute describes the percentage of the ABS pattern configured in the aggressor cell that is used by the mobiles served by the victim cell. Thereby, it is possible to get feedback related to the usefulness of the ABS pattern.

The ABS information enables the victim cell to schedule mobiles during either almost blank or ordinary subframes. To support scheduling and link adaptation in general, the base station can configure the mobiles to report channel state information. These may be averaged over some subframes, and this can be misguiding due to the ABS patterns considered by the aggressor cell. The channel state will be very different if averaged over only the ABSs, only the non-ABSs or some of both. To avoid this, the base station can configure two subframe patterns for channel state information – *MeasSubframeSet1* and *MeasSubframeSet2*. The mobile is thus requested to only consider subframes in the respective set when averaging channel state information. This means that channel state information is reported for both these sets. Victim mobiles can thereby be scheduled with link adaptation that considers channel state information feedback aggregated only over protected resources.

The benefit of ABS is not restricted to only control and data channel protection. It also enables mobiles to detect weak neighboring cells during the almost blank macro

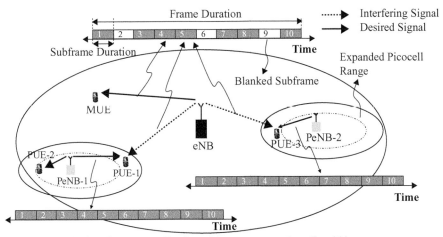

(a) Time-domain interference coordination with open access picocells within macro coverage.

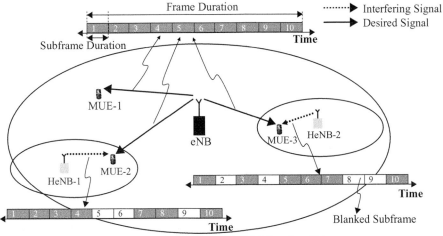

(b) Time-domain interference coordination with CSG HeNB cells within macro coverage.

Figure 7.6 Use of ABSs in heterogeneous networks for time-domain interference coordination (reproduced with permission from IEEE©).

subframes. This is possible by instructing the mobile to measure on neighbor cells only during the protected subframes. It can also be relevant to monitor the serving cell during such subframes. For flexibility, the radio resource control toolbox [24] enables configuration of one ABS pattern (*measSubframePatternNeigh*) to use when evaluating the neighboring cells, and another ABS pattern (*measSubframePatternPcell*) when monitoring the serving cell. Any kind of ABS pattern considered by the mobile is referred to as measurement resource restriction. Such separation is important in order to fully exploit the benefits of ABS [50].

Fig. 7.6 provides an example of how ABS can be configured for mitigating inter-cell interference in the HCN scenarios discussed in Fig. 7.1. In Fig. 7.6(a), ABS is configured in a macrocell to protect range-expanded PUEs. Hence, the victim UEs are PUE-1 and PUE-3. In order to protect them, eNB configures certain subframes as ABS. The subframe selection can be arbitrary, such as subframes 2, 6, and 9 in Fig. 7.6(a). Then, the picocells schedule their range-expanded UEs PUE-1 and PUE-3 in the subframes corresponding to these ABSs of the macrocell. The subframe selection could also consider behavior of the retransmission protocol in *medium access control* (MAC). In LTE it is based on eight parallel stop and wait *hybrid automatic repeat requests* (HARQs), meaning that, if reception of a data unit in a resource block at subframe S fails, then it is negatively acknowledged and retransmitted at subframe $S + 8$. This means that if subframe S is protected, also $S + 8k, k = 1, 2, \ldots$ should be protected as well [51].

In Fig. 7.6(b), a macrocell and CSG femtocell coexistence scenario is considered, and the use of ABS to mitigate intercell interference to victim MUEs such as MUE-2 and MUE-3 is demonstrated [52–54]. HeNB-1 and HeNB-2 are configured to not schedule any transmission (other than reference signals etc.) in certain subframes for allowing protection of victim MUEs. For example, in Fig. 7.6(b), subframes 5, 6, and 9 are left blank in HeNB-1, while subframes 2, 4, 8, and 9 are left blank in HeNB-2. Then, victim MUEs may be scheduled in macrocell resources overlapping with the ABSs of the HeNBs; i.e., MUE-2 may be scheduled in subframes 5, 6, and 9, while MUE-3 may be scheduled in subframes 2, 4, 8, and 9. There is no scheduling restriction for MUE-1.

The ABS pattern may be dynamically changed (as fast as every 40 ms [55]) for picocells through the X2 interface in LTE. In case of CSG femtocells, X2 support is less likely, and therefore the ABS reconfigurations cannot be that dynamic due to the absence of the X2 interface between the eNBs and the HeNBs. For example, through the management system, different ABS patterns (known both to macrocells and femtocells) may be used at different times of the day [55]. Moreover, in [52], three different solutions are proposed in order for macrocells to know the ABS pattern in femtocells. (1) A single ABS pattern is configured for all femtocells in a macrocell coverage area. However, this approach prevents the adaptation of femtocells to the traffic variations in their coverage area. (2) A configured ABS pattern is signaled in the system information of femtocells. This approach increases flexibility, but also increases HeNB signaling. (3) MUEs may identify the ABS pattern through UE measurements, and report the identified pattern to the macrocell. This approach reduces HeNB signaling, but increases eNB and UE complexity. Nevertheless, regardless of whether a static or a dynamic approach is used, the number of ABSs can be optimized in order to accommodate different number of UEs and their *quality of service* (QoS) requirements [53, 54].

7.6.2   Almost blank subframes for range-expanded picocells

This section considers the specific case of picocells with RE due to their significance, and investigates in more detail why time-domain ICIC is useful in this case and how it can be implemented. From the results in Section 7.2, while the benefits of RE are obvious for the UL communications (i.e., UL interference mitigation), its merits for the DL communications are not so obvious. Time-domain ICIC has the potential to let the network benefit from RE in the DL, and improve the performance of range-expanded UEs.

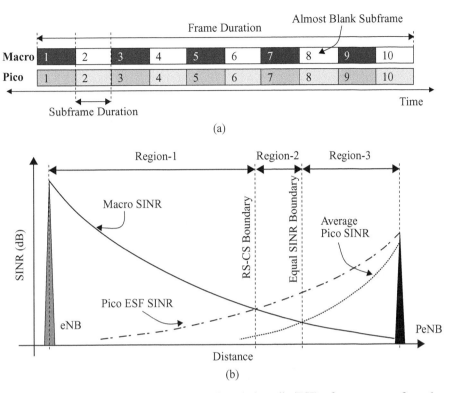

Figure 7.7 (a) Frame structure for the macrocells and picocells (ESFs of macro are configured as ABSs). (b) Macro/pico downlink SINRs and corresponding cell boundaries for equal SINR cell selection and RS-CS (reproduced with permission from IEEE©).

In a considered example [56] the ABS pattern in the macrocell is configured as in Fig. 7.7(a). This means that all *even subframes* (ESFs) are ABS, which means that MUEs are always scheduled in *odd subframes* (OSFs). This implies that the ESFs of PUEs in the PeNBs are protected from macrocell interference, and that the range-expanded PUEs may be scheduled within the ESFs in order to improve their SINRs. Thus the PeNBs may schedule their UEs in all subframes, but only the ESFs are protected. Both the macrocell and the picocell should therefore define the two measurement subframe sets for channel state information as OSFs and ESFs respectively. Thereby, their respective schedulers get the relevant feedback from the mobiles.

How aggressively to expand the range of picocells in the presence of ICIC is an important design issue. The cell-selection rule as in (7.1) based on received signal strength is agnostic to any kind of subframe blanking. However, if the handover is based on received signal quality, then the measurements can be impacted by the ABS patterns. This can be controlled by configuring measurement subframe patterns to be considered by the mobile when evaluating serving and neighbor cells. For example, the macrocell measurement subframe pattern can be configured to the non-ABSs, which are the OSFs, and the picocell measurement subframe pattern can be the ABSs, which are the ESFs. The difference between the picocell SINR values observed at ESFs and the average SINR over all subframes is illustrated in Fig. 7.7(b).

This *resource-specific cell selection* (RS-CS) method corresponds to a cell-selection rule for the UEs as follows:

$$\hat{i} = \arg \max_{i \in \mathcal{C}} \left\{ \tilde{\gamma}_i^{(\mathrm{dB})} \right\}, \tag{7.14}$$

and the SINRs are measured differently for the macrocells and picocells as

$$\tilde{\gamma}_i^{(\mathrm{dB})} = \begin{cases} \gamma_{i,\mathrm{OSF}}^{(\mathrm{dB})}, & \text{if } i \text{ is a macrocell} \\ \gamma_{i,\mathrm{ESF}}^{(\mathrm{dB})}, & \text{if } i \text{ is a picocell} \end{cases}. \tag{7.15}$$

It is important to note that the RS-CS as in (7.14) favors selection of picocells, inherently extending their coverage without using any dedicated REB.

7.6.3   Reduced-power subframes and UE interference cancellation

In Rel. 11, 3GPP has opened up another intercell interference management dimension in relation to time-domain ICIC. In particular, the definition of "almost blank" is generalized to also encompass non-zero or reduced-power transmissions during the ABS. Moreover, requirements on UE interference cancellation are discussed to investigate UE performance when canceling interference from full power reference symbols during ABS.

While ABS is very efficient to avoid aggressor interference to victim UEs, it also reduces the aggressor's performance, since some resources become unavailable. By blanking a certain portion of the resources, the throughput from the aggressor cell is reduced by the same amount. It is not obvious that this can be compensated by the increase in traffic uptake by the LPNs, which benefit from the interference reduction. The agreement [57], in which ABSs are generalized from completely blank data and control symbols to transmission of such symbols at a reduced power level, implies that the aggressing eNB may also be able to transmit information. The performance degradation of invoking ABS at the aggressing eNB is reduced significantly [11]. This is illustrated by Fig. 7.8, where two alternatives, zero-power and reduced-power ABS, are illustrated for the aggressing BS. Often, the term *reduced-power subframes* (RPSFs) is used for reduced-power ABS.

In RPSF, the *cell-specific reference symbols* (CRSs) and data symbols are transmitted at different power levels. The CRSs are used by the receiving UE as both phase reference and decoding threshold reference. The latter is important if the symbol power level itself carries information such as in high-order modulation and coding schemes, e.g., 16 QAM and higher QAM constellations. If the data symbol power offset relative to the CRS is unknown, then the transmission can either be based on constellations where only the phase carries information such as QPSK, or rely on transmission modes where additional UE-specific reference symbols (channel state information reference symbols in LTE) are transmitted at the same power as the data symbols. Another alternative is to signal the reduced-power offset to the UE [58].

A different dimension of interference suppression is via UE receiver optimization. One option is to consider the spatial properties of the interference. A typical example

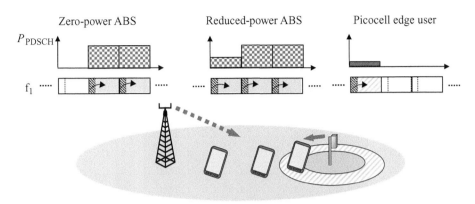

Figure 7.8 Use of zero-power ABS or reduced-power ABS at the aggressor MBS significantly reduces the interference to the expanded-range LPN UEs.

is *interference rejection combining* (IRC), a.k.a. *minimum mean square error* (MMSE) or optimal combining [59, 60]. This approach determines antenna combining weights based on the channel, the (spatial) noise and the interference covariance matrix, i.e., not only the interference power but also the spatial coloring of the interference is considered.

Furthermore, the 3GPP work item [61] focuses on UE mechanisms to cancel strong interferers, in particular from aggressing cells during ABS. Since the aggressing cell still transmits CRS, synchronization symbols, and some system information symbols at full power, UEs can detect and decode these symbols and estimate the channel, regenerate the signals, and subtract them from the received signal. With such cancellation in place, the UE becomes less sensitive to the MBS overlap, and deals with adverse DL interference conditions imposed by the RE. In [62], it is concluded that interference cancellation capable UEs should be able to support up to a 9 dB REB.

7.7 Performance evaluations

As seen in this chapter, interference can be dynamically managed with very different means. To examplify, some specific scenarios, configurations, and mechanisms are evaluated in this section. Special attention is on time-domain techniques together with power-based ICIC in Section 7.7.1, a capacity analysis in Section 7.7.2, and coverage and capacity analysis with various range expansion options in Sections 7.7.3 and 7.7.4, respectively. Finally, Section 7.7.5 evaluates impacts from ABS with non-zero data symbol power, UE interference cancellation, and FTP traffic models as addressed in 3GPP Rel.11.

7.7.1 Power-based and time-domain intercell interference coordination

In this section, the DL of an LTE HCN is simulated to test the different power-based ICIC schemes presented in Section 7.5. The scenario (Fig. 7.9) under scrutiny is a residential

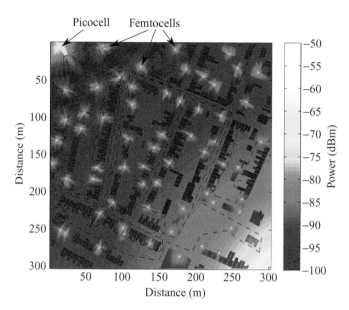

Figure 7.9 HCN simulated scenario based on an LTE-A network of 20 MHz with one macrocell transmitting at 46 dBm, one picocell transmitting at 30 dBm and 63 femtocells using up to 20 dBm (reproduced with permission from IEEE©). The dashed lines represent the routes followed by the eight MUEs.

area of size 300 m × 300 m in Luton, UK, containing 400 dwelling houses, of which 63 were selected to host a CSG femtocell (if we assume three network operators with equal customer shares, this corresponds to an approximate 50% femtocell penetration). These femtocells overlaid a macrocell, whose eNB was located 200 m south and 200 m east from the scenario's center (i.e., outside of Fig. 7.9). The femtocells were assumed to be fully loaded, thus using all frequency resources. Meanwhile, eight pedestrian UEs, using a *voice over IP* (VoIP) service, move along predefined paths according to a pedestrian mobility model with mean speed 1.1 m/s. The cells' transmit power is uniformly distributed among subcarriers, and a pedestrian UE carrying a VoIP service is considered to fall in outage if it cannot receive data signals (i.e., the UE SINR is smaller than −4 dB for a period of 200 ms).

Fig. 7.10 illustrates the SINR of a pedestrian UE when passing by the front doors of two different houses hosting a femtocell. It can be seen that, when no action is taken at the femtocells (no ICIC), the SINR of the pedestrian UE significantly falls due to cross-tier interference, thus resulting in UE outage. In contrast, when power-based ICIC is applied, the MUE SINR recovers and UE outages vanish. The performances of the different algorithms proposed in Section 7.5 are analyzed in the following. Note that an ICIC action occurs every time an eNB reports a low signal quality of one of its MUEs due to an overlaid HeNB, or every time an HeNB directly senses it. The results show that time-domain ICIC provides the best MUE protection, since those subframes overlapping with the ABSs of femtocells suffer from reduced inter-ference. In contrast, the different power-based ICIC schemes yield distinct levels of

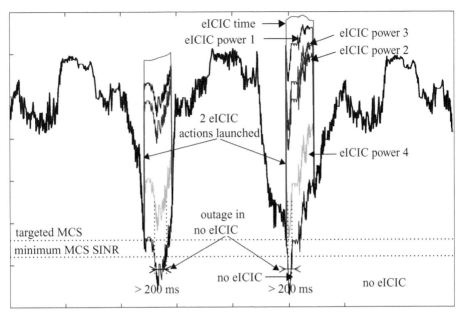

Figure 7.10 SINR versus time of a victim MUE when passing close to two houses hosting a CSG femtocell (MCS denotes the modulation and coding scheme) (reproduced with permission from IEEE©).

signal quality protection for the victim MUE. The behavior of these power-based ICIC schemes depends on their nature and tuning, but there is always a tradeoff between the performance of both the victim MUE and the aggressing HeNB, i.e., enhancing the victim MUE SINR always comes at the expense of reducing the transmit power of the HeNB and hence its throughput. Table 7.1, where simulation results are given, also shows this fact. The larger the average sum throughput of the eight MUEs, the smaller the average sum throughput of the femtocell tier. Moreover, in the fifth column, named "ICIC TP gain at a femto," the average throughput of a femtocell is given when an ICIC action is triggered on it. According to this performance indicator, the time-domain ICIC scheme provides the worst femtocell throughput performance, since it *switches off* the femtocell data channels (no data is carried in ABSs) to provide the best MUE signal quality protection. In contrast, taking no action at the femtocells results in the best femtocell throughput performance, at the expense of the worst MUE signal quality protection. In Table 7.1, we can also observe that the performance of power-based ICIC schemes (1) and (3) highly depends on the tuning of their parameters, i.e., α, β, and $SINR_{tar}$. If they are not fine tuned (ICIC power 1* and ICIC power 3*), a large number of UE outages occur. In contrast, power-based ICIC schemes (2) and (4) do not depend on this fine tuning, since they are able to adapt to each victim MUE situation considering either its path loss or SINR. Due to this fact, these two schemes can offer a "tailored" signal quality protection to victim MUEs, hence mitigating UE outages and recovering the maximum throughput at each femtocell. Moreover, in Table 7.1, we can also observe the average number of ICIC actions triggered in each femtocell every 10 minutes. When

Table 7.1 Performance comparison (600 s simulation).

ICIC methods	Number of MUE outages	Average ICIC TP gain at a femto [Mbps]	Average sum TP of pedestrian UEs [kbps]	Average sum TP of femtocell tier [Mbps]	ICIC actions femto · 10 min
no ICIC	267	73.32 (100%)	156.03	3974.25	–
ICIC time	0	0 (0%)	2158.82	2990.50	14.81
ICIC power 1 $\alpha = 1, \beta = 60$ dB	0	11.02 (15.03%)	1937.26	3153.88	14.81
ICIC power 1* $\alpha = 1, \beta = 75$ dB	25	46.49 (63.41%)	1139.20	3725.88	56.23
ICIC power 2	0	34.49 (47.03%)	1499.30	3558.75	20.80
ICIC power 3 $\text{SINR}_{\text{tar}}^{\text{FUE}} = 0$ dB	0	22.55 (30.75%)	1626.61	3333.75	17.47
ICIC power 3* $\text{SINR}_{\text{tar}}^{\text{FUE}} = 5$ dB	19	33.74 (46.02%)	1281.21	3520.75	47.52
ICIC power 4 $\text{SINR}_{\text{tar}}^{\text{MUE}} = 5$ dB	0	33.74 (66.05%)	1183.35	3751.13	39.78

* This ICIC method has not been properly tuned to avoid the number of UE outages. They are given for comparison purposes.

the power-based ICIC schemes (1) and (3) are not fine tuned (ICIC power 1* and ICIC power 3*), a large number of ICIC actions are triggered by the macrocell in the femtocells to attend to avoid MUE call drops. However, due to their inadequate tuning, these ICIC schemes cannot recover MUE SINRs to the desired value. The number of ICIC actions decreases when fine-tuning is performed. Power-based ICIC schemes (2) and (4) cast more ICIC actions than power-based ICIC schemes (1) and (3), when the latter are fine tuned. This is because the path losses and SINRs of victim MUEs continuously change when they move through the HeNBs' coverage, and hence HeNBs need to continuously update their transmit powers to prevent MUE outages according to MUEs' changing path loss or SINR conditions. Therefore, it can be stated that this tailored protection comes at the expense of larger backhaul signaling between cells.

7.7.2 Performance analysis for time-domain intercell interference coordination

Let us now analyze how the system capacity changes when using picocell deployments and time-domain ICIC. In more detail, we analyze system capacity as a function of REB and ICIC parameters such as the duty cycle of ABSs. Consider an HCN scenario where the macrocell of interest is surrounded by co-channel macrocells, and each macrocell also includes within its coverage N_p co-channel picocells. Let $\rho = [\rho_1, \ldots, \rho_{N_u}]$ and $\psi = [\psi_1, \ldots, \psi_{N_u}]$ denote the SINR sets of UEs when they are connected to the macrocell and the *strongest picocell*, respectively, where N_u is the total number of UEs within the coverage area of a macrocell and its picocells. Let $\delta_i = \rho_i - \psi_i$ denote the difference between macrocell and picocell DL SINRs for the ith UE, and $\delta = [\delta_1, \ldots, \delta_{N_u}]$. Without loss of generality, δ may be arranged as an ordered set in terms of δ_i, i.e., $\delta_i \leq \delta_{i+1}, \forall i$.

Then, the total number of PUEs can be expressed as

$$N_{\text{pue}} = \arg \max_i \delta_i \,, \quad \text{such that } \delta_i < 0 \,. \tag{7.16}$$

Without RE, all UEs with $i \leq N_{\text{pue}}$ are PUEs, while UEs with $i > N_{\text{pue}}$ are MUEs. Hence, if the *cumulative distribution function* (CDF) of δ_i is denoted by $F(\delta_i)$, then, without RE, we may write $N_{\text{pue}} = F(0)N_{\text{u}}$.

Let us now consider the case with RE, where a REB of λ (in dB) is added to the measured SINRs of PUEs prior to cell selection. Then, the total number of PUEs for the given REB of λ can be written as

$$N_{\text{pue}} = \arg \max_i (\delta_i + \lambda), \quad \text{such that } \delta_i - \lambda < 0 \,. \tag{7.17}$$

Alternatively, (7.17) may also be written as $N_{\text{pue}} = F(\lambda)N_{\text{u}}$, which captures the number of PUEs as a continuous function of the REB values.

Considering round-robin scheduling and a flat-fading channel, the sum capacity (for unit bandwidth) of all UEs connected to the desired macrocell can be written as a function of the REB values as

$$C_{\text{macro}}(\lambda) = \sum_{i=N_{\text{pue}}+1}^{N_{\text{u}}} \frac{1}{N_{\text{u}}\big(1 - F(\lambda)\big)} \log_2(1 + \rho_i) \,. \tag{7.18}$$

In contrast, the sum capacity (for unit bandwidth) of all the UEs connected to picocells within the desired macrocell can be written as a function of the REB values as

$$C_{\text{pico}}(\lambda) = \sum_{i=1}^{N_{\text{pue}}} \frac{N_{\text{p}}}{N_{\text{u}}F(\lambda)} \log_2(1 + \psi_i) \,. \tag{7.19}$$

Given (7.18) and (7.19), the sum capacity of all the UEs in macrocells/picocells can be written as

$$C_{\text{tot}}(\lambda) = C_{\text{macro}}(\lambda) + C_{\text{pico}}(\lambda) \,. \tag{7.20}$$

With RE, expanded-range PUEs may observe unfavorable SINRs, especially for large λ. In order to improve their performance, ICIC techniques based on macrocell ABSs are considered in the 3GPP, as discussed earlier in the chapter. Let $0 \leq \beta < 1$ denote the duty cycle of the ABSs in the macrocell. For example, Fig. 7.7(a) shows a scenario with $\beta = 0.5$, which considers that the macrocell schedules no MUEs in ESFs, but only in OSFs. This implies that ESF PUEs (with SINRs $\psi_{i,\text{esf}}$) will be protected from co-channel macrocell interference, while OSF PUEs (with SINRs $\psi_{i,\text{osf}}$) will still observe co-channel macrocell interference.

For a given duty cycle of β, and considering that the MUEs are scheduled *only* in the non-coordinated subframes (NSFs) of macrocells (e.g., OSFs in Fig. 7.7(a)), the sum capacity of the macrocell tier for unit bandwidth can be written as

$$C_{\text{macro}}(\lambda) = \sum_{i=N_{\text{pue}}+1}^{N_{\text{u}}} \frac{1 - \beta}{N_{\text{u}}(1 - F(\lambda))} \log_2(1 + \rho_i) \,, \tag{7.21}$$

where it is assumed that, while calculating $F(\delta_i)$ for this ICIC scenario, coordinated subframe (CSF) SINRs (e.g., ESF SINRs in Fig. 7.7(a)) of PUEs are used, i.e., $\delta_i = \rho_i - \psi_{i,\mathrm{csf}}$.

In contrast, the sum capacity for the picocell tier depends on whether the PUEs are scheduled in the NSFs or CSFs, and can be written as

$$
C_{\mathrm{pico}}(\lambda, K) = \sum_{i=1}^{K} \underbrace{\frac{\beta N_{\mathrm{p}}}{K} \log_2 \left(1 + \psi_{g(i),\mathrm{csf}}\right)}_{C_{\mathrm{p}}(g(i),\lambda,K)}
$$

$$
+ \sum_{i=K+1}^{N_{\mathrm{pue}}} \underbrace{\frac{(1-\beta)N_{\mathrm{p}}}{N_{\mathrm{pue}} - K} \log_2 \left(1 + \psi_{g(i),\mathrm{nsf}}\right)}_{C_{\mathrm{p}}(g(i),\lambda,K)}, \tag{7.22}
$$

where $C_{\mathrm{p}}(g(i), \lambda, K)$ is the capacity of UE $g(i)$, K is the number of PUEs that are scheduled in the CSFs, and $g(i)$ is an ordering function of the UE indices for $i = 1, \ldots, N_{\mathrm{pue}}$, such that $\psi_{g(i),\mathrm{nsf}} \leq \psi_{g(i)+1,\mathrm{nsf}}$. Note that the NSF SINR of a PUE is always worse than its CSF SINR, and the range-expanded PUEs with low NSF SINRs should ideally be served in the CSFs of picocells. Therefore, the PUEs are sorted with respect to their NSF SINRs, and the K PUEs with worse NSF SINRs are scheduled in the CSFs of picocells.

In order to select K, we consider a simple max–min capacity scheduler that maximizes the minimum capacity of PUEs as follows (see e.g., Theorem-1 in [63])

$$
K = \arg \max_{\tilde{K}} \left\{ \min_i C_{\mathrm{p}}\big(g(i), \lambda, \tilde{K}\big) \right\}, \tag{7.23}
$$

where $i = 1, 2, \ldots, N_{\mathrm{pue}}$, $\tilde{K} = 1, \ldots, N_{\mathrm{pue}} - 1$. A scheduler as in (7.23) efficiently assigns the victim PUEs to protected subframes considering overall fairness of PUEs. Given (7.21) and (7.22), $C_{\mathrm{tot}}(\lambda)$ can be calculated similarly as in (7.20), i.e., the no ICIC scenario.

7.7.3 Coverage analysis for time-domain intercell interference coordination and range expansion

Consider a simulation scenario as in Fig. 7.11 with different numbers of PeNBs, i.e., $N_{\mathrm{p}} = 3, 9$, where PeNBs are placed with uniform intervals around a ring of radius $R_{\mathrm{p}} = 400$ m, and an eNB is located at the center of the ring. Unless otherwise stated, simulation parameters such as path-loss models, minimum distance constraints between BSs and UEs, etc., are based on HCN simulation assumptions in [28]. In order to average out the impact of UE distribution, a grid of $10\,\mathrm{m} \times 10\,\mathrm{m}$ is considered, and a UE is placed on each grid point (excluding the regions that do not satisfy minimum distance constraints). Note that it is straightforward to modify the results in this section using different UE distributions and different PeNB locations.

For the above-described simulation scenario, the coverage areas of the eNB and PeNBs for $N_{\mathrm{p}} = 3, 9$ with different REB values and RS-CS (see Section 7.6.2) are shown in Fig. 7.11. It is assumed that same REB value is used for all picocells in the network. For

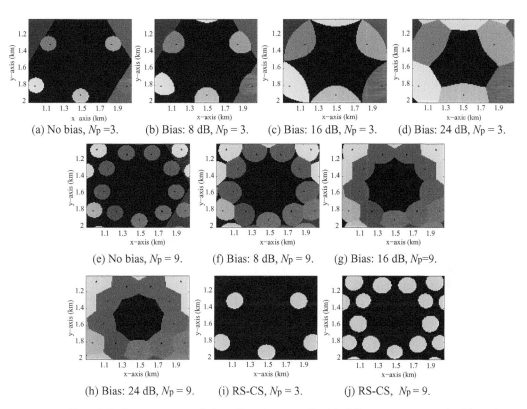

Figure 7.11 Coverage areas of picocells and macrocell with different range expansion bias values and different N_p (reproduced with permission from IEEE©). Coverage areas of picocells with RS-CS are also shown in Fig. 7.11(i) and Fig. 7.11(j).

smaller REB, picocells have small coverage areas. As the REB increases, the footprints of picocells become larger, which is more visible in Fig. 7.11 for smaller N_p. For sufficiently large REB values,[3] the picocell coverage areas eventually converge to the natural UL boundaries illustrated in Fig. 7.2, which can be represented by a Voronoi diagram with eNBs and PeNBs as center points (see, e.g., Fig. 7.11(h)).

With regard to RS-CS results in Fig. 7.11(i) and Fig. 7.11(j), the picocell coverage areas are adaptively determined using (7.14), without using any explicit REB value. The two different colors in the footprint of a picocell represent the areas where UEs are scheduled in ESFs (outer region) and OSFs (inner region), as shown in Fig. 7.7. The overall coverages of picocells resemble those of Fig. 7.11(c) (16 dB bias) for $N_p = 3$, and Fig. 7.11(g) (8 dB bias) for $N_p = 9$. This implies that, for larger N_p (where ESF interference from neighboring picocells will be larger), RS-CS expands less aggressively the picocells' coverage.

---

[3] In particular, when the REB value fully compensates for the transmit power difference (16 dB) and the antenna gain difference (9 dB) between the eNB and the PeNB, i.e., for REB= 25 dB considering the parameter set in Chapter 3.

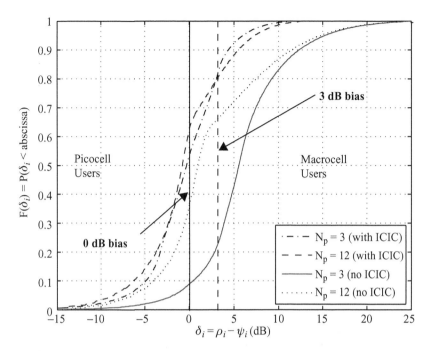

Figure 7.12 SINR difference CDFs, $F(\lambda)$, for different N_p, with/without ICIC (reproduced with permission from IEEE©). Note that CDF curves are independent from the REB values and cell associations of UEs. For a given CDF, the number of UEs that are offloaded from macrocell to picocells can be captured as a function of the REB.

7.7.4   Capacity analysis for time-domain intercell interference coordination and range expansion

For the above-described simulation scenario, the CDFs $F(\delta_i)$ for different configurations, with/without ICIC, are shown in Fig. 7.12. For any REB value of λ dB, $F(\lambda)$ shows the ratio of UEs that are associated with a picocell, and $1 - F(\lambda)$ shows the ratio of UEs that are associated with the macrocell of interest. Therefore, these CDFs fully characterize the number of UEs that are associated with picocells and the macrocell of interest for a given REB value, which is challenging to capture analytically. A vertical REB line divides the UEs into two sets; UEs to the left of the REB line are PUEs, while the UEs to the right of the REB line are MUEs.

When ICIC is applied, Fig. 7.12 verifies that $\rho_i - \psi_{i,\mathrm{esf}}$ becomes smaller due to the better ESF SINRs of PUEs. Therefore, coverage areas of picocells (and hence the number of PUEs) increase with ICIC as shown in Fig. 7.11, even when no REB is applied. For example, Fig. 7.12 shows that 60–65% of the UEs are connected to picocells in the case of ICIC and without range expansion. Moreover, if REB values over 5 dB are used with ICIC, over 90% of UEs get connected to picocells. As a result, even though UEs are scheduled in CSFs, picocells get overloaded and therefore the edge-UE capacity gets degraded for large REB values as will be shown in Fig. 7.14(b).

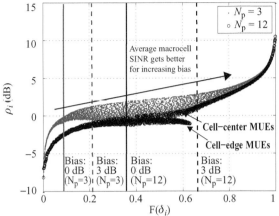

(a) Macrocell SINRs; all points to the right of a bias line are the UEs
connected to a macrocell for the corresponding REB.

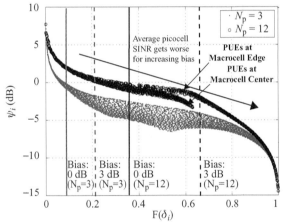

(b) Picocell SINRs; all points to the left of a bias line are the UEs
connected to the strongest picocell for the corresponding REB.

Figure 7.13 (a) Macrocell SINRs and (b) picocell SINRs, as a function of $F(\delta_i)$ (reproduced with permission from IEEE©).

Fig. 7.13(a) and Fig. 7.13(b) show the SINR distributions (without ICIC) of the set of grid points corresponding to the CDFs $F(\delta_i)$ in Fig. 7.12 for $N_p = 3, 12$. With increasing REB, the average MUE SINR gets better, while the average PUE SINR gets worse. For large N_p, two SINR clusters can be observed, corresponding to UE locations that are closer to the macrocell edge and the macrocell center. For $N_p = 3$, the SINRs are relatively more uniform due to the large distance between the picocells.

Fig. 7.14 shows the sum capacity, fifth-percentile UE capacity, and 50th-percentile UE capacity, for different N_p, with and without ICIC ($\beta = 0.5$). Fig. 7.14(a) shows that sum capacity with ICIC is slightly lower than sum capacity with no ICIC. The fifth-percentile UE capacities, which can be seen to represent the overall fairness of the system, are shown in Fig. 7.14(b) as a function of REB values. Results show that a critical REB that

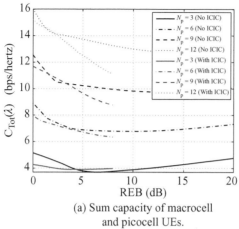

(a) Sum capacity of macrocell
and picocell UEs.

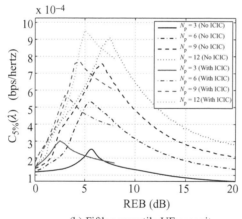

(b) Fifth-percentile UE capacity.

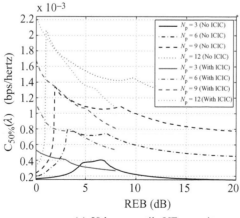

(c) 50th-percentile UE capacity.

Figure 7.14 Sum capacity, fifth-percentile capacity, and 50th-percentile capacity of all UEs, for different N_p, with and without ICIC (reproduced with permission from IEEE©).

Table 7.2 Percentage of the UEs that are served by a picocell.

	RE = 0 dB	RE = 6 dB	RE = 12 dB
Config. 1, Model-1	27%	42%	59%
Config. 1, ITU	55%	65%	75%

maximizes the fifth-percentile UE capacity exists for all scenarios. From the fairness perspective, while it is preferable to use REB values of the order of 6–8 dB with no ICIC, REB values of 3–5 dB are preferable with ICIC. In either case, after a certain REB value, picocells become overloaded, and the capacities of cell-edge PUEs degrade. For most REB values, fifth-percentile UE capacities with ICIC are better than those with no ICIC. The 50th-percentile UE capacities in Fig. 7.14(c) show a similar behavior to the fifth-percentile UE capacities in Fig. 7.14(b) for no ICIC and for increasing REB, and are maximized for a certain REB value. The 50th-percentile UE capacity improves sharply with the REB first (while UEs are still associated with the macrocell), and starts dropping slowly after a certain REB value (after the 50th-percentile UE becomes a PUE). With ICIC, on the other hand, the 50th-percentile UE capacity monotonically degrades with REB values. The reason for this is that, due to already expanded picocell range, the 50th-percentile UE is already a PUE (see Fig. 7.12), and its capacity is degraded as more UEs are offloaded to picocells with larger REB values.

7.7.5 Reduced-power ABS and UE interference cancellation

This section[4] evaluates some aspects of zero-power ABS and reduced-power ABS ICIC. A more complete evaluation is provided in [11], where additional simulation configurations are considered as well. In this case, the simulations are based on the 3GPP model 1 and ITU channel models (see Chapter 3), and are based on LTE deployments with macrocells and picocells. Moreover, picocells are deployed outdoors, four per macrocell, and they operate at 0.5×2 W. The same REB is applied to all picocells, and the same static ABS patterns are considered for all macrocells. The static ABS pattern ratios are optimized for each considered scenario, corresponding to a genie guided ratio determination operation that may be considered to give an upper bound on the performance of ABS.

A central system performance aspect is the percentage of the UEs that are served by picocells; see Table 7.2, where it is interesting to see that the numbers are different for different channel models. An observation is therefore that the need for RE is also different, and that picocells can offload considerable traffic from macrocells in some environments even without range expansion. In general, picocells offload more traffic in simulations based on ITU models compared with simulations based on 3GPP models.

Another aspect is to evaluate the costs and benefits of implementing ABS, and then considering either no data scheduled during ABS or data scheduled during ABS at

---

[4] The authors acknowledge Chrysostomos Koutsimanis and Lars Lindbom for providing the simulation evaluations in Section 7.7.5.

Table 7.3 Relative system performance in terms of fifth percentile user throughput for different time-domain and power-based ICIC configurations compared with no ICIC (no ABS and No RE) in the rightmost column. Simulated scenarios: (1) config. 1, 3GPP model 1, and (2) config. 1, ITU model.

Scenario	Load	ABS power	RE 0 dB	RE 6 dB	RE 12 dB	No ABS 1W MBS	No ABS [Mbps]
(1)	10%	0	−17%	−30%	−28%	33%	8.6
(2)	10%	0	−30%	−27%	−32%	16%	19.9
(1)	50%	0	−51%	−32%	−16%	78%	3.7
(2)	50%	0	−60%	−31%	−31%	63%	4.8
(1)	10%	>0	37%	36%	33%	33%	8.6
(2)	10%	>0	31%	26%	18%	16%	19.9
(1)	50%	>0	−5%	49%	78%	78%	3.7
(2)	50%	>0	−15%	40%	73%	63%	4.8

reduced power from the macrocells. The benefits are due to the reduced interference from macrocells to users served by picocells, and the costs are due to the reduced availability of resources in macrocells. The power reduction when applied is equal to the considered REB. For example, with a 46 dBm macro and REB = 12 dB, the data channel transmit power would be 34 dBm during reduced power ABS. The interference contributions during ABS are only associated with the data transmission power, thus reflecting the performance of a UE that is able to cancel all CRS interference. Traffic models also have a strong impact on the results, where full buffer traffic models have a tendency to exaggerate the behavior or some features compared with expected behavior in reality. Therefore, a non-full buffer model is considered: in this case, FTP model 1 (see Chapter 3).

The system performance is represented by the fifth percentile user throughputs. To facilitate comparisons with the baseline system performance with no ABS and no RE, the system performance is presented as the relative system performance to the baseline (fifth percentile user throughput difference divided by the baseline fifth percentile user throughput). Thus, negative percentages mean worse system performance compared with no ABS and no RE, and positive means better system performance. The results are summarized in Table 7.3, where the rightmost column represents the reference system performance with no ABS. The three columns labeled RE 0, 6 and 12 dB represents the relative system performance considering RE and ABS. In addition, one unconventional case is evaluated where also the macrocells are using the same power as the pico to further illustrate that the main costs are associated with not transmitting data during ABS from macrocells. The relative system performance for that case is presented in the column labeled No ABS, 1 W MBS.

The different rows represent different combinations of scenarios (1 or 2), load levels (low 10% and high 50% in terms of average radio resource utilization in the macro layer in the baseline configuration), and ABS data transmission powers (zero and non-zero). When considering non-zero ABS, the macrocell schedules users with favorable radio conditions during such reduced power subframes.

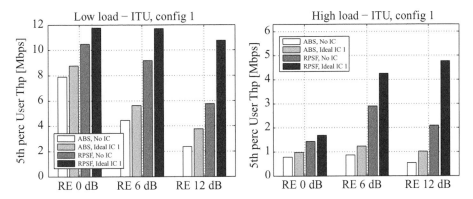

Figure 7.15 Comparison of costs and benefits of ABS, RPSF, and UE interference cancellation.

Some important observations are as follows.

- The cost of zero-power ABS dominates over the benefits, and operation without ABS is always better in the considered scenarios with the studied traffic model. Some small gains were observed in a few of the observed cases in [11], but in general, for non-full buffer traffic, there seems to be no or very limited gain with zero-power ABS. The main reason is the high cost of not transmitting any data during ABS.
- Instead, significant improvements are achieved when reducing the macrocell transmit power to be the same as the picocell transmit power at 1 W, thereby avoiding the UL/DL imbalance. However, this mainly indicates that the simulated scenarios do not capture all the aspects of a real network, where the high macrocell power is needed to provide full coverage over the intended service area.
- Reduced (but non-zero) transmit power ABS improves the system performance significantly in comparison with zero-power ABS for the considered scenarios with the studied FTP traffic model. Clearly, it is very beneficial to transmit data even at reduced power compared with not transmitting at all.

To increase realism, the impact from CRS to CRS collisions has been modeled via link-level simulations as a mapping of the impact to the effective SINR. Furthermore, Fig. 7.15 illustrates the gains of canceling only the CRS interference from the strongest interfering cell in the case of zero-power and reduced-power ABS. From the results, we can see that there are significant benefits with UEs capable of canceling CRS interference, in particular with large REBs, since the strongest interferer becomes more dominant.

As discussed above, some higher-order modulation schemes need a received power reference, typically from the pilot signals. Either the reduced-power offset is signaled to the UE, or only modulations where the pilot is only used as phase reference, such as QPSK, are available during reduced-power ABS. Fig. 7.16 provides a cell-edge performance comparison of zero-power ABS, on the one hand, and reduced-power ABS, on the other hand, with different modulation limitations. From the results, we can see that excluding 64 QAM does not have a significant impact on cell-edge performance, and RPSF limited to QPSK is still better than zero-power ABS. This is further analyzed in [58].

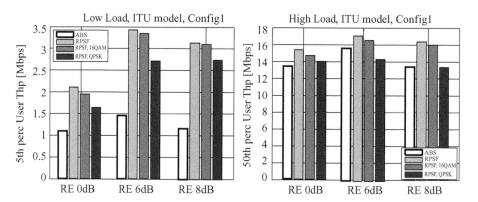

Figure 7.16 Comparison of costs and benefits of ABS, RPSF, and UE interference cancellation with different modulation limitations during RPSF.

7.8   Summary and conclusions

This chapter surveys the most recent advances as well as more traditional mechanisms for dynamic interference management. Some ICIC techniques developed for macrocell only scenarios and rather homogenous deployments have merits also for heterogenous networks. However, the more recent ICIC techniques are tailored to new HCN challenges, where UEs are not always being served by the most favorable BS, radio-propagation-wise, in the DL or the UL. This is due to the transmit power differences between the MBS and the LPN, and the implementation of RE and CSG access.

The dedicated focus is on frequency-domain, power-based, and time-domain interference management, and the most recent discussions in 3GPP are addressed. Moreover, performance comparisons and assessments indicate some design guidelines for dynamic interference management. The simulated scenarios are LTE specific, involving picocells and CSG HeNBs, and are aligned with the 3GPP scenarios. However, the discussion is applicable to any wireless network in general.

Copyright notices

Figure 7.12, 7.13 and 7.14 © 2011 IEEE. Reprinted, with permission, from I. Guvenc, Capacity and fairness analysis of heterogeneous networks with range expansion and interference coordination. *IEEE Communications Letters*, **15**:10 (2011).

References

[1] A. Lozano, D. Cox and T. Bourk, Uplink–downlink imbalance in TDMA personal communication systems. In *Proceedings of the IEEE International Conference on Universal Personal Communications (ICUPC)*, Florence (1998), vol. 1, pp. 151–155.

[2] W. Lee and D. Lee, The impact of front end LNA on cellular system. In *Proceedings of the IEEE Vehicular Technology Conference (VTC)*, Boston, MA (2000), vol. 5, pp. 2180–2184.

[3] D. López-Pérez, A. Valcarce, G. de la Roche and J. Zhang, OFDMA Femtocells: a roadmap on interference avoidance. *IEEE Communications Magazine*, **47**:9 (2009), 41–48.

[4] G. de la Roche, A. Valcarce, D. López-Pérez and J. Zhang, Access control mechanisms for femtocells. *IEEE Communications Magazine*, **48**:1 (2010), 33–39.

[5] Qualcomm, *Importance of Serving Cell Selection in Heterogeneous Networks (R1-100701)*, (Valencia, 2010).

[6] NTT DOCOMO, *Performance of eICIC with Control Channel Coverage Limitation (R1-103264)*, (Montreal, 2010), 3GPP TSG RAN WG1 Meeting-61.

[7] Kyocera, *Potential Performance of Range Expansion in Macro–Pico deployment (R1-104355)*, (Madrid, 2010), 3GPP TSG RAN WG1 Meeting-62.

[8] Motorola, *On Range Extension in Open-Access Heterogeneous Networks (R1-103181)* (Montreal, 2010), 3GPP TSG RAN WG1 Meeting-61.

[9] Huawei, *Evaluation of Rel-8/9 Techniques and Range Expansion for Macro and Outdoor Hotzone (R1-103125)* (Montreal, 2010), 3GPP TSG RAN WG1 Meeting-61.

[10] Alcatel-Lucent, *DL Pico/Macro Het Net Performance: Cell Selection (R1-102808)* (Montreal, 2010), 3GPP TSG RAN WG1 Meeting-61.

[11] Ericsson, *System Performance Evaluations on feICIC (R1-113482)* (Zhuhai, China, Oct. 2011).

[12] J. Wigard, T. Nielsen, P. Michaelsen and P. Morgensen, On a handover algorithm in a PCS1900/GSM/DCS1800 network. In *Proceedings of the IEEE Vehicular Technology Conference (VTC)*, Houston, TX (1999), vol. 3, pp. 2510–2514.

[13] A. Tolli, I. Barbancho, J. Gomez, and P. Hakalin, Intra-system load balancing between adjacent GSM cells. In *Proceedings of the IEEE Vehicular Technology Conference (VTC)*, Orlando, FL (2003), vol. 1, pp. 393–397.

[14] H.-S. Jo, Y. J. Sang, P. Xia, and J. G. Andrews, Outage probability for heterogeneous cellular networks with biased cell association. In *Proceedings of the IEEE Global Telecommunications Conference (GLOBECOM)*, Houston, TX (2011).

[15] R. Madan, J. Borran, A. Sampath, N. Bhushan, A. Khandekar, and T. Ji, Cell association and interference coordination in heterogeneous LTE-A cellular networks. *IEEE Journal on Selected in Areas Communications (JSAC) – Special Issue on Cooperative Communications in MIMO Cellular Networks* (Oct. 2010).

[16] D. Ghosh and C. Lott, Uplink–downlink imbalance in wireless cellular networks. In *Proceedings of the IEEE International Conference Communications (ICC)*, Glasgow (2007), pp. 4275–4280.

[17] K. Azarian, C. Lott, D. Ghosh and R. A. Attar, Imbalance issues in heterogeneous DO networks. In *Proceedings of the IEEE International Workshop on Femtocell Networks (FEMnet)*, Miami, FL (Dec. 2010).

[18] S. Landstrom, H. Murai and A. Simonsson, Deployment aspects of LTE pico nodes. In *Proceedings of the IEEE International Workshop on Heterogeneous Networks (HETnet)*, Kyoto (2011).

[19] K. Balachandran, J. H. Kang, K. Karakayali and K. Rege, Cell selection with downlink resource partitioning in heterogeneous networks. In *Proceedings of the IEEE International Workshop on Heterogeneous Networks (HETnet)*, Kyoto (2011).

[20] K. Okino, T. Nakayama, C. Yamazaki, H. Sato and Y. Kusano, Pico cell range expansion with interference mitigation toward LTE-Advanced heterogeneous networks. In *Proceedings of the IEEE International Workshop on Heterogeneous Networks (HETnet)*, Kyoto (2011).

[21] D. Lopez-Perez and X. Chu, Outage probability for heterogeneous cellular networks with biased cell association. In *Proceedings of the IEEE International Conference Computer Communications Networks (ICCCN)*, Maui (2011).

[22] I. Guvenc, Capacity and fairness analysis of heterogeneous networks with range expansion and interference coordination. *IEEE Communications Letters*, **15**:10 (2011), 1084–1087.

[23] M. Vajapeyam, A. Damnjanovic, J. Montojo, T. Ji, Y. Wei, and D. Malladi, Downlink FTP performance of heterogeneous networks for LTE-Advanced. In *Proceedings of the IEEE International Workshop on Heterogeneous Networks (HETnet)*, Kyoto (2011).

[24] 3GPP, *Evolved Universal Terrestrial Radio Access (E-UTRA); Radio Resource Control (RRC); Protocol Specification (TS 36.331)* (Dec. 2010).

[25] 3GPP, *Evolved Universal Terrestrial Radio Access (E-UTRA); Physical Layer Procedures (TS 36.213)*.

[26] A. Simonsson and A. Furuskar, Uplink power control in LTE – overview and performance: principles and benefits of utilizing rather than compensating for SINR variations. In *Proceedings of the IEEE Vehicular Technology. Conference (VTC)* (Sep. 2008), pp. 1–5.

[27] C. U. Castellanos, D. L. Villa, C. Rosa, K. I. Pedersen, F. D. Calabrese, P. H. Michaelsen and J. Michel, Performance of uplink fractional power control in UTRAN LTE. In *Proceedings of the IEEE Vehicular Technology Conference (VTC)* (Singapore, May 2008), pp. 2517–2521.

[28] 3GPP, *Evolved Universal Terrestrial Radio Access (E-UTRA); Further Advancements for E-UTRA Physical Layer Aspects (TS 36.814)* (Mar. 2010).

[29] G. Fodor, C. Koutsimanis, A. Racz, N. Reider, A. Simonsson and W. Muller, Intercell interference coordination in OFDMA networks and in the 3GPP long term evolution system. *Journal of Communications*, **4**:7 (2009), 445–453.

[30] 3GPP, *Evolved Universal Terrestrial Radio Access (E-UTRA); X2 Application Protocol (X2AP) (TS 36.423)*.

[31] *Summary of the Description of Candidate eICIC Solutions*. 3GPP Standard Contribution (R1-104968) (Aug. 2010).

[32] Nokia Siemens Networks, *On Advanced UE MMSE Receiver Modeling in System Simulations*. 3GPP Standard Contribution (R1-111031) (Feb. 2011).

[33] R. Beck and H. Panzer, Strategies for handover and dynamic channel allocation in microcellular mobile radio systems. In *Proceedings of the IEEE Vehicular Technology Conference (VTC)*, San Francisco, CA (1989), vol. 1, pp. 178–185.

[34] S. Kuek, W.-C. Wong, R. Vijayan and D. Goodman, A predictive load-sharing scheme in a microcellular radio environment. *IEEE Transactions on Vehicular Technology*, **42**:4 (1993), 519–525.

[35] M. Almgren, L. Bergstrom, M. Frodigh and K. Wallstedt, Channel allocation and power settings in a cellular system with macro and micro cells using the same frequency spectrum. In *Proceedings of the IEEE Vehicular Technology Conference (VTC)*, Atlanta, GA (1996), vol. 2, pp. 1150–1154.

[36] A. Fiorini and R. De Bernardi, Load sharing methods in a WCDMA macro multi-carrier scenario. In *Proceedings of the IEEE Vehicular Technology Conference (VTC)*, Orlando, FL (2003), vol. 2, pp. 816–820.

[37] M. Kazmi, O. Sjobergh, W. Muller, J. Wierok and B. Lindoff, Evaluation of inter-frequency quality handover criteria in E-UTRAN. In *Proceedings of the IEEE Vehicular Technology Conference (VTC)*, Anchorage, AK (2009), pp. 1–5.

[38] E. Dahlman, S. Parkvall and J. Sköld, *4G LTE/LTE-Advanced for Mobile Broadband* (Academic Press, 2011).

[39] Z. Shen, A. Papasakellariou, J. Montojo, D. Gerstenberger and F. Xu, Overview of 3GPP LTE-Advanced carrier aggregation for 4G wireless communications. *IEEE Communications Magazine*, **50**:2 (2012), 122–130.

[40] A. Szufarska, K. Safjan, S. Strzyz, K. Pedersen and F. Frederiksen, Interference mitigation methods for LTE-Advanced networks with macro and HeNB deployments. In *Proceedings of the IEEE Vehicular Technology Conference (VTC)*, San Francisco, CA (2011), pp. 1–5.

[41] Qualcomm (Rapporteur), *DL Interference Solutions for Carrier Based Hetnet* (email discussion summary, R3-120483) (Cabo San Lucas, 2011).

[42] Ericsson (Rapporteur), *TP on UL Interference Solutions for Carrier Based Hetnet* (email discussion summary, R3-120482) (Cabo San Lucas, 2011).

[43] J. Shapira, Microcell engineering in CDMA cellular networks. *IEEE Transactions Vehicular Technology*, **43**:4 (1994), 817–825.

[44] Ericsson, *UL Power Control in Hotzone Deployments* (R1-102619) (Montreal, 2010).

[45] R. Combes, Z. Altman, M. Haddad and E. Altman, Self-optimizing strategies for interference coordination in OFDMA networks. In *Proceedings of the IEEE International Conference Communications (ICC)*, Kyoto (2011), pp. 1–5.

[46] CATT (Rapporteur), *Summary of the Description of Candidate eICIC Solutions* (R1-104968) (Madrid, 2010).

[47] CMCC, *New Work Item Proposal: Enhanced ICIC for Non-CA Based Deployments of Heterogeneous Networks for LTE*. 3GPP Standard Contribution (RP-100383) (Mar. 2010).

[48] *Comparison of Time-Domain eICIC Solutions*. 3GPP Standard Contribution (R1-104661) (Aug. 2010).

[49] D. Lee, H. Seo, B. Clerckx, E. Hardouin, D. Mazzarese, S. Nagata and K. Sayana, Coordinated multipoint transmission and reception in LTE-Advanced: deployment scenarios and operational challenges. *IEEE Communications Magazine*, **50**:2 (2012), 148–155.

[50] A. Damnjanovic, J. Montojo, J. Cho, H. Ji, J. Yang and P. Zong, UE's role in LTE advanced heterogeneous networks. *IEEE Communications Magazine*, **50**:2 (2012), 164–176.

[51] CATT, *Evaluations of RSRP/RSRQ Measurement* (R4-110284) (Austin, TX, Jan. 2011).

[52] Samsung, *CSI Measurement Issue for Macro–Femto Scenarios*. 3GPP Standard Contribution (R1-106051) (Nov. 2010).

[53] InterDigital Communications, LLC, *eICIC Macro–Femto: Time-Domain Muting and ABS*. 3GPP Standard Contribution (R1-105951) (Jacksonville, FL, Nov. 2010).

[54] Ericsson, *Details of Almost Blank Subframes*. 3GPP Standard Contribution (R1-105335) (Xian, 2010).

[55] A. Damnjanovic, J. Montojo, Y. Wei, T. Ji, T. Luo, M. Vajapeyam, T. Yoo, O. Song and D. Malladi, A survey on 3GPP heterogeneous networks. *IEEE Wireless Communications*, **18**:3 (2011), 10–21.

[56] I. Guvenc, M.-R. Jeong, I. Demirdogen, B. Kecicioglu and F. Watanabe, Range expansion and inter-cell interference coordination (ICIC) for picocell networks. In *Proceedings of the IEEE Vehicular Technology Conference (VTC)* (Sep. 2011), pp. 1–6.

[57] Alcatel-Lucent, Ericsson, *et al.*, *CR on ABS Definition* (R2-111701) (Taipei, 2011).

[58] Ericsson, ST-Ericsson, *On Signalling Support for Reduced Power ABS* (R1-121749) (Jeju, 2012).

[59] J. Winters, Optimum combining in digital mobile radio with cochannel interference. *IEEE Journal on Selected Areas in Communications*, **2**:4 (1984), 528–539.

[60] Y. Ohwatari, N. Miki, T. Asai, T. Abe, and H. Taoka, Performance of advanced receiver employing interference rejection combining to suppress inter-cell interference in LTE-Advanced downlink. In *Proceedings of the IEEE Vehicular Technology Conference (VTC)*, San Francisco, CA (Sep. 2011), pp. 1–7.

[61] On *Signalling Support for Reduced Power ABS* (RP-111369) (2011).

[62] DOCOMO (Rapporteur), *LS on feICIC* (R1-114468) (San Francisco, 2011).

[63] N. Ksairi, P. Bianchi, P. Ciblat, and W. Hachem, Resource allocation for downlink cellular OFDMA systems part I: Optimal allocation. *IEEE Transactions on Signal Processing*, **58**:2 (2010), 720–734.

8   Uncoordinated femtocell deployments

David López-Pérez, Xiaoli Chu and Holger Claussen

8.1   Introduction

Nowadays, 50% of phone calls and 70% of data services are carried out indoors [1]. For this reason, one may expect that operators' networks are optimized to provide good indoor coverage and capacity for voice, video, and high-speed data services. However, surveys have shown that 45% of households and 30% of businesses experience poor indoor coverage [2]. This poor indoor coverage may lead to reduced subscriber loyalty and increased subscriber churn, which may significantly affect operators' revenues. As a consequence, vendors and operators are developing new solutions to address the indoor coverage problem.

A straightforward solution to enhance indoor coverage would be to increase the number of outdoor *macrocell base stations* (MBSs). Deploying a larger number of MBSs with a reduced cell radius may provide improved network coverage and capacity, but this approach is too expensive due to the high cost associated with MBSs. Moreover, this approach presents challenges in terms of site acquisition due to municipality and people's concerns about MBS towers [3]. It is also very difficult to achieve high indoor signal quality when providing coverage from outdoors due to wall attenuation losses. Therefore, providing indoor coverage from outdoors is not the best solution.

As a result, indoor solutions such as *distributed antenna systems* (DASs) and picocells have become attractive alternatives to provide services in indoor hotspots, e.g., shopping malls and office buildings. These operator-deployed solutions improve in-building coverage, enhance signal quality, offload traffic from outdoor MBSs, and allow high-data-rate services due to the fact that transmitters are closer to receivers. Although DASs and picocells may be more cost effective than increasing the number of MBSs, they are still too expensive to be used in scenarios such as small offices or homes, since operators have to plan, install, and maintain these solutions [4].

Due to the market pressures for improved indoor coverage and capacity, the cellular industry is developing new indoor solutions. Among the new technologies, the *femtocell access points* (FAPs) or *Home NodeBs* (HNBs) have gained a lot of momentum, because they have the potential to provide low-cost indoor coverage and capacity extension [5]. FAPs are low-cost, low-power, small-size *base stations* (BSs) initially designed to extend indoor coverage where it is limited or unavailable. On the air interface, FAPs provide coverage of a certain cellular standard, e.g., *Universal Mobile Telecommunication System* (UMTS), *High Speed Packet Access* (HSPA) or *Long Term Evolution* (LTE), while on

Table 8.1 Differences between macrocells and femtocells.

Features	Macrocell	Femtocell
Design target	Outdoors	Indoors
Cost	High	Low
Number of deployed cells per area	Small	Large
Number of connected UEs per cell	Large	Small
Physical access to the BS	Operators only	Users only
Access to the cell	Open access	Open, closed or hybrid access
Can be turned on/off at any time?	no	yes
Can be moved at any time?	no	yes (geographic restrictions may exist)

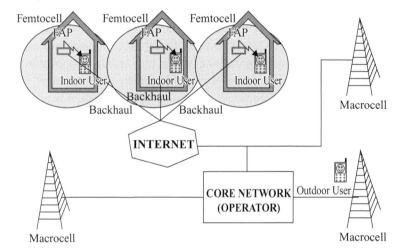

Figure 8.1 Femtocell architecture.

the backhaul, they are connected to the operator's network via a broadband connection such as optical fiber and *digital subscriber line* (DSL) (see Fig. 8.1). Table 8.1 gives an overview of the main differences between macrocells and femtocells.

Unlike macrocell, DAS, and picocell networks, femtocell networks are likely deployed in a similar manner to *Wireless Fidelity* (WiFi) systems, i.e., individual users will deploy their own FAPs in an *ad hoc* manner in unplanned locations without any operator supervision. This decentralized deployment will reduce both *capital expenditure* (CAPEX) and *operational expenditure* (OPEX). However, as a main difference, it is important to note that FAPs operate in licensed bands using a cellular standard, while WiFi *access points* (APs) operate in unlicensed bands. Moreover, WiFi APs are based on a simpler technology, which is not able to guarantee tailored *user equipment* (UE) quality of service.

The use of femtocells offers advantages to both users and operators. Users may benefit from improved signal quality due to the short distance between transmitter and receiver, resulting in enhanced capacity and reliability, together with reduced transmit power consumption in indoor environments. In this way, UE throughputs may be increased and UE battery life may be extended. In the meanwhile, operators may benefit from a larger network capacity due to improved spectral efficiency and spatial reuse. Moreover, since indoor traffic will be carried over the user-provided *Internet Protocol* (IP) backhaul, femtocells may help operators to manage the exponential traffic growth within macrocells through data offloading. Since users will deploy and maintain FAPs, operators' expenses can also be significantly reduced [4].

However, the above mentioned femtocell benefits may not be easy to realize in all scenarios. Although quite some work has been carried out on femtocells, there are still technical challenges that the cellular industry must face before successfully deploying a wide femtocell network over existing macrocells, while optimizing the use of the spectrum (e.g., co-channel deployments) [6]. In this chapter, we shall pay special attention to the management of intercell interference between the macrocell and femtocell tiers in co-channel deployments, as well as the interference between neighboring femtocells. Intercell interference in co-channel deployments could counteract the above mentioned benefits of femtocells, and degrade the overall network performance [7].

Most femtocell networks' challenges arise from the individualistic nature of femtocells and the uncertainty on the number and locations of FAPs. Due to the *ad hoc* deployment of FAPs by individual users, operators must adopt new approaches rather than conventional network planning and optimization methods to guarantee proper network operation. Signaling overhead and delay issues over the user-provided IP backhaul make it difficult, if not impossible, to perform a centralized planning of user-deployed femtocell networks [8]. As a result, FAPs must be self-organizing small BSs that can integrate the processes of planning, configuration, and optimization in a set of built-in autonomous functionalities. The self-organizing techniques allow FAPs to monitor their radio environment and fine tune their parameters according to network, traffic, and channel fluctuations, while reducing human involvement and enabling plug-and-play operation [6].

8.2   Femtocell market

According to an ABI Research survey [9] in early 2009, there would have been 102 million UEs of FAP products in 32 million FAPs worldwide by 2011. However, according to a recent Telecoms & Media survey [10], only 2.3 million FAPs are currently in use worldwide, much less than the forecast of 32 million FAPs. ABI Research anticipated this slower-than-expected adoption of femtocells by operators in its forecasts of April and November 2009, where they predicted that by the end of that year there would be only 790 000 and 350 000 FAP shipments, respectively. A significant adjustment of about 55% was performed in only seven months.

In 2010, femtocell deployments took off. A revised forecast from ABI Research in August 2010 anticipated that, by the end of that year, there would be around 1 million FAP shipments, indicating that in 2010 femtocell rollouts would be more than doubled. All major operators in the USA, e.g., AT&T, Verizon, and Sprint, had had femtocell offerings. Vodafone in the UK and China Unicom in China had also had femtocell services by then.

More recently, as forecasted by ABI Research, the number of FAP shipments was expected to grow even faster from 2.3 million FAPs in 2011 to more than 54 million FAP shipments in 2015. Informa Telecoms & Media has also predicted that there will be around 49 million FAPs and 114 million UEs accessing cellular networks via femtocells worldwide by 2014. Such a large market growth is expected due to two main facts: (1) eight of the top ten mobile operator groups (which are AT&T Group, Vodafone Group, France Telecom Group, NTT DOCOMO Group, Sprint, Telefonica Group, Deutsch Telecom Group, and Verizon Wireless) have already been offering femtocell services [11], and (2) the significant reduction of the FAP price, which is now less than 100 USD, makes it possible to even provide users with FAPs for free. Softbank in Japan was the first operator to offer FAPs for free. As a result, in the vast majority of developed markets, the number of FAPs is already larger than that of MBSs [11].

The promising future of femtocells is also reflected and supported by the quick standardization of femtocells in the *3rd Generation Partnership Project* (3GPP) and the *Institute of Electrical and Electronics Engineers* (IEEE) fora, the growing membership of the Small Cell Forum, which now includes 74 vendors and 63 mobile operators worldwide, and the large FAP portfolio provided by major vendors such as Alcatel-Lucent, Ericsson, and Huawei.

8.3   Femtocell deployment scenarios

FAPs can be installed at locations where UEs experience unsatisfactory radio coverage or have higher data rate requirements than can be offered by the existing macrocell network. In home-femtocell scenarios, FAPs help to improve the coverage of indoor UEs that suffer from poor reception due to high wall penetration losses. In this case, one single FAP is used to cover a household. In enterprise-femtocell scenarios, several FAPs are deployed to serve indoor demands in a large building or enterprise.

Corresponding to the above two deployment scenarios, there are two major FAP use cases. The first use case is for home environments, where only a few authorized UEs are allowed to connect to a home FAP. Home FAPs will be deployed by users, creating a need for decentralized interference coordination. In the home scenario, the main source of interference to femtocells is the outdoor macrocell network. In order to overcome macro-to-femto and femto-to-macro interference, operators may divide their spectrum into different bands so that some bands are used only by macrocells and others by femtocells (i.e., orthogonal deployment). However, orthogonal deployments may lead to a low spectral efficiency, and thus an inefficient use of the scarce and expensive spectrum. As a result, co-channel macrocell–femtocell deployments, where macrocells and femtocells share the spectrum resources, may be preferred, but they require efficient intercell interference coordination mechanisms, as will be shown in Section 8.7. The second use case is for enterprises, where a large number of authorized UEs are allowed to connect to several enterprise FAPs. Enterprise FAPs can be deployed by users or operators, and interfemtocell interference may become significant since multiple FAPs coexist within the same household/building. In order to optimize the performance of enterprise FAPs in co-channel macrocell–femtocell deployments, decentralized intercell interference coordination and load balancing techniques will be of importance.

Femtocells will also be components of *heterogeneous cellular networks* (HCNs), as indicated in Chapter 1, where the main target is to bring the network closer to UEs, not only at home or indoors, but also outdoors [12]. Outdoor FAPs, similar to enterprise FAPs, could be user deployed or operator deployed, the latter allowing for planned roll-outs and letting FAPs operate as low-transmit-power picocells (see Chapter 7). For instance, operators may deploy FAPs at macrocell edges to effectively improve cell-edge coverage and capacity.

Deployment scenarios and access mechanisms to femtocells (e.g., open, privately, or semi-publicly accessible FAPs) may also define the strategy used for carrier utilization [13], which has a significant impact on the overall network performance (as discussed in Chapter 4 and Chapter 6). In co-channel deployments, when the access method blocks the use of femtocell resources to a subset of UEs within its coverage area, this subset of UEs become potential sources/victims of interference. Hence, the use of closed-access FAPs makes the problem of interference management complex in co-channel deployments. Fig 7.1, presented in the previous chapter, depicts a challenging scenario of interference coordination, where *macrocell user equipments* (MUEs) move nearby a house hosting a closed-access FAP. In the *uplink* (UL), MUE (a) interferes the UL of the closed-access FAP, since MUE (a) uses a high transmit power to compensate for the high path loss to its distant serving MBS. In the *downlink* (DL), the closed-access FAP interferes the DL of MUE (b), since MUE (b) is connected to a distant MBS and hence suffers from a low received signal strength. The use of open-access FAPs solves these interference problems in co-channel deployments (see Chapter 7), since UEs can connect to the cell with the most favorable radio conditions, but may bring security and resource-sharing concerns to FAP owners. Moreover, when high-mobility UEs move across areas with dense open-access FAP deployments, the number of triggered handovers and associated signaling overhead become an critical issue too, as will be shown in Chapter 9. The use of closed- and open-access FAPs is discussed in more detail in Section 8.7.

Hybrid-access mechanisms can be seen as a trade-off between closed-access and open-access mechanisms. For hybrid access, the fraction of resources for sharing must be carefully tuned to avoid making femtocell subscribers feel that they are paying for a service that is exploited by non-subscribers. The impact of hybrid access on femtocell subscribers may be minimized through intelligent scheduling or economic advantages (e.g., reduced bills) [14]. In order to illustrate the effect of limiting the number N_r of shared *resource blocks* (RBs) for hybrid access in an *orthogonal frequency division multiple access* (OFDMA) network, Fig. 8.2 shows the simulated average throughput and UE outage predictions [14], where system-level simulations were based on Monte Carlo methods, the frequency band was divided into eight RBs, N_r ($0 \leq N_r \leq 8$) RBs were shared among served non-subscribers, and a *first-in first-out* (FIFO) scheme was used to schedule non-subscribers. The results indicate that reducing the amount of shared resources minimizes the impact of hybrid access on subscribers, and sharing a

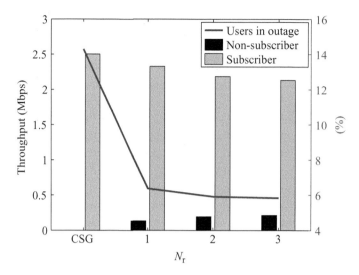

Figure 8.2 Average throughput per UE in a small residential scenario where each of 30% of the houses contains an OFDMA femtocell with three subscribers that request an intense data service. There are five outdoor non-subscribers each requesting a throughput between 80 kbps and 450 kbps for video service (reproduced with permission of IEEE©).

small amount of resources (e.g., $N_r = 1$) is sufficient to significantly reduce the outage probability of non-subscribers. Larger N_r values do not improve the outage probability of non-subscribers, but increase their average throughput.

8.4   The Small Cell Forum

In July 2007, the Small Cell Forum – formerly the Femto Forum – was founded to promote femtocell standardization. The Small Cell Forum supports, promotes, and helps drive the wide-scale adoption of small-cell technologies to improve coverage, capacity, and services delivered by cellular networks. Today, most vendors and operators around the world are members of the Small Cell Forum. Although the Small Cell Forum initially focused on UMTS femtocell standardization, due to the large interest of the mobile industry in LTE, the Small Cell Forum is now also focusing on LTE femtocells, and gives recommendations for *Home evolved NodeBs* (HeNBs). In the following, we summarize the work of the Small Cell Forum on data collection and interference management, which have been identified as the main challenges in femtocell deployments.

Because the Small Cell Forum focuses on the standardization of HeNBs, the solutions proposed for collecting data are based on network monitoring modes implemented at the HeNB [15]. Other possible solutions for collecting data, such as UE measurement reports, have been studied in 3GPP standardization and have been discussed in Chapter 6.

According to the Small Cell Forum, an HeNB switched on in network monitoring mode should behave like a UE, and should be able to scan the frequency spectrum and decode DL signals from neighboring cells, and thereafter perform the optimization of its own *radio frequency* (RF) parameters. The network monitoring mode should be executed at least at booting time and optionally during cell operation. For how long the network monitoring mode should be used, and how often it should be executed, are questions left up to vendor and operator implementation. A more regular network monitoring mode operation would allow the HeNB to perform a more powerful RF parameter optimization, as more reliable information would be available. However, the enhanced sensing may come at the expense of reduced performance, because HeNBs running in the network monitoring mode may need, e.g., to synchronize and measure over different frequency carriers other than the one used for transmission, as will be explained in the following.

The network monitoring mode operation can be classified into three types, depending on the amount of information that the HeNB is able to gather from the radio environment.

- *Basic carrier detection*: In this case, an HeNB scans either the whole frequency spectrum or a given list of candidate frequencies, and measures the *reference signal strength indicator* (RSSI) of each scanned carrier, so as to discern which carriers are being used by other cells.
- *LTE cell identification and power measurement*: In this case, given a list of candidate frequencies or *physical layer cell identities* (PCIs), the HeNB synchronizes with neighboring cells and identifies them. First, the *primary synchronization signals* (PSSs) and *secondary synchronization signal* (SSS) of neighboring cells are decoded to get synchronization with them. If both PSS and SSS are successfully derived, the cell is identified and its PCI is retrieved. Then, the *reference signal received power* (RSRP) of each identified cell is measured, and the list of neighboring cells is built.
- *LTE cell information retrieval*: In this last form of network monitoring mode, information from the sensed cells may be decoded. In this case, the sensing HeNB will decode the *primary broadcast channels* (PBCHs) (which contain their master information blocks) and the *physical downlink shared channels* (PDSCHs) (which contain the secondary information blocks) of neighboring cells, thus allowing the sensing HeNB to access key information from the sensed cells, such as the number of antennas, channel bandwidth, and proprietary information of neighboring cells.

The Small Cell Forum has also put a lot of effort into interference characterization and management [16], with main focuses on defining possible interference scenarios and proposing techniques/guidelines to mitigate intercell interference. In the *physical* (PHY) layer, the proposed schemes are mainly based on power control, carrier selection, and multi-antenna techniques. In the case of power control, it is suggested that, at booting time, HeNBs should adapt their maximum transmit power so that their signals are stronger than those of the closest neighboring cells at a targeted cell radius. In the *medium access control* (MAC) layer, the proposed schemes are mainly based on radio resource allocation, e.g., RB assignment, which can be performed in both the frequency and time domains due to the flexibility of the LTE subframe structure. Moreover, the Small Cell Forum has also investigated interference mitigation schemes that rely on an

X2 interface between eNBs and HeNBs, or between HeNBs themselves. However, the 3GPP has not standardized the X2 interface for femtocells yet, due to capacity and delay concerns of the user-provided backhaul.

8.5   Backhaul

Due to the increasing demand for cellular network capacity, the backhaul connection may quickly become the bottleneck in the near future. Traffic generated by macrocells is typically routed from the *radio access network* (RAN) to the core network through the lu-b interface. Femtocells are much closer to UEs, e.g., indoors or on the street, but further away from existing backhaul infrastructure designed for macrocells, making it harder for femtocells to reach the core network than for macrocells. Moreover, femtocell backhaul traffic is generally lighter but much burstier than that of macrocells. For these two reasons, the traffic from femtocells can be more efficiently handled when it is aggregated and routed over a separate gateway, via the lu-h interface using an additional network node, such as the *home gateway* (HGW) (see Chapter 4), which can accommodate a large number of FAPs.

DSL has been widely used for residential or enterprise broadband. Its current data-only requirement makes it suitable for femtocell backhaul [17]. Unlike the backhaul system of a well planned and centrally managed macrocell network, the broadband IP-based backhaul of femtocells is likely provided by a third-party entity. For example, DSL backhaul connections would typically be paid for by the primary femtocell subscribers [18]. In a home served by a DSL connection, a stand-alone FAP is connected to the DSL modem via a home router, while an integrated FAP includes the home router and the DSL modem within the same physical device [18].

While existing macrocell networks provide latency guarantees, current IP backhaul networks are not equipped to provide delay resiliency. The IP backhaul requires *quality of service* (QoS) for delay-sensitive traffic and for providing service parity to macro-cells [19]. QoS provisioning depends on the relationship between the wireless service provider and the wireline service provider. Lack of agreement between the two parties may cause problems, except in cases where the wireline backhaul provider is in a tight strategic relationship with or is the same company as the cellular operator [19]. There are three major types of potential relationship between the wireless service provider and the wireline service provider [18]: (1) common wireless and wireline operator, i.e., the same operator owns the network path from the femtocell to the *femto gateway* (FGW), and can thus manage the wireline access network to guarantee the QoS required by femtocell services; (2) separate wireless and wireline operators with a *service level agreement* (SLA), in which case, although the wireline access provider is different from the wireless operator, femtocell traffic is provided the required QoS within the access network based on an SLA between the two different operators; and (3) separate wireless and wireline operators without an SLA, in which case the wireline operator has no incentive to guarantee QoS for femtocell traffic, but simply provides best effort services in the access network. In the Internet, the peer-to-peer SLAs are typically best effort with little or no standardized tools to guarantee inter-provider SLAs [20].

The IP backhaul should provide sufficient capacity to avoid creating traffic bottle-neck [19]. For the case of separate wireless and wireline operators without an SLA, when the traffic load in the access network is less than the network capacity, best effort service may still provide good enough QoS, while if traffic demands come close to or beyond the access network capacity, then the QoS of delay-sensitive traffic will be degraded. In the latter case, there are a number of potential solutions for femtocells to achieve the desired QoS [18]. Firstly, if an FAP is placed between the home router and the DSL modem, then the femtocell can prioritize its traffic over other traffic generated by the networked devices in the home, e.g., by using dynamic path bandwidth measurement techniques to compute the available bandwidth periodically, so as to adapt to the changing network conditions. Secondly, if a femtocell finds that the available bandwidth toward the FGW is lower than a certain threshold, then UEs served by the femtocell can be handed over to the macrocell network, as long as there is a macrocell that can provide a satisfactory signal quality to these UEs. Thirdly, if there are multiple FGWs available, a femtocell can be assigned to another FGW when the access link to the current one is congested.

Considering that improved voice coverage has been a main driver for femtocells, the sharing of backhaul between femtocells and WiFi would also be important [19]. Femtocells require deployment in locations with limited space availability. A compact backhauling solution, in conjunction with smaller and lighter equipment, is essential. Consumer femtocell backhauling should work in a plug-and-play manner.

The following backhaul solutions for FAP deployments may be used:

- copper wire xDSL
- traditional microwave (6-42 GHz)
- sub 6 GHz microwave – either point-to-point or point-to-multipoint
- light licensed millimeter wave (E-band) microwave (70–80 GHz)
- unlicensed millimeter wave (60 GHz)
- cable

where every technology is best suited for specific scenarios, and multiple solutions will be used in any operator network. For example, copper wire xDSL is the usual backhaul solution in home scenarios, while microwave solutions may only be used for operator-deployed enterprise or outdoor FAPs.

8.6 Synchronization and localization

Proper operation of femtocells require both network synchronization and FAP location identification. Femtocells need to synchronize with existing macrocells and neighboring femtocells to facilitate seamless handover and interference management, especially in *time division duplexing* (TDD) networks. For instance, femtocells require synchronization to align received signals to minimize multiple-access interference, ensure a tolerable carrier offset, and hand over UEs to and from macrocells [19]. Regulatory agencies and cellular standards impose strict time, frequency, and location requirements on FAPs.

Synchronization and localization methods can be categorized into two types [21]: (1) network-based methods, such as IEEE 1588 *Precision Time Protocol* (PTP) over

IP [22], *Network Time Protocol* (NTP) (*Request for Comments* (RFC) 1305), and *Simple Network Time Protocol* (SNTP) (RFC 2030); and (2) air-interface-based methods, such as *Global Positioning System* (GPS) and air interface snooping.

IEEE 1588 [22] is used in *Worldwide Interoperability for Microwave Access* (WiMAX) to deliver timing information from a centralized synchronization server to all femtocells through the wired backhaul. In this case, each FAP measures the timing difference between itself and the synchronization server, and then corrects its clock [23]. IEEE 1588 PTP uses users' available backhaul connection for synchronization. Despite the universal availability of PTP, the protocol suffers from a few drawbacks [24]. First, the PTP algorithm determines the FAP location based on the coarse location estimates derived from the IP address associated with the Internet connection, which may not even be available information for all FAPs and/or operators. Second, the PTP algorithm assumes symmetry in the UL and DL packet delay times, but asymmetry is often present in residential Internet connections where bandwidth is heavily shared and packet congestion may occur. Such asymmetry coupled with variations in packet delay caused by network jitter degrades the accuracy of femtocell timing synchronization. Moreover, FAPs connect to the operator's network by users' IP connections, and there can be a number of intermediate nodes between a femtocell and the synchronization server [23]. This fact also affects the PTP accuracy. Other approaches, such as NTP and SNTP, are not accurate enough to be used for synchronization of an OFDMA system [21].

GPS is an established global synchronization scheme for BSs, but the GPS solution can only be used when an outdoor GPS antenna is available, because *line of sight* (LOS) needs to be ensured for communications between the GPS device and multiple satellites. Equipping femtocells with GPS for timing synchronization and location determination relies on maintaining stable satellite reception and keeping costs low. FAPs are typically installed indoors or on the street level, where GPS signals are usually weak and thus difficult to acquire. Specifically, indoor femtocells suffer from a severe penetration loss of GPS signals. The acquisition performance of GPS receivers can be enhanced by using data bits, coarse time and position estimates provided by the cellular network [25]. Receivers equipped with Enhanced GPS can acquire and track GPS signals with a carrier-to-receiver noise density ratio, C/N_0, of 14 dB Hz [26], e.g., by coherently integrating to recover enough GPS signal power [27].

A tightly coupled opportunistic navigation (TCON) scheme [28], which fuses the GPS signal and the signal of opportunity at the carrier-phase level, was proposed to extend the penetration of GPS-assisted femtocells in indoor environments. Due to cost and size constraints, most femtocells use low quality *temperature controlled oscillators* (TCXOs) for their local clocks. Instead of replacing the inexpensive TCXO with a more stable but more expensive *oven controlled oscillator* (OCXO), the TCON scheme uses a non-GPS signal of opportunity, to correct phase errors at the GPS receiver [28]. Candidate signals of opportunity for TCON include iridium communication satellite signals, which are modulated using a stable OCXO on-board the satellite and have C/N_0 values typically above 60 dBHz, *high-definition television* (HDTV) signals, which have a stable clock and a power advantage of about 50 dB over GPS, and cellular signals, which can penetrate buildings and use highly stable and GPS synchronized oscillators in their BSs [28].

Snooping synchronization channels (i.e., preambles) from nearby MBSs is also a simple solution to femtocell synchronization. In the air-interface-based network synchronization algorithm [29], densely deployed femtocells achieve a timing consensus by leveraging existing synchronization signals broadcast by a macrocell or neighboring femtocells. The timing of an FAP is then updated according to a convex combination of the current timing of itself and timings of selected neighboring FAPs.

FAP location requirements are motivated by three major concerns, which are spectrum use, operator control, and emergency caller location identification [19]. For spectrum use, femtocell operators must ensure that *femtocell user equipments* (FUEs) do not transport or activate FAPs in areas where the operator does not have a licensed spectrum. For operator control, if a user takes a femtocell out of its licensed area or out of its country, operators may wish to impose additional charges on the user or disable the femtocell to avoid losing revenue from roaming fees. The concern of emergency caller location identification leads to the most stringent requirement on FAP location.

Regulations on FAP localization have not been well defined. The USA *Federal Communications Commission* (FCC) has passed the Enhanced 911 (E911) legislation. Phase I of the E911 legislation mandates that an operator must be able to provide the location of the cell site receiving the 911 call [30]. A recent FCC Request for Comments suggests that the FCC would like FAP location accuracy to be roughly equivalent to MBS location accuracy, or within about 10 meters [31]. Among existing synchronization technologies for femtocells, only GPS is able to deliver location accuracy within 10 meters [24]. Phase II of the E911 legislation mandates that operators must not determine the FAP location, but the caller's coordinates, and be accurate to within 50 meters for 67% of emergency calls and 150 meters for 95% of calls. Since the femtocell coverage radius is typically short enough, the FAP location is an adequate proxy for that of the UE served by it. Thus for most femtocells, if the required accuracy of FAP location is met, so is the UE location accuracy.

8.7   Interference mitigation in femtocell networks

As indicated earlier, intercell interference is among the most urgent challenges to successful femtocell rollouts. This section presents a survey of cutting edge intercell interference mitigation approaches that can be used in uncoordinated FAP deployments. We will pay special attention to carrier allocation strategies, together with power-, antenna-, load-balancing-, and subcarrier-based intercell interference mitigation approaches.

8.7.1   Carrier allocation strategies

In a multi-carrier cellular network, such as UMTS or LTE, the adopted intercell interference mitigation techniques depend upon the selected carrier allocation strategies, which are impacted by the deployment scenarios and access mechanisms to femtocells. Three different carrier allocation strategies have been considered [32].

- *Separated carriers:* The simplest carrier allocation strategy for femtocell deployments is to use separate carriers for macrocells and femtocells, respectively. This approach completely avoids cross-tier interference, but may significantly reduce spectral efficiency per area. Both closed- and open-access femtocell deployments are possible in this strategy.

- *Shared carriers:* The most efficient carrier allocation scheme in terms of spectral efficiency per area is the co-channel deployment, where macrocells and femtocells can fully reuse all available carriers. However, co-channel operation has the disadvantage of intercell interference between macrocells and femtocells, which requires enhanced intercell interference mitigation algorithms. Due to such strong cross-tier interference, the deployment of closed-access femtocells may not be an option in this case.
- *Partially shared carriers:* As a tradeoff between the separated and shared carrier approaches, macrocells may use all available carriers, while femtocells may use only some of them. In this way, MUEs can always be allocated to the clean carriers to avoid interference from femtocells. For example, when an MUE moves near a closed-access femtocell to which it does not have access, it can hand over to a macrocell carrier that is not used by the femtocell and thus avoid interference. Both closed- and open-access femtocell deployments are possible. In order to minimize the number of mobility events caused by handovers between the shared carriers and the clean macrocell carriers, UEs could be allocated to carriers taking their mobility into account, i.e., static UEs on the shared carrier, highly mobile UEs on the clean carrier.

Shared carrier (co-channel) deployments

In the following, we pay particular attention to shared deployments, since they have the potential for a better spatial reuse.

A centralized DL frequency planning for shared deployments of OFDMA-based macrocells and femtocells was proposed in [33]. However, as plug-and-play devices, the number and locations of active FAPs would be hardly known to operators, and thus a centralized DL frequency planning would be difficult to realize [34]. Moreover, the proposed approach does not achieve a frequency reuse factor of 1, since frequency resources used by macrocells are not reused by femtocells and vice versa. A similar but distributed shared carrier approach was proposed in [35]. Cross-tier interference is avoided by assigning orthogonal spectrum resources within one carrier to the macrocell and femtocell tiers through coordinated sub-channel partitioning, while femtocell-to-femtocell interference is mitigated by allowing each femtocell to access only a random subset of sub-channels assigned to the femtocell tier. Although this approach can avoid cross-tier interference in a distributed manner, operators may still choose to deploy both macrocell and femtocell networks in a co-channel manner with a frequency reuse of 1, due to considerations of spectrum availability and network infrastructure [36, 37]. In other words, due to its better spectral efficiency per area and since some operators may only own one carrier, they may prefer co-channel deployments with a frequency reuse of 1 in both tiers.

In co-channel deployments with a frequency reuse of 1 in both tiers, the different transmit powers used by MBSs and FAPs, in conjunction with potentially densely deployed closed-access FAPs, may create dead spots where reliable DL services cannot be guaranteed to either MUEs or FUEs [38]. For example, the DL of a femtocell that is close to an MBS may be disrupted by macrocell DL transmissions due to the much higher transmit power used by the MBS, while the DL of an MUE that is far away from

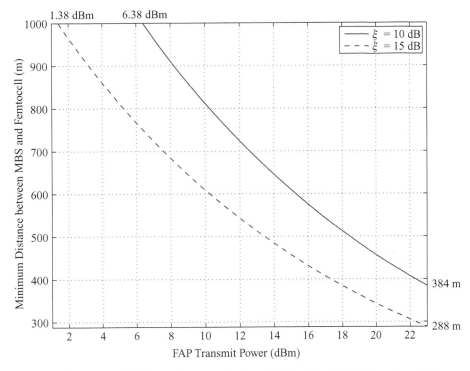

Figure 8.3 $d_{\text{FM,min}}$ versus FAP transmit power, for $\xi = 10\,\text{dB}$ and $\xi = 15\,\text{dB}$ (reproduced with permission of IEEE©).

its serving MBS may be blocked by the DL transmissions of nearby femtocells [32, 37]. The latter case is depicted in Fig. 7.1 in the previous chapter. As a result, co-channel operation may need to grant open access to femtocells in order to prevent excessive interference for UEs of the same operator located close to a privately owned FAP.

In [38], fundamental limits of co-channel OFDMA-based macrocell–femtocell deployments were analyzed using DL outage probabilities (based on similar principles as in Chapter 5). The DL outage probability analysis accounted for path loss, lognormal shadowing, and Rayleigh fading, and decomposed the femtocell DL outage probability into two parts corresponding to femtocell DL outages caused by strong macro-to-femto interference alone and those caused by composite macro-and-femto interference. The analysis also removed the conventional assumption that all FAPs transmit at the same power (also shown in [41]), and instead embraced the dynamics of transmit power used by different FAPs. Different FAPs may use different total transmit powers depending on their distance from the macrocell. The analysis also allowed different *signal to interference plus noise ratio* (SINR) targets and outage probability constraints for MUEs and FUEs. Based on the analysis, the authors derived analytical expressions of the minimum distance that a co-channel femtocell has to keep away from an MBS, and distance-dependent upper and lower bounds on FAP transmit power for maintaining reliable macrocell and femtocell DL services. Simulation results have shown that a necessary condition for a co-channel femtocell to meet its DL outage probability constraint is to be at least $d_{\text{FM,min}}$ m in distance from the MBS. Fig 8.3 illustrates $d_{\text{FM,min}}$ versus

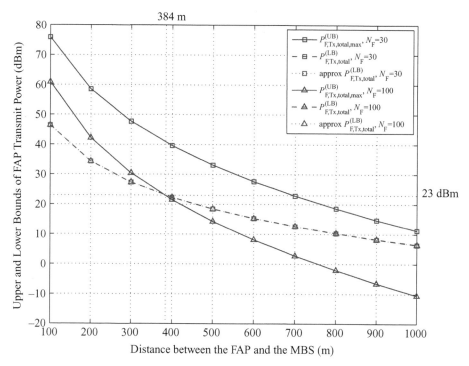

Figure 8.4 Upper bound of maximum FAP transmit power, lower bound of FAP transmit power, and approximate lower bound of FAP transmit power versus distance of an FAP from the MBS, for $N_F = 30$ and 100, and $\xi = 10\,dB$ (reproduced with permission of IEEE©).

the FAP transmit power, for wall penetration losses of $\xi = 10\,dB$ and $\xi = 15\,dB$, FUE SINR target of 10 dB, and FUE outage probability constraint of 0.1. Please refer to [38] for detail about the simulation scenario and parameters. For either value of ξ considered, $d_{FM,min}$ decreases with the increase of FAP transmit power. For a given FAP transmit power, $d_{FM,min}$ is reduced at a higher value of ξ, indicating that indoor co-channel femtocells can be deployed closer to an MBS when the wall-partition loss is higher.

Fig. 8.4 illustrates the upper and lower bounds of FAP transmit power required for keeping the macrocell DL outage probability (i.e., the probability of MUE SINR being lower than 5 dB) below 0.1 and the femtocell DL outage probability (i.e., the probability of FUE SINR being lower than 10 dB) below 0.1, versus the FAP's distance from the MBS, for the number of femtocells N_F per macrocell being 30 and 100 [38]. Both the upper and lower bounds of FAP transmit power decrease with the FAP's distance from the MBS. At a given distance, the upper bound of maximum FAP transmit power reduces significantly with the increase of N_F, while the lower bound of FAP transmit power does not change much with N_F. In this particular scenario, at distances shorter than 384 m, the required lower bound of FAP transmit power is larger than 23 dBm, which is the maximum transmit power of an FAP. This means that in this particular example FAPs cannot be deployed at distances less than 384 m from the MBS for a wall penetration loss of 10 dB.

The above DL outage probability analysis (which is in line with Chapter 5) helps to explain the fundamental limits of co-channel deployments of macrocells and femtocells, and has been extended to other scenarios, such as spectrum-sharing multiple-antenna macrocells and femtocells [40], sectorized macrocells [41, 42], frequency reused macrocells [50], etc. Deriving these fundamental limits has shed new light on the understanding of HCN deployments, and facilitates the development of practical schemes for interference mitigation such as the ones presented in the following.

8.7.2   Power-based techniques

As shown above, the tuning of the maximum power of pilot and data channels is of critical importance to guarantee a proper femtocell coverage for a given quality of service, particularly when open-access femtocells reuse the same frequency band as an existing macrocell network. This is because the transmit power defines the femtocell coverage area and has an impact on the interference, handover signaling, and dropped call rate. FAPs that are close to an MBS need to transmit at a higher maximum transmit power than those far away from the MBS to achieve a given target cell radius. Moreover, in order to mitigate cross- and co-tier interference, FAPs must dynamically tune their maximum transmit power in both control and data channels, according to the changing conditions of the radio environment, e.g., neighboring femtocells switching on/off, passing UEs, etc. The self-optimization of FAP maximum transmit power helps to (1) adapt the femtocell coverage to the household premise and thus reduce the interference created towards MUEs passing by and (2) mitigate the attempts of macro-to-femto handover by MUEs.

8.7.2.1   Power tuning for interference mitigation

In [41], the authors proposed a baseline approach for ensuring a given open-access femtocell coverage radius and mitigating intercell interference, where the maximum transmit power of the FAP is initialized based on the radio distance from the umbrella MBS, which is estimated using macrocell received signal strength measurements.

In the DL, both the pilot transmit power (that defines the cell range) and the maximum data channels transmit power (to limit interference) must be configured. The proposed approach consists of a power control algorithm for pilot and data channels, which ensures a constant femtocell coverage. Each FAP sets its power to a value that on average is equal to the power received from the closest MBS at a target femtocell radius, e.g., according to the features of the household. In this way, a constant femtocell radius is guaranteed regardless of the physical distance from the FAP to the MBS. The power received from the closest MBS can be estimated based on average channel modeling, or using path loss measurements at the femtocell target radius. The received macrocell DL power can also be derived using the built-in sensing capability of the femtocell (i.e., network monitoring mode) or UE measurement reports presented in Section 8.4.

In the UL, where the macrocell suffers from femtocell interference, the UE transmit power is upper bounded by a value so that the aggregate interference of all FUEs to the closest MBS is limited to a predefined value, thereby ensuring that the UL of the macrocell is not degraded significantly. In other words, a maximum interference

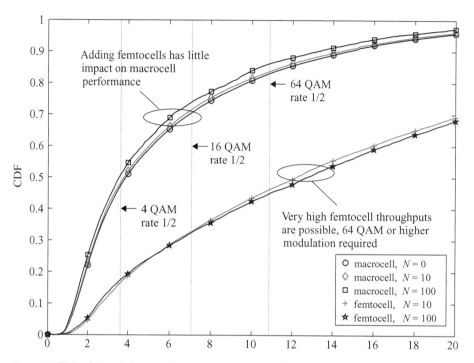

Figure 8.5 CDF of downlink throughputs for locations within the central macrocell for both macrocells and femtocells.

allowance is set for each macrocell, and the interference allowance is shared among all FUEs located within the coverage of the macrocell. This technique can be further improved by sharing the interference allowance only among active FUE UL connections, and by dynamically computing the interference allowance according to the current interference conditions of the macrocell.

Simulations results for open-access UMTS femtocells are presented in Fig. 8.5, where the targeted femtocell radius is set to 10 m and UEs associate with the FAPs providing the strongest pilot signal. The results show that when the above discussed transmit power control is used, the drop in macrocell DL throughput as a result of the additional interference caused by 10 or 100 deployed open-access FAPs is only minimal. This can be explained by the low FAP transmit power and the strong signal falloff at the femtocell boundary, in most cases, due to wall separations. This result is essential for successful co-channel femtocell deployments, since any significant degradation in coverage or capacity of the existing macrocells would be unacceptable. We can also see from Fig. 8.5 that, due to high indoor SINRs, the theoretically achievable femtocell throughput is very high. The high indoor SINRs are a result of the short distances between UEs and the FAP, and the wall separation that protects indoor femtocells from outdoor macrocell interference. Fig. 8.5 also shows that increasing the number of FAPs from 10 to 100 does not significantly affect the DL femtocell throughput due to the small cell size and a typically strong wall separation between any two houses. Therefore, when open access to the

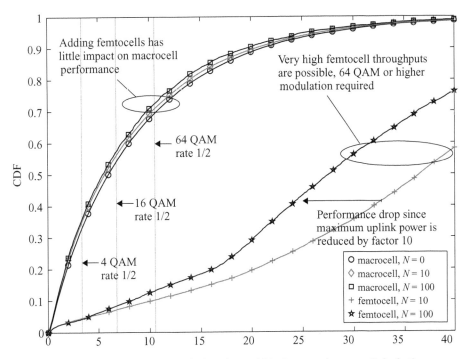

Figure 8.6 CDF of uplink throughputs for locations within the central macrocell for both macrocells and femtocells.

femtocells is granted, large numbers of FAPs can be deployed in the same frequency band of existing macrocell networks without causing a significant DL performance degradation of the macrocells.

Similarly to the DL, the degradation in macrocell UL performance resulting from the addition of 10 or 100 active FAPs is also minimal, as shown in Fig. 8.6. This is a result of the proposed power control in the UL for FUEs that limits the aggregate interference caused to MUEs to a predefined level. Despite the maximum FUE UL power limitation, results also show that the achievable throughput of FUEs is still very high, as a result of high FUE SINR due to the short distance from the serving FAP and the wall protection from the outdoor macrocell. When the number of FAPs is increased from 10 to 100 per macrocell sector, the maximum allowed interference per FUE is reduced by a factor of 10, and the SINR can drop by up to 10 dB, resulting in a reduction in femtocell UL throughput. However, since the femtocell UL throughput is very high, such a reduction is not problematic.

As a conclusion, with open-access and strongest cell selection, indoor femtocell deployments do not worsen the overall interference conditions or change the SINR statistics. This invariance property has also been supported by theoretical research presented in [42], and observed in real-world systems by Nokia Siemens [43] and Qualcomm [44]. It provides optimism that femtocell deployments do not need to compromise the integrity of the existing macrocell network.

8.7.2.2 Power tuning for handover failure mitigation

From the mobility management point of view, widespread deployment of co-channel macrocell and open-access femtocells can result in an increase in the number of macrocell–femtocell handover requests by underlay MUEs (i.e., UEs moving outdoors and connected to the macrocell), due to the leakage of the femtocell pilot signal outside the house that it is deployed in. In order to minimize the number of such handover requests, the femtocell pilot transmit power needs to be adjusted so that leakage outside the house is reduced. In [41], the authors provided simulation studies to obtain a large-scale insight into the effect of implementing pilot power auto-configuration schemes for open-access femtocells in residential scenarios. It was assumed that the pilot power is one-tenth of data power for both macrocells and femtocells, which is a common practice in UMTS. Handover probabilities are calculated for users walking past the front of a house with a femtocell at a speed of 1 meter per second (m/s), at a distance of 1 meter from the house boundary (see Fig. 8.7(a) and (b)). Three pilot power auto-configuration schemes are examined. Scheme A uses a fixed transmit power level that allows 90% of the femtocells within a macrocell coverage area to obtain a target cell radius of 10 m (in free space). In Scheme B, the femtocell pilot is adjusted to obtain a target cell radius of 10 m if it can, while limited by its maximum pilot power, as previously described in Section 8.7.2.1. In Scheme C, when there are active FUEs (e.g., in a call), the maximum pilot power is set to the same level as in Scheme B. When all UEs are in idle mode, the maximum pilot power is reduced by 10 dB. Fig. 8.7(c) illustrates the simulated handover probabilities as a function of the FAP distance from the footpath as shown in Fig. 8.7(a) and (b) in a UMTS network. One example of the resulting coverage for an FAP deployed at the center of the house using Scheme B is shown in Fig. 8.7(a). The complementary areas covered by the macrocell for this example are shown in Fig. 8.7(b). When Scheme A is implemented, according to Fig. 8.7(c), the resulting MUE handover probability per call is significantly high. This suggests that using a fixed transmit power would not be feasible in this scenario due to the very high probabilities of handover and thus call drop (it is estimated that the call drop probability during a handover event is 3.96% of the handover probability). When using Scheme B and Scheme C, the average handover probabilities per call are significantly reduced, which reduces the chance of call dropping due to handover. Scheme C provides the best performance because the FAP transmits at a very low power, when all UEs are in idle mode. The proposed approach fine tunes the pilot power of FAPs, and it can be directly used in LTE FAPs too.

The above presented method provides a baseline approach for initializing the maximum transmit power of open-access femtocells, and ensures a given femtocell coverage radius. However, ensuring a given femtocell coverage radius is not enough for adequate network performance, since households may differ in size and users can locate their FAPs at any positions within their households. Taking this into account, in [45], the authors proposed a more dynamic approach to further refine network performance through coverage adaptation. In this approach, each femtocell sets its maximum pilot power to a value that maximizes its coverage and minimizes on average the total number of unwanted events of attempts of passing and indoor UEs to connect to the femtocell. The unwanted

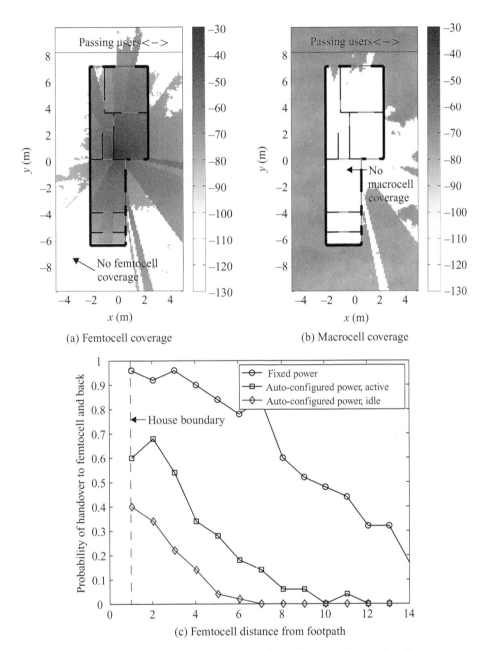

Figure 8.7 Handover probabilities versus the distance from the FAP, and examples of coverage around a house with an FAP inside.

events are defined as those handovers or attempts to hand over in which the UE connects to the femtocell and immediately hands over to a macrocell or another femtocell. Such unwanted events could be caused either by passing UEs that continuously hand over from an open femtocell to another or to the umbrella macrocell when moving across a residential area, or by non-subscribers trying to connect to closed-access femtocells. In contrast, wanted events are those allowed handovers that last longer than a predefined time period. In the approach proposed in [45], the FAP counts the numbers of *unwanted* and *wanted* mobility events. If the number of unwanted events is larger than a given threshold, then the femtocell reduces its pilot power. If the number of unwanted events is smaller than a given threshold, then the femtocell increases its pilot power. These thresholds must be tuned according to the scenario to achieve an optimal performance. Once the pilot power is decreased or increased, the mobility event counters are reset. Using this technique, the femtocell coverage shrinks when the leakage of power from indoors to outdoors causes unwanted mobility events, or enlarges when there are no passing UEs around the femtocell premises. Since the proposed approach fine tunes the pilot power of FAPs, it can be used in both UMTS and LTE FAPs.

The above two discussed power-based techniques [41, 45] can be combined to mitigate intercell interference, core network signaling overhead due to unwanted mobility events, and call drops. Even for FAPs located in unsuitable locations, these two self-optimization methods, which may result in insufficient coverage (i.e., the femtocell coverage may shrink too much), are still preferable from an operator's viewpoint. This is because they would encourage a re-deployment of the FAP to a better location by the users, resulting in improved coverage for UEs and a reduced number of mobility events in the network.

8.7.3   Antenna-based techniques

The performance of power-based techniques can be further improved by additionally tuning the antenna pattern of FAPs. When using a single omnidirectional antenna, the FAP may only self-optimize its maximum pilot power in order to minimize its impact on the macrocells and other femtocells. However, in order to reduce the number of unwanted mobility events, an FAP may have to significantly reduce its pilot power, thus compromising the femtocell coverage and resulting in an inadequate performance. In order to solve this issue, multiple antenna elements can be installed in the FAP to create different antenna patterns that can be used to adapt the femtocell coverage to the household features. However, it is not easy to integrate multiple antenna elements in the FAP due to the tight size and price constraints required to successfully commercialize FAPs. Therefore, the multiple antenna elements for FAPs must be of reduced volume and cost. The system handling the array of antenna elements must also be of low complexity. Accordingly, simple antenna switching systems are preferred to complex beam forming.

In [46], the authors presented an FAP architecture, where four low-size low-cost antennas are installed in the FAP (see Fig. 8.8). In this architecture, no more than two antennas are used simultaneously at any time in order to keep the impedance mismatch low. The use of one or the combination of two antennas generates 10 different antenna patterns (see Fig. 8.9). The target of the self-optimization procedure is to select the FAP transmit pilot power and antenna pattern that maximizes its coverage and minimizes on average the total number of handover attempts. During operation, the femtocell counts

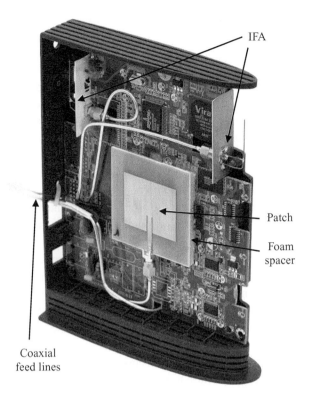

Figure 8.8 Prototype of an FAP with 4 installed small-size low-cost antennas (reproduced with permission of IEEE©).

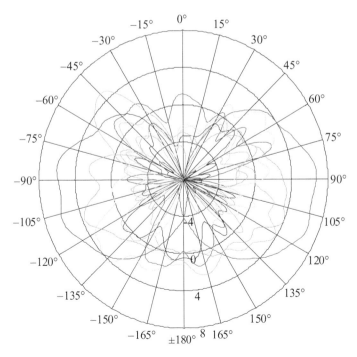

Figure 8.9 Measured antenna patterns resulting from the selection of one or a combination of two antennas (10 patterns in total) (reproduced with permission of IEEE©).

the numbers of unwanted and wanted mobility events as in [45], and then periodically collects information about the coverage performance of all available antenna patterns based on path-loss measurements. These measurements can be derived using the built-in sensing capability of the femtocell or from measurement reports sent by FUEs, as indicated in previous sections. Every time the pilot power is decreased or increased triggered by the mobility event counters, the antenna patterns are re-evaluated and the best combination of the pilot power and antenna pattern is selected.

Simulations were performed to compare the proposed multi-antenna solution with a single-antenna solution. A UMTS scenario was used for simulations, but the proposed multi-antenna solution can also be directly used in LTE FAPs. The simulation results show that the flexibility provided by the multiple-antenna patterns allows shaping the femtocell coverage such that the number of mobility events can be significantly reduced compared to a single-element-antenna solution. The reductions depend on where the FAP is deployed in the house, and has been estimated to be up to 39%. In addition to the reduced number of mobility events, the indoor coverage can be improved compared to a single-antenna solution with a similar sophisticated coverage self-optimization. It is shown that the most significant improvements in coverage are achieved when the FAP is deployed in an unsuitable location close to a footpath, where the flexibility in terms of gain patterns allows better coverage of the rest of the house without increasing the number of unwanted mobility events and the associated core network signaling. Improvements in average indoor coverage have been estimated to be up to 18%. Simulation results also show that antenna patterns with the highest spatial diversity (or those with the smallest pattern similarity) are most often used. Therefore, as a design rule, it is beneficial to achieve as diverse patterns as possible.

8.7.4   Load-balancing-based techniques

Residential femtocell deployments make use of a single FAP to provide coverage in a home to serve a small group of registered UEs. In such scenarios, as shown in previous sections, the aim of femtocell coverage optimization is to ensure that the power leakage is minimized while indoor coverage is maximized in order to reduce the increased core network mobility signaling. In contrast, for open-access femtocell deployments in enterprises, a group of FAPs may be deployed, where the individual FAPs need to collaborate to jointly provide service coverage. Therefore, the requirements for coverage optimization for the deployment of enterprise femtocells differ significantly from residential femtocells, and some of the previously presented solutions may not apply.

In [47], the authors presented a decentralized joint coverage optimization algorithm that runs individually in each FAP, and works towards achieving UE load balancing and minimization of coverage holes and overlap, thus resulting in interference mitigation. The pilot power dictates the femtocell coverage area, and its increment or decrement causes the femtocell coverage area to increase or shrink, respectively. This also results in a change in the number of UEs connected to the respective FAP. The coverage gap represents the area where the received pilot signal from the FAP is below a specified threshold, whereas the coverage overlap denotes the region where coverage areas of two or more femtocells overlap. Some degree of coverage overlap can be advantageous in terms of smooth UE handovers. However, excessive overlap may translate to greater interference to neighboring cells. Thus, the problem of distributed femtocell coverage optimization, by means of updating the femtocell pilot power only, consists of satisfying the following three objectives:

- to balance the UE load amongst the N collocated femtocells for efficient radio resource utilization (prevent overloading or underutilization);
- to minimize radio coverage holes or gaps;
- to achieve the first two objectives with minimum pilot transmit power possible, i.e., to reduce the coverage overlap with the neighboring femtocells.

Some degree of coverage overlap can be advantageous in terms of smooth UE handovers; however, excessive overlap may translate to greater interference to neighboring cells.

The proposed decentralized joint coverage optimization algorithm in [47] solves this multi-objective problem entailing conflicting objectives: increasing the coverage of a femtocell would reduce the number of coverage holes, but doing so may increase the FUE load, and/or increase the coverage overlap with its neighbors. By acquiring the relevant information from its neighbors, the algorithm periodically adjusts the FAP's pilot power to share the system UE load evenly across the network, and minimize coverage holes and overlap. Compared to the fixed pilot power allocation, simulation results show a performance improvement of 18% in terms of femtocell supported traffic and 80% in terms of pilot transmit power reduction. A UMTS scenario was used for simulations, but the proposed multi-antenna solution can also be directly used in LTE FAPs.

8.7.5   Frequency-based techniques

In addition to the previously presented techniques, the flexible OFDMA-based PHY layer of LTE provides more opportunities for intercell interference mitigation and network performance improvements through dynamic orthogonalization, sub-band scheduling, and adaptive fractional frequency reuse strategies [6, 48–50]. In essence, all these strategies tackle intercell interference mitigation through fine RB and transmit power allocations within each carrier. However, due to the lack of X2 interfaces in femtocell networks, where operators cannot rely on femto-to-femto communications in most scenarios, it is desired that each FAP thus make its own radio resource allocation decisions independently.[1]

In general, each cell performs scheduling to decide how RBs are to be allocated to UEs and how much transmit power is to be applied in each RB, so that the network

---

[1] Note that 3GPP Release-8 and Release-10 *intercell interference coordination* (ICIC) techniques presented in Chapter 7 may apply to enterprise and outdoor FAPs provided that a reliable X2 interface can be used. This is not usually the case in FAP deployments, where the backhaul is usually user provided through an IP connection.

capacity is enhanced. The radio resource allocation problem is complex because UEs may have different quality of service demands and may experience various channel conditions in different RBs. Moreover, there are scheduling constraints, e.g., when more than one RB is allocated to a UE, all these RBs must use the same *modulation and coding scheme* (MCS) [51].

In [52], the authors introduced a simple self-organization rule based on minimizing the DL transmit power per femtocell, and proposed a distributed resource allocation algorithm taking realistic resource allocation constraints into account. Following the proposed algorithms, a distributed femtocell network is able to converge into an efficient resource reuse pattern. In more detail, the proposed self-organization rule is defined as: *each cell assigns MCSs, RBs, and transmit power to UEs independently, while minimizing the cell total transmit power across all used RBs and meeting its UEs' throughput demands.* There are two reasons to minimize the total transmit power per cell.

1. A cell that aims at minimizing its own transmit power mitigates intercell interference to neighboring cells, because less power is allocated to those UEs with good channel conditions or lower throughput demands.
2. A cell that aims at minimizing its own transmit power tends to allocate those RBs that are not being used by its neighboring cells, because less transmit power is required for a less interfered and/or faded RB to get a targeted SINR.

Following this rule, a cell tends to allocate UEs that are closer to the FAP or have lower data-rate requirements (hence requiring lower MSCs as well as transmit power) to RBs that are used by cell-edge UEs in neighboring cells.

Simulation results in an open-access femtocell enterprise scenario show that, compared to existing radio resource allocation techniques in the literature, the proposed distributed approach is able to significantly decrease the number of UE outages, increase the average number of simultaneously transmitting UEs in the network (by around 15%), and enhance the average network sum throughput (by around 12%). This is because the proposed self-organization rule achieves intercell interference coordination without the need to assign orthogonal RBs among neighboring cells. Instead, it allows all cells to allocate all RBs to their UEs in an intelligent manner. A cell that minimizes its own transmit power assigns low transmit power to those RBs allocated to UEs having good channel conditions or with low data-rate demands. Thus, neighboring cells will 'see' low interference in such RBs and will allocate them to UEs having bad channel conditions or with high data-rate demands, thereby improving spatial reuse. In order to illustrate the self-organizing feature, Fig. 8.10 shows the transmit power allocated by three neighboring femtocells at a given time in eight available RBs. We can see that each femtocell tends to allocate higher power levels in RBs in which neighboring femtocells assign lower power levels and vice versa. In this way, the same RB can be dynamically reused in neighboring femtocells.

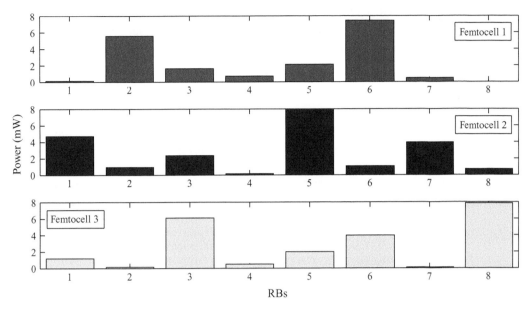

Figure 8.10 RB allocation in three neighboring femtocells.

8.8   Summary

In this chapter, we have provided a detailed review of uncoordinated femtocell deployments. We have covered the evolving femtocell market and deployment scenarios. The Small Cell Forum has also been introduced and self-organizing femtocell networks have also been presented for perspective. Special attention has also been paid to femtocell backhaul together with synchronization and localization issues, which are especially critical due to the less strict backhaul requirements as compared to other HCN small cells. Moreover, power-based, antenna-based, and frequency-domain techniques for interference mitigation in uncoordinated femtocells have been reviewed in detail as the main block of this chapter.

Copyright notices

Figures 8.7, 8.8 and 8.9 © 2009 IEEE. Reprinted, with permission, from H. Claussen and F. Pivit, Femtocell coverage optimization using switched multi-element antennas. In *IEEE Proceedings*, 2009.

References

[1] G. Mansfield, Femtocells in the US market – business drivers and consumer propositions. In *FemtoCells Europe* (London: ATT, 2008).

[2] J. Cullen, Radioframe presentation. In *FemtoCells Europe* (London, 2008).

[3] J. Zhang and G. de la Roche, *Femtocells: Technologies and Deployment* (Wiley, 2010).

[4] V. Chandrasekhar, J. Andrews and A. Gatherer, Femtocell networks: a survey. *IEEE Communications Magazine*, **46**:9 (2008), 59–67.

[5] D. Knisely, T. Yoshizawa and F. Favichia, Standardization of femtocells in 3GPP. *Communications Magazine, IEEE*, **47**:9 (2009), 68–75.

[6] D. López-Pérez, A. Valcarce, G. de la Roche and J. Zhang, OFDMA femtocells: a roadmap on interference avoidance. *IEEE Communications Magazine*, **47**:9 (2009), 41–48.

[7] Femto Forum, *Interference Management in UMTS Femtocells*. Technical Report (2008).

[8] D. López-Pérez, I. Guvenc, G. de la Roche, M. Kountouris, T. Quek and J. Zhang, Enhanced intercell interference coordination challenges in heterogeneous networks. *Wireless Communications, IEEE*, **18**:3 (2011), 22–30.

[9] ABI Research. http://www.abiresearch.com

[10] Informa Telecoms & Media. http://www.informatandm.com/section/home-page/

[11] Femto Forum (press releases). http://www.femtoforum.org

[12] D. López-Pérez, I. Güvenç, G. de la Roche, M. Kountouris, T. Q. Quek and J. Zhang, Enhanced inter-cell interference coordination challenges in heterogeneous networks. *IEEE Wireless Communications Magazine*, **18**:3 (2011), 22–31.

[13] D. López-Pérez, A. Valcarce, A. Ladênyi, G. de la Roche and J. Zhang, Intracell handover for interference and handover mitigation in OFDMA two-tier macrocell femtocell networks. *EURASIP Journal on Wireless Communications and Networking*, Article ID 142629 (2010). DOI: 10.1155/2010/142629

[14] G. de la Roche, A. Valcarce, D. López-Pérez and J. Zhang, Access control mechanisms for femtocells. *IEEE Communications Magazine*, **48**:1 (2010), 33–39.

[15] FemtoForum, *LTE Network Monitor Mode Specification* (2010).

[16] FemtoForum, *Interference Management in OFDMA Femtocells* (2010).

[17] S. Chia, M. Gasparroni and P. Brick, The next challenge for cellular networks: backhaul. *IEEE Microwave Magazine*, **10**:5 (2009), 54–66.

[18] D. Calin, H. Claussen and H. Uzunalioglu, On femto deployment architectures and macrocell offloading benefits in joint macro femto deployments. *IEEE Communications Magazine*, **48**:1 (2010), 26–32.

[19] V. Chandrasekhar, J. G. Andrews and A. Gatherer, Femtocell networks: a survey. *IEEE Communications Magazine*, **46**:9 (2008), 59–67.

[20] H. Raza, A brief survey of radio access network backhaul evolution: part I. *IEEE Communications Magazine*, **49**:6 (2011), 164–171.

[21] R. Y. Kim, Jin Sam Kwak and K. Etemad, WiMAX femtocell: requirements, challenges, and solutions. *IEEE Communications Magazine*, **47**:9 (2009), 84–91.

[22] IEEE Instrumentation and Measurement Society, *IEEE 1588 Standard for a Precision Clock Synchronization Protocol for Networked Measurement and Control Systems*. IEEE Std 1588 (2008).

[23] J. Yoon, J. Lee and H. S. Lee, Multihop based network synchronization scheme for femtocell systems. In *IEEE PIMRC 2009*, Tokyo (2009).

[24] Femtocell synchronization and location, a Femto Forum topic brief. *Femto Forum Whitepapers*, **1**:15 (2010).

[25] R. Rowe, P. Duffett-Smith, M. Jarvis and N. Graube, Enhanced GPS: the tight integration of received cellular timing signals and GNSS receivers for ubiquitous positioning. In *IEEE/ION Position, Location and Navigation Symposium* (2008), pp. 838–845.

[26] *SiRFstarIV GSD4t Datasheet* (Cambridge: CSR).

[27] Rosum Corporation, *In-Building Location, Timing, and Frequency Coverage Analysis of A-GPS and TV-GPS for Femtocell Applications*, Whitepaper (2009).

[28] K. M. Pesyna Jr., K. D. Wesson, R. W. Heath Jr. and T. E. Humphreys, Extending the reach of GPS-assisted femtocell synchronization and localization through tightly-coupled opportunistic navigation. In *IEEE GlobeCom 2011 Workshop on Femtocell Networks*, Houston, Texas (2011).

[29] S. Lien, H. Lee, S. Shih, P. Chen and K. Chen, Network synchronization among femtocells. In *IEEE GlobeCom 2011 Workshop on Femtocell Networks*, Houston, Texas (2011).

[30] FCC Docket No. 94-102, *Revision of the Commission's Rules to Ensure Compatibility with Enhanced 911*.

[31] FCC Docket No. 07-114, *In the Matter of Wireless E911 Location Accuracy Requirements*.

[32] J. D. Hobby and H. Claussen, Deployment options for femtocells and their impact on existing macrocellular networks. *Bell Labs Technical Journal*, **13** (2009), 145–160.

[33] D. López-Pérez, G. de la Roche, A. Valcarce, Á. Jüttner and J. Zhang, Interference avoidance and dynamic frequency planning for WiMAX femtocells networks. In *11th IEEE Singapore International Conference on Communication Systems (ICCS)* (Guangzhou, China, 2008), pp. 1579–1584.

[34] D. López-Pérez, Á. Ladányi, A. Jüttner and J. Zhang, OFDMA femtocells: a self-organizing approach for frequency assignment. In *IEEE Personal, Indoor and Mobile Radio Communications Symposium (PIMRC)*, Tokyo (2009).

[35] V. Chandrasekhar and J. G. Andrews, Spectrum allocation in tiered cellular networks. *IEEE Transactions on Communications*, **57**:10 (2009), 3059–3068. http://arxiv.org/abs/0805.1226

[36] L. T. W. Ho and H. Claussen, Effects of user-deployed, co-channel femtocells on the call drop probability in a residential scenario. In *IEEE 18th International Symposium on Personal, Indoor and Mobile Radio Communications (PIMRC 2007)*, Athens (2007), pp. 1–5.

[37] H. Claussen, Co-channel operation of macro- and femtocells in a hierarchical cell structure. *International Journal of Wireless Information Networks*, **15**:3 (2008), 137–147.

[38] X. Chu, Y. Wu, D. López-Pérez and X. Tao, On providing downlink services in collocated spectrum-sharing macro and femto networks. *IEEE Transactions on Wireless Communications*, **10**:12 (2011), 4306–4315.

[39] H. Claussen, L. T. W. Ho and L. G. Samuel, An overview of the femtocell concept. *Bell Labs Technical Journal*, **3**:1 (2008), 221–245.

[40] V. Chandrasekhar, M. Kountouris and J. Andrews, Coverage in multi-antenna two-tier networks. *IEEE Transactions on Wireless Communications*, **8**:10 (2009), 5314–5327.

[41] J. Y. Wu, X. Chu and D. López-Pérez, Downlink outage probability of co-channel femtocells in hierarchical 3-sector macrocells. *IEEE Communications Letters*, **16**:5 (2012), 698–701.

[42] H. Dhillon, R. Ganti, F. Baccelli and J. Andrews, Modeling and analysis of K-tier downlink heterogeneous cellular networks. *IEEE Journal on Selected Areas in Communications*, **30**:3 (2012), 550–560.

[43] A. Ghosh, N. Mangalvedhe, R. Ratasuk, B. Mondal, M. Cudak, E. Visotsky, T. Thomas, J. Andrews, P. Xia, H. Jo, H. Dhillon and T. Novlan, Heterogeneous cellular networks: from theory to practice. *IEEE Communications Magazine*, **50**:6 (2012), 54–64.

[44] A. Damnjanovic, J. Montojo, Y. Wei, T. Ji, T. Luo, M. Vajapeyam, T. Yoo, O. Song and D. Malladi, A survey on 3GPP heterogeneous networks. *IEEE Wireless Communications*, **18**:3 (2011), 10–21.

[45] H. Claussen, L. T. W. Ho and L. G. Samuel, Self-optimization coverage for femtocell deployments. In *Wireless Telecommunications Symposium*, ser. 24–26, California (2008), pp. 278–285.

[46] H. Claussen and F. Pivit, Femtocell coverage optimization using switched multi-element antennas. In *IEEE International Conference on Communications*, Dresden (2009).

[47] I. Ashraf, H. Claussen and L. T. Ho, Distributed radio coverage optimization in enterprise femtocell networks. In *IEEE International Conference on Communications (ICC)* (May 2010), pp. 1–6.

[48] G. Fodor, C. Koutsimanis, A. Racz, N. Reider, A. Simonsson and W. Muller, Intercell interference coordination in OFDMA networks and in the 3GPP Long Term Evolution system. *Journal of Communication*, **4**:7 (2009), 445–453.

[49] K. Sundaresan and S. Rangarajan, Efficient resource management in OFDMA femto cells. In *Proceedings of the Tenth ACM International Symposium on Mobile Ad Hoc Networking and Computing*, ser. MobiHoc '09 (New York: ACM, 2009), pp. 33–42. http://doi.acm.org/10.1145/1530748.1530754

[50] T. Novlan, R. Ganti, A. Ghosh and J. Andrews, Analytical evaluation of fractional frequency reuse for OFDMA cellular networks. *IEEE Transactions on Wireless Communications*, **10**:12 (2011), 4294–4305.

[51] E. Dahlman, S. Parkvall, J. Sköld and P. Beming, *3G Evolution: HSPA and LTE for Mobile Broadband*, 2nd edn (Elsevier, 2008).

[52] D. López-Pérez, X. Chu, A. V. Vasilakos and H. Claussen, Minimising cell transmit power: towards self-organized resource allocation in OFDMA femtocells. In *Proceedings of the ACM SIGCOMM 2011 Conference*, ser. SIGCOMM '11 (New York: ACM, 2011), pp. 410–411. http://doi.acm.org/10.1145/2018436.2018494

9   Mobility and handover management

Huaxia Chen, Shengyao Jin, Honglin Hu, Yang Yang, David López-Pérez, Ismail Güvenç and Xiaoli Chu

9.1   Introduction

Compared with current cellular networks, next generation mobile networks are expected to encompass more sophisticated features, including the support of higher data transmission rates and *user equipment* (UE) mobility, location management, diversified service levels, etc. In order to accommodate these requirements, the *3rd Generation Partnership Project* (3GPP) is devoted to the standardization of *Long Term Evolution* (LTE) and LTE-Advanced systems, which have been recognized as major candidates for the *fourth-generation* (4G) mobile networks. In LTE/LTE-Advanced systems, the network structure will be heterogeneous. How to maintain and improve mobility, *handover* (HO), and location management, while avoiding user experience deterioration, is a challenging task. In this chapter, we will study the mobility management challenge and illustrate advanced mobility management schemes.

In LTE/LTE-Advanced systems, the factors that make mobility, HO, and location management a challenging task are as follows

- The rapid evolution of cellular networks results in the coexistence of multiple *radio access technologies* (RATs), e.g., *Global System for Mobile Communications* (GSM), *Universal Mobile Telecommunication System* (UMTS) and LTE/*System Architecture Evolution* (SAE). This demands optimized cooperation among multiple RATs to enable UEs to roam from one RAT to another.
- The introduction of *low-power nodes* (LPNs) largely increases the total number of *base stations* (BSs), making the network structure and interference conditions more intricate. Thus, traditional *mobility load balancing* (MLB) and mobility management schemes need to be revisited to suit the new *heterogeneous cellular network* (HCN) architecture.
- The complexity of LTE/LTE-Advanced systems leads to a large number of network parameters. Therefore, efforts need to be made in defining proper key performance indicators and developing optimization techniques for mobility management in various scenarios.

Due to the unevenly distributed and time variant traits of mobile UEs and their traffics, traffic loads are usually unbalanced among different cells, resulting in a low resource utilization efficiency. In order to solve this problem, while minimizing cell reselections and HOs, mobility parameters in each cell need to be carefully and dynamically optimized according to cell traffic loads. Improper selection of mobility parameters may result in *radio link failures* (RLFs), unnecessary HOs and/or *handover failures* (HOFs), which will degrade user experience and network performance.

Mobility robustness optimization (MRO) encompasses the dynamic optimization of mobility parameters, tuning active mode HO and idle mode cell reselection parameters to ensure satisfactory mobility performance. MRO aims to provide enhanced user experience as well as network capacity, through reducing RLFs, unnecessary HOs and/or HOFs. While RLFs that do not lead to HOFs are often "invisible" to UEs, RLFs caused by incorrect HO parameter settings have a combined passive impact on both user experience and network capacity. Therefore, the primary objective of MRO is to reduce HO-related RLFs. Moreover, sub-optimal configuration of HO parameters may lead to service performance degradation, even if it does not result in RLFs. One example is the incorrect setting of HO hysteresis parameters, which may result in ping-pongs. Therefore, the secondary objective of MRO is to reduce the waste of network resources caused by ping-pongs.

In this chapter, we will study mobility robustness, HO management, and MLB in LTE/LTE-Advanced systems, and present MRO algorithms under both the *Radio Resource Control* (RRC) connected and idle modes. Under the RRC-connected mode, we will introduce the hard HO procedure in LTE and the MRO use case for UEs. Based on the mobility status of UEs, we propose a novel MRO scheme, which assigns different HO hysteresis parameters for UEs traveling at different velocities. Simulation results show that the proposed scheme greatly increases the HO success rate. Under the RRC-idle mode, we will present cell reselection procedures for UEs. Due to the conflict between mobility robustness and MLB management, which leads to the mismatching between HO and cell reselection boundaries, we also consider a negotiation procedure among cells in the process of optimizing cell reselection parameters. Simulation results show that the proposed negotiation procedure helps to balance cell traffic loads and decrease the number of ping-pongs. Moreover, in HCN scenarios, we identify the technical challenges in mobility management, and evaluate the mobility performance under 3GPP Release-10 enhanced *intercell interference coordination* (ICIC) features such as *almost blank subframes* (ABSs). Then, we present a mobility-based ICIC (MB-ICIC) scheme, in which LPNs configure coordinated resources so that macrocells can schedule their high-mobility UEs in these resources without co-channel interference from LPNs.

9.2   Mobility management in RRC-connected state

In [1], two mobility-related UE states are defined, which are RRC-connected and RRC-idle. The UE actions under the RRC-connected state will trigger HO procedures, while the RRC-idle state will trigger cell reselection procedures. In this section, we look into mobility management issues in RRC-connected mode.

UE mobility management in RRC-connected state is implemented through UE HO procedures. Key performance indicators that are widely used to evaluate HO performance include HO complete rate, HOF rate and ping-pong rate. In order to optimize mobility robustness, we first analyze the HO mechanism and define HO complete, HOF and ping-pong events. Then, we review the existing MRO schemes by referring to standardization efforts and introduce an improved scheme.

9.2.1   Overview of the handover procedure in LTE systems

In cellular networks, the HO is the process through which a mobile UE communicating with one BS is switched to another BS during a call or data service [2–4]. HO is an essential function in UE mobility management. Roaming UEs could be handed over among different RATs, different frequencies, and/or different cells. In this chapter, we only consider the intra-RAT and intrafrequency HO, where the source *evolved NodeB* (eNB) and the target eNB use the same RAT and the same frequency and communicate through an X2 interface [5, 6], with the *Mobility Management Entity* (MME) kept unchanged.

Based on the connectivity between the UE and the source eNB, before the UE connects to the target eNB, we can define two types of HO. One is hard HO, where the UE has cut connection with the source eNB before establishing connection with the target eNB. The other is soft HO, where the UE keeps the connections with both the source and target eNBs during the HO process. Since LTE systems are based on a flat architecture without *Radio Network Controllers* (RNCs), and adopt direct internode interfaces instead, LTE systems do not support macro-diversity combining methods. In other words, soft HO is made impossible by the decentralization of the network. Hence, LTE systems adopt hard HOs only, which will be considered hereafter.

The decision to HO from one cell to another is based on various criteria that take into account channel degradation, Erlang capacity, and blocking considerations. In LTE systems, event triggered *measurement reports* (MRs) based on *reference signal received power* (RSRP) and *reference signal received quality* (RSRQ) measurements have been specified. The eNBs may make HO decisions based on these MRs. Filtering of RSRP and RSRQ samples, *handover hysteresis margin* (HOM) and *time-to-trigger* (TTT) mechanisms are also provided to support efficient HO decisions and to avoid frequent HOs. For an intra-MME HO, the entire HO process will be mostly confined between the source eNB and the target eNB, although there may be interactions with the MME to switch the user plane.

The HO process can typically be divided into four stages: measurements, processing, decision, and execution. HO measurements are RSRP measurements used as a basis for HO decisions. RSRP measurements are taken in the downlink by UEs. Processing of RSRP measurements is also performed at the UE to filter out the effects of fast fading and Layer 1 measurement/estimation imperfections. The processed measurements are then reported back to the source eNB in a periodic or event-triggered manner through MRs. Based on the processed HO measurements and when a certain criterion is met, the HO is initiated. In the following, we first introduce the HO triggering procedure, and then describe the HO procedure as well as the message exchange flow between the source eNB and the target eNB in both the control plane and the UE plane.

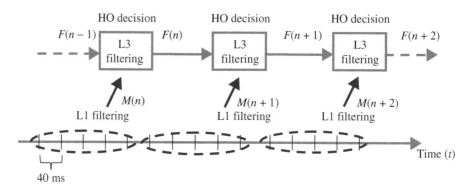

Figure 9.1 L1 and L3 filtering procedures. RSRP is measured over one subframe (1 ms and, e.g., six RBs) every 40 ms and recorded as $RSRP_{L1}(l)$. L1 filtering performs averaging over every 200 ms to provide $M(n) = \frac{1}{5} \sum_{k=0}^{4} RSRP_{L1}(l - k)$. Finally, L3 filtering performs averaging over every 200 ms to obtain $F(n) = (1 - a)F(n - 1) + a10 \log_{10}\{M(n)\}$, where a is the L3 filter coefficient.

9.2.1.1 Handover triggering procedure

In LTE systems, RSRP measurements and their processing are performed by the UE in Layer 1 and Layer 3, as shown in Fig. 9.1 [7]. RSRP is estimated as the average received power of all resource elements comprising the common reference signal in a subframe. In order to mitigate the effects of shadowing and fast fading from RSRP measurements, the UE also averages over several DL RSRP samples. Such linear averaging is performed in Layer 1, and hence is known as L1 filtering. For a typical setup (as shown in Fig. 9.1), in order to obtain an L1 filtered HO measurement, RSRP samples may be taken every 40 ms, and then averaged over every five samples.

The L1 filtered HO measurement is updated every HO measurement period (e.g., 200 ms) at the UE through a first-order *infinite impulse response* (IIR) filter in Layer 3, namely L3 filtering, as defined in Fig. 9.1 [7]. A typical L3 filtering period is 200 ms. Since successive log-normal shadowing samples are spatially correlated, it would be preferred for the L3 filtering period to be adaptive to the degree of shadowing correlation in the common reference signal. For example, shadowing would be less correlated for higher-mobility UEs, and thus it would be better to have a shorter L3 filtering period.

In order to trigger the HO procedure, the UE will send an MR to the serving cell if the L3 filtered HO measurement meets an event entry condition. In LTE, there are eight types of event entry condition (see [8], Section 5.5.4):

- Event A1: server becomes better than threshold;
- Event A2: server becomes worse than threshold;
- Event A3: neighbor becomes offset better than server;
- Event A4: neighbor becomes better than threshold;
- Event A5: server becomes worse than threshold1 and neighbor becomes better than threshold2;
- Event A6: neighbor becomes offset better than secondary server (this condition applies to carrier aggregation configurations);

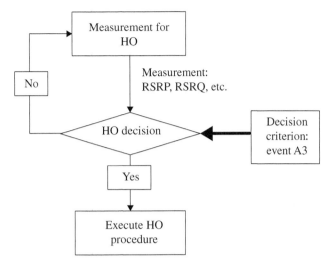

Figure 9.2 Overall HO process in LTE systems.

- Event B1: inter-RAT neighbor becomes better than threshold;
- Event B2: server becomes worse than threshold1 and inter-RAT neighbor becomes better than threshold2.

Events A1–A6 are used for intra-LTE system report triggering, and events B1 and B2 are used for intersystem report triggering. In LTE systems, MRs exchanged among eNBs operating in the same frequency band are triggered by event A3, which will be the focus of our studies in the following. Fig. 9.2 illustrates the HO process, while Fig. 9.3 illustrates event A3 [9].

The inequality of the event A3 entry condition is given by

$$M_{n_j} > M_s + Hyst_s - CIO_{s,n_j}, \tag{9.1}$$

where M_s is the filtered RSRP of the serving cell, not taking into account any offset, M_{n_j} is the filtered RSRP of the neighboring cell n_j, not taking into account any offset, $Hyst_s$ is the hysteresis parameter, and CIO_{s,n_j} is the cell-specific offset of the neighbor cell n_j with respect to the serving cell s [10].

Once the event A3 entry condition (9.1) is met, the UE starts the time to trigger timer. In order to mitigate fading effects, only when the event A3 entry condition is satisfied through the whole TTT time window, the UE reports event A3 via a MR to the serving cell, which may trigger an HO.

Optimization of the hysteresis $Hyst_s$ and/or TTT parameters could be used to mitigate HOFs and ping-pongs caused by too frequent HOs, and thus reduce the signaling overhead. For example, small values of TTT may lead to too early HOs, increasing ping-pongs, while large TTT may result in too late HOs, increasing HOFs. Different from $Hyst_s$ and TTT, which are used to adjust parameters in the serving eNB, *cell individual offset* (CIO) is used to regulate the target cell chosen by the UE. CIO is an offset that will be applied by the UE to the measurement results of neighboring cells before it

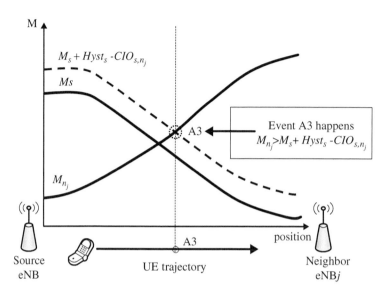

Figure 9.3 Illustration of event A3 entry condition.

determines whether or not an event A3 has occurred. From Fig. 9.3, we can observe that, if the values of $Hyst_s$ and CIO_{s,n_j} are improperly set, the event A3 entry condition may be met too late or too early, causing too late HO or too early HO, respectively.

9.2.1.2   Handover procedure and message flow

Control plane

In the control plane, detailed intra-MME/Serving Gateway HO procedures are illustrated in Fig. 9.4, where the HO measurement phase is depicted from step 0 to step 3, the HO preparation phase from step 4 to step 7, the HO execution phase from step 8 to step 11, and the HO complete phase from step 12 to step 18 [1]. In the following, let us describe the intra-MME/Serving Gateway HO procedure in more detail.

0 The UE context within the source eNB contains information regarding roaming restrictions, which are provided either at connection establishment or at the last *timing advance* (TA) update.

1 The source eNB configures the UE measurement procedures according to the area roaming restriction information. Measurements provided by the source eNB may assist the function controlling the UE's connection mobility.

2 The UE is triggered to send a MEASUREMENT REPORT by the rules set by the system, e.g., if the event A3 entry condition is met.

3 The source eNB decides whether or not to hand over the UE based on the MEASUREMENT REPORT and radio resource management information.

4 The source eNB sends a HANDOVER REQUEST message to the target eNB, passing necessary information to prepare for HO. The UE X2/S1 signaling references enable the source eNB to address the target eNB and the *Evolved Packet Core* (EPC). The

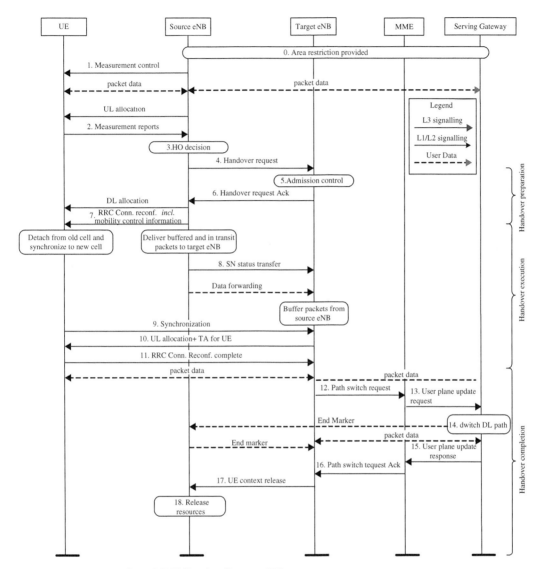

Figure 9.4 Intra-MME/Serving Gateway HO.

E-UTRAN radio access bearer (E-RAB) context includes necessary *radio network layer* (RNL) and *transport network layer* (TNL) address information, and *quality of service* (QoS) profiles of the E-RABs.

5 Admission control may be performed by the target eNB depending on the received E-RAB QoS information to increase the likelihood of a successful HO (if necessary resources can be granted by the target eNB). The target eNB configures the required resources according to the received E-RAB QoS information, and reserves a *cell radio network temporary identifier* (C-RNTI) and optionally a *random access channel* (RACH) preamble.

6 The target eNB prepares HO in Layer 1 and Layer 2, and sends a HANDOVER REQUEST ACKNOWLEDGE message to the source eNB. The HANDOVER REQUEST ACKNOWLEDGE message includes a transparent container to be sent to the UE and an RRC message to perform the HO. The container includes a new C-RNTI and the target eNB security algorithm identifiers for the selected security algorithms. Moreover, it may include a dedicated RACH preamble, and possibly some other parameters, e.g., access parameters, *system information blocks* (SIBs), etc. If necessary, the HANDOVER REQUEST ACKNOWLEDGE message may also include RNL/TNL information for the forwarding tunnels.

7 The target eNB generates the RRC request to perform the HO, i.e., the RRCConnectionReconfiguration request, which includes the mobilityControlInformation to be sent by the source eNB to the UE. The source eNB performs the necessary integrity protection and ciphering of the RRCConnectionReconfiguration request, and sends a HANDOVER COMMAND message to the UE. The UE receives the HANDOVER COMMAND message and thus the RRCConnectionReconfiguration request with necessary HO parameters (i.e., new C-RNTI, target eNB security algorithm identifiers, optionally dedicated RACH preamble, target eNB SIBs, etc.). The UE is instructed by the source eNB to perform the HO. The UE does not need to delay the HO execution for delivering the *hybrid automatic repeat request* (HARQ)/*automatic repeat request* (ARQ) responses to the source eNB.

8 The source eNB sends the *serial number* (SN) STATUS TRANSFER message to the target eNB to convey the uplink *Packet Data Convergence Protocol* (PDCP) SN receiver status and the DL PDCP SN transmitter status of E-RABs, for which PDCP status preservation applies (i.e., for *Radio Link Control* (RLC) *acknowledged mode* (AM)). The *uplink* (UL) PDCP SN receiver status includes at least the PDCP SN of the first missing Uplink Service Data Unit (UL SDU), and may also include a bitmap with the UL SDUs that the UE needs to retransmit to the target cell, if there are any such SDUs. The DL PDCP SN transmitter status indicates the next PDCP SN that the target eNB shall assign to new SDUs, which do not have a PDCP SN yet. The source eNB may omit sending this message if none of the E-RABs of the UE needs to be treated with PDCP status preservation.

9 After receiving the RRCConnectionReconfiguration message including the mobilityControlInformation, the UE performs synchronization to the target eNB and accesses the target cell via RACH, following a contention-free procedure if a dedicated RACH preamble was indicated in the mobilityControlInformation, or following a contention-based procedure if no dedicated preamble was indicated (refer to Chapter 6 for more information on RACH procedures). The UE derives target eNB specific keys and configures the selected security algorithms to be used in the target cell.

10 The target eNB responds with UL allocation and TA.

11 When the UE has successfully accessed the target cell, in order to confirm that the HO is completed for the UE, the UE sends the RRCConnectionReconfigurationComplete acknowledgment along with an UL buffer status report, whenever possible, to the target eNB through a HO COMPLETE message. The target eNB verifies the

C-RNTI sent in the RRCConnectionReconfigurationComplete acknowledgment, and can start sending data to the UE.

12 The target eNB sends a PATH SWITCH message to inform the MME that the UE has changed its serving cell.

13 The MME sends an UPDATE USER PLANE REQUEST message to the Serving Gateway.

14 The Serving Gateway switches the downlink data path to the target cell, and sends one or more "end marker" packets on the old path to the source eNB, and then releases any UE-plane/TNL resources towards the source eNB.

15 The Serving Gateway sends an UPDATE USER PLANE RESPONSE message to the MME.

16 In response to the PATH SWITCH message, the MME confirms with the PATH SWITCH ACKNOWLEDGE message.

17 The target eNB sends the UE CONTEXT RELEASE message after the PATH SWITCH ACKNOWLEDGE message is received from the MME. By sending the UE CONTEXT RELEASE message, the target eNB informs the source eNB of the successful completion of the HO and triggers the release of resources by the source eNB.

18 Upon reception of the UE CONTEXT RELEASE message, the source eNB releases control-plane resources associated to the UE context, while any ongoing data forwarding may continue.

User plane

In the UE plane, the following principles are taken into account to avoid data loss and support seamless data service during the HO process.

1 During HO preparation, UE-plane tunnels can be established between the source eNB and the target eNB. There is one tunnel established for UL data forwarding and another one for DL data forwarding for each E-RAB.

2 During HO execution, UE data can be forwarded from the source eNB to the target eNB. The forwarding may take place in a service- or deployment-dependent manner.
   • Forwarding of DL UE data from the source to the target eNB should take place if packets are received at the source eNB from the EPC or the source eNB buffer has not been emptied.

3 During HO completion:
   • The target eNB sends a PATH SWITCH message to inform the MME that the UE has gained access. The MME then sends a USER PLANE UPDATE REQUEST message to the Serving Gateway, and the UE-plane path is switched by the Serving Gateway from the source eNB to the target eNB.
   • The source eNB should continue forwarding UE-plane data if packets are received at the source eNB from the Serving Gateway or the source eNB buffer has not been emptied.

The HO procedure in LTE systems could also support seamless data transmission through the following special characteristics. During the execution of the HO procedure,

UEs will cut the connection with the source eNB and become synchronized to the target eNB. Simultaneously, the source eNB will forward the DL UE data to the target eNB, so that when the HO procedure is completed, the target eNB could transmit the original data to the UE.

These procedures imply that in LTE systems, the lossless HO is achieved through packet forwarding from the source eNB to the target eNB. Nevertheless, in the HO execution process, an HOF is likely to occur if high interference is experienced or parameters are improperly set when the UE has cut its connection with the source eNB but has not established a new connection with the target eNB. Most problems associated with HOFs or sub-optimal system performance can be associated with either too early or too late HOs, provided that the required fundamental network coverage exists.

9.2.2   Handover failures and ping-pongs

In 2G/3G systems, HO parameters are set manually by network operators, which is time consuming and is not robust to channel fluctuations. Following such manual configuration, when the network environment varies quickly, HOFs are likely to occur, due to too late HOs, too early HOs, and/or HOs to wrong cells.

In order to solve these problems and optimize the HO procedure, 3GPP has introduced MRO in LTE as part of *self-organizing network* (SON) as is discussed in Chapter 6. The main goal of MRO is to reduce the HO related RLFs and to avoid the waste of network resources caused by unnecessary HOs and ping-pongs. In the following, we relate to the discussion in Section 6.4.4 where HOF types are discussed together with MRO requirements. Here, further aspects of RLF, HOF, and ping-pong definitions and classifications are discussed together with an improved scheme to enhance mobility performance.

9.2.2.1   Handover failure

For the purpose of modeling and classifying HOFs, the HO procedure is divided into three states according to [12]:

1 State 1: before the event A3 entering condition is satisfied;
2 State 2: after the event A3 entering condition is satisfied, but before the HO COMMAND is successfully received by the UE;
3 State 3: after the HO COMMAND is received by the UE, but before the HO COMPLETE is successfully sent by the UE.

Moreover, the evaluation criterion of RLF [13, 14] is based on the model illustrated in Fig. 9.5 [15], where Q_{out} is the wideband *signal to interference plus noise ratio* (SINR) threshold for an out-of-sync event and Q_{in} is the wideband SINR threshold for an in-sync event. T310 is a timer that is triggered when the number of out-of sync events have reached a certain value. N310 is the threshold for the number of out-of-sync events to start the T310 timer. N311 is the threshold for the number of in-sync events to abort the T310 timer. In Fig. 9.5, x-axis represents the time domain, and the y-axis represents the variation of SINR values. In order to mitigate fading effects, when a UE tracks RLFs,

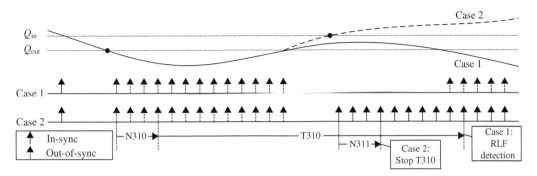

Figure 9.5 RLF detection model.

Q_{out} is monitored with a 200 ms window and Q_{in} is monitored with a 100 ms window (as specified in [13]). Both windows are updated once per frame, i.e., once every 10 ms, with the measured wideband *channel quality indicator* (CQI) value. If a UE detects that its average wideband CQI value is lower than Q_{out}, it will report an out-of-sync event. Thereafter, if it detects that its average wideband CQI value is higher than Q_{in}, it will report an in-sync event. When the out-of-sync event has been reported N310 times, the eNB will start the T310 timer, which is the time limit to decide whether an RLF occurs. If the in-sync event is detected less than N311 times when the T310 timer expires, an RLF occurs, otherwise, the T310 timer is aborted.

An HO failure occurs if an RLF occurs in state 2 or state 3 [12].

1 In state 2: when the UE is attached to the source cell, an HOF is counted if one of the following criteria is met: (1) timer T310 has been triggered or is running when the HO COMMAND is received by the UE; (2) RLF is declared in state 2. For monitoring RLF in state 2, the wideband CQI is measured once every 10 ms, and wideband CQIs are filtered by a linear filter with a sliding window of 200 ms (i.e., 20 samples).

2 In state 3: after the UE is attached to the target cell, an HOF is counted if the target cell filtered average wideband CQI is less than Q_{out} at the end of the HO execution time. For monitoring RLF in state 3, the wideband CQI should be measured at least twice every 40 ms (i.e., the HO execution time) and averaged over these samples.

The HOF rate is defined as the ratio of the number of HOFs to the total number of HO attempts.

Moreover, depending on when RLF occurs during the HO procedure, an HOF can be classified as too late HO, too early HO, or HO to wrong cell; see also Section 6.4.4.

Too late HO
In Fig. 9.6, there are two overlapping cells. Cell A is the source cell and cell B is the target cell. A UE is located in cell A and served by eNB A. Suppose that the UE moves from eNB A to eNB B. If the UE signal has been significantly degraded during the HO process, causing RLF to occur, or if the RLF happens during the HO process due to bad radio link quality, the UE will search and establish its radio link to a proper eNB after

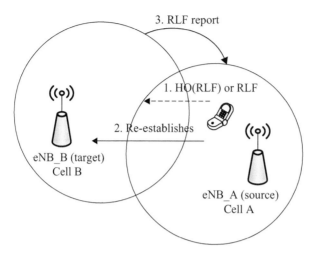

Figure 9.6 Typical scenario and detection mechanism for too late HO.

RLF. If the UE chooses to re-establish the radio link with eNB B, this RLF is caused by a too late HO. Thus, the features of a too late HO can be summarized as follows.

- An RLF occurs in the source cell before the HO is initiated or during the HO procedure.
- The UE re-establishes the radio link connection with the target cell.

The detection mechanism of too late HOs is shown in Fig. 9.6 and explained in the following. If a UE experiences RLF in eNB A and re-establishes its radio link with eNB B, since the RLF occurs before eNB A receives the HO MR message, eNB A is usually unconscious of the too late HO. Therefore, it is necessary for eNB B to send the RLF event message to eNB A after re-establishing radio links, so that eNB A could adjust the related HO parameters according to the already incurred too late HO. The RLF event message contains *failure cell ID* (FCI), *re-establish cell ID* (RCI), and C-RNTI.

Too early HO
In Fig. 9.7, there are two overlapping cells. Cell A is the source cell, and cell B is the target cell. A UE is located in cell A and served by eNB A. Suppose the UE moves from eNB A to eNB B. If after a successful HO from eNB A to eNB B the UE experiences an RLF, the UE will again search for an eNB to provide service. If the UE re-establishes its radio link with eNB A, this RLF is caused by a too early HO. Thus, the features of a too early HO can be summarized as follows.

- An RLF occurs shortly after a successful HO from a source cell to a target cell or during the HO process.
- The UE re-establishes the radio link connection with the source cell.

When a UE experiences an RLF after a successful HO and re-establishes its radio link with eNB A, eNB A should report the RLF event to eNB B. However, eNB B may recognize the RLF event report as an indication of a too late HO. Thus, a timer has to be

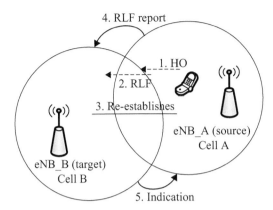

Figure 9.7 Typical scenario and detection mechanism for too early HO (reproduced with permission from IEEE©).

used at eNB B to differentiate between too late and too early HOs. If a UE experiences an RLF after successfully handing over to eNB B in time T, it will be classified as a too early HO, otherwise as a too late HO.

The detection mechanism of too early HOs is shown in Fig. 9.7 and explained as follows. If eNB B has sent the UE CONTEXT RELEASE message to eNB A for the completion of an incoming HO for the same UE within the last TstoreUEcntxt seconds, then eNB B should send an RLF INDICATION message indicating a too early HO event to eNB A.

Handover to wrong cell
In Fig. 9.8, there are three overlapping cells. Cell A is the source cell, cell B is the target cell, and cell C is a cell other than the source cell or the target cell. A UE is located in cell A and served by eNB A. Suppose the UE moves from eNB A to eNB B. If the HO parameters are improperly set, even though the HO triggering events are correct, the UE may still establish a radio link to the wrong cell after an RLF. If the UE establishes a radio link to eNB C after experiencing an RLF event in eNB B, eNB C will report the RLF event to eNB B, and eNB B will forward this RLF event to eNB A. This is a typical HO to a wrong cell event. It can be observed that HO to a wrong cell is similar to the too early HO, i.e., choosing an improper target cell. The features of HO to a wrong cell can be summarized as follows.

- An RLF event occurs shortly after a successful HO from the source cell to the target cell.
- The UE re-establishes the radio link connection with a cell other than the source cell and the target cell.

HO to a wrong cell and cell reselection optimization are related to the setting of the corresponding HO parameters. It is widely recognized that the HO process consumes a large amount of network resources. Unnecessary HOs are liable to occur when HO

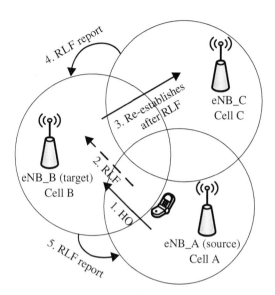

Figure 9.8 Typical scenario and detection mechanism for HO to a wrong cell.

parameters are improperly set, making UEs miss the best timing for HO and causing large waste of network resources.

9.2.2.2   Ping-pong

Whether or not a ping-pong occurs is determined by the time duration that a UE stays connected to a cell directly after an HO, namely time-of-stay, and the identities of the prior-to-source, source, and target cells. The time-of-stay starts when the UE sends an HO complete message to a cell, and ends when the UE sends an HO complete message to another cell. If a UE has a time-of-stay less than a threshold T_p, then the HO that terminates this time-of-stay is considered as an unnecessary HO. An unnecessary HO is considered as a ping-pong if the prior-to-source and target cells are the same cell, where the prior-to-source cell is the cell to which the UE was connected before handing over to the source cell.

The ping-pong rate is defined as the ratio of the number of ping-pongs to the total number of successful HOs (excluding HOFs). Recommended T_p is 1 second [12].

9.2.3   Improved schemes for mobility management in RRC-connected state

In this section, we study the relationship between UE mobility and HO parameter setting, and present an improved HO scheme based on UE mobility features. Simulations are presented to prove the effectiveness of the proposed scheme. It is suggested in [11] that MRO uses the following HO parameters as a basis for mobility enhancement: hysteresis (Hyst), TTT, CIO and cell reselection parameters. In this section, we focus on the optimization of Hyst parameters.

Different solutions have been proposed to improve HO performance [16–18]. In order to mitigate HOFs, in MRO, eNBs will analyze the RLF report from each UE, determine the type of RLF and then dynamically adjust the corresponding HO parameters. For too

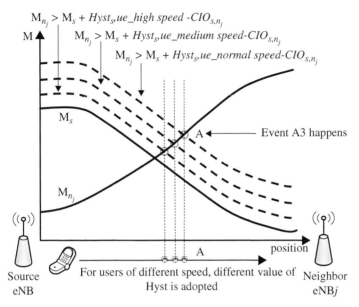

$M_{n_j} > M_s + Hyst_s, ue\_high\ speed\ -CIO_{s,n_j}$

$M_{n_j} > M_s + Hyst_s, ue\_medium\ speed-CIO_{s,n_j}$

$M_{n_j} > M_s + Hyst_s, ue\_normal\ speed-CIO_{s,n_j}$

Event A3 happens

position

For users of different speed, different value of Hyst is adopted

Source eNB   Neighbor eNBj

Figure 9.9 Illustration of the improved event A3 entry condition.

late and too early HO events, the major problem is the mismatch between the adjustment of HO parameters and the dynamics of mobile UEs. This problem can be mitigated by adopting different sets of HO parameters for different levels of UE mobility, so that eNBs could choose the proper set of HO parameters according to the UE mobility. Details of this scheme are provided in the following.

9.2.3.1   Improved scheme for HO parameter optimization based on UE mobility

In today's networks, a set of HO parameters is defined in each eNB, and all UEs of the same eNB use the same HO parameters, $Hyst_s$ and CIO_s, regardless of their mobility. However, it would be better if HO parameters were closely related to UE mobility features. For example, in a cell, the values of $Hyst_s$ that well suit low-velocity pedestrian UEs may not be suitable for high-velocity vehicular UEs, causing too late HOs or even RLFs. Based on the velocity levels classified in [19], we propose a new definition of $Hyst_s$, which takes into account UE velocity and supports three levels of $Hyst_s$, namely Hyst for low velocity, Hyst for medium velocity, and Hyst for high velocity. eNBs will inform UEs of the three Hyst values, so that UEs can choose the proper Hyst values according to their velocities. Accordingly, the inequity of event A3 entry condition (9.1) changes to

$$M_{n_j} > M_s + Hyst_{s,ue\_velocity} - CIO_{s,n_j} \qquad (9.2)$$

where the parameter $Hyst_{s,ue\_velocity}$ contains one of three values: $Hyst_{s,ue\_high\_velocity}$, $Hyst_{s,ue\_medium\_velocity}$, and $Hyst_{s,ue\_low\_velocity}$. This new event A3 entry condition is illustrated in Fig. 9.9.

As shown in Fig. 9.9, we choose low-velocity UEs as the baseline to lower the HO threshold of medium-velocity and high-velocity UEs. The adjustment of Hyst values

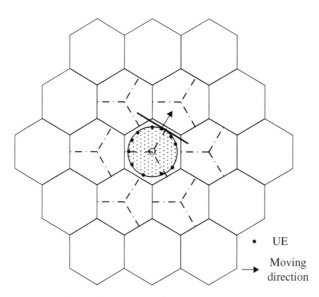

•   UE

→   Moving
    direction

Figure 9.10 Simulation scenario for mobility parameter optimization (reproduced with permission from IEEE©).

could enable high-velocity UEs to start HO procedures earlier, avoiding too late HOs. Moreover, the adjustment of Hyst values should be above a given threshold to avoid triggering ping-pongs. Although we only take into account the velocity feature of UEs, other UE features that have an influence on HO performance could also be considered in the the analysis.

9.2.3.2   Simulations and performance evaluation

We consider a macrocell system with two tiers of interfering macrocells as shown in Fig. 9.10. Each macrocell contains three sectors. The *intersite distance* (ISD) is 500 m. We consider path loss and shadow fading. UEs are initially randomly distributed on a circle around the center point of the scenario, with a radius that is 70% of the hexagonal side length. Each UE moves radially towards the center of the circle with a velocity randomly selected from three velocity levels, which are low, medium, and high. In the simulation, we compare the performance of two schemes, with and without the proposed UE velocity dependent Hyst. We set the values of TTT and CIO to be 0.3 s and 0 dB, respectively. The key mobility performance indicator used is the RLF probability.

In order to simplify the simulation and show the necessity of HO optimization based on UE velocities, we abridge the HO process to Fig. 9.11. At the beginning, the UE is served by eNB A and RLF detection is performed following the procedure described in previous sections. Once event A3 occurs, the HO procedure is initiated. If the Hyst parameter is set too small, a too early HO may occur. Thus, we select a short time period following a completed HO as an observation window. If no RLF occurs in the observation window, then this HO is confirmed to be successful; otherwise, an HOF is counted. The detailed parameter settings are given in Table 9.1.

Table 9.1 Simulation setup for mobility parameter optimization.

Parameter	Value
Intersite distance (ISD)	500 m
NLOS path loss model	$PL(dB) = 35.63 + 35\log_{10}(R)$ where R is the distance between the UE and the eNB (m)
Shadow fading compensation	6 dB
eNB antenna gain	$A(\theta) = -\min\left[12\left(\frac{\theta}{\theta_{3dB}}\right)^2, A_m\right]$, where $-180° \leq \theta \leq 180°$, $A_m = 20$ dB, $\theta_{3dB} = 70°$
eNB transmit power	43 dBm
RLF detection	$Q_{in} = -6$ dB, $Q_{out} = -8$ dB, N310 = 1, N311 = 1, T310 = 1 s
HO delay	0.234 s
HO check time	2 s
Number of UEs at a low velocity	3333
Number of UEs at a medium velocity	3333
Number of UEs at a high velocity	3334
Simulation time	500 s
TTT	0.3 s
CIO	0 dB
UE velocity (v)	Low velocity: $v < 5$ m/s; medium velocity: 5 m/s $< v < 30$ m/s; high velocity: $v > 30$ m/s

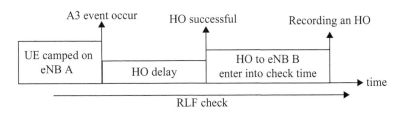

Figure 9.11 Simplified HO confirmation process (reproduced with permission from IEEE©).

In Fig. 9.12, we can observe that, for UEs with different velocity levels, Hyst values after optimization are different, resulting in a low RLF probability. For low-velocity UEs, the lowest probability of RLF is achieved when the Hyst value is set at 4 dB, whereas for medium-velocity UEs and high-velocity UEs, it is better to set Hyst values to 3 dB and 2 dB, respectively. Hence, in order to attain low RLF probability for every UE, we should set an appropriate Hyst value for each UE according to its velocity level. Using the respective best Hyst values for the three UE velocity levels found in Fig. 9.12, we evaluate the mobility performance of the improved HO scheme, and compare it with a conventional scheme that uses the same Hyst value (e.g., 3 dB) for all UEs. Table 9.2 shows that in the conventional algorithm, with the same Hyst value applied to all UEs with possibly different velocities, the probability of RLF is 4.3%, while in the improved scheme the probability of RLF is 1.3%. Hence, considering the feature of UE velocity in the selection of Hyst parameters, the probability of RLF could be reduced by almost

Table 9.2 Performance evaluation of the improved scheme.

Algorithm	HO times	RLF times	Number of RLF times after HO	RLF probability
Conventional algorithm (same Hyst$_s$ value 3 dB)	9568	398	34	4.32%
Optimized algorithm (best Hyst$_s$ values based on Fig. 9.12)	9872	96	32	1.28%

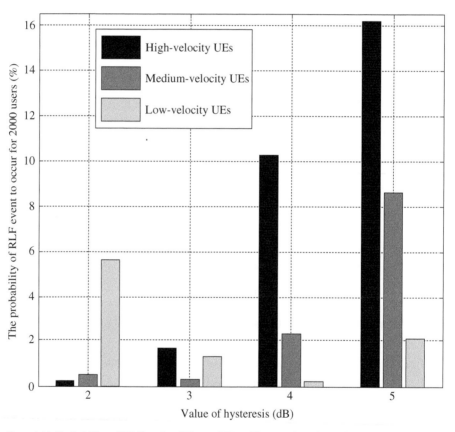

Figure 9.12 Probability of RLF under different UE and hysteresis values of the improved scheme.

70%. In practice, the adjustment of Hyst$_s$ values could be performed through a step by step tracking, normally starting from a small step length (e.g., 0.5 dB to 1 dB), until the requirement of MRO is fulfilled. Furthermore, with the development of new radio technologies, the parameter setting scheme could also dynamically adjust Hyst$_s$ values, taking mobility state estimation approaches and the influence of other parameters into account to obtain a better mobility performance.

9.3 Mobility management in RRC-idle state

The mobility management in the RRC-idle state is implemented through cell selection/reselection. Note that the cell selection/reselection parameter setting problem is not an isolated problem, but one largely integrated with the HO parameter setting problem. If they are not coordinated, system overhead will increase due to unnecessary HOs. Moreover, MLB may also be performed by adjusting HO parameters [11]. Therefore, the setting of cell reselection parameters should also take into account the amendments made by MLB, matching both the reselected parameters and the HO parameters. In this section, we will first introduce the mobility management mechanisms in the RRC-idle state, and then analyze them by referring to standardization efforts.

9.3.1 Overview of cell selection/reselection procedure

In LTE systems, the cell selection/reselection procedures and algorithms are much more agile than those in GSM systems, increasing the complexity of mobility optimization. In the following, we will review in detail the procedures and criteria of cell selection/reselection in LTE systems.

9.3.1.1 Cell selection and the S criterion

When a UE has selected a proper *public land mobile network* (PLMN), it will initiate the cell selection process. The cell selection criterion S is used and described as follows [15]:

(1) UEs calculate S_{rxlev} and S_{qual}, which are measurement evaluations for the signal quality of a cell when the UE is in the cell selection phase, and can be modelled as follows:

$$S_{rxlev} = Q_{rxlevmeas} - (Q_{rxlevmin} + Q_{rxlevminoffset}) - P_{compensation}, \quad (9.3)$$

$$S_{qual} = Q_{qualmeas} - (Q_{qualmin} + Q_{qualminoffset}), \quad (9.4)$$

where

$$P_{compensation} = \max(P_{MAX} - P_{PowerClass}, 0), \quad (9.5)$$

and the meanings of these parameters are described in Table 9.3.

(2) Then, UEs check the values of S_{rxlev} and S_{qual}. If $S_{rxlev} > 0$ and $S_{qual} > 0$, this cell is considered as a suitable cell for cell selection, otherwise, the cell is omitted from selection.

Based on whether the UE has stored information about each cell, such as carrier frequencies and cell parameters, the cell selection process can be divided into two procedures, namely "Initial Cell Selection" and "Stored Information Cell Selection" [19].

- The initial cell selection procedure requires no a priori knowledge of which *radio frequency* (RF) channels are carriers. The UE shall thus scan all RF channels in the

Table 9.3 Parameters for cell selection criterion S.

Parameter	Description
S_{qual}	Cell selection quality value (dB). Applicable only for FDD cells.
S_{rxlev}	Cell selection received signal power level value (dBm).
$Q_{qualmeas}$	Measured cell received signal quality value. The quality of the received signal expressed in CPICH Ec/N0 (dB) for FDD cells. Applicable only for FDD cells.
$Q_{rxlevmeas}$	Measured cell received signal power level value. This is the received signal, CPICH RSCP for FDD cells (dBm) and P-CCPCH RSCP for TDD cells (dBm).
$Q_{qualmin}$	Minimum required received signal power quality level in the cell (dB). Applicable only for FDD cells.
$Q_{qualminoffset}$	Compensation value of $Q_{qualmin}$ only when periodic search for a higher-priority PLMN is implemented.
$Q_{rxlevmin}$	Minimum required received signal power level in the cell (dBm).
$Q_{rxlevminoffset}$	Compensation value of $Q_{rxlevmin}$ only when periodic search for a higher-priority PLMN is implemented [20].
UE_TXPWR_MAX_RACH	Maximum transmit power level a UE may use when accessing the cell on RACH (dBm).
P_{MAX}	Maximum radio frequency (RF) output power of the UE (dBm) [21].
$P_{compensation}$	max(UE_TXPWR_MAX_RACH $- P_{MAX}$, 0) (dB).

system bands. For each identified carrier, the UE finds the strongest cell, and thereafter selects the strongest cell considering all carriers.

- The stored information cell selection procedure requires stored information of cell carriers and optionally of cell parameters. Once the most suitable cell is found among the stored ones, it should be selected. If no suitable cell is found, the initial cell selection procedure will be started again. It should be noted that, during the cell selection procedure, UEs should be informed of the priorities of different RATs and carriers through system information or dedicated signaling, but this informing process is not included in the cell selection procedures.

9.3.1.2   Cell reselection and the R criterion

Cell reselection is based on the above described cell selection procedure, i.e., the considered UE has camped on a cell and starts the cell reselection evaluation procedure. The cell reselection procedure is mostly impacted by the network environment. When the signal quality in the serving cell deteriorates, the considered UE will try to camp on a new cell that provides a better signal quality.

The priority management is an important part of the cell reselection process. Cell reselection priorities, which indicate which cells should be considered first in the reselection process, are parameters configured by the network side for each carrier and provided to UEs through the broadcast channel. The span of cell reselection priority values is from 0 to 7. The larger the value, the higher the priority of the corresponding carrier. In the same RAT, all cells on the same carrier have the same priority, while cells

Table 9.4 Parameters for cell reselection criterion R.

Parameter	Unit	Definition
$Q_{\text{meas,s}}$	dBm	Measured RSRP value of the serving cell.
$Q_{\text{meas,n}}$	dBm	Measured RSRP value of the neighboring cell.
Q_{Hyst}	dB	Hysteresis value of the serving cell for cell reselection.
$Q\text{offset}_{\text{frequency}}$	dB	Frequency compensation for the carrier of the measured neighboring cell. Only used for neighboring cells with different frequencies. The default value is zero.
Q_{offset}	dB	1 When the calculation of R values is for the same frequency, if $Q\text{offset}_{s,n}$ is effective, then $Q_{\text{offset}} = Q\text{offset}_{s,n}$, otherwise, use the default value 0. 2 When the calculation of R values is for different carrier frequencies, if $Q\text{offset}_{s,n}$ is effective, then $Q_{\text{offset}} = Q\text{offset}_{s,n} + Q\text{offset}_{\text{frequency}}$, otherwise, $Q_{\text{offset}} = Q\text{offset}_{\text{frequency}}$.

on different carriers may have different priorities. In addition to the system broadcast channel, cell reselection priority could also be sent to UEs through the RRC connection release message. If cell reselection priorities are provided in the RRC connection release message, then the UE shall ignore all priorities provided in the system broadcast channel. This mechanism ensures that the network has the highest authority in the cell reselection procedure.

After the determination of cell reselection priorities, UEs will choose measurement triggering rules and cell reselection decision rules for cells of different priorities. The lower the priority, the harsher the reselection decision rule, leaving less chance for UEs to reselect a better cell. This is to ensure that UEs get the best services from the network. When multiple cells are available, the UE will choose to camp on that with the highest priority. In the following, we will further discuss the cell reselection rules by using criterion R as an example.

Criterion R is the cell reselection criterion for cells with the same priority, including cells on the same serving carrier or interfrequency carriers (but not inter-RAT cells, since the reselection priorities for the inter-RAT cells are different). Criterion R can be described as follows.

(1) First, UEs calculate all R values of the cells that fulfill criterion S, where the R values are calculated as

$$R_s = Q_{\text{meas,s}} + Q_{\text{Hyst}}, \tag{9.6}$$

$$R_n = Q_{\text{meas,n}} - Q_{\text{offset}}, \tag{9.7}$$

and

$$Q_{\text{offset}} = Q\text{offset}_{\text{frequency}} + Q\text{offset}_{s,n}, \tag{9.8}$$

where the meanings of these parameters are described in Table 9.4.

(2) Then, UEs sort the cells in descending order according to their R values. The cell with the largest R value is considered as the best cell [19].

Based on the calculated R values, for cells whose cell reselection priority is the same as the serving cell, UEs will trigger the cell reselection procedure if the following two conditions are met.

- The R_s value of the serving cell is lower than the R_n value of the neighboring cell during a time interval.
- More than 1 s has elapsed since the UE camped on the current serving cell.

Moreover, the R value of the current serving cell may be biased. This is to avoid too frequent cell reselections among the serving cell and its neighboring cells. Note that the R_n value calculated in (9.7) is also applicable when multiple cells with the same priority are available for cell reselection. In this case, UEs should calculate the R_n values for all neighboring cells and choose the cell with the largest R_n for the cell reselection.

9.3.1.3   Interaction among cell selection/reselection and HO parameters

Cell selection/reselection and HO parameters have a significant impact on mobility performance. If these two sets of parameters are not coordinated, once a UE moves from RRC-idle to RRC-connected mode, an HO may be triggered, costing a large amount of network resources. Thus, the configuration of cell selection/reselection parameters must take the HO parameters into account. Moreover, MLB is also implemented through cell selection/reselection and HOs. In [11], the aim of MLB is defined as to avoid cell overload as indicated by some *key performance indicators* (KPIs), so as to balance the system load and minimize the number of HOs. Most of the MLB algorithms take the loads of the neighboring cells as an input, and provide the output by reconfiguring cell selection/reselection and HO parameters. In [22–27] optimization algorithms are presented taking into account mobility parameters, load factors, UE features, and diverse MLB scenarios. In [27], it is proposed to increase the threshold of some mobility parameters when the load of the target cell is higher than that of the current one, making it harder for the UE to HO to the target cell. In [28], a self-optimized MLB algorithm is proposed for LTE systems, based on choosing the best offset parameters according to cells' loads.

9.3.2   Improved schemes for mobility management in RRC-idle state

The mobility management in the RRC-idle state needs to take the matching between cell selection/reselection and HO parameters into account, as well as the impact of MLB on HO parameters, so as to reduce the number of HOFs and ping-pongs. In this section, we will first discuss the necessity for cell negotiation to optimize cell reselection parameters, then propose a cell negotiation scheme to optimize cell reselection parameters, and evaluate the performance of the proposed scheme through system-level simulations.

A large number of proposals for the adjustment of HO parameters have been submitted to 3GPP. In [9–30], it is indicated that the modification of HO parameters in one cell will influence HOs in neighboring cells, leading to ping-pongs and deteriorated system performance. In [9], possible HO parameter settings are analyzed in detail, and

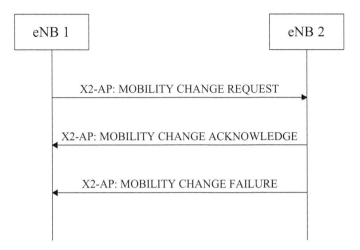

Figure 9.13 Mobility parameter modification procedures.

methods to differentiate between suitable and non-suitable HO parameter settings are provided. These works have shown that the configuration of HO parameters must be negotiated among the involved cells so as to avoid frequent HOs. In [1], a mobility setting modification process is designed to negotiate the change of HO parameters in a coordinated manner among neighboring cells, in order to increase the system capacity. As shown in Fig. 9.13 [1], when eNB 1 needs to modify its HO parameters, it will send a MOBILITY CHANGE REQUEST message via the X2 application protocol (AP) to negotiate with its neighboring cells the modification of HO parameters [6]. If eNB 2 agrees with this modification, it will reply to eNB 1 with a MOBILITY CHANGE ACKNOWLEDGE message, indicating that the negotiation is successful. Otherwise, eNB 2 will reply with a MOBILITY CHANGE FAILURE message, indicating that the negotiation has failed, and thus eNB 1 cannot modify its HO parameters.

Despite the efforts made in 3GPP for the intelligent adjustment of HO parameters, the negotiation of cell reselection parameters among neighboring cells has not been regulated yet. As a result, when eNB 1 sends a MOBILITY CHANGE REQUEST to eNB 2 to modify cell reselection parameters, eNB 1 and eNB 2 may or may not change their cell reselection parameters, or may even soon select new ones to cope with changing channel conditions. Moreover, since current standards have no regulation on the relationship between cell reselection and HO parameters, eNBs are free to choose various algorithms to modify cell reselection and HO parameters. For example, eNB 1 may change its Hyst value to 3 dB after successfully negotiating with eNB 2. Then, eNB 1 and eNB 2 will both have their Hyst value changed to 3 dB. At the same time, eNB 1 and eNB 2 may independently modify their own cell reselection parameters, respectively, e.g., with eNB 1 changing Q_{hyst} to 1 dB and eNB 2 changing Q_{hyst} to 2 dB. As shown in Fig. 9.14, such uncoordinated modification of cell reselection parameters may cause mismatch between cell reselection and HO parameters, leading to mismatch between cell reselection boundaries of neighboring cells, and potentially ping-pongs. Furthermore, the mismatch between cell reselection and HO parameters might also

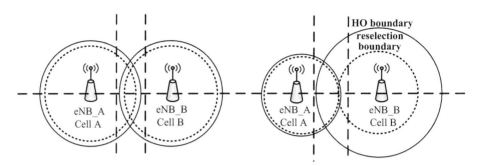

Figure 9.14 Illustration of the mismatched adjustment of the intercell mobility parameters.

cause UEs to initiate the HO process when entering the RRC-connected mode from the RRC-idle mode. Therefore, intercell negotiation is necessary during the setting of cell reselection parameters.

9.3.2.1   Improved scheme for cell reselection parameter optimization based on mobility load balancing methods

Conventionally, MLB is achieved by handing over cell-edge UEs from relatively high-load cells to low-load cells [31–34]. Meanwhile, in order to avoid wasting system resources or deteriorating user experience, it is necessary to reduce the number of cell reselections and HOs during MLB.

Before designing optimization schemes for MLB, we first need to determine a proper objective for the optimization of cell reselection and HO parameters. For high-velocity UEs, the major problems to be solved are how to choose the proper target cell and when to start the HO procedure. For example, when a mobile UE moves with high velocity from one cell to another, the received signal strength from the source eNB will deteriorate quickly. Without a timely HO, the high-velocity UE is likely to experience RLF, as shown earlier in this chapter. Therefore, the purpose of HO optimization for high-velocity UEs is different from that for MLB. The former implements HO to maintain basic radio links, while the latter implements HO to optimize system capacity. Hence, high-velocity UEs should not be involved in MLB. We recommend that priority is given to low-velocity UEs when targeting MLB. More specifically, the lower the velocity, the higher the priority.

Based on previous analysis, we propose a negotiation scheme for cell reselection parameter optimization based on MLB methods and UE mobility states. The procedure of the proposed scheme for cell reselection parameter negotiation is as follows.

1 The serving eNB measures the load in its cell and exchanges load information with other cells through the X2 interface.
2 eNBs decide whether or not to initiate intercell MLB procedures based on their MLB algorithms.
3 If the decision of implementing MLB procedures is made, the thresholds of cell reselection and HO parameters are negotiated among involved cells to mitigate ping-pongs.

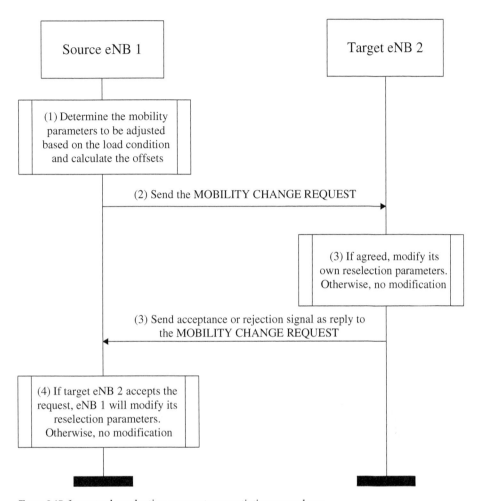

Figure 9.15 Improved reselection parameter negotiation procedures.

Fig. 9.15 illustrates a detailed implementation of the proposed cell reselection parameter negotiation scheme, which is summarized as follows.

(1) Source eNB 1 determines the mobility parameters to be modified, e.g., Q_{offset} and Q_{Hyst}, based on the load conditions of itself and neighboring cells. Then, eNB 1 calculates the variations to be imposed in neighboring cells on the determined mobility parameters.

(2) Source eNB 1 sends a MOBILITY CHANGE REQUEST message to target eNB 2, containing the calculated parameter variations.

(3) Target eNB 2 accepts or rejects the MOBILITY CHANGE REQUEST by sending the corresponding response to eNB 1. If eNB 2 accepts the request, it will adjust its mobility parameters based on the received parameter variations calculated by eNB 1. Otherwise, eNB 2 does not do any modification.

Table 9.5 Parameter settings for simulations of the mismatching problem.

Parameter	Value
Number of users	10 000
UE distribution	Uniform
ISD	1732 m
NLOS path loss model	$PL(dB) = 35.63 + 35\log_{10}(R)$
Shadow fading	6 dB
eNB antenna gain (horizontal plane)	$A(\theta) = -\min\left[12\left(\frac{\theta}{\theta_{3dB}}\right)^2, A_m\right]$, where $-180 \leq \theta \leq 180$ in degrees, $A_m = 20$ dB, $\theta_{3dB} = 70$ degrees
eNB transmit power	43 dBm
Q_{Hyst}, $Hyst$	0 dB
cell-specific reference symbol (CRS)/HO offset	-10 dB to 10 dB
Cell reselection criterion	$RSRP_s + Q_{Hyst} < RSRP_n - CRS_{offset}$
HO criterion	$RSRP_s + Hyst < RSRP_n - HO_{offset}$

(4) Source eNB 1 receives the reply from eNB 2. If the reply is positive, eNB 1 will modify its mobility parameters based on the calculations in step (1). Otherwise, eNB 1 will not perform any modification either.

9.3.2.2   Simulations and performance evaluation

We evaluate system performance under two conditions: (1) when cell reselection parameters do not match among neighboring cells, and (2) when cell reselection parameters do not match with HO parameters within a cell. Moreover, we assess the effectiveness of the scheme proposed in the previous subsection by evaluating the probabilities of HO and ping-pong events when UEs enter the RRC-connected state from the RRC-idle state. We consider a system with 19 hexagonal cells, where all UEs remain static once deployed. If the reselection criterion is not fulfilled, UEs will remain camped on their current serving cells. The simulation parameter settings are given in Table 9.5.

Fig. 9.16 shows the rate of cell reselection caused by ping-pongs, where S-CRS is the serving cell reselection parameter offset and N-CRS is the neighboring cell reselection parameter offset. We assume that all UEs are served by the cell with the largest RSRP at the beginning of the simulation. In the analyzed scenario, there is no cell reselection failure, because coverage is assumed to be ensured. The UE will stay at its current serving cell when the cell reselection criterion is not met. It can be observed that the rate of cell reselection caused by ping-pongs is zero when the S-CRS offset of the serving cell is larger than Q_{Hyst}, because the UE will always select the strongest cell. A negative offset means that the UE will probably reselect the neighboring cell. As a result, when both S-CRS and N-CRS are near -10 dB, the ping-pong reselection rate reaches 52%. When the serving cell offset S-CRS is -4 dB and the neighbor cell offset N-CRS is -5 dB, the ping-pong reselection rate is around 10%. When both S-CRS and N-CRS are positive, the ping-pong reselection rate is always 0%.

Fig. 9.17 and Fig. 9.18 show the rate of immediate HO due to the mismatch between cell reselection and HO parameters, for $Q_{Hyst}/Hyst = 3$ dB and 0 dB, respectively.

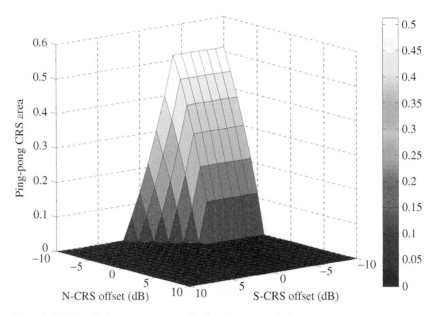

Figure 9.16 Rate of ping-pong cell reselection ($Q_{\text{hyst}} = 3$ dB).

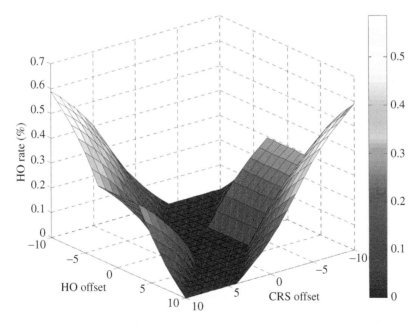

Figure 9.17 Rate of immediate HO when UE comes active ($Q_{\text{Hyst}}/\text{Hyst} = 3$ dB).

Considering all overlapping areas between the two cells, almost 59% of UEs will immediately HO to the neighboring cell when the difference between cell reselection and HO
offsets is 20 dB and $Q_{\text{Hyst}}/\text{Hyst} = 3$ dB. When the cell reselection offset is 0 dB, the HO
offset is 4 dB, and the $Q_{\text{Hyst}}/\text{Hyst} = 3$ dB, the immediate HO rate significantly decreases

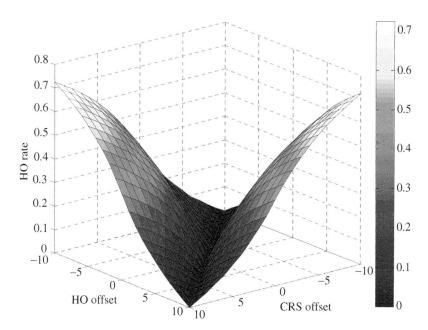

Figure 9.18 Rate of immediate HO when UE comes active ($Q_{\mathrm{Hyst}}/\mathrm{Hyst} = 0$ dB).

to 5.7%. As shown in Fig. 9.18, when $Q_{\mathrm{Hyst}}/\mathrm{Hyst} = 0$ dB, the maximum immediate HO rate increases.

The simulation results show that the negotiation of cell reselection parameters between two neighboring cells is necessary, and helps to mitigate the mismatch of cell reselection parameters between the two cells and the mismatch between cell reselection and HO parameters of a cell.

9.4   Mobility management in heterogeneous cellular networks

Among the technical challenges faced by operators to realize the potential coverage and capacity benefits of HCNs introduced in Chapter 1, the mobility management problem is of special importance. As already introduced in this chapter, the deployment of a large number of LPNs may increase the complexity of MRO and MLB further, since mobile UEs may trigger frequent HOs when they move across small coverage areas of LPNs.

In conventional homogeneous networks, mobile UEs typically use the same set of HO parameters (e.g., hysteresis margin, TTT), where some of them may be fine tuned according to UE velocity, as shown in Section 9.2.3. However, in HCNs (where macro-cells, picocells, femtocells, and relay nodes may have different coverage areas), using the same set of HO parameters for all cells and/or for all UEs may degrade mobility performance. For example, the use of range expansion in picocells will affect when and where the HO process is initiated, while *closed subscriber group* (CSG) LPNs may allow access to a group of registered UE only. Hence, in HCNs, there is a need for cell-specific

HO parameter optimization. For more details on range expansion and CSG access, please refer to Chapter 7 and Chapter 4, respectively. Moreover, high-mobility *macrocell user equipments* (MUEs) may run deep inside LPN coverage areas before the TTT optimized for macrocells expires, thus incurring HOFs due to degraded control channel quality. HOs performed for high-mobility MUEs could also be unnecessary, when they quickly pass through the small coverage areas of LPNs. These facts also impose the need for UE-specific HO parameter optimization.

Mobility management challenges in HCNs have attracted a lot of interest from the wireless industry, research community, and standardization bodies [12–37]. Indeed, a new study item (SI) "HetNet mobility enhancements for LTE" has recently been established in 3GPP RAN2 [12].

In the following, we identify technical challenges in mobility management. In more detail, we evaluate mobility performance in an HCN with 3GPP Rel-10 enhanced ICIC (eICIC) features such as ABSs. Then, we present a mobility-based ICIC (MB-ICIC) scheme, in which LPNs configure coordinated resources so that macrocells can schedule their high-mobility UEs in these resources without co-channel interference from LPNs. Without loss of generality, we will focus on the open-access picocell case.

9.4.1   Range expansion, almost blank subframes, and HO performance

In order to address the problems caused by the DL transmit power difference among eNBs and *pico evolved NodeBs* (PeNBs) in HCNs, cell selection methods that allow UE association with cells that do not provide the strongest DL RSRP are necessary. A widely considered approach is range expansion [38], in which a positive *range expansion bias* (REB) is added to the DL *reference signal strengths* (RSSs) of picocell pilot signals at UEs to increase picocells' DL coverage footprints. However, although range expansion is able to mitigate UL intercell interference and provide MLB in HCNs, it reduces the DL signal quality of *picocell user equipments* (PUEs) in the expanded region, since these PUEs are connected to cells that do not provide the strongest DL RSRP. In order to mitigate this DL signal quality degradation, Rel-10 eICIC ABS, in which no control or data signals but only reference signals are transmitted, can be used to mitigate DL intercell interference for range expanded picocells [38]. More specifically, the umbrella macrocell schedules ABSs, and the picocell schedules range-expanded PUEs in the subframes that overlap with the ABSs of the umbrella macrocell. For more details about range expansion and ABS, please refer to Chapter 7.

In order to illustrate HO behaviors in the presence of picocell range expansion, the coverage areas of a range-expanded picocell with an 8 dB bias for three different HO scenarios are depicted in Figs. 9.19(a)–9.19(c), where the eNB is located at (1500, 1500) m, the PeNB is located at (1500, 1650) m, UEs move at 3 km/h, eICIC is not implemented, and the HO preparation and execution time is assumed to be neglected. Simulation assumptions follow hotspot simulation recommendations in [12]. In Fig. 9.19(a), HOs are performed based on geometry only, while in Figs. 9.19(b) and 9.19(c), HOs are performed using a TTT of 160 ms and an event A3 offset of 2 dB. In Figs. 9.19(a) and 9.19(b), no effect of shadowing or fast fading is included, while in Fig. 9.19(c), the effect

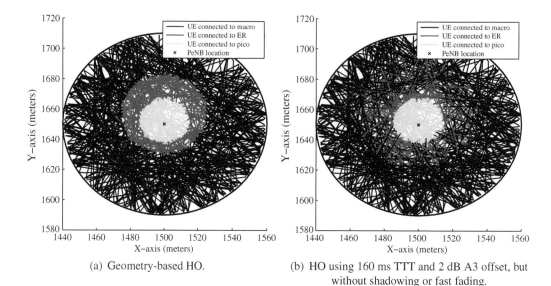

(a) Geometry-based HO.

(b) HO using 160 ms TTT and 2 dB A3 offset, but without shadowing or fast fading.

(c) HO using 160 ms TTT and 2 dB A3 offset, with shadowing.

Figure 9.19 3GPP Release-10 eICIC scenarios and handover behaviors in the presence of picocell range expansion (reproduced with permission from IEEE©).

of shadowing is considered. Comparing Fig. 9.19(b) with Fig. 9.19(a), we can see that MUEs invade the picocell expanded region but without handing over to the picocell due to the use of TTT. Comparing Fig. 9.19(c) with Fig. 9.19(b), we observe that due to shadowing effects, there is no clean-cut boundary of the picocell coverage area with or without range expansion, and UEs can connect to the picocell in a much wider area. This demonstrates the importance of considering channel variations in mobility management.

Fig. 9.20 also shows that, with range expansion at the picocell, the event A3 positions are pushed away from the PeNB location, thus increasing the picocell coverage area and

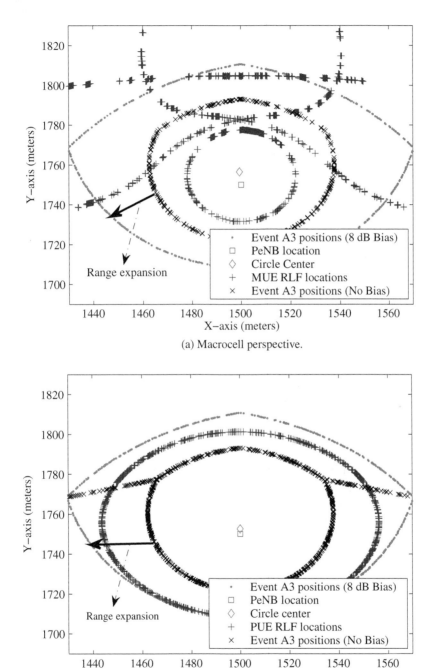

(a) Macrocell perspective.

(b) Picocell perspective.

Figure 9.20 Coverage areas of macrocell/picocell with and without range expansion, and RLF locations from macrocell and picocell perspectives ($Q_{out} = -8\,dB$) (reproduced with permission from IEEE©).

Table 9.6 HO parameter sets in [12] (reproduced with permission from IEEE©).

Profile	Set-1	Set-2	Set-3	Set-4	Set-5
TTT (ms)	480	160	160	80	40
A3 offset (dB)	3	3	2	1	−1
RSRP L3 Filter K	4	4	1	1	0

allowing for a better spatial reuse. In Fig. 9.20(a), since the gap between the event A3 entry condition boundary and the RLF boundary is larger with picocell range expansion, it is more likely that the TTT will expire before the UE SINR falls to Q_{out}. Picocell range expansion thus facilitates the macro-to-pico HO. In contrast, in Fig. 9.20(b), range expansion challenges the pico-to-macro HO, since the space between the event A3 boundary and the RLF boundary gets smaller, and it is thus more likely that the UE SINR falls to Q_{out} before the TTT expires. In Fig. 9.20(b), RLF occurs earlier than event A3, which indicates the need for eICIC to support expanded region picocells.

9.4.2   HCN mobility performance with 3GPP Release-10 eICIC

There is a tradeoff between HOFs and ping-pongs. Optimizing HO parameters to reduce HOFs would increase ping-pongs, and vice versa. For example, reducing TTT may decrease HOFs, but may increase ping-pongs. This makes HO optimization as shown in previous sections an intricate problem, which would be exacerbated by the large number of LPNs overlaid on macrocell networks in an HCN.

In order to illustrate the tradeoff between HOFs and ping-pongs, we have performed system-level simulations using simulation scenarios in [12]. In these simulations, a hexagonal eNB layout with 19 eNBs, 57 sectors, and an inter-eNB distance of 500 m were considered. Four PeNBs were randomly distributed within each eNB sector coverage area. Five different HO profiles were considered, as summarized in Table 9.6 [12], where the longest and shortest TTT durations are 480 ms in Set-1 and 40 ms in Set-5, respectively. For each simulation set, an L1 and L3 filtering period of 200 ms was used, along with full cell-loading. UEs were randomly distributed over the entire simulation scenario. Each UE moved along a straight line towards a randomly selected direction, and did not change direction until it hit the border of the simulation scenario. When a UE hit the simulation scenario border, it bounced back and moved towards another randomly selected direction. Simulation assumptions follow large-area simulation recommendations in [12]. Further details of the simulation and parameters are presented in Table 9.7 and [12].

Fig. 9.21 presents the simulated HOF and ping-pong rates, respectively, for two different HCN cases: (a) picocells without range expansion or eICIC; (b) picocells using an 8 dB bias for range expansion and implementing eICIC with ABSs configured at macrocells. The results confirm that reducing TTT decreases HOFs, but increases ping-pongs, and vice versa, for both cases considered. Among the five HO profiles, Set-3

Table 9.7 Simulation parameters (reproduced with permission from IEEE©).

Parameter	Macrocell
Carrier frequency	2.0 GHz
System bandwidth	10 MHz
Number of eNB/sectors	19/57, with 500 m ISD
eNB antenna patterns (TR 36.814)	3D pattern
PeNB antenna patterns (TR 36.814)	Omnidirectional pattern
eNB antenna tilt	15 degree
eNB antenna gain	15 dB
PeNB antenna gain	5 dB
UE antenna gain	0 dB
Macrocell path loss model	$128.1 + 37.6\log_{10}(R)$ dB
Picocell path loss model	$140.7 + 36.7\log_{10}(R)$ dB
Shadowing standard deviation	8 dB (macrocell), 10 dB (picocell)
Correlation distance of shadowing	25 m
Macrocell shadowing correlation	0.5 (1) between cells (sectors)
Picocell shadowing correlation	0.5 between cells
Transmit power	46 dBm (eNB), 30 dBm (PeNB)
Penetration loss	20 dB
Antenna configuration	1×2
Picocell range expansion bias	8 dB (whenever applicable)
Cell loading	100%
UE velocities	3, 30, 60, 120 km/h
UE noise figure	9 dB
Thermal noise density	-174d Bm/Hz
Channel model	Typical Urban (six rays)
HO metric	1 Rx for RSRP measurement
SINR metric	2 Rx, MRC and EESM
RSRP measurement bandwidth	25 RBs
L3 filter coefficient (a)	0.5
HO preparation (execution) delay	50 ms (40 ms)
Q_{out} (Q_{in})	-8 dB (-6 dB)
T310	1 s
Min. eNB–UE (PeNB–UE) distance	35 m (10 m)
Min. eNB–PeNB (PeNB–PeNB) distance	75 m (40 m)

with an intermediate TTT of 160 ms yields the best tradeoff. Moreover, we observe that using range expansion and implementing eICIC decreases HOFs, but increases ping-pongs. (This is also in line with the aforementioned tradeoff.) For UE velocities up to 30 km/h, Release-10 eICIC is shown to offer good HOF and ping-pong performance, but it becomes hard to simultaneously achieve good HOF and ping-pong performance for higher UE velocities, which fortunately are less likely for PUEs.

In terms of HOF rates, results are in line with [12], which indicate that UE velocity has a significant impact in HO performance with high-velocity UEs suffering a much higher HOF rate than low-mobility UEs. The case of picocells with no range expansion (Fig. 9.21(a)) results in worse performance. This is because the picocell coverage areas without range expansion are small, and thus MUEs may quickly run deep inside picocell

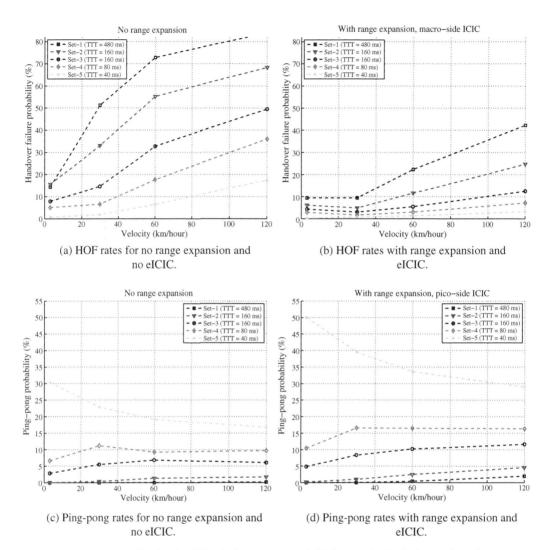

(a) HOF rates for no range expansion and no eICIC.

(b) HOF rates with range expansion and eICIC.

(c) Ping-pong rates for no range expansion and no eICIC.

(d) Ping-pong rates with range expansion and eICIC.

Figure 9.21 Simulated HOF and ping-pong rates (with four randomly deployed picocells per sector, with or without an 8 dB range expansion bias, and eICIC) (reproduced with permission from IEEE©).

coverage areas before the TTT expires, significantly degrading MUEs' SINR before the HO process is completed. When using picocell range expansion with pico eICIC (Fig. 9.21(b)), the number of HOFs is significantly reduced. Since the event A3 entry condition boundary is pushed away from the PeNB location by range expansion, the MUE's SINR when the macro-to-pico HO is initiated is improved, thus avoiding RLF before the TTT expires. Pico-to-macro HO is not an issue due to eICIC. In both cases, HOFs are alleviated with shorter TTTs, e.g., in Set-5, since UEs can be handed over faster.

In terms of ping-pong rates, results are also in line with [12], which indicate that ping-pong rates are relatively high for low-velocity UEs with parameter Set-5, while ping-pong rates are relatively low for low-velocity UEs with parameter Set-1 and parameter Set-2. The case of picocells with no range expansion (Fig. 9.21(c)) performs better than that of picocell range expansion with pico eICIC (Fig. 9.21(d)). When we increase the picocell coverage area through range expansion, cell selection oscillation caused by fading may occur in a larger area. As a consequence, the number of ping-pongs increases. In both cases, ping-pongs are alleviated with larger TTTs, e.g., in Set-1, since sufficient L1 and L3 filtering can be performed to mitigate fading and Layer 1 estimation imperfections.

9.4.3 Mobility-based intercell interference coordination for HCNs

As shown in previous sections, in co-channel deployments of macrocells and picocells, high-mobility MUEs are likely to be victim UEs, because they may not be able to connect soon enough to a picocell due to the TTT constraint, even when the picocell provides better link quality, and they may experience RLFs before the HO process is completed. In the following, we propose a mobility-based intercell interference coordination (MB-ICIC) scheme that combines HO parameter optimization with eICIC, so as to reduce the HOF and/or ping-pong rates. On the one hand, in order to protect high-mobility UEs from HOFs (which have been shown to be a bigger issue for them than ping-pongs), we propose that picocells release the use of certain resources (e.g., subframes 4 and 8 in Fig. 9.22(a)), so that macrocells can schedule their high-mobility UEs in these resources without co-channel interference from picocells. In a more general setting, resources can be coordinated in time (e.g., subframes), frequency (e.g., component carriers), code (e.g., spreading codes in CDMA systems), and space (e.g., beam direction) domains. On the other hand, in order to protect low-mobility UEs from ping-pongs (which have been shown to be a bigger issue for them than HOFs), we propose the use of large TTTs for low-mobility UEs. In addition, and following the traditional Rel-10 enhanced ICIC approach, macrocells may also leave certain resources blank (e.g., subframes 2, 6, and 9 in Fig. 9.22(a)), so that picocells can schedule their range-expanded PUEs in these resources [38]. This MB-ICIC depends upon the detection of UE mobility state, i.e., low mobility or high mobility.

HOF rates under the proposed MB-ICIC are shown in Fig. 9.22(b). With MB-ICIC, macrocells allocate high-velocity UEs in coordinated subframes (i.e., ABSs of picocells), so that their macro-to-pico HOs as well as strong pico-to-macro interference are avoided. In this way, the number of RLFs and thus HOFs is significantly reduced. However, such a performance improvement comes at the expense of releasing some resources from the picocells. In order to minimize the throughput loss of picocells, only high-velocity MUEs (e.g., >60 km/h) are assigned to ABSs of picocells, whereas low-velocity MUEs are handled through HO parameter optimization, i.e., using large TTTs (e.g., Set-2) to suppress ping-pongs. It may also be possible to semi-dynamically adjust the duty cycle of ABSs at the picocell based on the percentage of high-mobility UEs within a given time window, so as to minimize throughput loss of the picocell.

Ping-pong rates under the proposed MB-ICIC are shown in Fig. 9.22(c). The rate of ping-pong is also significantly reduced by MB-ICIC. This is because HOs for high-mobility UEs (e.g., >60 km/h) are avoided through cooperative radio resource

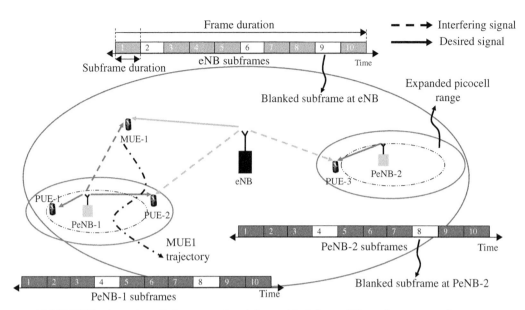

(a) MB-ICIC, where picocells leave certain subframes blank for mobility enhancement of high-velocity MUEs.

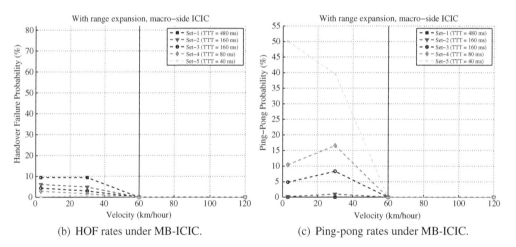

(b) HOF rates under MB-ICIC.   (c) Ping-pong rates under MB-ICIC.

Figure 9.22 Scenario for the proposed MB-ICIC, and simulated HOF and ping-pong rates (with four randomly deployed picocells per sector, 8 dB range expansion bias, and MB-ICIC) (reproduced with permission from IEEE©).

management, while HOs for low-mobility UEs go through the standard HO procedure but with long TTTs (e.g., Set-2), which reduce ping-pongs.

In order to allow a fair performance comparison, we quantify the gain of using the proposed MB-ICIC with respect to the case of picocell range expansion and pico eICIC for Set-3 (TTT of 160 ms), which has been shown to provide the best HOF and ping-pong tradeoff performance. When the UE velocity is 60 km/h, the HOF and ping-pong gains offered by the proposed MB-ICIC are around 5.5% (by comparing Fig. 9.22(b) with

Fig. 9.21(b)), and 10% (by comparing Fig. 9.22(c) with Fig. 9.21(d)), respectively. The gains are larger for high-velocity UEs. For example, at 120 km/h, the HOF and ping-pong gains provided by the proposed MB-ICIC in comparison with the same reference case are around 13% and 12%, respectively.

9.5   Conclusion

In this chapter, we have reviewed mobility and handover management in both RRC-connected and RRC-idle states in LTE systems. In the RRC-connected state, based on the analysis of HO procedures, we focus on the settings of HO parameters and introduce a scheme to assign different hysteresis values to UEs with different velocity levels. The effectiveness of this scheme is verified through simulations, which have shown that the probability of RLF could be significantly reduced. In the RRC-idle state, we present the cell selection/reselection procedures and the conflict between MLB and CRS parameter adjustment. Based on the dynamic parameter adjustment scheme of MLB, a negotiation mechanism is introduced to solve two parameter mismatching problems. One is the CRS parameter mismatch between neighboring cells, and the other is the mismatch between CRS and HO parameters. Simulation results have demonstrated the necessity of an intercell negotiation process during the setting of CRS and HO parameters. Finally, mobility management in HCNs is investigated, and a mobility-based ICIC technique is introduced, which decreases the number of handover failures and ping-pongs caused by high-velocity UEs.

Copyright notices

References

[1] 3GPP TR 36.300, *Technical Specification Group Radio Access Network; Evolved Universal Terrestrial Radio Access (E-UTRA) and Evolved Universal Terrestrial Radio Access Network (E-UTRAN); Overall Description*, v10.2.0, technical report (2011).

[2] D. Singhal, M. Kunapareddy, V. Chetlapalli, V. B. James and N. Akhtar, LTE-Advanced: handover interruption time analysis for IMT-A evaluation. In *International Conference on Signal Processing, Communication, Computing and Networking Technologies (ICSCCN)* (2011), pp. 81–85.

[3] L. K. L. Bajzik, P. Horvath and C. Vulkan, Impact of intra-LTE handover with forwarding on the user connections. In *16th Mobile and Wireless Communications Summit* (2007), pp. 1–5.

[4] Y. Yuan and Z. Chen, A study of algorithm for LTE intra-frequency handover. In *2011 International Conference on Computer Science and Service System (CSSS)* (2011), pp. 1986–1989.

[5] 3GPP TR 36.420, *LTE; Evolved Universal Terrestrial Radio Access Network (E-UTRAN); X2 general aspects and principles*, v10.0.0, technical report (2011).

[6] 3GPP TR 36.423, *LTE; Evolved Universal Terrestrial Radio Access Network (E-UTRAN); X2 Application Protocol (X2AP)*, v10.0.0, technical report (2011).

[7] M. Anas, F. D. Calabrese, P. E. Ostling, K. I. Pedersen and P. E. Mogensen, Performance analysis of handover measurements and layer 3 filtering for UTRAN LTE. In *Proceedings of the IEEE International Symposium On Personal, Indoor, Mobile Radio Communications (PIMRC)*, Athens, Greece (2007), pp. 1–5.

[8] TS 36.331, *Radio Resource Control; Protocol Specification*, 3GPP Technical Report, v.10.4.0 (2011).

[9] 3GPP TSG RAN WG3 Meeting 64, R3-091294, Alcatel-Lucent, *Exchange of Handover Parameters Directly between eNBs*, technical report (2009).

[10] 3GPP TR 36.331, *LTE; Evolved Universal Terrestrial Radio Access (E-UTRA); Radio Resource Control (RRC); Protocol specification*, v9.3.0, technical report (2010).

[11] 3GPP TR 36.902, *LTE; Evolved Universal Terrestrial Radio Access Network (E-UTRAN); Self-Configuring and Self-Optimizing Network (SON)Use Cases and solutions*, v9.3.0, technical report (2011).

[12] TR 36.839, *Mobility Enhancements in Heterogeneous Networks*, 3GPP Technical Report, v.0.2.0 (2011).

[13] 3GPP TS 36.133, *Requirements for Support of Radio Resource Management*, v10.1.0, technical report (2010).

[14] 3GPP TS 36.508, *Common Test Environments for User Equipment (UE) Conformance Testing*, v9.3.0, technical report (2010).

[15] 3GPP TSG RAN WG1 Meeting 56bis, R1-091578, NTT DOCOMO, *Evaluation Model for Rel-8 Mobility Performance*, technical report (2009).

[16] 3GPP TSG RAN WG3 Meeting 65, R3-091764, Huawei, *Source Cell Detection and Correction of Too Early Handover*, technical report (2009).

[17] 3GPP TSG RAN WG3 Meeting 65bis, R3-092523, Motorola, *Procedure Support for MRO*, technical report (2009).

[18] 3GPP TSG RAN WG3 Meeting 65bis, R3-092448, Ericsson, *Solutions to Support Mobility Robustness Ericsson*, technical report (2009).

[19] 3GPP TS 36.304, *User Equipment (UE) Procedures in IDLE Mode*, v10.0.0, technical report (2010).

[20] 3GPP TS 23.122, *NAS Functions Related to Mobile Station (MS) in Idle Mode*, v10.2.0, technical report (2010).

[21] 3GPP TS 36.101, *User Equipment (UE) Radio Transmission and Reception*, v10.1.1, technical report (2011).

[22] A. Lobinger, S. Stefanski, T. Jansen and I. Balan, Load balancing in downlink LTE self-optimizing networks. In *IEEE 71st Vehicular Technology Conference (VTC 2010 Spring)* (2010), pp. 1–5.

[23] H. Hu, J. Zhang, X. Zheng, Y. Yang and P. Wu, Self-configuration and self-optimization for LTE networks. *IEEE Communications Magazine*, **48**:2 (2010), 94–100.

[24] O. Yilmaz, Self-optimization of coverage and capacity in LTE using adaptive antenna systems, Master's thesis, Aalto University (2010).

[25] D. Lee, G. Gil and D. Kim, A cost-based adaptive handover hysteresis scheme to minimize the handover failure rate in 3GPP LTE system. *EURASIP Journal on Wireless Communications and Networking*, **2010**:6 (2010), 2–8.

[26] H. Zhang, X. Wen, B. Wang, W. Zheng and Z. Lu, A novel self-optimizing handover mechanism for multi-service provisioning in LTE-Advanced. In *International Conference on Research Challenges in Computer Science* (2009), pp. 221–224.

[27] K. Son, C. Song and G. Veciana, Dynamic association for load balancing and interference avoidance in multi-cell networks. *IEEE Transactions on Wireless Communications*, **8**:7 (2009), 3566–3576.

[28] SOCRATES, Self-optimisation and self-configuration in wireless networks. In *European Research Project*. http://www.fp7-socrates.eu

[29] 3GPP TSG RAN WG3 Meeting 63bis, R3-090794, Huawei, *Handover Parameters Tuning for Load Balancing*, technical report (2009).

[30] 3GPP TSG RAN WG3 Meeting 63bis, R3-090911, CATT, *Discussion of Load Balancing Parameter and Procedure*, technical report (2009).

[31] S. V. Hanly, An algorithm for combined cell-site selection and power control to maximize cellular spread spectrum capacity. *IEEE Journal on Selected Areas in Communications*, **13** (1995), 1332–1340.

[32] S. K. Das, S. K. Sen, R. Jayaram and P. Agrawal, A distributed load balancing algorithm for the hot cell problem in cellular mobile networks. In *The Sixth IEEE International Symposium on High Performance Distributed Computing* (1997), 254–263.

[33] R. Nasri and Z. Altman, Handover adaptation for dynamic load balancing in 3GPP Long Term Evolution systems. In *Proceedings of MoMM2007* (2007), pp. 145–153.

[34] H. L. A. Schroder and G. Nunzi, Distributed self-optimization of handover for the Long Term Evolution. *Lecture Notes In Computer Science*, **5343** (2008), 281–286.

[35] Samsung, *Mobility Support to Pico Cells in the Co-channel HetNet Deployment* (2010), 3GPP Standard Contribution (R2-104017).

[36] D. López-Pérez, I. Güvenc and X. Chu, Mobility enhancements for heterogeneous wireless networks through interference coordination. In *Proceedings of the IEEE International Workshop on Broadband Femtocell Technologies (Co-Located with IEEE WCNC)*, Paris (2012).

[37] D. López-Pérez, I. Güvenc and X. Chu, Theoretical analysis of handover failure and ping-pong rates for heterogeneous networks. In *Proceedings of the IEEE International Workshop on Small Cell Wireless Networks (Co-Located with IEEE ICC)*, Ottawa (June 2012).

[38] D. López-Pérez, I. Güvenç, G. de la Roche, M. Kountouris, T. Q. Quek and J. Zhang, Enhanced inter-cell interference coordination challenges in heterogeneous networks. *IEEE Wireless Communications Magazine*, **18**:3 (2011), 22–31.

10 Cooperative relaying

Jing Xu, Jiang Wang and Ting Zhou

Relaying is a well known technique to transmit signals from a source to a destination through one or several intermediate nodes (i.e., *relay nodes* (RNs)) without using increased power at the source [1–4]. In the past decade, many research efforts on relay technologies have been made to improve the cell coverage, enhance the transmission reliability, and increase the system throughput. More recently, RNs have become an important component in a *heterogeneous cellular network* (HCN) to provide service improvement and coverage extension at hotspots and cell edges. *Layer three* (L3) RN, which works as an independent *base station* (BS) except for the use of the wireless backhaul link, is specified in *3rd Generation Partnership Project* (3GPP) Release 10 to realize flexible network deployment and increase network throughput without any additional infrastructure.

In terms of data forwarding, four types of relay have been widely studied, which are *amplify-and-forward* (AF), *demodulate-and-forward* (DMF), *decode-and-forward* (DCF), and *estimate-and-forward* (EF). Because there is no baseband signal processing function, an AF-relay-based wireless network is cost efficient. The main disadvantage of the AF relay is that the received noise and interference would also be forwarded to the destination. To mitigate the received noise and interference at an RN, the DMF relay and EF relay have been proposed to perform some simple signal processing according to the constellation used. With the decoding operation performed at the RN, a DCF relay can regenerate the source-transmitted signal perfectly if the received signal is decoded correctly. As described in the WINNER project [5], a *user equipment* (UE) conventionally attaches to the destination through a single path, while in cooperative relaying a UE will be connected to the destination through multiple paths. With the soft information estimation and forwarding in cooperative relaying, an EF relay can improve the performance even if the received signal is not decoded correctly. With respect to the protocol architecture, the AF relay is a *layer one* (L1) relay, and the DMF relay and the EF relay are *layer two* (L2) relays. The DCF relay can be an L2 relay or L3 relay depending on the forwarding position in the protocol architecture [6, 7].

In Section 10.1, different relay functions are introduced, such as AF, DMF, and the corresponding link adaptation. The relay system architecture in 3GPP LTE-Advanced is introduced in Section 10.2. Then cooperative relaying with the EF relay function and the joint network-channel coding is presented in Section 10.3. Finally we draw conclusions for this chapter in Section 10.4.

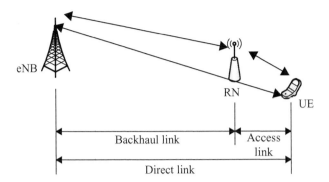

Figure 10.1 An illustration of a relay enhanced cellular network.

10.1 Relay function

In this section, we focus on memoryless single-path two-hop schemes, with AF relay, DMF relay, and DCF relay. Based on the analysis of the output *signal to noise ratio* (SNR), we present the principle of link adaptation for the DMF relay and investigate the spectral efficiency of AF relay, DMF relay, and DCF relay.

10.1.1 AF and DMF relay

A relay enhanced cellular network is shown in Fig. 10.1, where an RN and a UE are located at the cell edge. There are three types of link in this scenario: the direct link refers to the connection from the *evolved NodeB* (eNB) to the UE, the backhaul link refers to the link between the eNB and the RN, and the access link is from the RN to the UE. The eNB and the RN have the power constraints P_{eNB} and P_{RN}, respectively.

An AF RN performs the power scaling operation and forwards the received signal to the UE. The same modulation and coding scheme is adopted in the single-path two-hop transmissions. The received signal at the AF RN is given by

$$y_{\text{RN}} = h_{\text{BR}}\sqrt{P_{\text{eNB}}}x + v_{\text{RN}}, \tag{10.1}$$

where h_{BR} is the channel coefficient of the link from the eNB to the RN, x is the transmitted symbol with a normalized transmission power, the modulation scheme can be *binary phase-shift keying* (BPSK), *quadrature phase-shift keying* (QPSK), *16-quadrature amplitude modulation* (QAM) or 64QAM, and v_{RN} is the complex *additive white Gaussian noise* (AWGN) with the variance σ_{RN}^2. The received signal at the UE from the RN can be expressed as

$$\begin{aligned} y_{\text{UE}} &= h_{\text{RU}}a(h_{\text{BR}}\sqrt{P_{\text{eNB}}}x + v_{\text{RN}}) + v_{\text{UE}} \\ &= h_{\text{RU}}ah_{\text{BR}}\sqrt{P_{\text{eNB}}}x + h_{\text{RU}}av_{\text{RN}} + v_{\text{UE}}, \end{aligned} \tag{10.2}$$

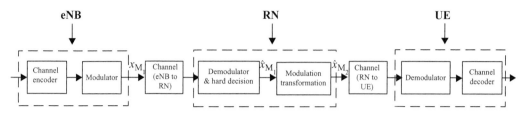

Figure 10.2 Demodulation-and-forward relay.

where $\alpha = \sqrt{\dfrac{P_{RN}}{P_{eNB}|h_{BR}|^2 + \sigma_{RN}^2}}$ is the power amplifier factor, h_{RU} is the channel coefficient of the link from the RN to the UE, v_{UE} is the complex AWGN with the variance σ_{UE}^2, and the power of the transmitted symbol is normalized to unity.

The received SNR or *end-to-end* (E2E) SNR at the UE is given by

$$\text{SNR}_{AF} = \frac{P_{eNB}\,|h_{RU}\alpha h_{BR}|^2}{|h_{RU}\alpha|^2\,\sigma_{RN}^2 + \sigma_{UE}^2}. \tag{10.3}$$

If the E2E SNR or the *channel state information* (CSI) of the two hops is known at the eNB, the scheduler can perform link adaptation for the AF relay transmission. From the perspective of the UE, the backhaul link and the access link could be jointly treated as a single equivalent link. The equivalent SNR is given by

$$\text{SNR}_{AF} = \frac{\text{SNR}_b\,\text{SNR}_a}{1 + \text{SNR}_b + \text{SNR}_a}, \tag{10.4}$$

where SNR_a and SNR_b denote the SNR of the access link and the backhaul link, respectively.

In contrast to the AF relay, the DMF relay can perform modulation transformation to adjust the transmission data rate of the access link as shown in Fig. 10.2. In Section 10.1.3, an efficient and robust modulation transformation scheme will be proposed. To simplify the description, we assume that modulation schemes for the backhaul link and the access link are M_1 and M_2, respectively, and the coding rate is C for both links. In other words, modulation scheme M_1 after the hard decision will be transformed into modulation scheme M_2. To analyze the effect of the modulation transformation, the received signal at the RN is rewritten as

$$y_{RN} = h_{BR}\sqrt{P_{eNB}}x_{M_1} + v_{RN}, \tag{10.5}$$

where x_{M_1} denotes the transmitted symbol with modulation scheme M_1 from the eNB. After the hard decision at the symbol level and the modulation transformation, the transmitted symbol from the RN can be expressed as

$$\sqrt{P_{RN}}\hat{x}_{M_2} = \sqrt{P_{RN}}(\rho_{M_2}x_{M_2} + (\hat{x}_{M_2} - \rho_{M_2}x_{M_2})), \tag{10.6}$$

where x_{M_2} is the virtually transmitted symbol with the modulation scheme M_2 from the eNB, $\hat{x}_{M_2}$ is the output of the modulation transformation, $\rho_{M_2} = E[x_{M_2}(\hat{x}_{M_2})^*]$ is the normalized correlation coefficient [8, 9] between $\hat{x}_{M_2}$ and the virtually transmitted symbol x_{M_2} from the eNB, and $()^*$ denotes the conjugate operation. The detail on how to

Table 10.1 The relationship between modulation order and modulation scheme.

Modulation scheme	BPSK	QPSK	16QAM	64QAM
q_i $(i = 1, 2)$	1	2	4	6

obtain ρ_{M_2} will be presented in Section 10.1.3. We can see that in (10.6) the first term is the desired signal, while the second term is the interference caused by the hard decision and the modulation transformation at the RN. Moreover, these two terms satisfy

$$E[x_{M_2}(\hat{x}_{M_2} - \rho_{M_2}x_{M_2})^*] = 0. \tag{10.7}$$

From (10.7) we can see that the second term of (10.6) does not contain any information of the virtually transmitted symbol x_{M_2} from the eNB. The received signal at the UE can be expressed as

$$y_{UE} = h_{RU}\sqrt{P_{RN}}\hat{x}_{M_2} + v_{UE}. \tag{10.8}$$

Substituting (10.6) into (10.8), the received signal at the UE can be rewritten as

$$y_{UE} = \sqrt{P_{RN}}h_{RU}\rho_{M_2}x_{M_2} + \sqrt{P_{RN}}h_{RU}(\hat{x}_{M_2} - \rho_{M_2}x_{M_2}) + v_{UE}. \tag{10.9}$$

The received SNR at the UE can be expressed as

$$SNR_{DMF} = \frac{\left|\rho_{M_2}h_{RU}\right|^2 P_{RN}}{\left(1 - \rho_{M_2}^2\right)\left|\rho_{M_2}h_{RU}\right|^2 P_{RN} + \sigma_{UE}^2}. \tag{10.10}$$

10.1.2 Throughput comparison

In this section, the throughput performance of the AF relay, DMF relay, and DCF relay is compared. We assume that the bandwidth is definite and a time unit is normalized to unity. One time unit is split into two parts: the first part (α) is used for the backhaul link transmission, and the second part (β) is used for the access link transmission.

For the AF relay, the time unit should be split 50–50 between the backhaul link and the access link, i.e.,

$$\alpha = \beta = \tfrac{1}{2}. \tag{10.11}$$

Thus, the E2E spectral efficiency of the AF relay is given by

$$SP_{AF} = \tfrac{1}{2}\log_2\left(1 + SNR_{AF}\right). \tag{10.12}$$

The DMF relay could choose the feasible modulation constellation of the access link according to its channel quality. Let the modulation order of the backhaul link and the access link be q_1 and q_2, respectively. The relationship between the modulation order and the modulation scheme is given in Table 10.1.

For the DMF relay, α and β should be determined as follows:

$$\alpha = \frac{q_2}{q_1 + q_2}, \quad \beta = \frac{q_1}{q_1 + q_2}. \tag{10.13}$$

Considering the backhaul link and the access link as a single equivalent link for the UE, the E2E spectral efficiency of the DMF relay is given by

$$SP_{DMF} = \beta \log_2 (1 + SNR_{DMF}). \tag{10.14}$$

One constraint on SP_{DMF} is that SP_{DMF} could not exceed the maximum spectral efficiency of a modulation scheme, i.e.,

$$SP_{DMF} \leq \frac{q_1 q_2}{q_1 + q_2}. \tag{10.15}$$

Another constraint on SP_{DMF} is that SP_{DMF} could not exceed the maximum spectral efficiency of the backhaul link, i.e.,

$$SP_{DMF} \leq \alpha \log_2 (1 + SNR_b). \tag{10.16}$$

Then, the spectral efficiency of the UE served by the DMF relay is given by

$$SP_{DMF} = \min \left\{ \beta \log_2 (1 + SNR_{DMF}), \frac{q_1 q_2}{q_1 + q_2}, \alpha \log_2 (1 + SNR_b) \right\}. \tag{10.17}$$

In order to maximize the spectral efficiency of the DCF relay, we set

$$\alpha + \beta = 1, \tag{10.18}$$

$$R_b \alpha = R_a \beta, \tag{10.19}$$

where R_b and R_a are the normalized spectral efficiency of the backhaul link and the access link, respectively, and can be calculated as

$$R_b = \log_2 (1 + SNR_b),$$
$$R_a = \log_2 (1 + SNR_a). \tag{10.20}$$

From (10.18), (10.19) and (10.20), we obtain that

$$\alpha = \frac{R_a}{R_a + R_b}, \beta = \frac{R_b}{R_a + R_b}. \tag{10.21}$$

The E2E spectral efficiency of the DCF relay is therefore given by

$$SP_{DCF} = R_b \alpha = R_a \beta = \frac{R_b R_a}{R_a + R_b}. \tag{10.22}$$

The E2E spectral efficiency of a cell-edge UE served by the AF relay, DMF relay, and DCF relay is plotted in Fig. 10.3. The SNR of the backhaul link is assumed to be 20 dB better than that of the direct link, and the SNR of the access link is assumed to be 10 dB better than that of the direct link. The 5%-tile and 10%-tile of the SNR *cumulative distribution function* (CDF) are also plotted in Fig. 10.3 to illustrate the spectral efficiency of the cell-edge UE. The spectral efficiency of the DMF relay is higher than that of the AF relay, and is close to that of the DCF relay in low-SNR regions. Note that the delay and processing complexity of the DMF relay are lower than those of the DCF relay.

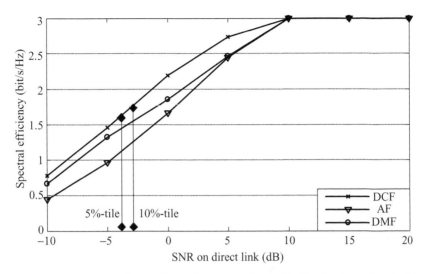

Figure 10.3 Spectral efficiency of AF relay, DMF relay, and DCF relay (reproduced with permission from IEEE©).

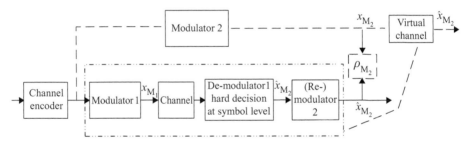

Figure 10.4 Virtual channel for modulator 1, the channel fading, the hard decision, and the modulation transformation.

10.1.3 Link adaptation of DMF relay

From (10.10), we can see that the E2E SNR of the DMF relay depends on the modulation schemes of the backhaul and access links. This can be utilized to perform link adaptation for the DMF relay. The optimal modulation schemes and coding rate are given by

$$(M_1{}^*, M_2{}^*, C^*) = \arg\max_{M_1, M_2, C}(1 - \mathrm{BLER}(\mathrm{SNR}_{\mathrm{DMF}}(M_1, M_2), C))\frac{L_{\mathrm{TB}}}{T_1 + T_2}, \quad (10.23)$$

where $\mathrm{BLER}(\mathrm{SNR}_{\mathrm{DMF}}(M_1, M_2), C)$ denotes the E2E *block error rate* (BLER), C is the coding rate, L_{TB} is the size of a transport block, and T_1 and T_2 are the time duration of the transmission for the backhaul link and the access link, respectively.

As shown in Fig. 10.4, the virtual channel is defined to represent effects of the modulator 1, the channel fading between the eNB and the RN, the hard decision at the symbol level, and the modulation transformation. The introduction of the virtual channel makes it easier to perform the link adaptation, although it is difficult to derive

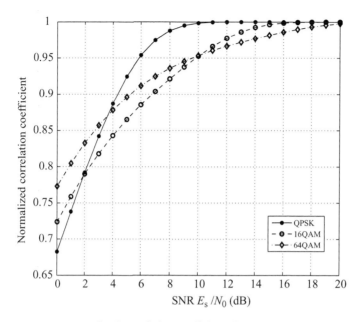

Figure 10.5 Normalized correlation coefficients for QPSK, 16QAM, and 64QAM.

the closed form expression of the normalized correlation coefficient over the virtual channel. Numerical solutions to $\rho_{M_2} = f(\rho_{M_1})$ over the AWGN channel obtained using the Monte Carlo method are shown in Figs. 10.6–10.9. Note that in Fig. 10.8 and Fig. 10.9, the curve for M_2 being BPSK and that for M_2 being QPSK almost overlap with each other. The normalized correlation coefficients for QPSK, 16QAM, and 64QAM over a complex AWGN channel are shown in Fig. 10.5. Given the normalized correlation coefficient of the modulation scheme and $\rho_{M_2} = f(\rho_{M_1})$, the spectral efficiency of a candidate transmission scheme can be determined. For example, the spectral efficiency of the transmission scheme (QPSK, 16QAM, 1/2) can be computed as follows.

1. With the CSI of the backhaul link and the modulation scheme QPSK ($M_1 = 4$), the normalized coefficient ρ_{QPSK} over the first-hop channel can be obtained based on Fig. 10.5.
2. The normalized correlation coefficient ρ_{16QAM} for modulation scheme 16QAM over the virtual channel can be obtained using Fig. 10.7 as a lookup table.
3. The E2E SNR of DMF relay can be computed from (10.10).
4. Given the E2E SNR, 16QAM used for the access link and the coding rate $C = 1/2$, BLER(SNR_{DMF}(QPSK, 16QAM), 1/2) can be obtained with the simulation specification in Table 10.2.
5. The spectral efficiency of the transmission scheme (QPSK, 16QAM, 1/2) is given by $(1 - \text{BLER}(\text{SNR}_{\text{DMF}}(\text{QPSK}, 16\text{QAM}), 1/2))\frac{L_{\text{TB}}}{T_1 + T_2}$.

Following the above procedure, the spectral efficiency of each candidate transmission scheme can be obtained, and the transmission scheme with the highest spectral efficiency

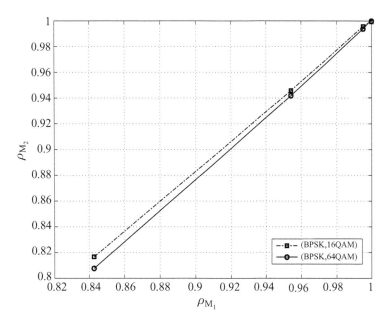

Figure 10.6 Normalized correlation coefficient of the second hop when BPSK is used in the first hop.

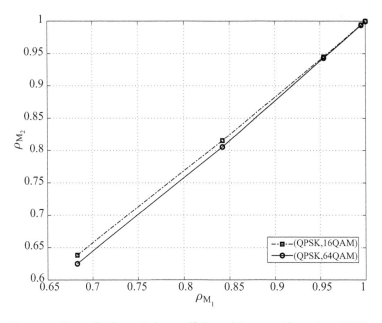

Figure 10.7 Normalized correlation coefficient of the second hop when QPSK is used in the first hop.

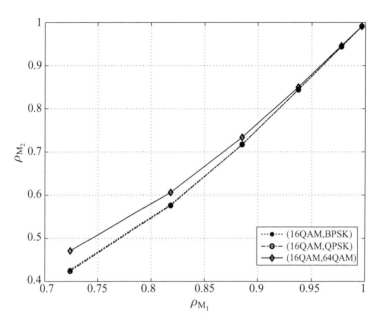

Figure 10.8 Normalized correlation coefficient of the second hop when 16QAM is used in the first hop.

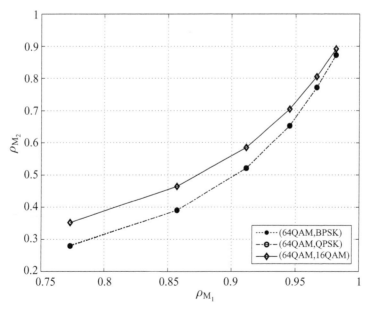

Figure 10.9 Normalized correlation coefficient of the second hop when 64QAM is used in the first hop.

Table 10.2 Simulation specification.

Channel model	Block fading channel
Channel encoder/channel decoder	Turbo code/max-log-MAP eight iteration
Modulation	BPSK, QPSK, 16QAM , 64QAM
Code rate	1/3, 1/2, 2/3 and 3/4
Transport block size (bits)	240
Hybrid automatic repeat request (HARQ)	No
Channel estimation	Perfect

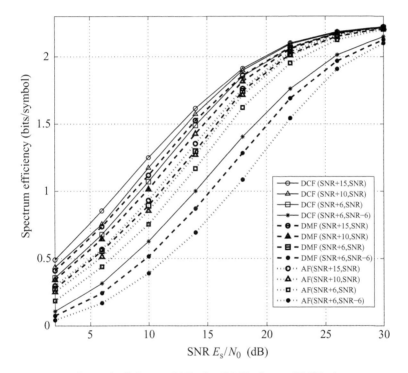

Figure 10.10 Spectral efficiency of AF relay, DMF relay, and DCF relay.

is selected. Simulation results are shown in Fig. 10.10 and Table 10.3, where SNR_1 and SNR_2 denote the SNR of the backhaul link and the access link, respectively, and (SNR_1, SNR_2) are given as functions of the abscissa SNR in the legends of Fig. 10.10. From the simulation results in Fig. 10.10 and Table 10.3, we can see that significant performance gain can be achieved by the DMF relay over the AF relay scheme. There exists a performance gap between the DMF relay and the DCF relay.

10.2   Relay architecture in LTE-Advanced

10.2.1   Interface and architecture

In LTE-Advanced systems, an RN is connected to its *donor eNB* (DeNB) via the Un interface, which is a radio interface modified from the Uu interface in *Evolved UTRAN*

Table 10.3 Spectral efficiency (bits/symbol).

(SNR$_1$, SNR$_2$)	AF	DMF	DCF
(16,4)	0.39	0.51	0.63
(16,10)	0.76	0.89	1.07
(20,10)	0.86	1.04	1.17
(25,10)	0.93	1.12	1.25
(12,0)	0.17	0.24	0.32
(12,6)	0.44	0.56	0.68
(16,6)	0.51	0.64	0.76
(21,6)	0.57	0.74	0.86

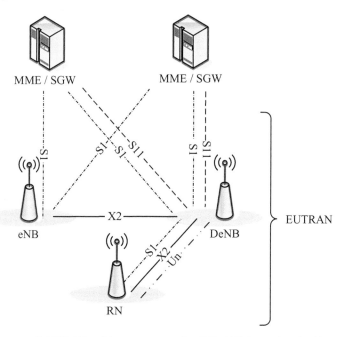

Figure 10.11 E-UTRAN architecture supporting RNs [10] (reproduced with permission from 3GPP).

(E-UTRAN). The architecture for supporting RNs is shown in Fig. 10.11, where the RN terminates the S1, X2, and Un interfaces, while the DeNB provides the S1 and X2 proxy functionality between the RN and other network nodes such as eNBs, *Mobility Management Entities* (MMEs), and *Serving Gateways* (SGWs).

The S1 and X2 proxy functionality includes passing UE-dedicated S1 and X2 signalling messages, as well as *GPRS Tunnel Protocol* (GTP) data packets, between the S1 and X2 interfaces associated with the RN and those associated with other network nodes. Due to the proxy functionality, the DeNB appears as an MME in *S1 for the control plane* (S1-MME), an eNB in X2, and an SGW in *S1 for the user plane* (S1-U) to the RN.

In addition to the eNB functionality, the RN also supports a subset of the UE functionality in order to connect to the DeNB via a wireless link. In the RN startup procedure

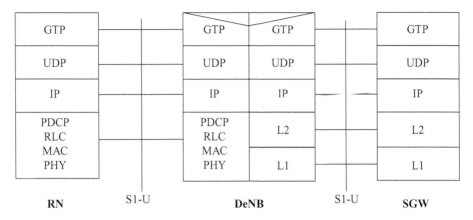

Figure 10.12 S1 user plane protocol stack for supporting RNs [10] (reproduced with permission from 3GPP).

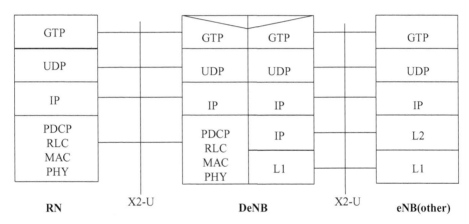

Figure 10.13 X2 user plane protocol stack for supporting RNs [10] (reproduced with permission from 3GPP).

described in Section 6.3.2, the DeNB also embeds and provides the SGW or *Packet Data Network Gateway* (PGW) like functions, which include creating a session, managing *Evolved Packet System* (EPS) bearers for the RN, and terminating the S11 interface towards the MME that is serving the RN.

10.2.2   Protocol stack

The S1 user plane protocol stack is depicted in Fig. 10.12. There is a GTP tunnel per UE bearer, spanning from the SGW/PGW of the UE to the DeNB. The DeNB switches to another GTP tunnel, going from the DeNB to the RN in a one-to-one mapping.

The X2 user plane protocol stack for supporting RNs is shown in Fig. 10.13. There is a GTP forwarding tunnel associated with each UE EPS bearer subject to forwarding, spanning from another eNB to the DeNB. The DeNB switches to another GTP tunnel, going from the DeNB to the RN in a one-to-one mapping.

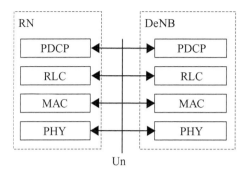

Figure 10.14 User plane protocol stack for supporting RNs [10] (reproduced with permission from 3GPP).

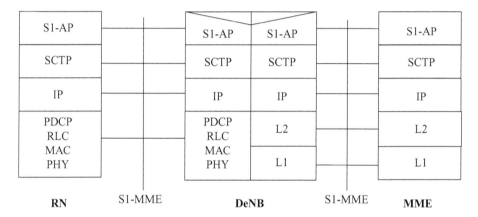

Figure 10.15 S1 control plane protocol stack [10] (reproduced with permission from 3GPP).

The S1 and X2 user plane packets are mapped to radio bearers over the Un interface. The mapping can be based on *QoS class identifiers* (QCIs) of UE EPS bearers. UE EPS bearers with similar *quality of service* (QoS) can be mapped to the same Un radio bearer. The RN connects to its DeNB via the Un interface using same radio protocols and procedures as a UE connecting to an eNB. The user plane protocol stack is shown in Fig. 10.14.

The control plane protocol architecture of S1 interface is shown in Fig. 10.15. *S1 Application Protocol* (S1AP) messages are exchanged between the MME and the DeNB, and between the DeNB and the RN. Upon the DeNB receiving S1AP messages, it translates UE *identifiers* (IDs) between the two interfaces by modifying the S1AP UE IDs in the message but leaving other parts of the message unchanged. This operation corresponds to an S1AP proxy mechanism. The S1AP proxy operation would be transparent to the MME and the RN. That is, from the perspective of MME, the UE is connected to the DeNB. While from the perspective of the RN, the RN is connected to the MME directly. S1AP messages encapsulated by *Stream Control Transmission Protocol* (SCTP)/*Internet Protocol* (IP) are transferred over an EPS data bearer of the RN.

S1 interface relations and signaling connections are shown in Fig. 10.16. There is one S1 interface relation between the RN and the DeNB and one between the DeNB

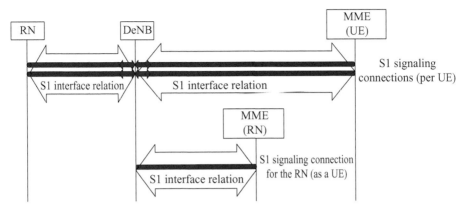

Figure 10.16 S1 interface relations and signaling connections [7] (reproduced with permission from 3GPP).

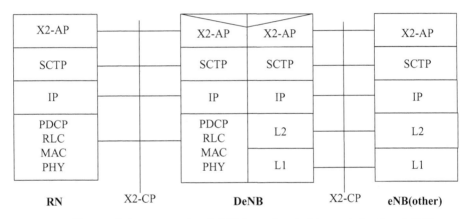

Figure 10.17 X2 control plane protocol stack [10] (reproduced with permission from 3GPP).

and the MME (serving the UE). S1 signaling connections are processed by the DeNB (as indicated by the arrows in Fig. 10.16). Note that the RN has to maintain only one S1 interface (to the DeNB), while the DeNB maintains one S1 interface to each MME in the respective MME pool. Note also that there is an S1 interface relation and an S1 signaling connection corresponding to the RN (as a UE), going from the DeNB to the MME that is serving the RN.

The X2 control plane protocol stack for supporting RNs is shown in Fig. 10.17. There is a single X2 interface relation between each RN and its DeNB, while the DeNB may have X2 interface relations with neighboring eNBs. The DeNB processes and forwards all X2 messages from the RN to other eNBs for UE-dedicated procedures. The processing of an *X2 Application Protocol* (X2AP) message modifies S1/X2AP UE IDs, transport layer address, and GTP *tunnel endpoint identifiers* (TEIDs), but leaves other parts of the message unchanged.

The control plane protocol stack when the RN acts as a UE is shown in Fig. 10.18. In fact, the Un interface is similar to the Uu interface, and the same *radio resource control* (RRC) protocol is used. One difference is the need for subframe configuration

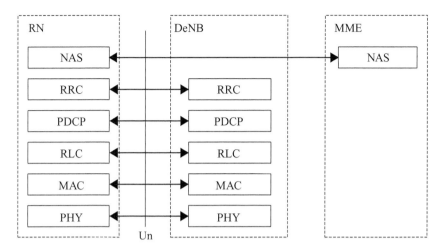

Figure 10.18 Control plane protocol stack for supporting RNs [10] (reproduced with permission from 3GPP).

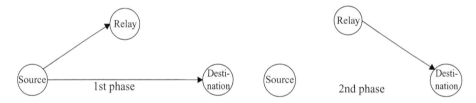

Figure 10.19 Cooperative relaying scenario.

over Un. Basically, if there is no cross-talk between the Uu and Un interfaces, the RN can operate these two simultaneously; otherwise, a subframe configuration is needed to operate these two in time division.

10.3   Cooperative relaying

10.3.1   Introduction

To improve the performance of relaying systems, cooperative schemes between the source and RNs have been developed [11–13], leading to cooperative relaying systems. A basic cooperative relaying scenario is shown in Fig. 10.19, where the source node broadcasts signals to both the RN and the destination node in the first phase, and the RN forwards the received and already processed signals to the destination node in the second phase. The destination combines the signals received in the two phases to decode the information.

For example, in the source node, an information bit sequence $\mathbf{u} = [u_0, u_1, \ldots, u_{K-1}]$ is passed through a binary encoder to produce a codeword $\mathbf{b} = [b_0, b_1, \ldots, b_{N-1}]$. The codeword is then passed through an m-order modulator (i.e., every m coded bits

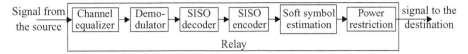

Figure 10.20 Block diagram of EF relaying.

correspond to one symbol) to produce a transmit symbol sequence $\mathbf{s} = [s_0, s_1, \ldots, s_{(N/m)-1}]$, assuming N is an integer multiple of m. We apply the notation $\mathbf{b}^k = [b_0^k, b_1^k, \ldots, b_{m-1}^k]$ to represent the m code bits corresponding to the symbol s_k, where k denotes the symbol index and $k \in \{0, 1, \ldots, (N/m) - 1\}$.

Assuming that channel fading coefficients are fixed during a transmission block, signals received by the RN and the destination in the first phase are given respectively by

$$y_{\mathrm{sr},k} = h_{\mathrm{sr}} s_k + n_{\mathrm{sr},k}, \tag{10.24}$$

$$y_{\mathrm{sd},k} = h_{\mathrm{sd}} s_k + n_{\mathrm{sd},k}, \tag{10.25}$$

where h_{sr} and h_{sd} are fading coefficients between the source and the RN and between the source and the destination, respectively, and $n_{\mathrm{sr},k}$ and $n_{\mathrm{sd},k}$ are zero-mean complex Gaussian white noise processes with the variances σ_{sr}^2 and σ_{sd}^2, respectively.

The RN then operates on the received signal sample $y_{\mathrm{sr},k}$ and forwards the processed signal $x_{\mathrm{r},k}$ to the destination in the second phase. The signal received by the destination in the second phase is given by

$$y_{\mathrm{rd},k} = h_{\mathrm{rd}} x_{\mathrm{r},k} + n_{\mathrm{rd},k}, \tag{10.26}$$

where h_{rd} is the fading coefficient between the RN and the destination, and $n_{\mathrm{rd},k}$ is a zero-mean complex Gaussian white noise process with the variance σ_{rd}^2. After receiving both signals from the source and the RN, the destination equalizes and demodulates $y_{\mathrm{sd},k}$ and $y_{\mathrm{rd},k}$ separately, combines the soft information at the bit level, and then decodes the information originally transmitted by the source.

10.3.2   Cooperative EF relay

Both the DCF relay and the EF relay perform demodulating, decoding, re-encoding, remodulating, and forwarding the regenerated signal to the destination, thus supporting hop-by-hop *adaptive modulation and coding* (AMC) at RNs. The main difference is that the DCF relay performs the hard decision after the decoding and uses the binary information to re-encode and remodulate, while the EF relay performs re-encoding and remodulating in a soft manner. Consequently, if imperfect decoding occurs, the DCF relay loses information due to the erroneous hard decision and causes error propagation through re-encoding. In contrast, the EF relay enables both regenerating the signal and keeping the soft information.

The EF relay operation is shown in Fig. 10.20, where the *single-input single-output* (SISO) re-encoding is similar to the SISO decoding and can be implemented following the *Bahl–Cocke–Jelinek–Raviv* (BCJR) algorithm in [14] and [15]. The difficult part of the EF relay is how to transmit the soft information, i.e., how to transform the soft information into digital symbols. The EF relaying scheme in [16] involves a soft information forwarding scheme for the distributed turbo coding based on the system model in Fig. 10.20. It employs a SISO convolutional re-encoder and a soft symbol calculation in the RN. The soft symbol calculation uses the *a posteriori probability* (APP) of a coded bit from the SISO decoder or the SISO re-encoder, and calculates the soft symbol in BPSK (where a binary bit 0 is mapped to 1 and a bit 1 is mapped to -1) as follows:

$$\tilde{s}_k = P_{s|\mathbf{y}_{sr}}(s_k = 0| \mathbf{y}_{sr}) \times 1 + P_{s|\mathbf{y}_{sr}}(s_k = 1| \mathbf{y}_{sr}) \times (-1). \tag{10.27}$$

Moreover, the following equivalent noise model is proposed in [16] for the soft estimation:

$$\tilde{s}_k = s_k(1 - \tilde{n}_k), \tag{10.28}$$

where s_k is the exact symbol transmitted by the source, and $\tilde{n}_k$ is the equivalent noise with its mean and variance estimated as [16]

$$\mu_n = \frac{1}{L}\sum_{k=1}^{L} |\tilde{s}_k - s_k|, \tag{10.29}$$

$$\sigma_n^2 = \frac{1}{L}\sum_{k=1}^{L} (1 - \tilde{s}_k s_k - \mu_n), \tag{10.30}$$

where L is the number of symbols used in the estimation.

If the transmit power of an RN is restricted to be P_{RN}, then the signal transmitted by the RN is given by $x_{r,k} = \beta \tilde{s}_k$, where $\beta = \sqrt{\frac{P_{RN}}{E[|\tilde{s}_k|^2]}}$. The EF relaying scheme in [16] is sub-optimal, as it can only forward the received signal in BPSK modulation, which results in a low spectral efficiency and is not practical for future wireless communication systems.

The optimal EF relaying scheme in [17] employs the *maximum likelihood* (ML) algorithm to estimate soft symbols of any required modulation from the coded soft information, under the assumption of perfect channel estimation at the RN. When receiving $y_{sr,k}$, the RN performs channel equalization, soft demodulation, and soft decoding to get the APP or *likelihood ratio* (LR) of coded bits, which are then used to estimate soft symbols of a required modulation to be transmitted by the RN. Soft estimates of forwarded symbols are denoted by $\mathbf{z} = [z_0, z_1, \ldots, z_k, \ldots]$, where each symbol z_k is modeled as

$$z_k = s_k + n_{sr,O}, \tag{10.31}$$

where $n_{sr,O}$ is the zero-mean complex AWGN at the relay output with the equivalent variance $\sigma_{sr,O}^2$, which is much smaller than the input noise variance $\sigma_{sr,I}^2$ due to the decoding gain.

The relationship between the APPs of code bits and those of corresponding symbols is given by

$$P_{\mathbf{b}^k|\mathbf{z}}\left(b_i^k = q|\mathbf{z}\right) = \sum_{\substack{x \in S, \\ b_i^k(x)=q}} P_{s_k|\mathbf{z}}\left(s_k = x|\mathbf{z}\right), \tag{10.32}$$

where S is the candidate constellation set of the considered modulation, $b_i^k = q$ indicates that s_k takes the constellation point x that has $b_i^k = q$, q takes the value of 0 or 1 for the binary information, and every m bits (for $i = 0, 1, \ldots, m - 1$) are mapped to a symbol of the considered modulation.

Applying the Bayes expression [18] and assuming an equal probability of each symbol being transmitted, one can get the expression of the LR of the coded bit conditioned on the symbols $\mathbf{z}$ as follows:

$$
\begin{aligned}
\mathrm{LR}_i^k &= \frac{P_{\mathbf{b}^k|\mathbf{z}}\left(b_i^k = 1|\mathbf{z}\right)}{P_{\mathbf{b}^k|\mathbf{z}}\left(b_i^k = 0|\mathbf{z}\right)} = \frac{\displaystyle\sum_{\substack{x \in S \\ b_i^k(x)=1}} P_{s_k|\mathbf{z}}\left(s_k = x|\mathbf{z}\right)}{\displaystyle\sum_{\substack{x \in S \\ b_i^k(x)=0}} P_{s_k|\mathbf{z}}\left(s_k = x|\mathbf{z}\right)} \\[2em]
&= \frac{\displaystyle\sum_{\substack{x \in S \\ b_i^k(x)=1}} \exp\left(-\frac{\left(z_{k,\mathrm{R}}-x_\mathrm{R}\right)^2+\left(z_{k,\mathrm{I}}-x_\mathrm{I}\right)^2}{2\sigma_{\mathrm{sr,O}}^2}\right)}{\displaystyle\sum_{\substack{x \in S \\ b_i^k(x)=0}} \exp\left(-\frac{\left(z_{k,\mathrm{R}}-x_\mathrm{R}\right)^2+\left(z_{k,\mathrm{I}}-x_\mathrm{I}\right)^2}{2\sigma_{\mathrm{sr,O}}^2}\right)},
\end{aligned}
\tag{10.33}
$$

where $\mathbf{b}^k = [b_0^k, b_1^k, \ldots, b_{m-1}^k]$ denotes the m bits corresponding to symbol z_k, and LR_i^k denotes the likelihood ratio for the bit b_i^k of the symbol z_k.

According to (10.33), one can estimate each symbol z_k from the LRs $[\mathrm{LR}_0^k, \mathrm{LR}_1^k, \ldots, \mathrm{LR}_{m-1}^k]$ of its m bits, which are the output of the SISO re-encoder. Then, we will have m equations while there is only one unknown variable z_k. Since the optimal solution to the overdetermined equation set is difficult to find, the *minimum mean square error* (MMSE) criteria is used to find a sub-optimal solution, and the resulting symbol estimation is given by

$$\tilde{z}_{k,\mathrm{MMSE}} = \arg\min_z \sum_{i=0}^{m-1} \left| \mathrm{LR}_i^k - \frac{\displaystyle\sum_{\substack{x \in S \\ b_i^k(x)=1}} f(z,x)}{\displaystyle\sum_{\substack{x \in S \\ b_i^k(x)=0}} f(z,x)} \right|^2, \tag{10.34}$$

where $f(z,x) = \exp\{-\frac{[\mathrm{Re}(z)-\mathrm{Re}(x)]^2+[\mathrm{Im}(z)-\mathrm{Im}(x)]^2}{2\sigma_{\mathrm{sr,O}}^2}\}$.

The estimated symbol $\tilde{z}_{k,\mathrm{MMSE}}$ contains the soft information of m bits. The result of (10.34) is a bit like an m-order modulator. The difference is that $\tilde{z}_{k,\mathrm{MMSE}}$ can be of any complex value, but a conventional modulation has fixed constellation points. The transmitted signal at the EF relay is similar to that of the AF relay and has a power

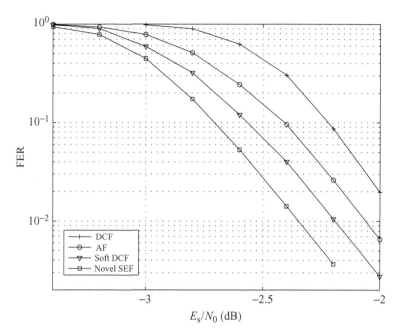

Figure 10.21 Comparison of FER.

restriction. The power restriction of the EF relay is implemented as follows:

$$x_{\mathrm{r},k} = \beta \tilde{z}_{k,\mathrm{MMSE}}, \tag{10.35}$$

where $\beta = \sqrt{P_{\mathrm{RN}} / \mathrm{E}[|\tilde{z}_{k,\mathrm{MMSE}}|^2]}$, and P_{RN} is the average power per symbol of the RN.

Fig. 10.21 compares the *frame error rate* (FER) performance of the following four schemes: AF, DCF, the soft information forwarding scheme for distributed turbo coding (denoted as Soft DCF) in [16], and the ML *soft symbol estimation and forward* (SEF) scheme (denoted as Novel SEF) in [17]. In the simulation of cooperative relay systems, the source encodes the data block with length $K = 1000$ bits, using a turbo encoder with the coding rate of 1/2, the constraint length of 3 and the generator polynomials in octal expression (15, 13). Both the source and the RN use BPSK. The AWGN channel is assumed. The SNR E_{s}/N_0 of the link from the RN to the destination is fixed at -1 dB, while E_{s}/N_0 from the source to the RN and that from the source to the destination are assumed to be the same and are indicated by as the horizontal axis. The RN forwards K systematic bits of a packet no matter whether it is decoded correctly or not. We can see from Fig. 10.21 that the SEF scheme outperforms AF relay, DCF relay, and Soft DCF relay, because it keeps the soft information and obtains the decoding gain at the RN. When the channel from the source to the RN is relatively bad, the DCF relay performs even worse than the AF relay due to the error propagation introduced by the RN. When the channel from the source to the RN gets better, the performance of the DCF relay gradually approaches and finally outperforms the AF relay in high-SNR regions.

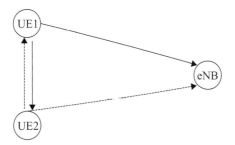

Figure 10.22 User cooperation scenario.

10.3.3   Joint network-channel coding for user cooperation

UEs can be used as mobile relays to provide an extra diversity gain. In the user cooperation, UEs pair up as partners to achieve the uplink spatial diversity by relaying each other's information [19, 20]. In particular, when UEs cannot support multiple antennas due to the size or other constraints, the user cooperation can still support multi-antenna features in a distributed manner. Many variations of the user cooperation have been proposed, such as simple repetition relaying [21], user coded cooperation [22], and distributed *space-time block coding* (STBC) [23]. However, these user cooperation schemes require extra time slots for cooperation, causing longer delays. The chained user cooperation scheme in [24] employs the joint network-channel coding to fully explore the spatial diversity and avoid using extra time slots for the cooperation. It has been shown to outperform the above conventional user cooperation schemes.

The scenario of the user cooperation is depicted in Fig. 10.22, where two source nodes, UE1 and UE2, work in cooperation to deliver their packets to the common destination eNB. UE1 and UE2 are partners of each other. Each source node receives and attempts to decode the packet sent by its partner. For example, if UE1 decodes UE2's information successfully, then part or all of this information that originated at UE2 will be relayed by UE1 in a future transmission to provide the eNB with spatial diversity. If UE1 fails to decode UE2's information, then UE1 will not relay UE2's information and will operate in a non-cooperative mode. With the DCF relay, there will be no spatial diversity at the eNB if the packets originated from UE2 are not received correctly by UE1. This indicates the importance of minimizing the packet error rate of the link between UE1 and UE2. Fortunately, the link between two not-far-from-each-other UEs is typically much better than that between either UE to the eNB in the uplink, and such uplink cooperative diversity can be achieved in most cases.

We investigate three different *time division multiplexing* (TDM) user cooperation schemes: non-cooperative, coded cooperation [22], and space-time coded cooperation [23]. Turbo code is considered in the investigation since it has been widely specified in standardizations for current and future wireless systems. Frame structures of three TDM schemes are shown in Fig. 10.23, where the horizontal axis represents time, $S_i(k)$ ($i \in \{1, 2\}$ and $k \in \{1, 2, \ldots\}$) refer to systematic bits, $P_i(k)$ refer to parity bits, and $P_i(k)'$ denote redundant parity bits. The subscript i indicates the UE index and k in the bracket indicates the block number.

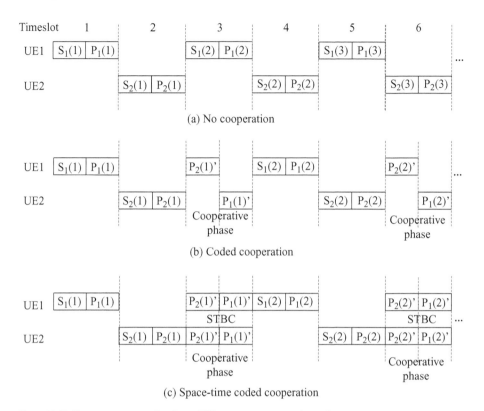

Figure 10.23 Frame structures for three different user cooperation schemes.

No cooperation
UE1 and UE2 transmit their own codewords in turn during each time slot as shown in Fig. 10.23(a).

Coded cooperation
In coded cooperation, each UE tries to relay the incremental redundancy for its partner, so that the codewords of each UE can be sent via two independent fading paths. As shown in Fig. 10.23(b), UE1 and UE2 first transmit their own codewords in turn and also listen to each other's transmissions. After decoding its partner's information, each UE encodes it and relays redundant parity bits. The information relaying takes an extra time slot, which is named the cooperative phase in Fig. 10.23.

Space-time coded cooperation
As shown in Fig. 10.23(c), the space-time coded cooperation scheme differs from the coded cooperation scheme only in that space-time coding is adopted in the cooperative phase. The two partners simultaneously transmit the redundant parity bits for each other using the space-time coding.

The above coded and space-time coded cooperation schemes can achieve spatial diversity and thus have better performance than the non-cooperative configuration. However, since these two user cooperation schemes treat the uplinks of two partners

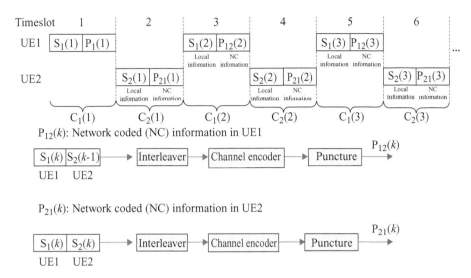

Figure 10.24 Frame structure of the joint network-channel coding scheme.

separately, two partners have to share available transmissions, and it is not possible to obtain the full diversity gain. Furthermore, they have to use an extra time slot for the cooperation, which increases the E2E transmission latency.

The joint network-channel coding based on turbo codes has been proposed for the user cooperation [24]. This network coding approach to the user cooperation has the main characteristic of network coding, i.e., the transmitting node constructs a combination of multiple data flows that it wishes to convey and transmits the combination as a whole, rather than simply routing data flows individually. The joint network-channel coding approach realizes the network coding through channel coding on the particularly combined multiple data flows. Resulting superimposed codewords from two partners can be viewed as a chain at the destination. The decoding at the destination is carried out by iterating between adjacent codewords from the two partners. In the following, operations of the network-channel coding based user cooperation are described.

Design for UEs as source nodes
The frame structure and the joint network-channel coding at the source node (i.e., UE) are depicted in Fig. 10.24, where $S_1(k)$ and $S_2(k)$ denote systematic bits of turbo codes of the kth block of UE1 and UE2, respectively, $P_1(1)$ denotes parity bits of the initial transmission that has no network coding, and $P_{12}(k)$ and $P_{21}(k)$ denote the kth block of the network coded information of UE1 and UE2, respectively. Invoking symmetry, we now provide a detailed description of the encoding operation at UE1 and the decoding operation at UE2.

Encoder at UEs
In the initial phase (i.e., time slot 1), UE1 first transmits its local codeword $C_1(1)$ (including systematic bits $S_1(1)$ and corresponding parity bits $P_1(1)$). Both the eNB and UE2 listen to it. During the time slot $(2k-1)$, for $k \in \{2, 3, \ldots\}$, UE1 must convey

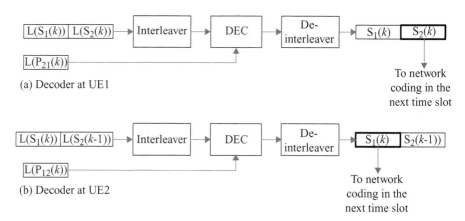

(a) Decoder at UE1

Figure 10.25 Joint network-channel decoder at the UEs.

its local information $S_1(k)$ while relaying UE2's information $S_2(k-1)$, assuming it has decoded $S_2(k-1)$ correctly. UE1 relays UE2's information $S_2(k-1)$ in the following way: UE1 first generates its new local information block $S_1(k)$, concatenates it with the information $S_2(k-1)$ of UE2, and then puts them through a pseudorandom interleaver and a channel encoder subsequently. The resulting parity bits realize the effect of network coding and are denoted as $P_{12}(k)$. The resulting codeword is denoted as $C_1(k)$. The pseudorandom interleaving facilitates iterative decoding at the destination node (i.e., eNB). It guarantees that, at the destination, the decoder for UE2 provides the decoder for UE1 with extrinsic information that is independent of the other information available to the decoder for UE1.

Decoder at UEs

In the initial phase (i.e., time slot 1), there is no network coding. During the time slot $(2k-1)$, for $k \in \{2, 3, \ldots\}$, when receiving the codeword $C_1(k)$ (containing $S_1(k)$ and $P_{12}(k)$) from UE1, the decoder at UE2 works as illustrated in Fig. 10.25, where the soft-decision decoding is assumed. The demodulator generates a *log-likelihood ratio* (LLR) for each bit, which is denoted, e.g., as $L(S_1(k))$ or $L(P_{12}(k))$. Since $S_2(k-1)$ is originated from UE2, UE2 knows the exact information of $S_2(k-1)$. UE2 can transform $S_2(k-1)$ into the LLR form $L(S_2(k-1))$ as follows:

$$
\begin{aligned}
L(S_2(k-1)) &= +D, \quad \text{if} \quad S_2(k-1) = 0 \\
L(S_2(k-1)) &= -D, \quad \text{if} \quad S_2(k-1) = 1
\end{aligned}
\tag{10.36}
$$

where D is a substantial positive integer.

Then UE2 concatenates $L(S_1(k))$ with $L(S_2(k-1))$, and puts them through a pseudorandom interleaver. The resulting sequence is put through a channel decoder and a de-interleaver subsequently. UE2 conveys the obtained information block of $S_1(k)$ and its new local information block $S_2(k)$ by the joint network-channel encoding.

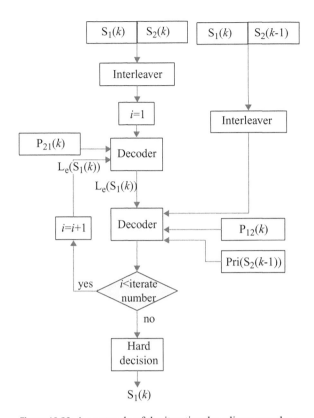

Figure 10.26 An example of the iterative decoding procedure.

Design for eNB as the destination node

The receiver at the destination node (i.e., eNB) makes decisions about all the packets generated by two partner UEs. Unlike partner UEs, the eNB has no a priori certainty of the received signal that it can use in the decoding. The decoder at the destination must exploit the redundancy provided by the channel code, because the joint network-channel encoding provides a close relationship between the adjacent codewords transmitted by the two partner UEs. By using the relationship, iterative decoding among the adjacent codewords with a soft-in-soft-out decoder can be employed.

In the initial phase, since there is no network coding, the eNB can decode the codeword $S_1(1)$ in a usual way. Without loss of generality, consider the decoding of $S_1(k)$ for $k \in \{2, 3, \ldots\}$ at the eNB. Note that both $C_1(k)$ and $C_2(k)$ carry the information about $S_1(k)$, but $C_1(k)$ is considered as the local information while $C_2(k)$ (in a interleaved form) is considered as relayed information. Since $C_2(k - 1)$ has already been processed, the extrinsic information about $S_2(k - 1)$ obtained from $C_2(k - 1)$ can be used as a priori information (denoted as $\text{Pri}(S_2(k - 1))$) in the processing of $C_1(k)$, while no a priori information is used for $S_1(k)$. Soft decoders processing $C_1(k)$ and $C_2(k)$ exchange extrinsic information about $S_1(k)$ (denoted as $L_e(S_1(k))$), with the assistance of an interleaver or de-interleaver, which is not shown in Fig. 10.26 for simplicity. After a

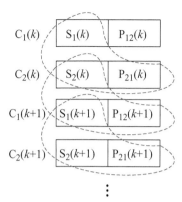

Figure 10.27 Multiple codeword blocks can be viewed as a chain.

fixed number of iterations, the decoder makes a decision about the desired information vector $S_1(k)$.

The decoding of UE2's packet $S_2(k)$ at the destination is similar, and makes use of both $C_2(k)$ and $C_1(k + 1)$. Multiple codewords can be viewed as a chain at the destination, as shown in Fig. 10.27.

Simulations have been carried out to compare the performance of four different approaches: no cooperation, coded cooperation [22], space-time coded cooperation [23], and the joint network-channel coded cooperation scheme [24]. In the simulations, a pair of source nodes try to convey the information to a common destination. Each packet consists of 1000 bits. A *cyclic redundancy check* (CRC)-12 code is used in the cooperative systems to identify decoding failures. All schemes use BPSK modulation. Each channel is independent and subject to AWGN plus block Rayleigh fading. The channel between a pair of UEs is assumed to be much better than that between either UE and the eNB, e.g., by 9 dB. Moreover, the following codes are incorporated in simulations.

- In the non-cooperative approach, i.e., a point-to-point reference system, the turbo code with the generator matrix [13   15] in the octal expression is used. The code is punctured to rate 1/2.
- The same turbo code in the octal expression [13   15] is used as the mother code for the coded cooperation approach. To leave the part of the transmission period available for relaying, this code is punctured to rate 2/3 for the transmission of the local information. The relayed information is constructed by re-encoding the recovered packet with the same code and then puncturing to rate 1/2, i.e., the whole equivalent rate of each packet is still 1/2. The destination node thus decodes a rate 1/2 code by combining the copy received from the originating UE and the relayed copy from its partner UE.
- The space-time coded cooperation scheme is different from the coded cooperation scheme only in that the originating UE and its partner UE simultaneously transmit the relayed information in a space-time coded manner.
- The joint network-channel coded cooperation scheme uses the rate 1/2 *recursive systematic convolutional* (RSC) code with the generator matrix [13   15] in the octal expression at the initiating UE. In subsequent time slots, an S-random interleaver and

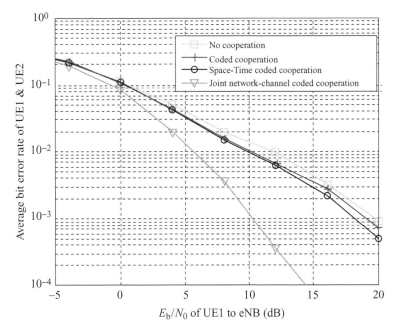

Figure 10.28 Bit error rate of the four different schemes.

a RSC code with the generator matrix [13   15] in the octal expression are used in the partner UEs for joint network-channel coding. Resulting parity bits are punctured to length K to be transmitted together with the locally generated information block. The equivalent code rate of each transmission is still 1/2. Iterative decoding with six iterations is adopted at the eNB to exploit the spatial diversity. Note that each iteration requires two BCJR algorithm calculations, which is the same as the turbo code used in the other three schemes, i.e., the decoding complexity is the same for all four schemes considered.

Fig. 10.28 depicts the *bit error rate* (BER) versus the average SNR per coded bit (E_b/N_0) in dB for the four schemes considered. The slope of BER curves shows the diversity gain provided by each scheme. The joint network-channel coded cooperation scheme achieves the full diversity gain. It outperforms the non-cooperative, coded cooperation, and space-time coded cooperation schemes by 6 dB, 4 dB, and 3.5 dB, respectively, for a BER of 10^{-2}.

10.4   Conclusion

In this chapter, we have analyzed the performance of AF relay, DMF relay, and DCF relay, and proposed a link adaptation scheme for the DMF relay. The performance comparison based on simulation results has shown that the DMF relay provides better performance than the AF relay at a relatively low implementation complexity. Then, 3GPP Release 10 relay system architectures in HCNs are introduced. In order to fully explore the capacity of cooperative relaying, the emerging EF relay is presented together with a new link adaptation scheme.

Acknowledgment

The work was supported by Shanghai Science and Technology Development funds (No 10QA1406300).

Copyright notices

References

[1] M. O. Hasna and M.-S. Alouini, End-to-end performance of transmission systems with relays over Rayleigh-fading channels. *IEEE Transactions on Wireless Communications*, **2**:6 (2003), 1126–1131.

[2] Y. Li and B. Vucetic, On the performance of a simple adaptive relaying protocol for wireless relay networks. In *IEEE Vehicular Technology Conference (VTC Spring 2008)* (2008), pp. 2400–2405.

[3] O. Sahin, E. Erkip and O. Simeone, Interference channel with a relay: models, relaying strategies, bounds. In *IEEE Information Theory and Applications Workshop* (2009), pp. 90–95.

[4] M. N. Khormuji and M. Skoglund, On instantaneous relaying. *IEEE Transactions on Information Theory*, **56**:7 (2010), 3378–3394.

[5] IST-4-027756 D3.5.2 V1.0, *Assessment of Relay Based Deployment Concepts and Detailed Description of Multi-Hop Capable RAN Protocols as Input for the Concept Group Work*, WINNER II.

[6] 3GPP R1-082024, Ericsson, *A Discussion on Some Technology Components for LTE-Advanced*, 3GPP TSG RAN WG1 Meeting 53 (2008).

[7] 3GPP TR36.806 V9.9.0, *Evolved Universal Terrestrial Radio Access (E-UTRA); Relay Architectures for E-UTRA (LTE-Advanced)*, 3GPP Technical Report (2010).

[8] Y. Yang, H. Hu, J. Xu and G. Mao, Relay technologies for WiMax and LTE-Advanced mobile systems. *IEEE Communications Magazine*, **47**:10 (2009), 100–105.

[9] H. Xiong, J. Xu and P. Wang, Frequency-domain equalization and diversity combining for demodulate-and-forward cooperative systems. In *IEEE International Conference on Acoustics, Speech and Signal Processing (ICASSP 2008)* (2008), pp. 3245–3248.

[10] 3GPP TS36.300 V10.4.0, *Evolved Universal Terrestrial Radio Access (E-UTRA) and Evolved Universal Terrestrial Radio Access Network (E-UTRAN); Overall Description*, 3GPP Technical Specification (2011).

[11] J. N. Laneman and G. W. Wornell, Distributed space-time coded protocols for exploiting cooperative diversity in wireless networks. In *IEEE Global Telecommunications Conference (GLOBECOM'02)* (2002), vol. 1, pp. 77–81.

[12] H. Muhaidat and M. Uysal, Cooperative diversity with multiple-antenna nodes in fading relay channels. *IEEE Transactions on Wireless Communications*, **7**:8 (2008), 3036–3046.

[13] D. S. Michalopoulos, A. S. Lioumpas, G .K. Karagiannidis and R. Schober, Selective cooperative relaying over time-varying channels. *IEEE Transactions on Communications*, **58**:8 (2010), 2402–2412.

[14] L. Bahl, J. Cocke, F. Jelinek and J. Raviv, Optimal decoding of linear codes for minimizing symbol error rate. *IEEE Transactions on Information Theory*, **20**:2 (1974), 284–287.

[15] S. T. Brink, Convergence behavior of iteratively decoded parallel concatenated codes. *IEEE Transactions on Communications*, **49**:10 (2001), 1727–1737.

[16] Y. Li, B. Vucetic, Y. Tang, Z. Zhou and M. Dohler, Practical distributed turbo coding through soft information relaying. In *IEEE 16th International Symposium on Personal, Indoor and Mobile Radio Communications (PIMRC 2005)*, vol. 4, pp. 2707–2711.

[17] T. Zhou, F. Wang, J. Xu and J. Lilleberg, Soft symbol estimation and forward scheme for cooperative relaying. In *IEEE 20th International Symposium on Personal, Indoor and Mobile Radio Communications (PIMRC 2009)*, pp. 3069–3073.

[18] T. Bayes and R. Price, An essay towards solving a problem in the doctrine of chances. By the late Rev. Mr. Bayer, communicated by Mr. Price, in a letter to John Canton, M. A. and F. R. S. *Philosophical Transactions of the Royal Society of London*, **53** (1763). DOI: 10.1098/rstl.1763.0053.

[19] A. Sendonaris, E. Erkip and B. Aazhang, User cooperation diversity. Part I. System description. *IEEE Transactions on Communications*, **51**:11 (2003), 1927–1938.

[20] A. Sendonaris, E. Erkip and B. Aazhang, User cooperation diversity. Part II. Implementation aspects and performance analysis. *IEEE Transactions on Communications*, **51**:11 (2003), 1939–1948.

[21] J. N. Laneman, G. W. Wornell and D. N. C. Tse, An efficient protocol for realizing cooperative diversity in wireless networks. In *IEEE International Symposium on Information Theory* (2001), p. 294.

[22] T. E. Hunter and A. Nosratinia, Cooperation diversity through coding. In *IEEE International Symposium on Information Theory* (2002), p. 220.

[23] M. Janani, A. Hedayat, T. E. Hunter and A. Nosratinia, Coded cooperation in wireless communications: space-time transmission and iterative decoding. *IEEE Transactions on Signal Processing*, **52**:2 (2004), 362–371.

[24] T. Zhou, J. Xu, F. Wang *et al.*, Chained user cooperation scheme employing joint network-channel coding. *High Technology Letters*, **18**:2 (2012), 167–172.

11 Network MIMO techniques

Gan Zheng, Yongming Huang and Kai-Kit Wong

11.1 Introduction

As the demand for high-rate wireless services increases, new techniques and architectures have emerged to increase their spectral efficiency and improve their reliability. During the last decade, *multiple-input multiple-output* (MIMO) or multiple-antenna technology has attracted much attention due to its ability to provide fast and reliable transmission without bandwidth expansion or increase in transmit power. For point-to-point MIMO systems, it has been shown that the capacity of an MIMO channel grows linearly with the minimum number of antennas at both ends [1]. For multi-user systems, MIMO can support *space-division multiple access* (SDMA) and provide multi-user diversity gain. MIMO has been a key to most modern wireless communication standards such as the *3rd Generation Partnership Project* (3GPP) *Long Term Evolution* (LTE) and LTE-Advanced, *Worldwide Interoperability for Microwave Access* (WiMAX) and IEEE 802.11n *Wireless Fidelity* (WiFi).

While the original LTE mainly considered capacity, *heterogeneous cellular networks* (HCNs), where macrocells are overlaid with *low-power nodes* (LPNs) such as picocells, femtocells, and relay nodes, have attracted lots of interest in LTE-Advanced to meet the explosive but unequal mobile data traffic demands. On the other hand, due to the scarcity of spectrum, full frequency reuse has been an attractive strategy considered in LTE-Advanced [2]. In conventional homogeneous macrocell cellular networks operating under the principle of *single-cell processing* (SCP), there is strong intercell interference (ICI), which can be treated as noise and becomes the major challenge that limits system performance in terms of both throughput and fairness. In particular, *user equipments* (UEs) at the cell edges suffer the most from ICI. In HCNs, using a co-channel deployment makes ICI a more pronounced problem. The macrocell can generate heavy interference to LPNs such as pico *evolved NodeBs* (eNBs) outdoors, femto eNBs at home, and relay nodes that extend the range of macrocells. Moreover, femto eNBs can severely interfere with macro or pico UEs that pass by. Therefore, there is a need for *intercell interference coordination* (ICIC) in LTE-Advance deployments. In LTE Releases 8 and 9, fractional frequency reuse is proposed to avoid high interference to cell-edge UEs; *relative narrowband transmit power* (RNTP) is exchanged between macro eNBs via a backhaul X2 interface to notify a neighboring *base station* (BS) whether the transmit power of a specific *physical resource block* (PRB) is above a certain threshold such that others can make good scheduling decisions. In LTE-Advanced Releases 10 and 11, due to

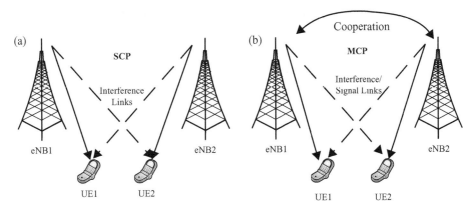

Figure 11.1 Illustration of SCP and MCP.

the introduction of multiple carrier components (CCs), more flexible frequency-domain and time-domain ICIC techniques have been proposed.

In this chapter, we introduce a new solution to deal with ICI in HCNs, namely, network MIMO or *multi-cell processing* (MCP). The basic idea is that neighboring BSs can form a cluster and cooperate to process the data to and from all UEs within the network. The history of BS cooperation can be dated back to the concept of "soft handover" in *code division multiple-access* (CDMA) systems [3]. In this strategy, a UE is connected to multiple BSs nearby, and at each time instance the best BS is selected to provide macrocell diversity. However, there are fundamental differences between soft handover and network MIMO. Soft handover aims to provide smooth handover and achieve macrocell diversity, while in network MIMO the purpose of multiple BS cooperation is to serve multiple UEs simultaneously for spatial multiplexing and ICI mitigation. We will demonstrate how MCP together with MIMO transmit beamforming techniques and power control can manage ICI to achieve a better tradeoff between system throughput and UE fairness in HCNs.

The remainder of this chapter is organized as follows. Section 11.2 analyzes the problems with existing SCP and reviews the benefits, capacity gains, and different types of MCP. Section 11.3 introduces application scenarios of MCP for HCNs. Section 11.4 proposes a distributed implementation of MCP and provides detailed analysis on BS signaling for macrocell scenarios. Section 11.5 proposes a distributed implementation of MCP for HCNs. Section 11.6 lists major network MIMO challenges in future investigations and Section 11.7 concludes this chapter.

11.2 General principles of network MIMO

11.2.1 Problems of single-cell processing

We first discuss a baseline system that uses SCP where each eNB individually serves its own UEs by treating interference from other cells as Gaussian noise; see Fig. 11.1(a).

Interference management is one of the most challenging issues for wireless communications, which limits the achievable spectral efficiency in cellular systems. In conventional cellular systems, this is achieved by frequency reuse with an SCP policy, meaning that UEs in a cell are served by the eNB located at the center of the cell only. An eNB is not supposed to take care of links to and from neighboring cells, which contains ICI. ICI is normally handled by careful frequency planning or power control and usually treated as background noise.

Because the need for high-rate wireless communications is ever growing and due to the scarcity of spectrum, more aggressive and even full frequency reuse has been proposed for future mobile communications standards such as LTE-Advanced [2]. In this context, SCP will experience strong ICI among neighboring cells. A major problem is that UEs at cell edges suffer from not only high signal attenuation, but also severe ICI. Therefore, they exhibit much poorer performance than interior UEs, which raises an alarm in UE fairness. Some interference mitigation techniques have been proposed to remedy this issue, such as dynamic frequency allocation, UE scheduling [4], and soft handover [5]. Although these techniques improve system performance, they do not fully utilize system resources. Interference is still treated as noise with no coordination among BSs.

11.2.2   Advantages of multi-cell processing

It is already known from the above discussion that ICI is the limiting factor for SCP and difficult to deal with. Recently, cooperative communication has received much attention due to its ability to improve system performance and benefit all participating units. The broad concept of cooperative communications means that two or more originally isolated units could cooperate to help one another to achieve better performance. Those units can be, e.g., a set of UEs a set of eNBs. The term "cooperation" has been used in a wide range of contexts, but in this chapter "cooperation" will be referred to as employing joint signal processing among multiple cells. In this context, interference can become a useful signal if some eNBs properly cooperate to jointly process received signals (e.g., in the *uplink* (UL)) and coordinately send messages (e.g., in the *downlink* (DL)) as shown in Fig. 11.1(b).

The cooperation paradigm of MCP was initially introduced by Wyner [6], in the early 1990s, as a novel network architecture for wireless cellular networks. In MCP, eNBs could be interconnected through high-speed delay-less error-free channels (e.g., optical fibres) to a central unit, which is capable of jointly processing all UEs' signals within the cellular system. This basic principle applies to both homogeneous and heterogeneous cellular networks, e.g., macrocells, picocells, femtocells, and relay nodes. The benefit of BS cooperation is a high capacity gain compared to conventional SCP cellular systems. This benefit comes from two aspects. Firstly, all cooperating eNBs share the same bandwidth and therefore the difficulty of frequency planning is avoided, while the increased bandwidth, if properly used, could result in more capacity. Secondly, eNB cooperation has the ability of converting harmful ICI into a useful information-bearing signal to improve capacity. Specifically, in the UL, the received power from UEs can be received by more than one adjacent eNB, and then sent to a *super base station* (SBS) as a hyper-receiver for joint detection [7]. Similarly, in the DL an SBS acts as a hyper-transmitter and jointly designs all UEs' signals through precoding, before transmission, as if the BSs form a large virtual antenna array to optimally transmit these signals [8]. MCP will benefit cell-edge UEs most and eventually provide an improved UE performance with respect to SCP.

MCP has also attracted much attention in industry and standardization bodies. In LTE standardization, there have been many proposals relating to MCP [9–13], In these proposals, though diverse definitions are used to describe MCP techniques, e.g., multi-cell MIMO, collaborative MIMO, network MIMO, coordinated multi-cell transmission, multipoint transmission/reception, coordinated multi-point, and multi-site collaborative MIMO, their basic principles are the same. In the standardization work of 3GPP LTE Release 11, the MCP technology, also termed "*coordinated multi-point* (CoMP)" is being intensively investigated as a study item. In this chapter, the terminologies of "network MIMO," "MCP" and "CoMP" are used interchangeably. Although promising, its suitability for next-generation cellular networks still needs a comprehensive evaluation because the advantage of MCP comes at the cost of increased computational complexity and a heavy burden of intercell information exchange including data and *channel state information* (CSI). The practical issues of MCP implementation will be discussed later.

11.2.3   Capacity results

In this section, we briefly review the capacity comparison between SCP and MCP for non-fading channels in both UL and DL. A comprehensive literature review on this topic can be found in [14].

We consider a cluster consisting of K eNBs, each with M antennas and serving M single-antenna UEs. The channel from eNB k to UE m in the jth cell is denoted by $\mathbf{h}_{mjk}$. We assume full cooperation among eNBs, i.e., all eNBs forming an SBS can share data and CSI. The corresponding result represents the highest capacity gain that can be achieved.

11.2.3.1   Uplink

We first consider the UL. The received signal y_k at eNB k is given by

$$\mathbf{y}_k = \sum_{j=1}^{K} \sum_{m=1}^{M} \mathbf{h}_{mjk} s_{mj}^u + \mathbf{n}_k, \tag{11.1}$$

where s_{mj}^u is the signal from UE m in the jth cell with average power P, and $\mathbf{n}_k$ is the independent and identically distributed (i.i.d.) Gaussian noise with an identity covariance matrix $\mathbf{I}$. In the following, we look into the capacities of SCP and MCP.

- For SCP, each eNB independently decodes its own signals treating ICI as noise. We can rewrite (11.1) as

$$
\mathbf{y}_k = \underbrace{\mathbf{H}_{kk}\mathbf{s}_k^u}_{\text{useful signal}} + \underbrace{\sum_{\substack{j=1 \\ j \neq k}}^{K} \mathbf{H}_{kj}\mathbf{s}_j^u}_{\text{ICI}} + \underbrace{\mathbf{n}_k}_{\text{noise}}, \tag{11.2}
$$

where $\mathbf{H}_{kj} \triangleq [\mathbf{h}_{1jk} \cdots \mathbf{h}_{Mjk}] \in \mathbb{C}^{M \times M}$ is the channel from all UEs in the jth cell to eNB k, and $\mathbf{s}_k^u \triangleq [s_{1k}^u \cdots s_{Mk}^u]^T \in \mathbb{C}^{M \times 1}$ is the signal vector sent from all M UEs in the kth cell where the superscript $(\cdot)^T$ denotes the transpose of a vector or matrix. Then, the capacity of the kth cell can be modeled as follows:

$$
C_{\text{SCP}}^u(k) = \log_2 \det\left(\frac{P\mathbf{H}_{kk}\mathbf{H}_{kk}^\dagger + P\sum_{j \neq k}\mathbf{H}_{kj}\mathbf{H}_{kj}^\dagger + \mathbf{I}}{P\sum_{j \neq k}\mathbf{H}_{kj}\mathbf{H}_{kj}^\dagger + \mathbf{I}} \right), \tag{11.3}
$$

where the superscript $(\cdot)^\dagger$ denotes the Hermitian transpose of a vector or matrix. Note that $P\sum_{j \neq k}\mathbf{H}_{kj}\mathbf{H}_{kj}^\dagger$ is the interference from all the other cells to eNB k. Then, the average UL capacity is given by

$$
\bar{C}_{\text{SCP}}^u = \frac{1}{K}\sum_{k=1}^{K} C_{\text{SCP}}^u(k). \tag{11.4}
$$

- For MCP, we stack all the received signals from all eNBs into a $KM \times 1$ vector

$$
\mathbf{y} = \mathbf{H}\mathbf{s}^u, \tag{11.5}
$$

where $\mathbf{H} \in \mathbb{C}^{KM \times KM}$ is constructed such that its (k, j)-submatrix is $\mathbf{H}_{kj}$ and $\mathbf{s}^u \triangleq [\mathbf{s}_1^u; \ldots; \mathbf{s}_K^u] \in \mathbb{C}^{KM \times 1}$ is a vector collecting all signals from the cells. Then, the average UL capacity (details to be discussed in the next section) is given by

$$
\bar{C}_{\text{MCP}}^u = \frac{1}{K}\log_2 \det\left(\mathbf{I} + P\mathbf{H}\mathbf{H}^\dagger\right). \tag{11.6}
$$

Clearly, in (11.3), cross terms $\mathbf{H}_{kj}\mathbf{H}_{kj}^\dagger, \forall j \neq k$ are all ICIs. However, in (11.6) they appear in the diagonal blocks of $\mathbf{H}\mathbf{H}^\dagger$. Therefore, they are also useful signals that can be used for joint detection instead of ICIs that limit the system performance.

The comparison of UL capacity between SCP and MCP is shown in Fig. 11.2, where a substantial performance gain provided by MCP over SCP is observed, especially at high *signal to noise ratios* (SNRs). For SCP, the system becomes interference limited and the capacity saturates from an SNR of around 15 dB. Under some benign environments of a *frequency division duplexing* (FDD) LTE trial system, average DL capacity improvement of 50% was reported in [15].

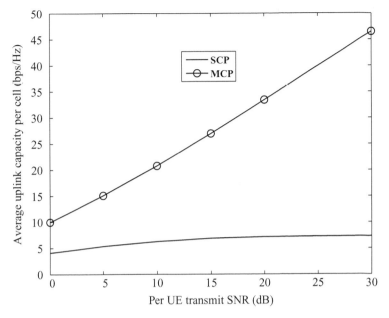

Figure 11.2 Per cell UL capacity comparison between SCP and MCP.

11.2.3.2 Downlink

In the DL channel, the received signal y_{mk} at UE m in the kth cell is given by

$$y_{mk} = \sum_{j=1}^{K} \sum_{n=1}^{M} \mathbf{h}_{mkj} \mathbf{s}_{nj}^{u} + n_{mk}, \qquad (11.7)$$

where $n_{mk} \sim \mathcal{CN}(0, 1)$ is the additive Gaussian noise.

For SCP, (11.7) can be rewritten as

$$y_{mk} = \mathbf{h}_{mkk} \mathbf{s}_{mk}^{u} + \underbrace{\sum_{\substack{n=1 \\ n \neq m}}^{M} \mathbf{h}_{mkk} \mathbf{s}_{nk}^{u}}_{\text{intra-cell interference}} + \underbrace{\sum_{j=1, j \neq k}^{K} \sum_{n=1}^{M} \mathbf{h}_{nkj} \mathbf{s}_{nj}^{u}}_{\text{ICI}} + \underbrace{n_{mk}}_{\text{noise}}. \qquad (11.8)$$

Assuming that each eNB has a power constraint of P, the capacity of cell k is given by

$$C_{\text{SCP}}^{d}(k) = \max_{\text{trace}(\Phi_k) \leq 1} \log_2 \det \left(\frac{P\mathbf{H}_{kk}\Phi_k\mathbf{H}_{kk}^{\dagger} + P\sum_{j \neq k} \mathbf{H}_{kj}\mathbf{D}_j\mathbf{H}_{kj}^{\dagger} + \mathbf{I}}{P\sum_{j \neq k} \mathbf{H}_{kj}\Phi_j\mathbf{H}_{kj}^{\dagger} + \mathbf{I}} \right), \qquad (11.9)$$

where $\Phi_k \succeq \mathbf{0}$ is the transmit covariance matrix of cell k without considering UEs in other cells, and $\mathbf{A} \succeq \mathbf{0}$ means that a matrix $\mathbf{A}$ is positive semi-definite. It can be seen that for SCP, although intracell interference can be managed by the eNB k via optimization of Φ_k, the ICI is simply treated as noise and becomes a limiting factor.

Meanwhile for MCP, using results of group-antenna power constraints originally derived in [16], the average DL capacity (its achievability to be discussed in the next section) is given by [17]

$$\bar{C}_{\text{MCP}}^{d} = \min_{\text{trace}(\mathbf{\Lambda}) \le \frac{1}{MP}} \max_{\text{trace}(\mathbf{D}) \le 1} \frac{1}{2} \log_2 \frac{\det(\mathbf{\Lambda} + \mathbf{HDH}^{\dagger})}{\det(\mathbf{\Lambda})}, \tag{11.10}$$

where we have assumed that each eNB has a power constraint of P, and $\mathbf{\Lambda}$ and $\mathbf{D}$ are positive diagonal matrices to be optimized. It was shown in [17] that the optimized average DL sum capacity $\bar{C}_{\text{MCP}}^{d}$ is exactly the same as the UL sum capacity $\bar{C}_{\text{MCP}}^{u}$, given $\mathbf{H}$ is block circulant.

11.2.4   Categories of network MIMO

Depending on the level of cooperation and the backhauling requirement, MCP can be roughly divided into the following categories.

11.2.4.1   Fully cooperated multi-cell MIMO

To fully exploit the potential of MCP, the cooperating eNBs should perform joint decoding from or coherent transmission to multiple UEs in the UL and DL, respectively, where each UE is simultaneously served by a cluster of eNBs, and both data and CSI are required to be shared among eNBs. Full cooperation requires an SBS connecting all eNBs via links without bandwidth or delay limits so that SBS can collect global information, including all UEs' data and CSI from all cooperating eNBs. In this case, the concept of "cell" vanishes, and the eNB cluster essentially becomes a single-cell multi-user multi-antenna system, and the link between any BS antenna and any UE may contain useful information. For practical transmission, existing multi-user precoding techniques (see, e.g., [16, 18–25]) can be directly applied to obtain a performance gain (see, e.g., [8, 26, 27]).

Full cooperation exploits all degrees of freedom provided by the channels and the system, so that it achieves the highest networked MIMO capacity. In the UL, the system capacity can be achieved by jointly decoding at the SBS using *successive interference cancellation* (SIC). This scheme first decodes the strongest data stream with the highest received *signal to interference plus noise ratio* (SINR); then the contribution of the strongest data stream is subtracted from the received signal. Thereafter, the second strongest data stream is decoded and canceled. The process continues until the last data stream is decoded. Similar ideas can be applied to the DL, while encoding the signal. Specifically, the data stream under the most favorable channel condition is encoded first, which will receive interference from other streams. Then, this data stream is known while encoding the next data steam, and its interference can be pre-canceled at the SBS. The last encoded stream sees no interference from the others since they are already known and canceled. This strategy is well known as *dirty-paper coding* (DPC) [28], and has been applied to precoding in multiuser MIMO, e.g., implementation of DPC with low complexity like Tomlinson–Harashima precoding [29] can be adopted.

From the viewpoint of practical implementation, some issues for the fully cooperative network MIMO should be considered.

- The optimal joint transmission in fully cooperative network MIMO is hard to realize in practice, because it requires full phase coherence among signals received from different BSs, which is usually impossible due to the difference in propagation delay.
- The requirement of UEs' data and CSI sharing among BSs brings a huge burden of backhaul links. This challenges the capability of the X2 interface among BSs.
- The realization of fully cooperative network MIMO usually needs a central unit to perform optimization, which is difficult for future and more complex HCNs, where operation tends to be distributed.

11.2.4.2 Coordinated network MIMO

The capacity gain brought by full intercell cooperation comes at the cost of high complexity and high demand on backhaul links (especially for data sharing), which is by no means practical both technically and economically. For example, if some eNBs are geographically well apart from others, joint processing of all UEs' data streams has little impact on the performance gain, but introduces extra complexity and backhauling. One possible way to reduce implementation complexity is that eNBs can cooperate with one another to serve only their own UEs. To be more specific, it does not require an SBS, i.e, eNBs do not need to share data or synchronization information with all other BSs, thereby greatly reducing the requirement on extra complexity and backhauling. To cooperate among themselves and mitigate interference, eNBs need to have CSI of their own UEs as well as UEs in cooperating cells, which refers to the interference links. Based on this CSI, coordinated beamforming and power control can be jointly optimized by all BSs by taking the interference to other cells into account while designing its own beamforming vectors. Limited information sharing among BSs, if possible, could improve overall performance further. Details on limited feedback will be provided later in Section 11.4 and Section 11.5. In addition, each BS can also have CSI of other interfering eNBs through its UEs, i.e., the CSI can be estimated at the UEs first then fed back to both their serving eNB and cooperating eNBs. Such CSI also facilitate distributed beamforming for interference channels [30].

Note that the performance of coordinated networked MIMO is strictly upper bounded by fully cooperated network MIMO, since the links from one eNB to all UEs in the other cells contain purely interference, and not every channel link is exploited to enhance system performance.

11.3 Application scenarios of network MIMO in HCN

The evolution of new-generation mobile communications demands a significant increase of spectral efficiency compared to current 4G systems. It is known that the spectral efficiency of a cellular network can be improved by increasing the cell density through cell splitting. However, its achievable gain would be significantly limited by severe intercell interference. Furthermore, deploying more macrocell BSs in an urban environment may be prohibitively expensive. Therefore, underlaying LPNs in macrocells' coverage, so

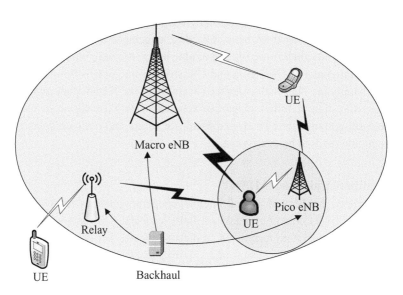

Figure 11.3 Application of MCP to HCN.

called HCN, has emerged as a cost-effective solution to increase spectral efficiency and
network capacity.

As shown in Fig. 11.3, multiple LPNs, including *remote radio head* (RRH), pico
eNB, femto eNB, and *relay node* (RN), may be underlaid in the coverage of macrocells.
This constitutes a two-layer coverage network. Previous study has shown that introducing LPNs within an existing macrocell network provides both coverage and capacity
improvement by offloading UEs from the macrocell to the underlaid LPNs. However,
because of the lower transmit power compared to the macro eNB, the footprint of the
LPN can be very small if a conventional access mechanism is used at UEs. This makes
the number of UEs offloaded from the macrocell limited in most scenarios of interest,
leading to only a marginal capacity improvement. To address this issue, an approach
known as cell range expansion can be used to increase the size of the LPN. It is realized by a modified UE access mechanism based on biased *reference signal received
power* (RSRP). However, this gain can only be achieved if the interference between
macrocells and underlaid LPNs, i.e., intertier interference, is addressed suitably. In particular, if a UE is located in the region of cell range expansion, it would suffer from
severe DL interference from the macro eNB, since it is not connected to the cell that provides the best DL received signal. Therefore, the development of intertier interference
addressing technology is necessary. Different ICIC schemes have emerged as promising
approaches, which address the intertier interference by coordinating the use of frequency
and time resources among cells. For more information on range expansion and ICIC,
the reader may refer to Chapter 7. Here, we focus on a more efficient approach, i.e.,
applying network MIMO to HCN.

Similar to the macrocell network, there are quite a few challenges while implementing inter-cell cooperation in HCN, such as backhauling requirements, coordinated UE

scheduling, CSI acquisition and synchronization, etc. Some of these issues are coupled together and need to be jointly considered while designing a network MIMO system. In the remainder of this section, we will review some of them.

11.3.1 Backhaul limit in HCN

The application of network MIMO in HCN needs a substantial amount of information exchange among the macro eNB and the LPNs. The fully cooperated MCP imposed demanding intercell information exchange due to the fact that it needs to share all UEs' traffic data and CSI. That is, it requires a backhaul link that is delay free and has unlimited capacity. Usually this can only be realized in some particular scenarios such as the macro–RRH or macro–pico heterogenous scenarios, where the macro eNB is connected to the LPN through a fibre-optical link. However, this kind of backhaul connection will significantly increase operational expenditures. Alternatively, backhaul connection based on LTE air interface has drawn much attention recently. For example, in the macro–RN heterogenous scenario, the connection between the macro eNB and the RN is usually set up via wireless backhaul, which can use the same frequency resource (in band) or a different frequency band (out-band) than the traffic links. In this case, the backhaul link is rate limited and is not delay free. How to consider this limitation in MCP optimization becomes a critical issue. So care must be taken in deciding what and how much quantized information (data, CSI) needs to be shared. The effects of limited-capacity backhaul on both UL and DL channels were studied in [31] for simplified cellular systems.

For the particular macro–femto heterogenous scenario, it is hard to realize an effective network MIMO due to the fact that its backhaul link is usually UE established through the Internet and will suffer from large delay. Therefore, the possible information exchange between the macro eNB and the femto eNB would be very limited. A common method is to coordinate the multi-cell transmissions from the perspectives of power control and frequency–time resource allocation.

Moreover, as wireless signals decay quickly as distance increases, it is not always necessary to jointly process far-apart UEs' data, and local MCP may be sufficient for interference mitigation. This motivates the clustered network MIMO to be discussed next.

Alternatively, coordinated MCP is another way to reduce backhaul requirement and manage ICI effectively since each eNB only serves its own UEs and does not need data sharing with other eNBs, thereby easing the major backhaul burden. The CSI may still be shared, but compared with data sharing the bandwidth requirement is much less.

11.3.2 Clustering mechanism for HCNs

The clustering of cooperating eNBs or cells plays a key role in the network MIMO system. From the implementation point of view, it is not always preferred to have multi-eNBs cooperating to serve multiple UEs, especially when UEs are far apart and the received interference from other cells is little. Therefore, only a limited number of cooperating cells may be used for coordinated transmission/reception. The first problem is how and when the cooperating cell cluster is formed. Several solutions have been proposed recently, including the network-centric, UE-centric, or combined methods [32]. Note that the network-centric method is performed on a static or semi-static basis. Thus, the variation of UE channel conditions cannot be fully exploited. In contrast, the UE-centric method is a dynamic clustering scheme and usually produces partially overlapped cell clusters, which achieves better performance but involves huge scheduling complexity. To achieve a tradeoff between the performance and the complexity of the previously presented method, a combined method uses a UE-centric clustering method within the given candidate clusters predetermined in a network-centric manner.

The clustering of cooperating eNBs has also been extensively investigated in the standardization bodies. As shown in the LTE-Advanced study item work for CoMP, there is an efficient approach to dynamically determine the coordinating cell cluster in a UE-centric manner [33–35]. Each UE reports a CoMP cell set based on RSRP measurements from each cell, where a non-serving cell is included in the reported CoMP set only if its corresponding RSRP is within a given offset from the RSRP of the serving cell, given by [33]

$$\text{cell } i \in \text{reporting CoMP set, if and only if } \text{RSRP}_i \geq \text{RSRP}_0 - \alpha_{\text{th}}, \qquad (11.11)$$

where RSRP_0 is the RSRP of the serving cell, RSRP_i is the RSRP of the ith cooperating cell reported by the UE, and α_{th} is the given given offset, whose value is usually set to be a small portion of RSRP_0 and all are in dBms (dBs). Then, CoMP UE scheduling can be performed among the UEs that report non-empty cooperating cell sets. If α_{th} is small, it is understood that the scheduled CoMP UEs see significant interference from the cooperating cells, such that a coordinated transmission among these cells promises to bring considerable gains over the non-coordinated single-cell transmission.

11.3.3   CSI sharing

CSI is essential for transmit precoding design even for single-cell multi-user MIMO systems. Cooperation of multi-cell MIMO brings more problems. Here, we focus on the DL where transmit CSI can only be obtained with the help of UEs.

In a general FDD system, the channel is estimated at the receiver, e.g., using pilots sent from the eNBs. Then each (participating) UE sends CSI to the eNB via a return channel. However, the return channel from a UE to eNBs is practically rate limited, so proper limited feedback mechanisms need to be designed to use the return channel effectively. For example, in LTE [36], a fixed codebook containing predetermined transmit beamforming vectors is used, which is known to both UEs and eNBs such that each UE only selects and feeds back the index of the most favorable beamforming vector. The main challenge is how to design a good codebook based on channel statistics and transmission schemes. This topic has been well studied in the single-cell multi-user MIMO, and interested readers are referred to [37] for more details. However, when applied to MCP, there are several noticeable differences. One is the fact that channel statistics are heterogeneous since the distances from one eNB to UEs within its cell and UEs within other cells are quite different. This needs to be measured and taken into account before channel estimation and feedback. Moreover, cooperative transmissions among eNBs determine what kind of beamforming vectors are desired, and this may differ from the objectives in a single cell. In addition, global CSI sharing introduces additional signaling overhead and the advantages of network MIMO are sometimes offset and questionable. Therefore, the extension of limited CSI feedback within the intercell cooperation is non-trivial and needs further investigation.

In the following, we consider a *time division duplexing* (TDD) system. In TDD, the DL CSI can be directly estimated at the eNB by exploiting the UL–DL reciprocity. In addition, it is less likely that UEs in different cells could collaborate to send pilots, so *time division multiple access* (TDMA) can be used to avoid interference for channel estimation. This strategy is well known in a single-cell scenario, while the difficulty of extending it to a multi-cell scenario is that one eNB cannot not get the CSI feedback from a particular UE if this UE is not served by the eNB. Effective protocols need to be carefully designed.

Both of the above cases involve CSI sharing, if global CSI is needed. If a central unit is present to coordinate the intercell cooperation and the backhaul capacity is infinite, we can assume that perfect global CSI is available and this can be obtained by feedback from UEs to all eNBs, and then to the central unit. If there is no such central unit, eNBs need to share CSI, which involves additional signaling, estimation errors, and delay. In [38], the authors considered the imperfect sharing of CSI with full data sharing, in which each eNB has different versions of the same CSI feedback from UEs due to the fact that the distances between eNBs and a UE in a cluster could be significantly different, which affects the quality of the available CSI. Despite the difference in CSI, the eNBs still cooperate to design the beamforming vectors via solving a team decision theory problem.

The codebook design for CSI feedback depends on the cooperation among eNBs, which is also true for general CSI estimation. While most schemes aim at achieving/ sharing full CSI at each eNB, this is not always necessary depending on the implementation. For example, in the proposed implementation in the following two sections, i.e., Section 11.4 and Section 11.5, only the availability of local CSI (meaning CSI from one eNB to all UEs within the cluster) rather than global CSI at each eNB is sufficient for the design of coordinated beamforming, which greatly reduces the signaling overhead. This can be well justified for a TDD system where the DL CSI corresponding to eNB k (i.e., $\mathbf{h}_{kk}$) can be directly estimated at the eNB by exploiting the UL–DL reciprocity. On the other hand, it is reasonable to assume that each eNB can estimate the crosstalk channels to all the other UEs in the same cluster. This can be explained by recognizing that usually a cell cluster is dynamically set up only if a UE inside the cluster sees significant interference from other cells in the same cluster. In other words, the channels between the eNB and the UEs in other cells and those between the eNB and its connected UEs are of similar strengths in a cluster. To do so, a coordinated training phase for the cell cluster is required such that each eNB can estimate the crosstalk channels together with the channel for its own UEs in predefined time slots, in which the training signals from different UEs are designed to be orthogonal in order to avoid mutual interference.

11.4   Distributed downlink coordinated beamforming for macrocell network

As discussed in the above section, from the perspective of implementation, the design of CoMP needs to fully take into account the issue of backhaul burden. Moreover, distributed implementation becomes more and more important with the air interface evolving. For these reasons, in this section, we focus on distributed CoMP with limited intercell coordination, starting by a review of some recent results in this direction. Here, we consider the DL of the homogeneous network, which could be macrocell, picocell, or femtocell. An application example in the HCN will be introduced in next section.

The achievable rates of DL CoMP using distributed implementation were discussed in [39] and [40]. Recently, there have been many emerging distributed CoMP algorithms, which can be grouped into two classes, i.e., distributed multi-cell joint processing and distributed multi-cell processing. Their main difference lies in that distributed multi-cell joint processing requires UEs' data sharing among the cooperating cells, while distributed multi-cell processing usually only requires CSI sharing. Typical distributed multi-cell joint processing strategies are discussed as follows. A distributed solution in TDD systems was devised based on a virtual SINR framework [41]. It was demonstrated that this strategy achieves a performance close to the Pareto bound, and requires no CSI sharing provided TDD based local CSI is available at each cooperating BS. The distributed realization of multi-cell joint processing can also be obtained by recasting the DL beamforming problem into a *linear minimum mean-square-error* (LMMSE) estimation problem [42]. Although distributed realization of this class of CoMP can be achieved, the required UE's data sharing among eNBs will cause heavy burden on the backhaul link, which prohibits its use in practical scenarios.

This motivates the development of distributed multi-cell processing, in which eNBs do not cooperate at the signal level but only jointly design transmit beamformers and power levels [43–46]. Aiming at reducing the required intercell information exchange, the UL–DL duality theory was widely used to design distributed coordinated processing. The main result of the duality is that, given the same total power, both UL and DL UEs can achieve the same rate/SINR region. In [43], the duality theory was first applied to solve the sum power minimization problem subject to SINR constraints. The design of distributed coordinated precoding without BS signaling using the duality theory is also possible if the criterion based on interference leakage to other cells is considered. Details of this strategy will be introduced in Section 11.4.3. In [47], an optimal distributed multi-cell beamforming scheme using a rate-fairness criterion was presented. In particular, it is aimed to maximize the minimum UEs' rate, which guarantees fairness between cells and UEs. To further reduce the required intercell information exchange, in Section 11.4.3.2 we also propose a more efficient distributed MCP algorithm without full CSI sharing.

11.4.1   System model and problem formulation

Let us consider DL transmissions, in a macrocell cluster, which is composed of B eNBs, each equipped with M transmit antennas. Coordinated beamforming and power control is used by the eNBs for serving K UEs. We assume that each eNB is serving at least one UE equipped with a single receive antenna, i.e., $K \geq B$, and its transmit signal is only intended for its own UEs. This means that there is no UE data sharing among cells, but the transmit beamformers from different eNBs need to be designed jointly based on limited intercell information exchange in order to suppress ICI.

Let $\mathcal{K}_b$ be the set of UEs served by eNB b, and s_k be the message to UE k with $\mathrm{E}\{s_k^2\} = p_k$, where $\mathrm{E}\{\cdot\}$ denotes expectation. Assuming $k \in \mathcal{K}_b$, the DL received signal of UE k is given by

$$y_k = \tilde{\mathbf{h}}_{kb}^{\dagger} \mathbf{w}_k s_k + \tilde{\mathbf{h}}_{kb}^{\dagger} \sum_{\substack{i \in \mathcal{K}_b \\ i \neq k}} \mathbf{w}_i s_i + \sum_{\bar{b} \neq b} \tilde{\mathbf{h}}_{k\bar{b}}^{\dagger} \sum_{j \in \mathcal{K}_{\bar{b}}} \mathbf{w}_j s_j + n_k, \tag{11.12}$$

where $\mathbf{w}_k$ is the unit-norm transmit beamforming vector at eNB k, $\tilde{\mathbf{h}}_{k,b}$ is the channel vector from eNB b to UE k, and n_k is the complex additive white Gaussian noise with zero mean and variance N_0, i.e., $n_k \sim \mathcal{CN}(0, N_0)$. Without loss of generality, we assume $N_0 = 1$. The received SINR for UE k is therefore given by

$$\mathrm{SINR}_k = \frac{p_k |\tilde{\mathbf{h}}_{kb}^{\dagger} \mathbf{w}_k|^2}{\sum_{\substack{i \in \mathcal{K}_b \\ i \neq k}} p_i |\tilde{\mathbf{h}}_{kb}^{\dagger} \mathbf{w}_i|^2 + \sum_{\bar{b} \neq b} \sum_{j \in \mathcal{K}_{\bar{b}}} p_j |\tilde{\mathbf{h}}_{k\bar{b}}^{\dagger} \mathbf{w}_j|^2 + 1}. \tag{11.13}$$

For notational convenience, we expand the set of B eNBs to a set of K *virtual eNBs* (VeNBs) by repeating the eBNs that serve more than one UE. Therefore, each eNB in the VeNB set serves one UE. The index of the VeNBs serving UE k is denoted as k, with the set of VeNBs $\{k \in \mathcal{K}_b\}$ corresponding to the actual eNB b. The channel vector from VeNB j to UE k is denoted by $\mathbf{h}_{kj}$. The relationship between the actual channel and the virtual channel is given by

$$\tilde{\mathbf{h}}_{kb} = \mathbf{h}_{kj}, \ \forall j \in \mathcal{K}_b. \tag{11.14}$$

Then, the above input–output model (11.12) can be rewritten as

$$y_k = \mathbf{h}_{kk}^{\dagger} \mathbf{w}_k s_k + \sum_{\substack{j=1 \\ j \neq k}}^{K} \mathbf{h}_{kj}^{\dagger} \mathbf{w}_j s_j + n_k, \tag{11.15}$$

and the received SINR for UE k can be concisely expressed as

$$\mathrm{SINR}_k = \frac{p_k |\mathbf{w}_k^{\dagger} \mathbf{h}_{kk}|^2}{\sum_{\substack{j=1 \\ j \neq k}}^{K} p_j |\mathbf{w}_j^{\dagger} \mathbf{h}_{kj}|^2 + 1}. \tag{11.16}$$

The objective of intercell coordination is to jointly optimize the beamformers $\{\mathbf{w}_k\}$ and $\{p_k\}$, using a certain criterion which usually can be expressed as a function of SINRs. In the following, we will focus on some predefined criteria and the corresponding optimizations.

11.4.2   Distributed multi-cell beamforming based on interference leakage

Maximization of the sum rate of the multi-cell system is a popular design criterion. It is, however, difficult to address since the SINR expression involves beamforming vectors and power levels of all UEs. Alternatively, it is possible to devise distributed algorithms for coordinated beamforming based on the *signal to leakage interference and noise ratio* (SLNR) [44, 48], where each eNB only exploits its local CSI. For the reason described in the previous subsection, we assume that each eNB in the cooperating cluster only serves a single UE, i.e., $K = B$. The expression of SLNR for UE k is given by

$$\mathsf{SLNR}_k = \frac{\frac{P_\mathrm{T}}{B}|\mathbf{w}_k^\dagger \mathbf{h}_{kk}|^2}{\frac{P_\mathrm{T}}{B}\sum_{\substack{j=1\\j\neq k}}^{K}|\mathbf{w}_k^\dagger \mathbf{h}_{jk}|^2 + 1}$$

$$= \frac{|\mathbf{w}_k^\dagger \mathbf{h}_{kk}|^2}{\mathbf{w}_k^\dagger \left(\sum_{\substack{j=1\\j\neq k}}^{K}\mathbf{h}_{jk}\mathbf{h}_{jk}^\dagger + \frac{P_\mathrm{T}}{B}\mathbf{I}\right)\mathbf{w}_k}, \tag{11.17}$$

where $\frac{P_\mathrm{T}}{B}$ is the power constraint at each BS. To maximize the SLNR for each UE, the optimized beamforming vector can be easily calculated as

$$\mathbf{w}_k = \frac{\left(\sum_{\substack{j=1\\j\neq k}}^{K}\mathbf{h}_{jk}\mathbf{h}_{jk}^\dagger + \frac{P_\mathrm{T}}{B}\mathbf{I}\right)^{-1}\mathbf{h}_{kk}}{\left\|\left(\sum_{\substack{j=1\\j\neq k}}^{K}\mathbf{h}_{jk}\mathbf{h}_{jk}^\dagger + \frac{P_\mathrm{T}}{B}\mathbf{I}\right)^{-1}\mathbf{h}_{kk}\right\|}. \tag{11.18}$$

The implementation of this algorithm requires no information exchange among cooperating cells, provided that TDD based local CSI is available at each BS. It has been shown that the above beamforming vector is able to attain the Pareto boundary of *multiple-input single-output* (MISO) interference channels [49]. However, this strategy is only asymptotically optimal in terms of the sum rate for the scenarios with two BSs, when the SNR goes to infinity. Although it is a practical and good heuristic approach, in general it cannot achieve the optimal solution. In what follows, we will provide an optimal distributed CoMP strategy in the sense of maximizing the minimum UEs' SINRs.

11.4.3   Distributed multi-cell beamforming based on max–min SINR

To guarantee a rate-fairness among the cooperating cells, here we focus on the maximization of the minimum UEs' SINRs in the design of distributed multi-cell beamforming. This metric not only allows us to reach any point on the rate-region boundary, but also provides fairness for cell-edge UEs. Mathematically, this can be expressed as

$$\max_{\{\mathbf{w}_k,p_k\}}\ \min_k\ \frac{\mathsf{SINR}_k}{\Gamma_k} \quad \text{s.t.} \quad \sum_{k=1}^{K} p_k \leq P_\mathrm{T}, \tag{11.19}$$

where P_T is a total transmit power constraint for the cooperating eNB cluster, and Γ_k is the SINR requirement for UE k. Note that though per-eNB individual transmit power constraints are more relevant in HCNs, the consideration of a simple total transmit power constraint provides mathematical tractability for the optimization problem, and helps to shed some insight on the optimal solution in macrocells networks. Since it is difficult to directly solve this DL optimization problem, alternatively, it is addressed by transforming it into a dual UL problem.

11.4.3.1 Dual optimization problem

UL–DL duality is a powerful tool to examine a DL problem by forming the equivalent virtual UL problem; see [20] for single-cell scenarios and [43, 46] for extensions to multi-cell scenarios. Specifically, here we consider the virtual UL where UE k transmits to its assigned eNB k, and suppose that the UL channel vector between UE j and eNB k is $\mathbf{h}_{jk}$. In what follows, the received beamforming vector in the virtual UL is given by $\mathbf{w}_k$, while the UL SINR for UE k is given by

$$\text{SINR}_k^{\text{UL}} = \frac{q_k |\mathbf{w}_k^\dagger \mathbf{h}_{kk}|^2}{\sum_{\substack{j=1 \\ j \neq k}}^{K} q_j |\mathbf{w}_k^\dagger \mathbf{h}_{jk}|^2 + 1}, \tag{11.20}$$

where q_k is the transmit power of UE k. Let us note that with the same total transmit power constraint $\sum_{k=1}^{K} p_k = \sum_{k=1}^{K} q_k \leq P_{\text{T}}$ and the same set of beamforming vectors $\{\mathbf{w}_k\}$, the same DL and UL SINR region for each UE can be achieved [43]. Therefore, to solve (11.19), it suffices to address the following virtual UL problem:

$$\max_{\{\mathbf{w}_k\}, \boldsymbol{q}} \min_k \frac{\text{SINR}_k^{\text{UL}}}{\Gamma_k} \quad \text{s.t.} \quad \sum_{k=1}^{K} q_k \leq P_{\text{T}}, \tag{11.21}$$

where $\boldsymbol{q} \triangleq [q_1, \ldots, q_K]$. The main advantage of UL formulation is that the beamforming vectors are decoupled and can be maximized on a per-cell basis, as can be seen by comparing (11.21) with (11.16). The corresponding DL solution can be easily retrieved from the virtual UL solution. Therefore, we can focus on the UL problem (11.21). Detailed signaling due to UL–DL conversion will be discussed in Section 11.4.4.

11.4.3.2 Algorithms and analysis

Regarding (11.21), in this section, we first review two previous solutions. The first one is for centralized processing while the second one is suitable for distributed implementation. We then propose a more efficient decentralized processing.

We rewrite (11.21) in a matrix form below by introducing an auxiliary variable γ as follows:

$$\max_{\gamma > 0, \boldsymbol{q} \geq 0, \{\mathbf{w}_m\}} \gamma \tag{11.22}$$

$$\text{s.t.} \quad \begin{cases} \dfrac{\mathbf{q}}{\gamma} = \mathbf{DF}\mathbf{q} + \mathbf{D}\sigma, \\ \mathbf{1}^{\mathsf{T}}\mathbf{q} \leq P_{\text{T}}, \end{cases} \tag{11.23}$$

where $\mathbf{1}$ is an all-one vector, and we have defined $\mathbf{D}$ and $\mathbf{F}$ as

$$\mathbf{D} \triangleq \mathrm{diag}\left(\frac{\Gamma_1}{\mathbf{w}_1^\dagger \mathbf{h}_{11}\mathbf{h}_{11}^\dagger \mathbf{w}_1}, \ldots, \frac{\Gamma_K}{\mathbf{w}_K^\dagger \mathbf{h}_{11}\mathbf{h}_{KK}^\dagger \mathbf{w}_K}\right),$$ (11.24)

$$[\mathbf{F}]_{i,j} = \begin{cases} \mathbf{w}_j^\dagger \mathbf{h}_{ij}\mathbf{h}_{ij}^\dagger \mathbf{w}_j, & \text{for } i \neq j, \\ 0, & \text{for } i = j. \end{cases}$$ (11.25)

Left multiplying both sides of (11.23) by $\mathbf{1}^\mathsf{T}$ results in

$$\frac{P_\mathrm{T}}{\gamma} = \frac{\mathbf{1}^\mathsf{T}\mathbf{q}}{C} = \mathbf{1}^\mathsf{T}\mathbf{DF}\mathbf{q} + \mathbf{1}^\mathsf{T}\mathbf{D1},$$ (11.26)

and an eigensystem can then be constructed as follows:

$$\frac{1}{\gamma}\begin{bmatrix}\mathbf{q}\\1\end{bmatrix} = \begin{bmatrix}\mathbf{DF} & \mathbf{D1}\\ \frac{1}{P_\mathrm{T}}\mathbf{1}^\mathsf{T}\mathbf{DF}\mathbf{q} & \frac{1}{P_\mathrm{T}}\mathbf{1}^\mathsf{T}\mathbf{Dq}\end{bmatrix}\begin{bmatrix}\mathbf{q}\\1\end{bmatrix}.$$ (11.27)

For notational convenience, let us define $\mathbf{\Psi} \triangleq \begin{bmatrix}\mathbf{DF} & \mathbf{D}\sigma\\ \frac{1}{P_\mathrm{T}}\mathbf{1}^\mathsf{T}\mathbf{DF}\mathbf{q} & \frac{1}{P_\mathrm{T}}\mathbf{1}^\mathsf{T}\mathbf{D}\sigma\end{bmatrix}$. Note that $\mathbf{\Psi}$ is a non-negative matrix and γ is strictly increasing with P_T. Therefore, there is a unique positive eigenvector associated with the maximal eigenvalue $\frac{1}{\gamma}$ (>0) [50]. No other eigenvector satisfies this positiveness constraint and the optimal balanced SINR can be expressed as follows:

$$\gamma^\mathrm{opt} = \max_{\{\mathbf{w}_k\}} \frac{1}{\lambda_\mathrm{max}(\mathbf{\Psi})},$$ (11.28)

where $\lambda_\mathrm{max}(\cdot)$ is the maximal eigenvalue of a matrix. Based on the above observation, in [20], a matrix-based algorithm was proposed to find the optimal UL power and beamforming solution, which is summarized in Algorithm 11.1.

Algorithm 11.1 CENTRALIZED MCP IN [20] ($\{\mathbf{h}_{jk}\}_{\forall j,k}, \{\Gamma_k\}_{\forall k}, P_\mathrm{T}$)

Initialize $\mathbf{q}$.
while not converge

do $\begin{cases} \text{Update } \{\tilde{\mathbf{w}}_m\} \text{ using} \\[4pt] \mathbf{w}_k = \dfrac{\left(\sum_{\substack{j=1\\j\neq k}}^{K} q_k \mathbf{h}_{jk}\mathbf{h}_{jk}^\dagger + \mathbf{I}\right)^{-1}\mathbf{h}_{kk}}{\left\|\left(\sum_{\substack{j=1\\j\neq k}}^{K} q_k \mathbf{h}_{jk}\mathbf{h}_{jk}^\dagger + \mathbf{I}\right)^{-1}\mathbf{h}_{kk}\right\|}, \quad \forall k. \\[12pt] \text{Update the power vector } \mathbf{q} \text{ by choosing } [\mathbf{q};1] \text{ as the eigenvector} \\ \text{associated with the maximal eigenvalue of } \mathbf{\Psi}. \end{cases}$

return ($\{\mathbf{w}_k\}, \boldsymbol{q}$)

Although Algorithm 11.1 converges very fast to the optimum solution, the main drawback is that, at each iteration, global CSI is needed and an SBS is necessary to collect the global CSI, perform the required optimization, and distribute the results to all cooperating UEs, which cannot be realized in a distributed manner.

Therefore, in the following, we revisit one of the possible algorithms that fit the distributed mechanism proposed in [47]. This algorithm relates the SINR balancing problem (11.21) to the following power minimization problem:

$$\min_{\{w_k\}, q} \sum_{k=1}^{K} q_k \text{ s.t. } \text{SINR}_k^{\text{UL}} \geq \gamma \, \Gamma_k, \forall k, \qquad (11.29)$$

with γ as a parameter. If we can find the maximum parameter γ_{max} that makes (11.29) feasible, then γ_{max} is also the optimum solution to (11.21). γ_{max} can be found via a bisection search as the outer iteration and repeatedly solving (11.29) in the inner iteration, for which the distributed algorithm in [51] can be applied. This algorithm is referred to as Algorithm 11.2, where the superscript (n) designates the given parameter at the nth iteration. The optimality of the obtained solution is guaranteed by the bisection search and the optimality of the power control algorithm in [51].

The main advantage of Algorithm 11.2 is that in each iteration, eNB k updates its own transmit power q_k and beamforming vector $\mathbf{w}_k$ by exploiting its local CSI $\{\mathbf{h}_{jk}\}_{j=1,...,K}$. For this, it needs all other UEs' transmit powers $q^{(n)}$, which can be obtained by information exchange among eNBs ($\{q^{(n)}\}_{\forall n}$ is the backhauling burden). Note that the required signaling is in proportion to the number of iterations and details will be discussed in Section 11.4.4.

Algorithm 11.2 DISTRIBUTED MCP IN [47] ($\{\mathbf{h}_{jk}\}_{\forall j,k}, \{\Gamma_k\}_{\forall k}, P_T$)

Initialize the lower bound γ_l and the upper bound γ_h.
while not converge

do $\begin{cases} \gamma = \frac{\gamma_l + \gamma_h}{2} \\ \text{Solve the power minimization problem with SINR constraint } \{\gamma \, \Gamma_k\} \\ \text{using the following algorithm} \\ \textbf{while} \text{ not converge} \\ \textbf{do} \begin{cases} q_k^{(n)} = \gamma \, \tilde{I}_k(q^{(n-1)}) \\ \triangleq \gamma \min_{\mathbf{w}_k} \dfrac{\sum_{j=1, j \neq k}^{K} q_j^{(n-1)} |\mathbf{w}_k^\dagger \mathbf{h}_{jk}|^2 + 1}{|\mathbf{w}_k^\dagger \mathbf{h}_{kk}|^2} \Gamma_k, \; \forall k \\ n = n + 1 \end{cases} \\ \textbf{if } \sum_{k=1}^{K} q_k \geq P_T \\ \quad \textbf{then } \gamma_h = \gamma \\ \quad \textbf{else } \gamma_l = \gamma \end{cases}$

return ($\{\mathbf{w}_k\}, q$)

Fast convergence of Algorithm 11.2 is crucial to reduce the amount of signaling among eNBs. Though distributed implementation of Algorithm 11.2 is possible without a central controller, it requires a two-layer iteration, which increases the amount of information exchange. In the following, we present a more efficient iterative algorithm of MCP, referred to as Algorithm 11.3, which differs from Algorithm 11.2 in that only one-layer iteration is needed. It uses similar distributed transmit power optimization as in [47] (originally proposed in [51]), but a normalization step is used to ensure the total power constraint. A similar algorithm was proposed in [25, (72–73)] and [52] for a single-cell system, but no detailed analysis on distributed implementation is available.

Algorithm 11.3 PROPOSED MCP($\{\mathbf{h}_{jk}\}_{\forall j,k}, \{\Gamma_k\}_{\forall k}, 1, P_{\mathrm{T}}$)

$\boldsymbol{q}^{(0)}$ = any vector with $\|\boldsymbol{q}^{(0)}\| = P_{\mathrm{T}}$, $\gamma^{(0)} = 1$, $n = 1$
while not convergence

$$\mathbf{do} \begin{cases} \tilde{q}_k^{(n)} = \gamma^{(n)} \mathcal{I}_k(\boldsymbol{q}^{(n-1)}) \triangleq \\[2mm] \qquad \gamma^{(n)} \min_{\mathbf{w}_k} \dfrac{\sum_{\substack{j=1 \\ j \neq k}}^{K} q_j^{(n-1)} |\mathbf{w}_k^\dagger \mathbf{h}_{jk}|^2 + 1}{|\mathbf{w}_k^\dagger \mathbf{h}_{kk}|^2} \Gamma_k \ \forall k \\[4mm] q_k^{(n)} = \dfrac{\tilde{q}_k^{(n)}}{\sum_{\forall j} \tilde{q}_j^{(n)}} P_{\mathrm{T}} \ \forall k \\[4mm] \gamma^{(n)} = \min_k \dfrac{\mathrm{SINR}_k(\boldsymbol{q}^{(n)})}{\Gamma_k} \\[3mm] n = n + 1 \end{cases}$$

return $(\{\mathbf{w}_k\}, \boldsymbol{q})$

In the following, we prove the convergence of Algorithm 11.3. First, we revisit the fixed-point iteration in [51] that finds the distributed solution to (11.29) with SINR targets $\{\gamma \Gamma_k\}$. Given the $(n-1)$th iteration, the update is simply given by

$$q_k^{(n)} = \mathcal{I}_k(\boldsymbol{q}^{(n-1)}), \ \forall k, \tag{11.30}$$

for $\mathcal{I}_k(\boldsymbol{q}) \triangleq \gamma \min_{\mathbf{w}_k} \dfrac{\sum_{\substack{j=1 \\ j \neq k}}^{K} q_j |\mathbf{w}_k^\dagger \mathbf{h}_{jk}|^2 + 1}{|\mathbf{w}_k^\dagger \mathbf{h}_{kk}|^2} \Gamma_k$, $\forall k$. It has been shown in [53] that, if $\mathcal{I}_k(\cdot)$ is a standard interference function, then given a feasible solution $\boldsymbol{q}^{(0)}$, the above iteration has the following properties:

(P1): $\boldsymbol{q}^{(n)}$ is component-wise monotonically decreasing;
(P2): $\boldsymbol{q}^{(n)}$ converges to the unique optimal solution q_k^*, with $q_k^* = \gamma \mathcal{I}_k(\boldsymbol{q}^*) \ \forall k$;
(P3): $\boldsymbol{q}^{(n)}$, for all n, are all feasible solutions.

Before proceeding, it is necessary to define a standard interference function. An interference function $\mathcal{I}(\boldsymbol{p})$ is said to be *standard*, if $\forall \boldsymbol{p} \geq \mathbf{0}$ and the following properties are satisfied [53].

- Positivity: $\mathcal{I}(\boldsymbol{p}) \geq \mathbf{0}$.
- Monotonicity: If $\boldsymbol{p} \geq \boldsymbol{p}'$, then $\mathcal{I}(\boldsymbol{p}) \geq \mathcal{I}(\boldsymbol{p}')$.
- Scalability: For all $\alpha > 1$, $\alpha \mathcal{I}(\boldsymbol{p}) > \mathcal{I}(\alpha \boldsymbol{p})$.

The following theorem and proof are drawn.

Theorem 11.1 *Algorithm 11.3 converges to the unique optimum of the SINR balancing problem (11.19).*

Proof. The proof has two parts. The first part is devoted to the proof of convergence, and the second part addresses the uniqueness and optimality of the fixed point after convergence.

Let us start with $\gamma^{(n)}$ (for $n \geq 1$), which is achievable for the transmit power vector $\boldsymbol{q}^{(n)}$. According to [51], $I_k(\boldsymbol{q}^{(n)})$ is a standard interference function. At the $(n+1)$th iteration, according to property (P1), $\tilde{q}_k^{(n+1)} \leq q_k^{(n)} \, \forall k$, and thus $\alpha \geq 1$ in the next step, which ensures that $\mathsf{SINR}_k(\boldsymbol{q}^{(n+1)}) > \mathsf{SINR}_k(\tilde{\boldsymbol{q}}^{(n+1)})$. Then, according to property (P3), $\mathsf{SINR}_k(\tilde{\boldsymbol{q}}^{(n+1)}) > \gamma^{(n)}\Gamma_k \, \forall k$, and as a consequence we have

$$\gamma^{(n+1)} = \min_k \frac{\mathsf{SINR}_k(\boldsymbol{q}^{(n+1)})}{\Gamma_k} \geq \min_k \frac{\mathsf{SINR}_k(\tilde{\boldsymbol{q}}^{(n+1)})}{\Gamma_k} > \gamma^{(n)}. \tag{11.31}$$

The balanced SINR $\gamma^{(n)}$ is thus increasing with (n), and since $\gamma^{(n)}$ is upper bounded by $\min_k \frac{P_T \|\mathbf{h}_{kk}\|^2}{1\Gamma_k}$ the algorithm converges to a fixed point $\boldsymbol{q}^{(\infty)}$.

In the following, we prove that the fixed point is also optimal. To do so, we see that $\boldsymbol{q}^{(\infty)}$ satisfies

$$q_k^{(\infty)} = \gamma^{(\infty)} I_k(\boldsymbol{q}^{(\infty)}) \, \forall k. \tag{11.32}$$

Therefore, $q_k^{(\infty)}$ is the unique solution to (11.29) with the normalized SINR constraints $\gamma^{(\infty)}$ according to property (P2). Clearly, the optimal transmit power in (11.29) is a monotonically non-decreasing function of $\gamma^{(\infty)}$. This implies that there is no solution $\boldsymbol{q}^*$ which provides a strictly higher normalized SINR $\gamma^* > \gamma^{(\infty)}$, while maintaining the power constraint $\sum_k q_k^* = \sum_k q_k^{(\infty)} = P_T$. This completes the proof. $\square$

Theorem 11.2 *Algorithm 11.3 has a linear convergence rate, while Algorithm 11.1 has a superlinear convergence rate.*

Proof. We first present the definition of linear and superlinear convergence. Suppose there is a sequence $y^{(n)}$, which converges to y^*, and

$$\lim_{n \to \infty} \frac{|y^{(n+1)} - y^*|}{|y^{(n)} - y^*|} = \mu. \tag{11.33}$$

If $\mu = 0$, $y^{(n)}$ is said to have superlinear convergence; and if $0 < \mu < 1$, $y^{(n)}$ is said to have linear convergence. We then proceed with the proof.

Algorithm 11.3 aims to solve the SINR balancing problem (11.21), while we start with the total transmit power minimization problem (11.29) with SINR constraints $\{\gamma^*\Gamma_k\}$, where γ^* is the optimal solution to the SINR balancing problem (11.21). In the last iteration of Algorithm 11.3, the power update satisfies

$$q_k = \gamma^* I_k(\boldsymbol{q}) \, \forall k, \tag{11.34}$$

which is the same as the fixed-point iteration in [51] to solve (11.29). Therefore, the fixed-point iterative algorithm used in (11.34) to solve (11.29) converges strictly faster than Algorithm 11.3, and it suffices to show the linear convergence of the transmit power update (11.34).

Assuming that beamforming vectors are fixed (the impact of beamforming will be studied later in this proof), the power sequence generated by (11.34) satisfies

$$\boldsymbol{q}^{(n)} - \boldsymbol{q}^* = \boldsymbol{\Psi} \left(\boldsymbol{q}^{(n-1)} - \boldsymbol{q}^* \right) \tag{11.35}$$

where we have defined

$$[\boldsymbol{\Psi}]_{kj} \triangleq \begin{cases} \gamma^* \Gamma_k \dfrac{|\mathbf{w}_k^\dagger \mathbf{h}_{jk}|}{|\mathbf{w}_k^\dagger \mathbf{h}_{kk}|}, & \text{if } k \neq j, \\ \\ 0, & \text{if } k = j, \end{cases} \tag{11.36}$$

as the coupling matrix. Since ϵ can be arbitrarily small and, according to [54], (11.35) has a linear convergence rate $\bar{\rho} \triangleq \rho(\boldsymbol{\Psi})$ only when $\bar{\rho} < 1$, where $\rho(\boldsymbol{\Psi})$ is the spectral radius which is the maximum of the absolute values of matrix eigenvalues. The smaller $\bar{\rho}$ is, the faster the algorithm converges.

In the following, we study the impact of adaptive beamforming, and show that the transmit power update dominates the convergence of the iteration. Let us denote the UL SINR in (11.20) as

$$\text{SINR}_k^{\text{UL}} = \max_{\mathbf{w}_k} \frac{q_k |\mathbf{w}_k^\dagger \mathbf{h}_{kk}|^2}{\sum_{\substack{j=1 \\ j \neq k}}^K q_j |\mathbf{w}_k^\dagger \mathbf{h}_{jk}|^2 + 1} \triangleq \max_{\mathbf{w}_k} \frac{\mathbf{w}_k^\dagger \mathbf{A}_k \mathbf{w}_k}{\mathbf{w}_k^\dagger \mathbf{B}_k \mathbf{w}_k}, \tag{11.37}$$

where $\mathbf{A}_k(\boldsymbol{q}) \triangleq q_k \mathbf{h}_{kk} \mathbf{h}_{kk}^\dagger$ and $\mathbf{B}_k(\boldsymbol{q}) \triangleq \sum_{\substack{j=1 \\ j \neq k}}^K q_j \mathbf{h}_{jk} \mathbf{h}_{jk}^\dagger + \mathbf{I}$. For brevity, we drop the UE index and rewrite it as

$$\text{SINR} = \max_{\mathbf{w}} \frac{\mathbf{w}^\dagger \mathbf{A}(\boldsymbol{q}) \mathbf{w}}{\mathbf{w}^\dagger \mathbf{B}(\boldsymbol{q}) \mathbf{w}}. \tag{11.38}$$

Clearly, SINR in (11.38) is the generalized eigenvalue of the matrix pair $(\mathbf{A}(\boldsymbol{q}), \mathbf{B}(\boldsymbol{q}))$ and the optimal $\mathbf{w}$ is the associated eigenvector. Similar to [55, Theorem 6.3.12], it can be shown that the change in $\boldsymbol{q}$ (and effectively in $\mathbf{A}(\boldsymbol{q})$ and $\mathbf{B}(\boldsymbol{q})$) dominates the change in $\mathbf{w}$. This proves that the above convergence comparison result based on the transmit power update with fixed beamforming vectors holds true even when the beamforming vectors are updated in each iteration, which completes the the first part of the proof.

In Algorithm 11.1, the transmit power update directly optimizes the objective function, while exactly satisfying all SINR constraints with equality, i.e.,

$$\mathbf{q} = (\mathbf{I} - \mathbf{DF})^{-1} \mathbf{Dq}. \tag{11.39}$$

Therefore, it is expected that starting with the same initial point, at each iteration, the power vector generated by Algorithm 11.1 is element-wise less than that generated by Algorithm 11.3, which has been formally proved in [56]. It is further shown that the transmit power sequence generated by Algorithm 11.1 satisfies

$$\limsup_{\substack{n \to \infty \\ x > 0}} \frac{\|\boldsymbol{\lambda}^{(n+1)} - \boldsymbol{\lambda}^*\|_x}{\|\boldsymbol{\lambda}^{(n)} - \boldsymbol{\lambda}^*\|_x} = 0. \tag{11.40}$$

This indicates that Algorithm 11.1 has superlinear convergence. □

Finally, in order to realize a distributed CoMP for the DL, the above solution to the dual optimization problem needs to be converted to the original problem. After we solve the problem (11.21), we have the optimal beamforming vectors $\{\mathbf{w}_m^{\text{opt}}\}$ and the maximal balanced SINR γ^{opt}. Then, the optimal DL power allocation vector, $\mathbf{p}_{\text{opt}}$, can be obtained by solving the following set of linear equations:

$$\text{SINR}_k = \gamma^{\text{opt}} \Gamma_k, \forall k, \tag{11.41}$$

where SINR_k is defined in (11.16).

Alternatively, we first define $\mathbf{D}^{\text{opt}}$ and $\mathbf{F}^{\text{opt}}$ as follows:

$$\mathbf{D}^{\text{opt}} \triangleq \text{diag}\left(\frac{\Gamma_1}{\mathbf{w}_1^{\text{opt}\dagger} \mathbf{h}_{11} \mathbf{h}_{11}^{\dagger} \mathbf{w}_1^{\text{opt}}}, \ldots, \frac{\Gamma_K}{\mathbf{w}_K^{\text{opt}\dagger} \mathbf{h}_{11} \mathbf{h}_{KK}^{\dagger} \mathbf{w}_K^{\text{opt}}} \right), \tag{11.42}$$

$$\left[\mathbf{F}^{\text{opt}} \right]_{i,j} = \begin{cases} \mathbf{w}_j^{\text{opt}\dagger} \mathbf{h}_{ij} \mathbf{h}_{ij}^{\dagger} \mathbf{w}_j^{\text{opt}}, & \text{for } i \neq j, \\ 0, & \text{for } i = j. \end{cases} \tag{11.43}$$

Then, the solution to the DL problem is given by [46, (19)–(21)]

$$\mathbf{p}_{\text{opt}} = \left[\frac{1}{\gamma^{\text{opt}}} (\mathbf{D}^{\text{opt}})^{-1} - \mathbf{F}^{\text{opt}} \right]^{-1} \mathbf{1}, \tag{11.44}$$

where $\mathbf{1}$ is the all-one vector.

11.4.4   Analysis of distributed implementation

11.4.4.1   Distributed implementation

Distributed implementation is especially important in CoMP because the backhauling burden has to be minimized as indicated in previous sections. Centralized processing such as data sharing in eNB clusters and exchange of CSI among eNBs is undesirable for most practical scenarios. As such, the non-distributed Algorithm 1.4.1, which links the optimal balanced SINR with the eigenvalue of the coupling matrix. is not feasible. In the following, we assume that each eNB can estimate the channels among itself and UEs in other cells of the same cluster, i.e., each eNB k knows $\{\mathbf{h}_{jk}\}_{j=1,2,\ldots,K}$. This can be justified by the fact that a multi-cell cluster is usually set up only if a UE in this cluster sees considerable interference from other cells. As a result, distributed CoMP can be achieved in two steps. First, a distributed algorithm solves the virtual UL problem, and then the virtual UL solution is converted into the DL solution. Note that the conversion is conducted only once, after the solution is found. In each step, eNB k exploits only its local CSI $\{\mathbf{h}_{jk}\}_{j=1,\ldots,K}$ and has limited information exchange with others.

11.4.4.2   Discussion on parameter exchange

The implementation of a distributed strategy depends largely on the required parameter exchange among eNBs in the two steps described above. To simplify the analysis, we assume that each BS serves only one UE, i.e., $B = K$. First, a distributed algorithm such as Algorithm 11.3 is used to solve the virtual UL problem (11.21), where in each iteration a length-K transmit power vector $\boldsymbol{q}$ has to be shared among eNBs. Therefore, the number of exchanged parameters is $N_I K$, where N_I is the total number of iterations. Thereafter, in the step of virtual UL–DL solution conversion [20], K^2 positive scalars are required to be shared among eNBs. As a result, the total number of exchanged parameters is $N_I K + K^2$, regardless of the number of transmit antennas M. As a comparison, the full CSI exchange, which would make possible a centralized CoMP optimization such as Algorithm 11.1, requires $2MK^2$ scalars to be shared, where the factor of 2 comes from the complex entries of the channel matrix. If the same quantization scheme is used, the percentage of required exchanged bits in Algorithm 11.3 (distributed) over that in Algorithm 11.1 (full CSI) exchange is given by $\eta = \frac{N_I + K}{2MK}$, showing a significant reduction in backhaul signaling.

A key factor to evaluate a distributed implementation is whether or not the iterative algorithm used to solve the virtual UL problem converges and how fast if it does (or how small N_I could be). In this regard, Algorithm 11.2 is not very efficient due to two-layer iterations, i.e., N_I should be the total number of inner iterations summed over all outer iterations. This makes Algorithm 11.3 more promising for distributed CoMP.

11.4.5   Simulation results

In this section, we provide simulation results to evaluate the SLNR scheme and Algorithms 11.1, 11.2 and 11.3. Specifically, Algorithm 11.3 is referred to as "the distributed CoMP algorithm." Let us consider a macrocell network composed of two or three eNBs with 1 km inter-eNB distance. As it is well recognized that CoMP usually brings more gain for cell-edge UEs than for cell-centre UEs, in this simulation setup, we assume that all UEs are located at cell edges. In particular, we consider a macrocell network where each UE has a distance of $d_1 = 350$ m from its serving BS and a distance of $d_2 = 550$ m from other cooperating BSs. To simulate the channel vectors, the channel model for a typical urban macro scenario with *non line of sight* (NLOS) is used. The channel vectors are generated using the formulation $\mathbf{h}_{ki} \triangleq \gamma_{ki} \tilde{\mathbf{h}}_{ki}$, where $\tilde{\mathbf{h}}_{ki}$ is the small-scale fading and is assumed to be zero-mean Gaussian distributed with a covariance matrix of $\mathbf{I}$, i.e., $\tilde{\mathbf{h}}_{ki} \sim \mathcal{CN}(\mathbf{0}, \mathbf{I})$, and γ_{ki} is the large-scale path loss, given by

$$\gamma_{ki} = \frac{\beta \chi_{ki}}{d_{ki}^l}, \tag{11.45}$$

in which β is a positive scaling factor, l is the path loss exponent (typically $l > 2$), d_{ki} is the distance between UE k and BS i, and χ_{ki} is lognormal shadow fading. Throughout our simulations, we set $\beta = 10^{-3.45}$, $l = 3.8$, which in dB gives $10 \log_{10}(\gamma_{ki}) = -38 \log_{10}(d_{ki}) - 34.5 + \mu_{ki}$, where μ_{ki} is the lognormal shadow fading and follows the normal distribution $\mathcal{N}(0, 8$ dB$)$. To be consistent with the parameter setup in 4G systems, we consider a 10 MHz bandwidth, and that each eNB transmits an average power of P_{BS} over the entire bandwidth. For each channel realization, BSs in the same cluster cooperate to consume a total power of $K P_{BS}$, the noise figure at each UE receiver is set to be 9 dB, and to show simulation results, we use P_{BS} to denote transmit SNR.

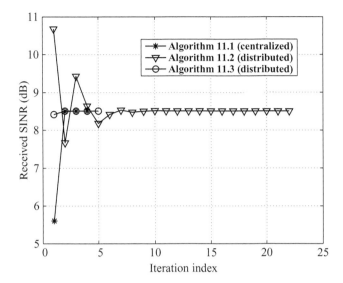

Figure 11.4 Convergence behavior of CoMP algorithms.

Fig. 11.4 compares the convergence behavior for the centralized Algorithm 11.1, the bisection search-based Algorithm 11.2 in [47] and the distributed Algorithm 11.3 for a solution accuracy of 10^{-6}, a typical channel with $(M, K) = (4, 2)$ and $P_T = 46$ dBm. Results reveal that all three algorithms converge to the same optimal SINR value. Algorithm 11.3 converges within a few iterations while the bisection search Algorithm 11.2 needs many more (which refers to the outer bisection search of γ in Algorithm 11.2), not to mention the additional required inner iterations. Thus, Fig. 11.4 confirms the advantage of Algorithm 11.3 over Algorithm 11.2 in being distributed, hence more suitable for CoMP.

Fig. 11.5 illustrates the average rate achieved by the worst UE in the distributed CoMP algorithm and the SLNR scheme with $(M, K) = (6, 3)$. Results show that the distributed CoMP significantly outperforms the SLNR scheme especially at high SNRs. In particular, the rate performance of the distributed CoMP with two iterations already shows an advantage over the SLNR scheme. Moreover, it is seen that the distributed CoMP has obtained near-optimal rate performance with around four iterations. This means that the ratio of required exchanged information in the proposed distributed CoMP scheme over direct CSI exchange is approximately given by $\zeta \approx \frac{N_I + K}{2MK} = \frac{7}{36}$, showing a significant reduction in backhaul signaling. This reduction will be further increased with M.

Fig. 11.6 illustrates the ratio of the worst UE's rate of the distributed CoMP to the optimal worst UE's rate, varying with the number of iterations. Results show that more than 90% of the optimal rate can be achieved by the distributed CoMP with three iterations, and almost 100% of the optimal rate can be attained with only six iterations. This suggests that the distributed CoMP algorithm is able to achieve a good tradeoff between performance and the amount of intercell information exchange.

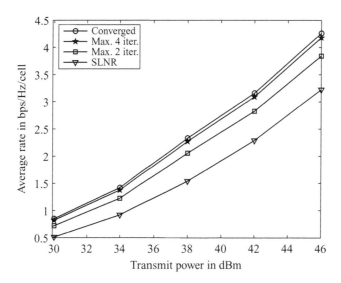

Figure 11.5 The average rate of the worst cell in the proposed multi-cell distributed beamforming system with $K = 3$.

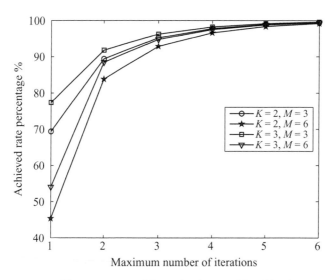

Figure 11.6 The percentage of the optimal rate, achieved by the worst cell in the proposed multi-cell distributed beamforming system after a fixed number of iterations, with $P_{BS} = 40$ dBm.

Fig. 11.7 illustrates the worst UE's rate of the distributed CoMP using a simple uniform linear quantization of the exchanged power parameters. Simulation results show that the performance loss of distributed CoMP resulting from four-bit parameter quantization is inappreciable, and a three-bit parameter quantization only exhibits a little penalty.

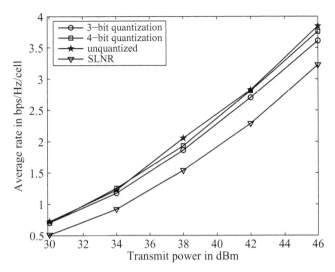

Figure 11.7 The average rate of the worst cell in the proposed multi-cell distributed beamforming system using quantized intercell parameter exchange with $M = 6$, $K = 3$, and a maximum of two iterations.

This implies that a simple uniform quantization is highly efficient for distributed CoMP, because the transmit power parameter is strictly included in the interval $[0, KP_{BS}]$.

11.5   Downlink coordinated beamforming applications in HCN

As discussed in Section 11.3, when CoMP is applied in HCN scenarios such as macro–pico and macro–RN networks, individual transmit power constraints at heterogenous eNBs need to be carefully considered. Note that, in the CoMP beamforming algorithms presented in Section 11.4, the scheme based on interference leakage can be directly used in HCNs, while the above Algorithms 11.1–3 only focus on a simple sum transmit power constraint, as revealed in [47], and thus the extension to the case of individual transmit power constraints in HCN is not straightforward. It would need to handle the uncertain noise level in the virtual UL problem, resulting in much slower convergence. Alternatively, this section will introduce an efficient realization of DL coordinated multi-cell beamforming, which considers individual power constraints and is applicable to HCN.

It is well known that the optimal rate region achieved by the coordinated multi-cell system is characterized by a Pareto boundary, where it is impossible to improve any rate without simultaneously decreasing at least one of the others. Most typical design criteria can be translated into reaching a certain point on this boundary. For instance, the sum rate maximization is equivalent to targeting at the rightmost point on the Pareto boundary (the rate being plotted on the x-axis). However, the sum rate maximizing problem is usually non-convex, and is hard to address. In the following, we consider both rate optimization and UE fairness. More specifically, the goal of the coordinated multi-cell beamforming design is to reach the particular point on the Pareto boundary that has the maximum worst-UE rate, i.e., the Pareto boundary with max–min optimality. Note that this objective function cannot be simply equivalent to maximizing the worst-UE rate, due to the fact that a simple max–min rate optimization usually achieves identical rate performance among multiple UEs, which cannot guarantee the Pareto optimality.

The proposed algorithm towards this goal can be implemented in two steps, i.e., a max–min optimization and a Pareto improvement, and it is proved that it is able to reach the desired point on the Pareto boundary for the case of two cooperating eNBs.

11.5.1   System model

Consider an HCN multi-cell DL system consisting of a single macrocell and multiple LPN cells such as picocells and relay nodes. As mentioned in Section 11.3, the footprint of LPNs is usually enlarged through cell range expansion, so as to increase the traffic offload from the macrocell network. As a result, UEs located in the region of cell range expansion would suffer from severe DL interference from the macrocell. To address this issue, we consider that these cells are allowed to perform coordinated transmission. Similar to the clustering approach in homogeneous multi-cell networks, it is efficient to dynamically determine the coordinating cell cluster in a UE-centric way [33–35], in which it is very possible that UEs in the region of cell range expansion report an expected CoMP cell set including the macrocell. Once the cell clustering and UE scheduling have been completed, a key issue remains to design the coordinated beamforming or precoding. For simplicity, assume that the cell cluster consists of B eNBs (henceforth in this section we do not distinguish between macro eNBs and LPNs), each serving at least one UE. Each eNB is equipped with M transmit antennas and each UE is equipped with a single antenna. The transmit signal from each eNB is only intended for its own UEs, meaning that no UE-data sharing between the eNBs is needed. However, the transmit beamformers from different eNBs need to be designed jointly to suppress the interference.

Let $\mathcal{K}_b$ denote the UE set served by eNB b, and s_k denote the information symbol intended for UE k, with $\mathbb{E}\{|s_k|^2\} = 1$, where $\mathbb{E}\{\cdot\}$ denotes the expectation operator. Let b_k denote the index of the eNB serving UE k, $\mathbf{w}_k \in \mathbb{C}^{M \times 1}$ denote the unit-norm beamforming vector (i.e., $\|\mathbf{w}_k\| = 1 \ \forall k$) for UE k, and p_k denote the transmit power for UE k. Each eNB has a transmit power constraint P_b, i.e., $\sum_{k \in \mathcal{K}_b} p_k \leq P_b$. Assuming $k \in \mathcal{K}_b$, the received signal of UE k is given by

$$y_k = \sqrt{p_k} \tilde{\mathbf{h}}_{kb}^{\dagger} \mathbf{w}_k s_k + \tilde{\mathbf{h}}_{kb}^{\dagger} \sum_{\substack{i \in \mathcal{K}_b \\ i \neq k}} \sqrt{p_i} \mathbf{w}_i s_i + \sum_{\bar{b} \neq b} \tilde{\mathbf{h}}_{k\bar{b}}^{\dagger} \sum_{j \in \mathcal{K}_{\bar{b}}} \sqrt{p_j} \mathbf{w}_j s_j + n_k, \qquad (11.46)$$

where $\dagger$ is the Hermitian transpose, n_k is the zero-mean complex white Gaussian noise with variance σ_k^2, and $\tilde{\mathbf{h}}_{kb}$ is the frequency-flat fading channel vector from eNB b to UE k. Note that although the above model appears to assume perfect symbol-to-symbol synchronization, this assumption can be removed in *orthogonal frequency division multiplexing* (OFDM) systems. This can be done if we consider a certain utility function, which is monotonically increasing with the SINR due to the fact that the SINR criterion only involves the transmit power of the interference.

The problem of interest is the joint optimization of the eNB beamforming vectors $\{\mathbf{w}_k\}_{\forall k}$ and the transmit power allocation $\{p_k\}_{\forall k}$. To emphasize capacity improvements on the basis of UE fairness, our objective here is to reach the particular point on the Pareto boundary of the rate tuples, which has the largest worst-UE rate and thus guarantees max–min rate fairness.

As for the CSI available at eNBs, we first consider a TDD system, where each eNB (say b) can directly estimate the CSI of its own UEs, i.e., $\{\mathbf{h}_{kb}\}_{k\in\mathcal{K}_b}$, by exploiting UL–DL reciprocity. It is also reasonable to assume that each eNB can estimate from the reverse link the crosstalk channels to other UEs in the same cluster. This is because a UE-centric clustering approach usually implies that the channels among each scheduled UE and multiple coordinating eNBs have similar orders of magnitude. For doing so, a coordinated training phase for the cell cluster might be required, in which the training signals from different UEs may be designed to be orthogonal in order to avoid mutual interference. Based on these assumptions, we define $\{\mathbf{h}_{kb}\}_{k=1}^K$ as local CSI at eNB b, which can be known without direct information exchange among eNBs.

11.5.2 Downlink multi-cell beamforming approaching Pareto optimality with max–min fairness

As indicated above, we aim to reach the particular point on the Pareto boundary that has the largest worst-UE rate, i.e., the Pareto optimality with max–min fairness. The developed scheme is realized with two steps of optimizations. The first step is to use the max–min criterion to maximize the worst rate achieved by the UEs. It is then followed by the second step, which realizes a Pareto improvement to further increase individual UEs' rates. It is proved that, for the case of two eNBs and two UEs, this scheme achieves the desired point on the Pareto boundary. Even in a general case with more than two UEs, it is also very possible to almost achieve the desired optimality. In the following, we describe this two-step optimization algorithm in detail.

11.5.2.1 Max–min optimization

To provide fairness among UEs in the cluster, a max–min optimization is first used, which maximizes the worst rate achieved by UEs. For notational convenience, we view the set of B eNBs as a set of K virtual eNBs (VBSs), each serving one UE. The index of the VBS serving UE k is denoted as k, with the set of VBSs $\{k \in \mathcal{K}_b\}$ corresponding to the actual eNB b. The channel vector from VBS i to UE k is denoted by $\mathbf{h}_{ki}$. Then, the SINR of UE k can be written as

$$\mathrm{SINR}_k^{\mathsf{DL}}(\mathbf{W}, \mathbf{p}) = \frac{p_k \mathbf{f}_k^\dagger \mathcal{H}_{kk} \mathbf{w}_k}{\sum_{\substack{i=1 \\ i \neq k}}^{K} p_i \mathbf{w}_i^\dagger \mathcal{H}_{ki} \mathbf{w}_i + 1}, \tag{11.47}$$

where $\mathbf{p} \triangleq [p_1, \ldots, p_K]^\mathsf{T}$ with the superscript T denoting transposition, and $\mathcal{H}_{ji} \triangleq \frac{\mathbf{h}_{ji}\mathbf{h}_{ji}^\dagger}{\sigma_j^2}$, and $\mathbf{W} \triangleq \mathrm{diag}(\mathbf{w}_1, \mathbf{w}_2, \ldots, \mathbf{w}_K)$ is the overall precoder or beamformer for the cluster of eNBs. As a consequence, the max–min optimization problem can be expressed as

$$\max_{\mathbf{W},\mathbf{p}} \min_{k} \frac{\text{SINR}_k^{\text{DL}}(\mathbf{W},\mathbf{p})}{\rho_k} \quad \text{s.t.} \sum_{k\in\mathcal{K}_b} p_k \leq P_b \ \forall b, \tag{11.48}$$

where ρ_k is a weighting factor and has the effect to steer the optimization in such a way that a larger factor for a cell will give rise to a higher SINR for that cell. Solving (11.48) is difficult because it is a non-convex problem. However, we see that (11.48) is highly related to the minimum transmit power problem:

$$\min_{\mathbf{W},\mathbf{p}} \|\mathbf{p}\|_1 \quad \text{s.t.} \begin{cases} \dfrac{\text{SINR}_k^{\text{DL}}}{\rho_k} \geq \gamma \ \forall k, \\[2ex] \displaystyle\sum_{k\in\mathcal{K}_b} p_k \leq P_b \ \forall b, \end{cases} \tag{11.49}$$

where γ is a preset target and $\|\cdot\|_1$ returns the 1-norm of an input vector. It is easily seen that the solution of (11.49) attains a set of equal weighted SINRs. If we denote the optimal weighted SINR of (11.48) as γ^* and its corresponding beamforming vectors and transmit power vector as $\mathbf{W}^*$ and $\mathbf{p}^*$, respectively, then it can be proved, see [25], that the solution to (11.49) with a target γ^* will produce the same optimal $\mathbf{W}^*$ and $\mathbf{p}^*$. As a result, if (11.49) is solved, (11.48) can be simultaneously solved by searching for a suitable target γ^*, for which the solution $\mathbf{p}^*$ of (11.49) has at least one active power constraint. Furthermore, as shown in [46], (11.49) can be transformed into a standard *second-order cone programming* (SOCP) problem, and hence can be solved using standard optimization packages, such as CVX. Based on this, a bisection search method can be used to efficiently search for the appropriate weighted SINR target γ^*, hence giving the solution of (11.48).

11.5.2.2   Pareto improvement

Note that the max–min optimization method above always gives a balanced weighted SINR. Numerical results show that, to achieve this max–min optimization, in most cases only a single eNB may transmit at full power, while other eNBs transmit with less power. This implies that it is possible to further improve the sum rate of the multi-cell beamforming system without losing the max–min optimality; such a behavior is referred to as Pareto improvement in game theory.

In the following, we focus on the case of $M \geq K$. To achieve an efficient Pareto improvement, we propose the following beamforming updating algorithm. Let $\{\mathbf{w}_k^*\}_{\forall k}$ denote the beamforming vectors obtained from the max–min optimization, and $\mathbf{p}^* = [p_1^*, \ldots, p_K^*]^T$ denote the corresponding transmit power vector. For eNB b, if $\sum_{i\in\mathcal{K}_b} p_i^* < P_b$ and $p_k^* = \min_{i\in\mathcal{K}_b} p_i^*$, then we propose to perform the following beamforming update:

$$\mathbf{w}_k^{\text{new}} = \frac{\sqrt{p_k^*}\mathbf{w}_k^* + \alpha_k e^{j\theta_k}\mathbf{h}_k^{\text{ZF}}}{\sqrt{P_b}}, \tag{11.50}$$

where $\theta_k = \angle(\mathbf{h}_{kk}^\dagger \mathbf{w}_k^*)$, α_k is a positive scalar such that $\|\mathbf{w}_k^{\text{new}}\| = 1$, and $\mathbf{h}_k^{\text{ZF}}$ is the projection of $\mathbf{h}_{kk}$ onto the complement of the column space of $\bar{\mathbf{H}}_k = [\mathbf{h}_{1k} \cdots \mathbf{h}_{k-1k} \; \mathbf{h}_{k+1k} \cdots \mathbf{h}_{Kk}]$, given by

$$\mathbf{h}_k^{\text{ZF}} = \left(\mathbf{I} - \bar{\mathbf{H}}_k (\bar{\mathbf{H}}_k^\dagger \bar{\mathbf{H}}_k)^{-1} \bar{\mathbf{H}}_k^\dagger \right) \mathbf{h}_{kk}. \tag{11.51}$$

The Pareto optimality of this method will be addressed in the following section.

11.5.3   Performance analysis

As shown in [40], the rate region of all UEs with respect to $\mathbf{W}$ and $\mathbf{p}$ is outer bounded by the so-called Pareto boundary, for which it is impossible to improve any of the rates without simultaneously decreasing at least one of the other rates. To achieve the rate-fair optimality, we need to reach the particular point on the Pareto boundary, which has the largest minimum rate ($\rho_k = 1$ for all k), i.e., the Pareto optimal point with the max–min rate optimality. The following analysis shows that the above two-step algorithm is able to achieve this goal. In particular, the result summarized in Theorem 1 shows that the proposed algorithm guarantees to achieve this max–min Pareto optimal point when $M \geq K = B = 2$. To do so, we first present the following two lemmas.

Lemma 11.1 *We denote the optimal max–min solution with $M \geq K = B = 2$ as r^*, $\mathbf{W}^*$, and $\mathbf{p}^*$. Without loss of generality, suppose that $p_1^* < P_1$ and $p_2^* = P_2$, i.e., eNB 2 uses full transmit power and eNB 1 uses less transmit power. If a Pareto improvement over the rate tuple (r^*, r^*) can be achieved, in which UE 1's rate improves while UE 2's rate remains,[1] then the updated beamforming vector for UE 1, $\mathbf{w}_1^{\text{new}}$ must generate no interference to UE 2, i.e., $\mathbf{w}_1^{\text{new}\dagger} \mathcal{H}_{21} \mathbf{w}_1^{\text{new}} = 0$.*

Proof. The proof is by contradiction. First, we let the beamforming vectors and the transmit power of UE 2 be fixed by $\mathbf{w}_1^{\text{new}}, \mathbf{w}_2^*$ and P_2, respectively. Then, we can express the SINRs (ρ_k omitted here) of the two UEs as functions of p_1, given by

$$\text{SINR}_1^{\text{DL}}(p_1) = \frac{p_1 \mathbf{w}_1^{\text{new}\dagger} \mathcal{H}_{11} \mathbf{w}_1^{\text{new}}}{P_2 \mathbf{w}_2^{*\dagger} \mathcal{H}_{12} \mathbf{w}_2^* + 1}, \tag{11.52}$$

$$\text{SINR}_2^{\text{DL}}(p_1) = \frac{P_2 \mathbf{f}_2^{*\dagger} \mathcal{H}_{22} \mathbf{w}_2^*}{p_1 \mathbf{w}_1^{\text{new}\dagger} \mathcal{H}_{21} \mathbf{w}_1^{\text{new}} + 1}. \tag{11.53}$$

Let us define $\Delta \text{SINR}(p_1) \triangleq \text{SINR}_1^{\text{DL}}(p_1) - \text{SINR}_2^{\text{DL}}(p_1)$. We observe that $\Delta \text{SINR}(p_1)$ is a monotonically increasing and continuous function. The assumption of Pareto improvement thus yields

$$\Delta \text{SINR}\left(p_1^{\text{new}}\right) > 0, \tag{11.54}$$

but it is easily seen that $\Delta \text{SINR}(0) < 0$. As a consequence, there must exist some $p_1' < p_1^{\text{new}}$ such that $0 < \Delta \text{SINR}(p_1') < \Delta \text{SINR}(p_1^{\text{new}})$. If we assume that $\mathbf{f}_1^{\text{new}\dagger} \mathcal{H}_{21} \mathbf{f}_1^{\text{new}} \neq 0$

---

[1] It is impossible to have both rates greater than γ^*, as γ^* is by definition the max–min SINR.

is true, then this implies that, if eNB 1 chooses to use transmit power p'_1, the resulting SINRs of both UEs will be greater than γ^*, which contradicts the result in which γ^* is the solution to the max–min optimization. Therefore, we must have $\mathbf{w}_1^{\text{new}\dagger}\mathcal{H}_{21}\mathbf{w}_1^{\text{new}} = 0$, which completes the proof.   □

Lemma 11.2 *For $M \geq K \geq 2$, both the Pareto optimal beamforming vectors and the max–min optimal beamforming vectors can be expressed as a linear combination of channels among the VBSs and all UEs, i.e., $\sum_{i=1}^{K} \zeta_{ik}\mathbf{h}_{ik}$ $\forall k$, where $\{\zeta_{ik}\}$ denotes some complex scalars.*

Proof. The proof for the Pareto optimal beamforming vectors is given in [40]. In the following, we only focus on the max–min optimal beamforming vectors, i.e., the solution to (11.48), and prove that

$$\mathbf{w}_k^* = \sum_{i=1}^{K} \zeta_{ik}\mathbf{h}_{ik}. \tag{11.55}$$

For convenience, the unit-norm constraint for the beamforming vector $\mathbf{w}_k^*$ has been removed. Our proof goes by contradiction.

When $M \geq K$, any $\mathbf{w}_k^* \in \mathbb{C}^M$ can always be expressed as

$$\mathbf{w}_k^* = \sum_{i=1}^{K} \lambda_{ik}\mathbf{h}_{ik} + \sum_{m=1}^{M-K} \eta_{mk}\mathbf{u}_{mk}, \tag{11.56}$$

where $\{\mathbf{u}_{mk}\}$ denotes an orthonormal basis for the orthogonal complement of the space spanned by $\{\mathbf{h}_{ik}\}_{\forall i}$, and $\{\lambda_{ik}, \eta_{mk}\}$ are some complex-valued scalars.

Suppose that for some $m = m'$, we have $\eta_{m'k} \neq 0$. Then, we define a vector for each k:

$$\tilde{\mathbf{w}}_k \triangleq \mathbf{w}_k^* - \eta_{m'k}\mathbf{u}_{m'k}. \tag{11.57}$$

The fact that we have $\mathbf{u}_{m'k}^\dagger\mathbf{h}_{ik} = 0$ $\forall i$ gives $|\mathbf{h}_{ik}^\dagger\mathbf{w}_k^*| = |\mathbf{h}_{ik}^\dagger\tilde{\mathbf{w}}_k|$ $\forall i$, meaning that the rate tuple achieved by the set of beamforming vectors $\{\mathbf{w}_1^*, \ldots, \mathbf{w}_k^*, \ldots, \mathbf{w}_K^*\}$ is equal to that achieved by $\{\mathbf{w}_1^*, \ldots, \tilde{\mathbf{w}}_k, \ldots, \mathbf{w}_K^*\}$. At the same time, it can be easily verified that

$$\|\tilde{\mathbf{w}}_k\|^2 < \|\mathbf{w}_k^*\|^2. \tag{11.58}$$

Thus, $\tilde{\mathbf{w}}_k$ will consume less transmit power than $\mathbf{w}_k^*$ but achieve the same rate. This contradicts the result that $\mathbf{w}_k^*$ is max–min optimal. Hence, we must have $\eta_{m'k} = 0$, or $\mathbf{w}_k^*$ must have the form in (11.55). This concludes the proof.   □

Based on the above observations, we present our main result in Theorem 1 as follows.

Theorem 11.3 *For $B = K = 2$ and $M \geq K$, if a Pareto improvement over the rate tuple (r^*, r^*) achieved by the max–min optimization is possible, the proposed beamforming updating method in (11.50) must achieve a maximum rate improvement. Hence, the proposed two-step optimization reaches the particular point on the Pareto boundary with the maximum worst-UE rate.*

Proof. We consider two separate cases. If the solution to the max–min optimization suggests that both eNBs use full transmit power, then it is impossible to have Pareto improvement. Therefore, we only need to focus on the other case, where one eNB uses full transmit power while the other one uses less transmit power. Without loss of generality, we assume $p_1 < P_1$ and $p_2 = P_2$. As shown in Lemma 11.1, the updated beamforming vector for UE 1 should satisfy $\mathbf{w}_1^{\text{new}\dagger}\mathcal{H}_{21}\mathbf{w}_1^{\text{new}} = 0$. This also implies that it is impossible for the SINR of UE 2 to outperform γ^*, and any interference to UE 1 will lead to a result in which UE 1's SINR becomes smaller than γ^*. As a by-product, it can be seen that $\mathbf{w}_1^{*\dagger}\mathcal{H}_{21}\mathbf{w}_1^* = 0$.

Based on these observations, if we denote the beamforming update as

$$\Delta\mathbf{w}_1 = \sqrt{p_1^{\text{new}}}\mathbf{w}_1^{\text{new}} - \sqrt{p_1^*}\mathbf{w}_1^*, \tag{11.59}$$

we must have $\mathbf{h}_{21}^\dagger \Delta\mathbf{w}_1 = 0$. Furthermore, it has been shown in Lemma 2 that both the Pareto optimal beamforming vector and the max–min optimal beamforming vector can be formulated as

$$\mathbf{w}_1 = \zeta_{21}\mathbf{h}_{21} + \zeta_{11}\mathbf{h}_{11} = \tilde{\zeta}_{21}\mathbf{h}_{21} + \zeta_{11}\mathbf{h}_1^{\text{ZF}}, \tag{11.60}$$

where $\tilde{\zeta}_{21} = \zeta_{21} + \zeta_{11}\frac{\mathbf{h}_{21}^\dagger\mathbf{h}_{11}}{\|\mathbf{h}_{21}\|}$. It follows from $\mathbf{w}_1^\dagger\mathcal{H}_{21}\mathbf{w}_1 = 0$ that $\tilde{\zeta}_{21} = 0$. This means that the beamforming update should be restricted in the direction of $\mathbf{h}_1^{\text{ZF}}$. Thus, the updated unit-norm beamforming vector $\mathbf{w}_1^{\text{new}}$ and the updated transmit power p_1^{new} can be written as

$$\sqrt{p_1^{\text{new}}}\mathbf{w}_1^{\text{new}} = \sqrt{p_1^*}\mathbf{w}_1^* + b_1\mathbf{h}_1^{\text{ZF}}, \tag{11.61}$$

where b_1 is a complex scalar. For the maximum Pareto improvment, the optimal transmit power allocation p_1^{new} and the scaling factor b_1 can be determined by solving the following problem:

$$\max_{\substack{b_1 \in \mathbb{C} \\ 0 < p_1^{\text{new}} \leq P_1}} \left|\sqrt{p_1^{\text{new}}}\mathbf{h}_{11}^\dagger\mathbf{w}_1^{\text{new}}\right|^2 \quad \text{s.t. } \|\mathbf{w}_1^{\text{new}}\| = 1. \tag{11.62}$$

From the triangle inequality, we have

$$\left|\sqrt{p_1^{\text{new}}}\mathbf{h}_{11}^\dagger\mathbf{w}_1^{\text{new}}\right| = \left|\sqrt{p_1^*}\mathbf{h}_{11}^\dagger\mathbf{w}_1^* + b_1\mathbf{h}_{11}^\dagger\mathbf{h}_1^{\text{ZF}}\right| \leq \left|\sqrt{p_1^*}\mathbf{h}_{11}^\dagger\mathbf{w}_1^*\right| + |b_1|\left|\mathbf{h}_{11}^\dagger\mathbf{h}_1^{\text{ZF}}\right|. \tag{11.63}$$

The equality holds only if the two terms on the right have the same phase. Hence, to achieve the maximum rate improvement, one necessary condition is that the phase of b_1 should be $\angle(b_1) = \angle(\mathbf{h}_{11}^\dagger\mathbf{w}_1^*) = \theta_1$, and the rate improvement is represented by the term $|b_1||\mathbf{h}_{11}^\dagger\mathbf{h}_1^{\text{ZF}}|$. On the other hand, $|b_1|$ and p_1^{new} should meet the constraint in (11.62) given by

$$\xi_1|b_1|^2 + \xi_2|b_1| + p_1^* - p_1^{\text{new}} = 0, \tag{11.64}$$

where $\xi_1 = |\mathbf{h}_1^{\text{ZF}}|^2$ and $\xi_2 = 2\sqrt{p_1^*}\text{Re}\{(\mathbf{w}_1^*)^\dagger\mathbf{h}_1^{\text{ZF}}e^{j\theta_1}\}$ with $\text{Re}\{\cdot\}$ extracting the real part of a complex-valued argument. It is easy to see that, if $p_1^* < p_1^{\text{new}}$, then (11.64) will always have a positive solution for $|b_1|$ with its value increasing with p_1^{new}. This implies that the choice of $p_1^{\text{new}} = P_1$ gives the greatest $|b_1|$, i.e., the maximum Pareto improvement. This concludes the proof.

So far, we have proved the desired Pareto optimality of the proposed two-step optimization method if $B = K = 2$. When extending to a general case $M \geq K > 2$, it is seen from Lemma 2 that both the beamforming vectors with the Pareto optimality and those with the max–min optimality can be written as

$$
\mathbf{w}_k = \underbrace{\sum_{\substack{i=1 \\ i \neq k}}^{K} \zeta_{ik}\mathbf{h}_{ik} + \zeta_{kk}\bar{\mathbf{H}}_k \left(\bar{\mathbf{H}}_k^{\dagger}\bar{\mathbf{H}}_k\right)^{-1} \bar{\mathbf{H}}_k^{\dagger}\mathbf{h}_{kk}}_{\mathbf{h}_k^{\text{I}}} + \zeta_{kk}\mathbf{h}_k^{\text{ZF}}, \quad \text{for } k \in \{1, \ldots, K\}. \quad (11.65)
$$

In general, $\mathbf{h}_{ik}^{\dagger}\mathbf{h}_k^{\text{ZF}} = 0 \ \forall i$ and $\mathbf{h}_{ik}^{\dagger}\mathbf{h}_k^{\text{I}} \neq 0$ with probability one. As a result, the beamforming update in (11.50) will not harm the rates of all the other UEs, while being able to improve its own rate. This means that our proposed two-step optimization method is also effective in a more general case $M \geq K > 2$.   □

11.5.4   Distributed implementation

The above two-step optimization scheme is by itself centralized as far as implementation is concerned. It is seen from (11.50) that the step of Pareto improvement can be performed distributively using local CSI. The problem lies in the distributed implementation of the max–min optimization. As seen from (11.48) and (11.49), the main difficulty for distributed implementation is to address the per-BS transmit power constraints in (11.49). It was revealed in [47] that, if the original DL optimization problem (11.49) is addressed by recasting it into a virtual UL problem, it will suffer from uncertain virtual noise level and is hard to achieve an efficient solution.

Alternatively, this problem can be effectively solved by recasting it into an approximate virtual UL problem. This can be realized by developing an approximate UL–DL duality, as shown in [57], in which it is further revealed that the use of an approximate dual problem for solving the original DL problem might result in some performance loss in terms of max–min fairness. However, a better sum rate performance could be achieved, as it would leave more room for the second step of rate improvement. For details please refer to the reference literature.

11.5.5   Simulation results

In this section, computer simulations are provided to evaluate the performance of the above schemes. Consider an HCN consisting of a macro eNB and an LPN such as a pico eNB or a RN, each serving a single UE. Both the macro eNB and the LPN are equipped with M antennas, and each UE is equipped with a single antenna. Assume that both UEs are located in or near the region of cell range expansion. In particular, in our configuration the distance from the macrocell UE to its serving eNB is randomly

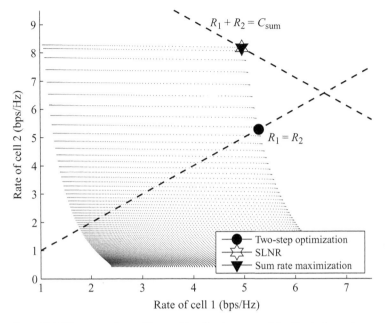

Figure 11.8 The rate points of the multi-cell schemes and the Pareto boundary under a random channel realization.

distributed between 400 and 450 m; the distance from the macrocell UE to the LPN is randomly distributed between 40 and 45 m. Similarly, the distance from the LPN UE to its serving LPN is randomly distributed between 35 and 50 m, while the distance from the LPN UE to the macro eNB is randomly distributed between 410 and 460 m.

Flat fading is assumed and the channel vector between BS (macro eNB or LPN) i and UE k is modeled as $\mathbf{h}_{ki} \triangleq \gamma_{ki} \mathbf{h}_{ki}^{(s)}$, where $\mathbf{h}_{ki}^{(s)}$ is the small-scale fading and is zero-mean Gaussian distributed with covariance $\mathbf{I}$, or $\mathbf{h}_{ki}^{(s)} \sim \mathcal{CN}(\mathbf{0}, \mathbf{I})$, and γ_{ki} is the large-scale path loss given by

$$\gamma_{ki} = \frac{\beta \mu_{ki}}{d_{ki}^l} \tag{11.66}$$

in which β is a scaling factor, l is the path loss exponent, and d_{ki} is the distance between BS i and UE k. In particular, we choose $\beta = 10^{-3.45}$, $l = 3.8$, which in dB gives $10\log_{10}(\gamma_{ki}) = -38\log_{10}(d_{ki}) - 34.5 + \mu_{ki}$, where the shadow fading follows a normal distribution $\mu_{ki} \sim \mathcal{N}(0, 8 \text{ dB})$. The noise figure at the UE is 9 dB. The transmit power constraint at the macro eNB is 46 dBm and that at the LPN is 30 dBm, assuming a 10 MHz bandwidth.

Figs. 11.8 and 11.9 illustrate the Pareto boundaries for a multi-cell coordinated beamforming system with $(M, K) = (4, 2)$, under two random channel realizations. The points are generated using the method in [40, Corollary 2]. The rate tuples achieved by the above two-step optimization schemes and the SLNR-based scheme are also shown. As we can see, the rate tuples of the two proposed schemes are both located on the Pareto boundary. In contrast to the SLINR scheme, as shown in Fig. 11.8, the rate tuple of the

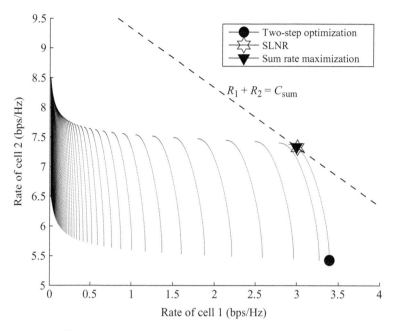

Figure 11.9 The rate points of the multi-cell schemes and the Pareto boundary under a random channel realization.

proposed centralized scheme locates exactly at the intersection point between the Pareto boundary and the line $R_1 = R_2$, while in the other case, shown in Fig. 11.9, the Pareto boundary has no intersection point with the line $R_1 = R_2$, implying that the max–min optimal point does not operate on the Pareto boundary, but the rate tuple of the two-step optimization scheme is still on the Pareto boundary, and is the closest point to the line $R_1 = R_2$. This means that the point achieved by the two-step optimization scheme has a worst rate no less than any other points on the Pareto boundary. Fig. 11.10 also shows the *cumulative distribution function* (CDF) of the worst-UE rate results achieved by the above schemes. It can be seen that at all outage levels, the two-step optimization scheme outperforms the SLNR based scheme, i.e., the two-step optimization scheme guarantees a good max-min fairness.

11.6   The road ahead of network MIMO in HCN

Having introduced the basic concepts of network MIMO and shown its possible applications to both macrocell and HCNs, a critical question is how to implement them in practical HCN. Based on the fact that eNBs can be macrocell BSs, picocell BSs or RNs, practical issues include (i) which eNBs need to cooperate and how to form a cluster in a reasonable way, (ii) how to maintain low signaling overhead among cooperating eNBs, while still enabling effective cooperation and coordination, and (iii) how to guarantee moderate upgrade on the current network infrastructures. In the following, we point out

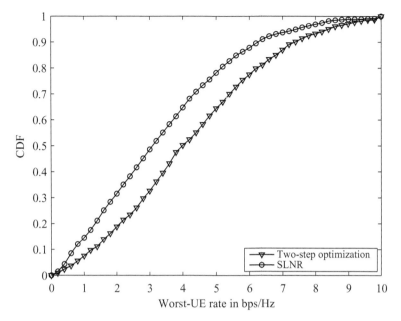

Figure 11.10 The CDF of the worst-UE rate achieved by the multi-cell beamforming schemes.

a number of concrete challenges that need to be tackled in the near future for successful MIMO implementations.

- Frequency and time synchronization is a problematic and complex issue, since in LTE there are different kinds of reference signals and the propagation delays in different cells are distinct.
- Approaches to acquire CSI at the transmitter side in FDD systems with minimum efforts, and design codebooks with effective transmit beamforming vectors, are required.
- One additional cost is the consumed energy due to the backhaul requirements and additional signal processing at eNBs. This should be taken into account for the future system design.
- Advanced UE will be crucial to network MIMO. Most current studies assume that UEs need no or moderate updates. However, in LTE-Advanced, there is also ongoing work to improve UE architecture. If UEs can have more powerful signal processing abilities, this will definitely facilitate the implementation of network MIMO.
- To design more effective and distributed schemes with per-cell power constraints than that discussed in Section 11.5.
- To improve distributed precoding based on SLNR without data, CSI, or common information sharing among eNBs, to achieve the lowest signaling overhead.
- To design adaptive frequency reuse schemes suitable for CoMP transmission. The authors in [58] proposed a cooperative frequency reuse scheme, which divides the cell-edge area of each cell into two types of zone, and defines a frequency reuse rule to support CoMP transmission for UEs in these zones.

11.7   Summary and conclusions

The ever-increasing demand for high-rate HCNs has called for more advanced interference management schemes. Network MIMO, also known as MCP or CoMP, combines MIMO and cooperation techniques to provide new means to manage interference more intelligently, while increasing the spectral efficiency. This chapter has discussed the advantages of MCP over SCP, and introduced different categories of eNB cooperation and their application scenarios to HCNs. As application examples, we have shown details of how the technique of distributed coordinated beamforming, which approaches the performance of centralized solutions, can be applied to both macrocell and HCN scenarios. Practical implementation issues such as backhaul requirement, eNB clustering, and CSI acquisition have been discussed. Future research directions are outlined, concluding that significant efforts are still needed to explore the potential of network MIMO in HCNs.

References

[1] I. E. Telatar, Capacity of multi-antenna gaussian channels. *European Transactions on Telecommunications*, **10**:6 (1999), 585–595.

[2] 3rd Generation Partnership Project (3GPP), *Further Advancements for E-UTRA: Physical Layer Aspects*, technical report 36.814 (2008).

[3] A. J. Viterbi, A. M. Viterbi, K. S. Gilhousen and E. Zehavi, Soft handoff extends CDMA cell coverage and increases reverse link capacity. *IEEE Journal on Selected Areas in Communication*, **12**:8 (1994), 1281–1288.

[4] S. H. Ali and V. C. M. Leung, Dynamic frequency allocation in fractional frequency reused OFDMA networks. *IEEE Transactions on Wireless Communications*, **8**:8 (2009), 4286–4295.

[5] A. J. Viterbi, A. M. Viterbi, K. S. Gilhousen and E. Zehavi, Soft handoff extends CDMA cell coverage and increases reverse link capacity. *IEEE Journal on Selected Areas in Communication*, **12**:8 (1994), 1281–1288.

[6] A. Wyner, Shannon-theoretic approach to a Gaussian cellular multiple-access channel. *IEEE Transactions on Information Theory*, **40**:6 (1994), 1713–1727.

[7] S. V. Hanly and P. A. Whiting, Information-theoretic capacity of multi-receiver networks. *Telecommunications Systems*, **1**:1 (1993), 1–42.

[8] S. Jing, D. Tse, J. Soriaga, J. Hou, J. Smee and R. Padovani, Multi-cell downlink capacity with coordinated processing, *EURASIP Journal on Wireless Communications and Networking* (2008), 586878.

[9] 3GPP R1-082499, *Multi-Cell MIMO with Distributed Inter-Cell Interference Suppression for LTE-A Uplink* (Alcatel Shanghai Bell, Alcatel Lucent, 2008).

[10] 3GPP R1-082325, *Inter-Cell Interference Management and Network MIMO* (Samsung, 2008).

[11] 3GPP R1-083192, *Network MIMO for Downlink Transmission in LTE-Advanced* (Qualcomm Europe, 2008).

[12] 3GPP R1-082501, *Collaborative MIMO for LTE-A Downlink* (Alcatel Shanghai Bell, Alcatel Lucent, 2008).

[13] 3GPP R1-082469, *LTE-Advanced-Coordinated Multipoint Transmission/Reception* (Ericsson, 2008).

[14] D. Gesbert, S. Hanly, H. Huang, S. Shamai, O. Simeone and W. Yu, Multicell MIMO coopera-tive networks: a new look at interference. *IEEE Journal on Selected Areas in Communication*, **28**:9 (2010), 1380–1408.

[15] V. Jungnickel, A. Forck, S. Jaeckel, F. Bauermeister, S. Schiffermueller, S. Schubert, S. Wahls, L. Thiele, T. Haustein, W. Kreher *et al.*, Field trials using coordinated multi-point transmission in the downlink. In *Personal, Indoor and Mobile Radio Communication Workshops (PIMRC Workshops), 2010 IEEE 21st International Symposium* (2010), pp. 440–445.

[16] W. Yu and T. Lan, Transmitter optimization for the multi-antenna downlink with per-antenna power constraints. *IEEE Transactions on Signal Processing*, **55**:6 (2007), 2646–2660.

[17] O. Somekh, B. M. Zaidel and S. Shamai, Sum rate characterization of joint multiple cell-site processing. *IEEE Transactions on Information Theory*, **53**:12 (2007), 4473–4497.

[18] O. Somekh, O. Simeone, Y. Bar-Ness, A. M. Haimovich and S. Shamai, Cooperative multicell zero-forcing beamforming in cellular downlink channels. *IEEE Transactions on Information Theory*, **55**:7 (2009), 3206–3219.

[19] Y. Huang, L. Yang and J. Liu, A limited feedback SDMA for downlink of multiuser MIMO communication system. *EURASIP Journal on Advances in Signal Processing* (2008), 947849.

[20] M. Schubert and H. Boche, Solution of the multiuser downlink beamforming problem with individual SINR constraints. *IEEE Transactions on Vehicular Technology*, **53**:1 (2004), 18–28.

[21] P. Viswanath and D. N. C. Tse, Sum capacity of the vector Gaussian broadcast channel and uplink–downlink duality. *IEEE Transactions on Information Theory*, **49**:8 (2003), 1912–1921.

[22] M. Schubert and H. Boche, A generic approach to QoS-based transceiver optimization. *IEEE Transactions on Communications*, **55**:8 (2007), 1557–1566.

[23] F. Rashid-Farrokhi, K. J. R. Liu and L. Tassiulas, Transmit beamforming and power control for cellular wireless systems. *IEEE Journal on Selected Areas in Communication*, **16**:8 (1998), 1437–1450.

[24] R. D. Yates, A framework for uplink power control in cellular radio systems. *IEEE Journal on Selected Areas in Communication*, **13**:7 (1995), 1341–1347.

[25] A. Wiesel, Y. C. Eldar and S. Shamai, Linear precoding via conic optimization for fixed MIMO receivers. *IEEE Transactions on Signal Processing*, **54**:1 (2006), 161–176.

[26] O. Somekh, B. M. Zaidel and S. Shamai, Sum rate characterization of joint multiple cell-site processing. *IEEE Transactions on Information Theory*, **53**:12 (2007), 4473–4497.

[27] W. Choi and J. G. Andrews, The capacity gain from intercell scheduling in multi-antenna systems. *IEEE Transactions on Wireless Communications*, **7**:2 (2008), 714–725.

[28] M. Costa, Writing on dirty paper. *IEEE Transactions on Information Theory*, **29**:3 (1983), 439–441.

[29] M. Tomlinson, New automatic equalizer employing modulo arithmetic. *Electronics Letters*, **7**:5 (1971), 138–139.

[30] J. Qiu, R. Zhang, Z. Luo and S. Cui, Optimal distributed beamforming for MISO interference channels. *IEEE Transactions on Signal Processing*, **59**:11 (2011), 5638–5643.

[31] O. Simeone, O. Somekh, H. V. Poor and S. Shamai, Downlink multicell processing with limited backhaul capacity. *EURASIP Journal on Advances in Signal Processing* (2009), 840814.

[32] 3GPP R1-090140, *Clustering for CoMP Transmission* (Nortel).

[33] 3GPP R1-111282, *Performance Evaluation of CoMP JT for Scenario 2* (Samsung, 2011).

[34] 3GPP R1-111290, *CoMP Phase 1 Evaluation Results* (ZTE, 2011).

[35] 3GPP R1-111277, *CoMP JT Evaluation for Phase I Homogenous Deployment* (Texas Instruments, 2011).

[36] S. Sesia, I. Toufik and M. Baker, eds., *LTE: the UMTS Long Term Evolution* (Wiley, 2009).

[37] D. J. Love, R. W. Heath, V. K. N. Lau, D. Gesbert, B. D. Rao and M. Andrews, An overview of limited feedback in wireless communication systems. *IEEE Journal on Selected Areas in Communication*, **26**:8 (2008), 1341–1365.

[38] R. Zakhour and D. Gesbert, Team decision for the cooperative MIMO channel with imperfect CSIT sharing. In *Information Theory and Applications Workshop, 2010*, San Diego, pp. 1–6.

[39] E. Larsson and E. Jorswieck, Competition versus cooperation on the MISO interference channel. *IEEE Journal on Selected Areas in Communication*, **26**:7 (2008), 1059–1069.

[40] E. Jorswieck, E. Larsson and D. Danev, Complete characterization of the Pareto boundary for the MISO interference channel. *IEEE Transactions on Signal Processing*, **56**:10 (2008), 5292–5296.

[41] E. Bjornson, R. Zakhour, D. Gesbert and B. Ottersten, Cooperative multicell precoding: rate region characterization and distributed strategies with instantaneous and statistical CSI. *IEEE Transactions on Signal Processing*, **58**:8 (2010), 4298–4310.

[42] B. L. Ng, J. Evans, S. Hanly and D. Aktas, Distributed downlink beamforming with cooperative base stations. *IEEE Transactions on Information Theory*, **54**:12 (2008), 5491–5499.

[43] H. Dahrouj and W. Yu, Coordinated beamforming for the multicell multi-antenna wireless systems. *IEEE Transactions on Wireless Communications*, **9**:5 (2010), 1748–1795.

[44] B. O. Lee, H. W. Je, I. Sohn, O.-S. Shin and K. B. Lee, Interference-aware decentralized precoding for multicell MIMO TDD systems. In *IEEE Global Communications Conference*, New Orleans, LA (2008).

[45] N. Hassanpour, J. Smee, J. Hou and J. Soriaga, Distributed beamforming based on signal-to-caused-interference ratio. In *Proceedings of the IEEE International Symposium on Spread Spectrum Techniques and Applications*, vol. 8, Bologna, Italy (2008), pp. 405–410.

[46] M. Bengtsson and B. Ottersten, Optimal and suboptimal transmit beamforming. In *Handbook of Antennas in Wireless Communications*, ed. L. C. Godara (CRC Press, 2001).

[47] Y. Huang, G. Zheng, M. Bengtsson, K. K. Wong, B. Ottersten and L. X. Yang, Distributed multicell beamforming with limited inter-cell coordination. *IEEE Transactions on Signal Processing*, **59**:2 (2011), 728–738.

[48] W. W. L. Hoy, T. Q. S. Quek, S. Sun and J. R. W. Heath, Decentralized precoding for multicell MIMO downlink. *IEEE Transactions on Wireless Communications*, **10**:6 (2011), 1798–1809.

[49] R. Zakhour and D. Gesbert, Coordination on the MISO interference channel using the virtual SINR framework. In *Proceedings of the International ITG Workshop on Smart Antennas*, Berlin (2009).

[50] W. Yang and G. Xu, Optimal downlink power assignment for smart antenna systems. In *Proceedings of the IEEE International Conference on Acoustics, Speech, and Signal Processing* (1998).

[51] S. Ulukus and R. D. Yates, Adaptive power control and multiuser interference suppression. *ACM Wireless Networks*, **4**:6 (1998), 489–496.

[52] C. W. Tan, M. Chiang and R. Srikant, Maximizing sum rate and minimizing MSE on multiuser downlink: optimality, fast algorithms, and equivalence via max–min SIR. In *Proceedings of the IEEE International Symposium on Information Theory*, Seoul, South Korea (2009), pp. 2669–2673.

[53] R. D. Yates, A framework for uplink power control in cellular radio systems. *IEEE Journal on Selected Areas in Communication*, **13**:7 (1995), 1341–348.

[54] C. D. Meyer, *Matrix Analysis and Applied Linear Algebra* (SIAM, 2000).

[55] R. Horn and C. Johnson, *Matrix Analysis* (Cambridge University Press, 1985).

[56] H. Boche and M. Schubert, A superlinearly and globally convergent algorithm for power control and resource allocation with general interference functions. *IEEE/ACM Transactions on Networking*, **16**:2 (2008), 383–395.

[57] Y. Huang, G. Zheng, M. Bengtsson, L. Y. K. K. Wong and B. Ottersten, Distributed multicell beamforming design approaching Pareto boundary with Max–Min fairness. *IEEE Transactions on Wireless Communications*, **11**:8 (2012), 2921–2933.

[58] J. Li, H. Zhang, X. Xu, X. Tao, T. Svensson, C. Botella and B. Liu, A novel frequency reuse scheme for coordinated multi-point transmission. In *IEEE 71st Vehicular Technology Conference*, Taipei (2010), pp. 16–19.

12 Network coding

Haishi Ning and Cong Ling

12.1 Introduction

The main motivation behind using network coding is the wireless broadcast nature, which means that every other node can potentially overhear the signal transmitted by one node. Conventionally, the overheard signal is treated as noise or interference, and thus completely ignored. However, as shown in [1], a smartly controlled interference can be used to greatly improve the total network throughput. While interference is harmful in a conventional perspective, if a node has previously transmitted or overheard the interference, its detrimental effects can be completely removed to increase the chance of conveying more information in a single transmission.

Network coding was initially proposed in [2] to achieve the multicast capacity of a single-session multicast network by permitting intermediate nodes to encode the received data in addition to traditional routing operations. For a single-session multicast network, it was shown in [3] that linear codes are sufficient to achieve the multicast capacity. A polynomial time algorithm for network code construction was proposed in [4]. The distributed random linear code construction approach in [5] was shown to be asymptotically valid given a sufficiently large field size. For a multiple-session network, it was shown in [6, 7] that linear network coding may be insufficient to achieve the multicast capacity. Moreover, finding a network coding solution for a network with multiple sessions was shown to be an NP-hard problem [8, 9]. Although optimal network coding solutions for multiple-session networks are generally unknown, simple network coding solutions are able to offer tremendous throughput improvements for wireless cooperative networks, which was famously demonstrated by [1, 10–12].

12.2 Coding opportunities in heterogenous cellular networks

It has been proved that previously well known code construction approaches for a single-session multicast network would be optimal for a multiple-session network, if and only if we can find an edge-disjoint subgraph for each session, which can at the same time support the multicast capacity for each session. In [13], random coding was applied to multiple sessions after transforming the network topology to construct pollution-free subgraphs. However, such qualified edge-disjoint subgraphs may not exist most of the time, because of the overlapping of edges that belong to multiple sessions. The benefits of overlapping among multiple sessions were studied in [14], where the authors proposed two metrics, overlap ratio and overlap width, to measure the benefits that a system can achieve by coding across multiple sessions. The main idea was to divide multiple sessions into different groups based on those two metrics and construct a linear network coding solution for each group. Simulation results in [14] showed that such a scheme can achieve about 30% higher throughput than coding within the same session.

Given the inherent difficulty of multiple-session network information flow problems, the authors in [15] considered network coding for only two unicast sessions and found conditions under which a linear network coding solution exists. The conditions under which there exists a linear network coding solution for two multicast sessions were proved in [16]. Since the butterfly network is well known to admit network coding gains, the authors in [17, 18] focused on decomposing the whole network into many butterfly-based structures and used linear programming to find a network coding solution.

Heterogeneous cellular networks (HCNs) are ideal to admit such simple yet great coding gains due to their dynamic structures with relay nodes, picocells, and femtocells overlaid on macrocells. To show its potential coding gain, we will propose several illustrative examples. Firstly, we restrict the coding operations to be linear for practically easy implementation, because this is often a sub-optimal yet tractable simplification. Second, our coding algorithm is built upon the routing layer and does not awake current sleeping nodes, i.e., if an intermediate node is made inactive by traditional routing algorithms, it remains inactive in our coding upon routing algorithm. In other words, we only use the active nodes, i.e., the nodes that need to transmit with or without network coding. This means that our algorithm uses network coding only on existing established routes to improve the transmission efficiency. The use of existing routes only makes our algorithm sub-optimal, because sometimes longer or more congested routes can actually provide more coding opportunities or higher network coding gains. However, such cross-layer approaches may need to consider the physical layer and the transport layer jointly and can make the analysis quite complicated. To our best knowledge, no practical scheme exists for general communication networks with arbitrary topology and multiple unicast sessions. Thus, we turn to a sub-optimal solution to build our coding upon routing algorithms between the network layer and the transport layer. The advantages of such sub-optimal solutions are similar to that of the *Open Systems Interconnection* (OSI) model, i.e., we can separately design either better routing algorithms or better coding opportunity detection algorithms to improve the network performance. Other advantages include easy implementation by inserting a coding layer between the network layer and the transport layer. This may be even more important in practice because the network performance can be improved by software modifications without replacing the expensive network core with network coding enabling devices.

In summary, our goal is to find linear network coding opportunity, gain, and solution upon routing for HCNs. Thus, routing is the underlying technology and we are aiming to explore possible linear coding opportunities along the existing routes to save bandwidth and improve efficiency without major change to the current network infrastructure.

12.2.1   An upper bound on coding gain without geometry consideration

Assume there are K unicast sessions, denoted by $f_1, f_2, \ldots, f_K$. Each unicast session has a unique source node S_k and a unique destination node D_k, for $k = 1, 2, \ldots, K$.

Each unicast session has its individual maxflow mincut bound C_k. This means that we can find a group of C_k edge-disjoint paths from S_k to D_k. There may be several groups of such paths, and the group with the minimum number of transmissions needs N_k transmissions. Assume the unicast sessions are ordered so that $N_1 \leq N_2 \leq \ldots \leq N_K$.

Denoting the set of intermediate nodes along the group of paths with the minimum number of transmissions as $\mathcal{R}_k$, we have

$$N_k = C_k + |\mathcal{R}_k|. \tag{12.1}$$

With only routing, it is obvious that the overall minimum number of transmissions needed for all the unicast sessions is

$$N_1 + N_2 + \cdots + N_K = (C_1 + |\mathcal{R}_1|) + (C_2 + |\mathcal{R}_2|) + \cdots + (C_K + |\mathcal{R}_K|)$$
$$= \sum_{i=1}^{K} C_i + \sum_{i=1}^{K} |\mathcal{R}_i|. \tag{12.2}$$

With network coding, we calculate the minimum possible number of transmissions, which corresponds to a coding gain upper bound, as follows:

1. For unicast session f_1 only, we need at least N_1 transmissions.
2. a. For unicast session f_2 given unicast session f_1 is completed, we need at least C_2 transmissions to inject the messages for unicast session f_2 into the network.
   b. Assume the C_1 messages emitted by S_1 and the C_2 messages emitted by S_2 can be overheard by D_2 and D_1 respectively through a genie without any extra transmission, then the nodes in the intersection of $\mathcal{R}_1 \cap \mathcal{R}_2$ can serve both unicast sessions by coding their messages together.
   c. Unicast session f_2 needs to use the nodes in $\mathcal{R}_2$. Now that the nodes in $\mathcal{R}_1 \cap \mathcal{R}_2$ have already been used, f_2 only needs $|\mathcal{R}_2 \backslash \mathcal{R}_1|$ more transmissions.
   d. Thus, for unicast session f_2, we need at least $C_2 + |\mathcal{R}_2 \backslash \mathcal{R}_1|$ transmissions.
3. Similarly, for unicast session f_k, the number of transmissions needed is at least

$$C_k + |\mathcal{R}_k \backslash \{\mathcal{R}_1, \mathcal{R}_2, \ldots, \mathcal{R}_{k-1}\}|, \tag{12.3}$$

where $k = 1, 2, \ldots, K$ and $\mathcal{R}_0$ is an empty set.

So, with network coding, the minimum possible number of transmissions needed is

$$N_1 + (C_2 + |\mathcal{R}_2 \backslash \mathcal{R}_1|) + (C_3 + |\mathcal{R}_3 \backslash \{\mathcal{R}_1, \mathcal{R}_2\}|)$$

$$+ \cdots + (C_K + |\mathcal{R}_K \backslash \{\mathcal{R}_1, \mathcal{R}_2, \ldots, \mathcal{R}_{K-1}\}|)$$

$$= (C_1 + |\mathcal{R}_1|) + (C_2 + |\mathcal{R}_2 \backslash \mathcal{R}_1|) + (C_3 + |\mathcal{R}_3 \backslash \{\mathcal{R}_1, \mathcal{R}_2\}|)$$

$$+ \cdots + (C_K + |\mathcal{R}_K \backslash \{\mathcal{R}_1, \mathcal{R}_2, \ldots, \mathcal{R}_{K-1}\}|)$$

$$= \sum_{i=1}^{K} C_i + (|\mathcal{R}_1| + |\mathcal{R}_2 \backslash \mathcal{R}_1| + |\mathcal{R}_3 \backslash \{\mathcal{R}_1, \mathcal{R}_2\}| + \cdots + |\mathcal{R}_K \backslash \{\mathcal{R}_1, \mathcal{R}_2, \ldots, \mathcal{R}_{K-1}\}|)$$

$$= \sum_{i=1}^{K} C_i + \left| \bigcup_{i=1}^{K} \mathcal{R}_i \right|. \tag{12.4}$$

Thus, we can upper bound the coding gain, which is defined as the ratio of the number of transmissions needed by routing and the number of transmissions needed by coding, as follows

$$\begin{aligned}
\eta &= \frac{\sum_{i=1}^{K} C_i + \sum_{i=1}^{K} |\mathcal{R}_i|}{\sum_{i=1}^{K} C_i + |\bigcup_{i=1}^{K} \mathcal{R}_i|} \\
&\leq \frac{\sum_{i=1}^{K} C_i + K|\bigcup_{i=1}^{K} \mathcal{R}_i|}{\sum_{i=1}^{K} C_i + |\bigcup_{i=1}^{K} \mathcal{R}_i|} \\
&\leq \frac{K|\bigcup_{i=1}^{K} \mathcal{R}_i|}{|\bigcup_{i=1}^{K} \mathcal{R}_i|} \\
&= K. \tag{12.5}
\end{aligned}$$

From the derivation of the coding gain upper bound, we observe that in order to achieve this upper bound, the following conditions need to be satisfied.

1. Each source's messages must be overheard by its unintended destinations.
2. All the unicast sessions should use the same intermediate nodes.
3. The number of intermediate nodes should be much larger than the sum of their individual maxflow mincut bounds.

Based on these observations, we now construct an example, which can achieve this coding gain upper bound asymptotically. As shown in Fig. 12.1, K unicast sessions share the same paths $(R_1, R_2, \ldots, R_N)$. With only routing, the minimum number of transmissions needed is $K(N + 1)$. With simple network coding, i.e., to mix the messages at R_1 and broadcast the mixed message at R_N, we only need $K + N$ transmissions. Thus, when the number of intermediate nodes is large, the coding gain is

$$\eta = \lim_{N \to \infty} \frac{K(N+1)}{K+N} = \lim_{N \to \infty} \frac{KN}{N} = K. \tag{12.6}$$

However, the achievable coding gain for an arbitrary network with multiple unicast sessions is related to its topology and connectivity.

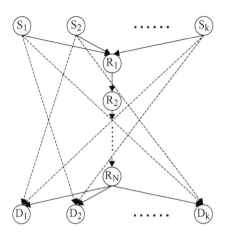

Figure 12.1 An example to asymptotically achieve the coding gain upper bound.

12.2.2   An upper bound on coding gain with geometry consideration

Lemma 12.1 *K unicast sessions can be coded together to provide a network coding gain if and only if their paths have at least one intersection node.*

Lemma 12.2 *With geometry consideration, the paths for the K unicast sessions have at most one intersection node.*

Lemma 12.3 *At most $K - 1$ transmissions can be saved if the paths for the K unicast sessions have one intersection node.*

Lemma 12.4 *K unicast sessions can be coded together to provide a network coding gain if and only if their routing solutions need at least $2K$ transmissions.*

Conjecture 12.1 *The coding gain upper bound with geometry consideration is 2.*

Proof. Assume there are K unicast sessions with distinct source and destination nodes. Following Lemma 12.4, if there is a network coding gain, routing solutions need $2K + \alpha$ transmissions. Following Lemma 12.1 and Lemma 12.2, the paths for the K unicast sessions have exactly one intersection node. Thus, by Lemma 12.3, the saving of transmissions is at most $K - 1$. So, the coding gain can be upper bounded by

$$
\begin{aligned}
\eta &\le \frac{2K + \alpha}{2K + \alpha - (K - 1)} \\
&= \frac{2K + \alpha}{K + \alpha + 1} \\
&\le \frac{2K}{K + 1} \\
&\le \frac{2K}{K} \\
&= 2.
\end{aligned}
\tag{12.7}
$$

$\square$

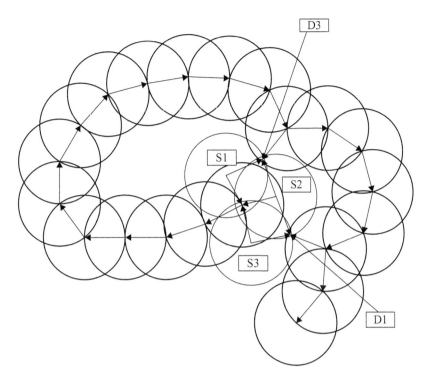

Figure 12.2 An example to show that K unicast sessions cannot have more than one intersection node.

12.2.3 Generalized butterfly network

Definition 12.1 *A level-K generalized butterfly network (level-K GBN) is defined as a network satisfying the following conditions.*

1. *There are K unicast sessions with K distinct source nodes and K distinct corresponding destination nodes.*
2. *The paths from the source nodes to their corresponding destination nodes decided by a chosen existing routing algorithm for these K unicast sessions have intersections at some common nodes.*
3. *There exist $(K-1)$ opportunistic paths from any subset of $(K-1)$ source nodes to each destination node, which do not pass through the intersections.*

We model the level-K GBN as a directed acyclic graph, $\text{GBN}_K = (\mathcal{V}, \mathcal{E})$, where $\mathcal{V}$ is the node set and $\mathcal{E}$ is the edge set. An edge e can be represented by an ordered node pair (x, y), where $x, y \in \mathcal{V}$, y is called the head of the edge, and x is called the tail of the edge. The messages can only be transmitted from x to y. The incoming edge set and the outgoing edge set of a node v are respectively defined as

$$\mathcal{E}_{\text{in}}(v) = \{(x, y) \mid (x, y) \in \mathcal{E}, y = v\},$$

$$\mathcal{E}_{\text{out}}(v) = \{(x, y) \mid (x, y) \in \mathcal{E}, x = v\}. \tag{12.8}$$

Let $\mathcal{S}$ denote the node set containing all source nodes, $\mathcal{D}$ denote the node set containing all destination nodes, $\mathcal{B}$ represent the node set containing the intersecting nodes of all unicast sessions of GBN_K, $\mathcal{R}$ represent the node set containing all the nodes involved in existing paths to accommodate the traffic demands for all unicast sessions, and $\mathcal{W}$ represent the node set containing all the nodes involved in the opportunistic paths for all the destination nodes. Then, $g = (K-1)\,|\,\mathcal{B}\,|$ and $p = |\,\mathcal{W} \setminus \mathcal{R}\,|$ are called the *network coding saving* and *network coding penalty*, respectively. The *saving of transmissions* (SoT) by using network coding compared to traditional routing is the network coding saving minus the network coding penalty, and can be written as

$$\text{SoT} = (g - p)^+, \tag{12.9}$$

where $(x)^+ = \max\{x, 0\}$. For a level-K GBN, network coding can save transmissions if SoT > 0; otherwise, network coding is not better than traditional routing.

Using graph-theoretic characterization, a level-K GBN is a directed acyclic graph $\mathrm{GBN}_K = (\mathcal{V}, \mathcal{E})$ containing K unicast sessions $(s_1 \rightarrow d_1), (s_2 \rightarrow d_2), \ldots, (s_K \rightarrow d_K)$, where $s_k \in \mathcal{S}$ and $d_k \in \mathcal{D}$ for $k \in \{1, \ldots, K\}$. The paths decided by a chosen existing routing algorithm for all unicast sessions travel through $b_i\ (\in \mathcal{B})$ for $i = 1, \ldots, |\,\mathcal{B}\,|$. For each destination, there are $(K-1)$ opportunistic paths from any subset of $(K-1)$ source nodes to it with edges chosen from $\mathcal{E}_{\mathrm{o}}\ (\in \mathcal{E} \setminus \mathcal{E}_{\mathrm{out}}(v))$, where $v \in \mathcal{B}$.

12.2.4   Necessary condition for network coding gain

Lemma 12.5 *K unicast sessions can be coded together to admit a network coding gain if these K unicast sessions can form a level-K GBN. In other words, a level-K GBN is necessary for K unicast sessions to be coded together to provide network coding gain.*

Proof. The proof is equivalent to showing that if we can apply network coding to these K unicast sessions, there must exist a level-K GBN. According to Definition 12.1, this is equivalent to showing that the following conditions must be satisfied.

1. There are K unicast sessions with K distinct source nodes and K distinct corresponding destination nodes.
2. The paths from the source nodes to their corresponding destination nodes decided by a chosen existing routing algorithm for these K unicast sessions have intersections at some common nodes.
3. There exist $(K-1)$ opportunistic paths from any subset of $(K-1)$ source nodes to each destination node, which do not travel through the intersection nodes.

First, we note that, if the existing paths for the multiple unicast sessions are disjoint from one another as shown in Fig. 12.3, then there is no network coding gain. This is obvious because, if there is no intersection among the existing paths, independent routing is sufficient to achieve the minimum number of transmissions, and network coding does not provide any gain. Thus, for a network to admit network coding gain, there must be at least two unicast flows intersecting with each other.

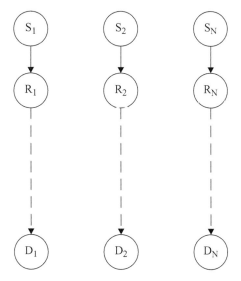

Figure 12.3 Edge disjoint multiple unicast flows.

Pick any K unicast flows, which can be coded together to admit network coding gain. The intersections mean these K unicast flows travel through the same set of nodes $\mathcal{B} = \{b_1, b_2, \ldots, b_{|\mathcal{B}|}\} \subset \mathcal{R}$. However, the network coding operation mixes the K unicast sessions' messages, i.e., the information conveyed by edges $e \ (\in \mathcal{E}_{\mathrm{out}}(v \in \mathcal{B}))$ is a function of the K unicast sessions' messages. The function is a linear function if we only consider linear network coding from practical implementation and low-complexity perspectives.

In order for each destination node to decode its desired information, there must be another $(K-1)$ opportunistic paths other than the existing path to construct a system of equations (to remove the interference). Such opportunistic paths should be independent of the main routes (and thus be independent of the intersections) in order to be able to help the destination nodes to construct non-degraded systems of equations (to remove the interference).

Based on the above argument, a network with multiple unicast sessions can be coded together to admit network coding gain if it satisfies all the conditions in Definition 12.1. Thus, it can now be concluded that if K unicast sessions can be coded together, then there must exist a level-K GBN. As a consequence, if there is no level-K GBN, not all the K unicast sessions can be coded together. □

Lemma 12.6 *The paths of two unicast sessions decided by traditional routing algorithms can only intersect once before reaching their destination nodes.*

Proof. We prove this theorem by using contradiction. Assume the two paths for two unicast sessions decided by traditional routing algorithms intersect more than once before reaching their destination nodes. Thus, after the first intersection, the two paths need to depart for the second intersection. However, traditional routing algorithms will choose the best possible path from the end of the first intersection to the start of the second intersection. This means that, between the end of the first intersection and the start of the second intersection, the two paths will travel through the same intermediate nodes. Thus, being apart after the first intersection is not possible, which contradicts the necessary condition for multiple intersections. So, we conclude that two unicast sessions' paths decided by traditional routing algorithms can intersect at most once before reaching their destination nodes.   □

Lemma 12.7 *A GBN can provide network coding gain if the network coding saving is larger than the network coding penalty.*

Proof. Consider a GBN satisfying all the conditions in Definition 12.1. As shown in Lemma 12.5, network coding gain comes from the congestion in the intersections (nodes in the node set $\mathcal{B}$) of the existing paths (nodes in the node set $\mathcal{R}$), which can be potentially resolved by coding several unicast sessions' messages together. The interference introduced by the network coding operation can be removed by the opportunistic paths (nodes in the node set $\mathcal{W}$) of the GBN. However, nodes in the opportunistic paths that are not used by the existing routes (nodes in the node set $\mathcal{W} \setminus \mathcal{R}$) must be used by paths for other unicast sessions decided by traditional routing algorithms. Otherwise, they are inactive and contradict the GBN conditions. As a result, in order to use network coding, these nodes in $\mathcal{W} \setminus \mathcal{R}$ must transmit some extra messages that are not needed in traditional routing algorithms. This is the penalty paid by using network coding.

Thus, the SoT using network coding compared to traditional routing, which is a characterization of the overall network coding gain, is the network coding saving minus the network coding penalty, and can be written as $\text{SoT} = (g - p)^+$.   □

12.2.5   Supporting examples

Example 12.1   A traditional butterfly network as shown in Fig. 12.4 is a level-2 GBN with the following definitions.

1. $\text{GBN}_2 = (\mathcal{V}, \mathcal{E})$ with two unicast sessions $S_1 \to D_1$ and $S_2 \to D_2$.
2. $\mathcal{S} = \{S_1, S_2\}$, $\mathcal{D} = \{D_1, D_2\}$, and $\mathcal{R} = \{S_1, S_2, D_1, D_2\}$.
3. $\mathcal{B} = \{R\}$, and $\mathcal{W} = \{S_1, S_2, D_1, D_2\}$.
4. $g = |\mathcal{B}| = 1$, and $p = |\mathcal{W} \setminus \mathcal{R}| = 0$.
5. $\text{SoT} = (g - p)^+ = 1$.

Example 12.2   A two-way exchange network as shown in Fig. 12.5 contains a level-2 GBN with the following definitions.

1. $\text{GBN}_2 = (\mathcal{V}, \mathcal{E})$ with two unicast sessions $A \to B$ and $B \to A$.
2. $\mathcal{S} = \{A, B\}$, $\mathcal{D} = \{A, B\}$, and $\mathcal{R} = \{A, R, B\}$.
3. $\mathcal{B} = \{R\}$, and $\mathcal{W} = \{A, B\}$.

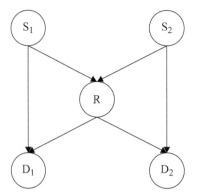

Figure 12.4 Traditional butterfly network.

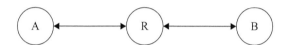

Figure 12.5 Two-way exchange network.

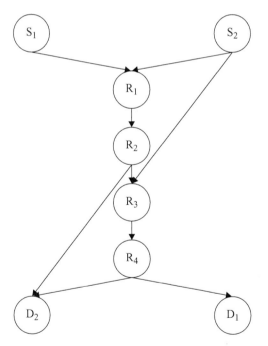

Figure 12.6 The grail network.

4. $g = |\mathcal{B}| = 1$, and $p = |\mathcal{W} \setminus \mathcal{R}| = 0$.
5. $\text{SoT} = (g - p)^+ = 1$.

Example 12.3 The grail network as shown in Fig. 12.6 is a level-2 GBN with the following definitions.

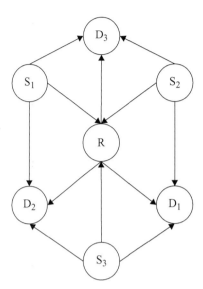

Figure 12.7 Three-user star network.

1. $GBN_2 = (\mathcal{V}, \mathcal{E})$ with two unicast sessions $S_1 \to R_3$ and $S_2 \to D_2$.
2. $\mathcal{S} = \{S_1, S_2\}$, $\mathcal{D} = \{R_3, D_2\}$, and $\mathcal{R} = \{S_1, S_2, R_1, R_2, R_3, R_4, D_1, D_2\}$.
3. $\mathcal{B} = \{R_1, R_2\}$, and $\mathcal{W} = \{S_2, R_3, R_4, D_2\}$.
4. $g = |\mathcal{B}| = 2$, and $p = |\mathcal{W} \setminus \mathcal{R}| = 0$.
5. $SoT = (g - p)^+ = 2$.

Example 12.4 The star network as shown in Fig. 12.7 is a level-3 GBN with the following definitions.

1. $GBN_3 = (\mathcal{V}, \mathcal{E})$ with three unicast sessions $S_1 \to D_1$, $S_2 \to D_2$, and $S_3 \to D_3$.
2. $\mathcal{S} = \{S_1, S_2, S_3\}$, $\mathcal{D} = \{D_1, D_2, D_3\}$, and $\mathcal{R} = \{S_1, S_2, S_3, R, D_1, D_2, D_3\}$.
3. $\mathcal{B} = \{R\}$, and $\mathcal{W} = \{S_1, S_2, S_3, R_1, R_2, R_3\}$.
4. $g = 2 |\mathcal{B}| = 2$, and $p = |\mathcal{W} \setminus \mathcal{R}| = 0$.
5. $SoT = (g - p)^+ = 2$.

Example 12.5 A network model with three source destination pairs is shown in Fig. 12.8, which contains a level-2 GBN with the following definitions.

1. $GBN_2 = (\mathcal{V}, \mathcal{E})$ with two unicast sessions $S_1 \to D_1$ and $S_2 \to D_2$.
2. $\mathcal{S} = \{S_1, S_2\}$, $\mathcal{D} = \{D_1, D_2\}$, and $\mathcal{R} = \{S_1, S_2, R_1, D_1, D_2\}$.
3. $\mathcal{B} = \{R_1\}$ and $\mathcal{W} = \{S_1, S_2, R_2, D_1, D_2\}$.
4. $g = |\mathcal{B}| = 1$ and $p = |\mathcal{W} \setminus \mathcal{R}| = 1$.
5. $SoT = (g - p)^+ = 0$.

Example 12.6 A wireless access network with two mobile *user equipments* (UEs) is shown in Fig. 12.9, which contains a level-2 GBN with the following definitions.

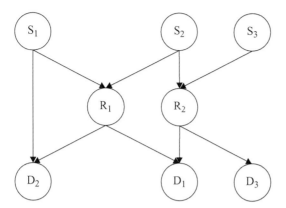

Figure 12.8 A network model with three source destination pairs.

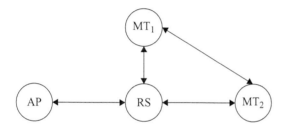

Figure 12.9 A wireless access network with two mobile UEs.

1. $GBN_2 = (\mathcal{V}, \mathcal{E})$ with two unicast sessions $AP \to MT_2$ and $MT_1 \to AP$.
2. $\mathcal{S} = \{AP, MT_1\}$, $\mathcal{D} = \{MT_2, AP\}$, and $\mathcal{R} = \{AP, RS, MT_1, MT_2\}$.
3. $\mathcal{B} = \{RS\}$, and $\mathcal{W} = \{AP, MT_1, MT_2\}$.
4. $g = |\mathcal{B}| = 1$, and $p = |\mathcal{W} \setminus \mathcal{R}| = 0$.
5. $SoT = (g - p)^+ = 1$.

Example 12.7 In the *point to point* (P2P) network shown in Fig. 12.10, the server has a download capacity of C_0, each of the n peers has an upload capacity of C_i ($i \in \{1, \dots, n\}$), and each peer's download capacity is assumed to be sufficiently large. This P2P network has been proved to have no network coding gain [19]. The basic idea is to first determine the throughput of traditional routing by solving a linear program, which finds explicit formulas in two cases based on the values of C_0 and $\sum_{i=1}^{n} C_i$. Then, for each case, network coding is shown to be unable to improve the throughput. We argue the same conclusion using our approach.

Note that our previous examples focus on directed wireless networks. For wired networks without the wireless broadcast advantage, the network coding gain should be defined as $g = (K - 1)(|\mathcal{B}| - 1)$, while the network coding penalty remains unchanged. Thus, no matter how many GBNs we can find, the number of intersecting nodes is no larger than unity. This means that, for such a P2P network, the network coding gain

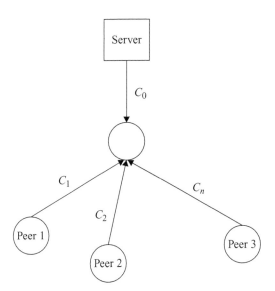

Figure 12.10 A P2P network.

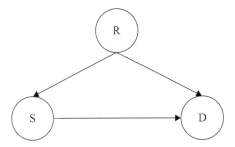

Figure 12.11 The single-relay network.

$g = (K - 1)(| \, \mathcal{B} \, | - 1) = 0$. Because we will later show that GBNs are the necessary element to admit network coding gain, we can conclude that such a P2P network does not have any coding advantage.

Example 12.8   The single-relay network and the single-cell cellular network models as shown in Fig. 12.11 and Fig. 12.12, respectively, do not contain any GBN, and thus do not have any network coding gain.

12.3   Efficiency and reliability

In order to analyze the efficiency and reliability performance of network coding in HCNs, we consider a special network topology called the *relay-aided X network* (RAXN). This network model is ubiquitous in practice because the source and destination nodes do not

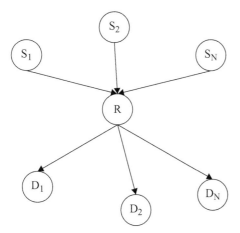

Figure 12.12 The single-cell cellular network.

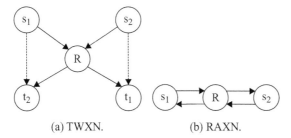

(a) TWXN.   (b) RAXN.

Figure 12.13 Illustrative channel models for the use of WNC in RAXN.

have to be the true communication end-users. It may happen as long as two traditional routing paths intersect at some point and share one or more intermediate relay nodes. Because of the shared use of resources, the throughput and reliability requirements are higher at the shared relay nodes. This motivates the development of new transmission strategies to meet the ever-increasing QoS requirements. Moreover, this network has combined features of primitive unit multiplexing topologies (e.g., broadcast channels and multiple-access channels) as well as properties of more complex interference channels. Thus, it requires interference management technologies to control the information flow in order to maximize the system performance.

In this section, we introduce a *wireless network coding* (WNC)-based partial interference cancelation strategy for RAXN. The main concept of WNC can be demonstrated using the *two-way exchange network* (TWXN) as shown in Fig. 12.13(a), where nodes s_1 and s_2 want to exchange information through the help of a relay node R. Without WNC, the conventional hop-by-hop transmission strategy needs four transmissions including $s_1 \rightarrow R$, $R \rightarrow s_2$, $s_2 \rightarrow R$, and $R \rightarrow s_1$. With WNC, only two transmissions are needed, i.e., $(s_1, s_2) \rightarrow R$ and $R \rightarrow (s_1, s_2)$.

Interference cancellation (IC) is often used in the decoding process of WNC to retrieve the desired signal [1, 11, 12, 20], by using a priori knowledge to reconstruct

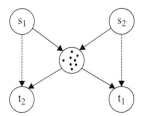

Figure 12.14 Relay-aided X network with a cluster of N relays.

the interference signal. The a priori information comes from either a node's previously transmitted information or its overheard information. While it is reasonable to assume a node's previously transmitted information to be perfect (without fading or noise), it is less so to assume the overheard information to be lossless.

As pointed out in [12], for a simple RAXN as shown in Fig. 12.13(b), WNC increases the network throughput at the expense of higher *bit error rate* (BER), because of imperfect overhearing. Since one can always trade a strategy's achievable diversity gain for its achievable multiplexing gain [21], it is not immediately clear whether WNC or the conventional hop-by-hop transmission strategy is fundamentally better. This question can be equivalently interpreted as follows: for an RAXN with imperfect overhearing, if we allow the conventional hop-by-hop transmission strategy to have a higher BER than that of WNC, can it transmit as fast as WNC? Or, if we force WNC to use more redundancy to reduce its BER to be the same as that of the conventional hop-by-hop transmission strategy, can its throughput still be higher? This fundamental problem needs to be investigated before the prevalent use of WNC.

Due to the problems discussed above, WNC for RAXN is not a straightforward extension of that for TWXN, due to imperfect wireless overhearing. Thus, the main objective of this section is to study how WNC can be used to improve the performance of RAXN with imperfect overhearing, and its fundamental performance in terms of the *diversity and multiplexing tradeoff* (DMT) as shown in Fig. 12.14. We propose a WNC-based partial IC strategy for RAXN, which could smartly use the imperfect overhearing to cancel part of the interference. The DMT analysis of the strategy will prove the fundamental superiority of WNC over the conventional hop-by-hop transmission strategy even with imperfect overhearing.

Notations: hereafter, we use $S = \{s_1, s_2\}$ to denote the two sources, $\mathcal{T} = \{t_1, t_2\}$ to denote the two destinations, and $R = \{R_1, R_2, \ldots, R_N\}$ to denote the N relays. We use x_{s_1} and x_{s_2} to denote the signals transmitted from the two sources, respectively, and x_{R_n} to denote the signal transmitted from the nth relay, for $n \in \{1, \ldots, N\}$. Similarly, y_{R_n} represents the received signal at the nth relay, and y_{t_1} and y_{t_2} denote the received signals at the two destinations, respectively.

Every node is constrained with an average energy E. All sources transmit independent information at the same rate R. The channel gains between the mth source and the nth relay, between the mth source and the kth destination, and between the nth relay and the kth destination are denoted by h_{s_m, R_n}, h_{s_m, t_k}, and h_{R_n, t_k}, respectively, where

$1 \leq m \neq k \leq 2$ and $n \in \{1, \ldots, N\}$. We assume that the physical links are all quasi-static flat Rayleigh-fading, i.e., the channel gains are constant during each frame but may change from one frame to the next.

We characterize the channels between the sources and their unintended destinations using the amount of information overheard by the destinations from their undesired sources in the first time slot. In order to evaluate the performance of our proposed WNC based partial IC strategy when the overheard information is imperfect, we assume that, in the first time slot, each destination can only decode part of its undesired source's information correctly, i.e., t_1 can decode R_{t_1} amount of information from s_2 correctly, and t_2 can decode R_{t_2} amount of information from s_1 correctly, where $0 \leq R_{t_1}, R_{t_2} \leq R$.

Let $d_1 = 1 - Ar_1$ and $d_2 = 1 - Br_2$ be two linear functions that denote the DMTs of two independent messages, where A and B are two constants, d_1 and d_2 are two diversity gains, and r_1 and r_2 are two multiplexing gains. The overall DMT is obtained by adding up the multiplexing gains subject to equal diversity gains, and can be written as

$$d = 1 - \frac{AB}{A+B}r. \tag{12.10}$$

12.3.1 Issues of naïve interference cancellation

Firstly consider the simple RAXN shown in Fig. 12.13(b), where s_1 and s_2 want to send independent information to t_1 and t_2, respectively, through the help of relay R. With WNC, the signalling is as follows.

1. In the first time slot, s_1 transmits x_{s_1} and s_2 transmits x_{s_2} simultaneously. The received signal at the other nodes are

$$y_{t_1,1} = h_{s_2,t_1}x_{s_2} + z_{t_1,1},$$

$$y_{t_2,1} = h_{s_1,t_2}x_{s_1} + z_{t_2,1},$$

$$y_{R,1} = h_{s_1,R}x_{s_1} + h_{s_2,R}x_{s_2} + z_{R,1}. \tag{12.11}$$

2. In the second time slot, R amplifies and forwards its previously received signal to both t_1 and t_2, whose received signals are

$$y_{t_1,2} = h_{R,t_1}\beta_R h_{s_1,R}x_{s_1} + h_{R,t_1}\beta_R h_{s_2,R}x_{s_2} + h_{R,t_1}\beta_R n_{R,1} + z_{t_1,2},$$

$$y_{t_2,2} = h_{R,t_2}\beta_R h_{s_1,R}x_{s_1} + h_{R,t_2}\beta_R h_{s_2,R}x_{s_2} + h_{R,t_2}\beta_R n_{R,1} + z_{t_2,2}, \tag{12.12}$$

where β_R denotes the scaling factor used at the relay to satisfy the energy constraint. In order to retrieve the desired signal at each destination, there are two possible approaches to perform IC using the overheard signal. Take destination t_1 for instance.

1. Without decoding the overheard information, IC can be done by

$$\tilde{y}_{t_1} = y_{t_1,2} - \frac{h_{R,t_1}\beta_R h_{s_2,R}}{h_{s_2,t_1}}y_{t_1,1}$$

$$= h_{R,t_1}\beta_R h_{s_1,R}x_{s_1} + \tilde{z}_{t_1}, \tag{12.13}$$

where $\tilde{z}_{t_1} = \frac{h_{R,t_1}\beta_R h_{s_2,R}}{h_{s_2,t_1}} z_{t_1,1} + z_{t_1,2}$. It is easy to verify that the accumulative noise variance is infinitely large, because $\frac{1}{x}$ is not integrable for an exponentially distributed variable x. This fact clearly prohibits the use of such an IC method.

2. With decoding the overheard information, IC can be done by

$$\tilde{y}_{t_1} = y_{t_1,2} - h_{R,t_1}\beta_R h_{s_2,R}\hat{x}_{s_2}$$

$$= h_{R,t_1}\beta_R h_{s_1,R}x_{s_1} + h_{R,t_1}\beta_R h_{s_2,R}(x_{s_2} - \hat{x}_{s_2}) + z_{t_1,2}, \qquad (12.14)$$

where $\hat{x}_{s_2}$ represents the estimated overheard information. Ideally, we would like the overhearing links to be reliable enough such that $x_{s_2} = \hat{x}_{s_2}$. However, due to wireless fading and noise corruption, it is highly possible for the decoded overheard information to be subject to errors, and it is not reasonable to assume more resources (e.g., coding or retransmission requests) to be used to ensure the overhearing reliability. Thus, such an IC method would lead to error propagation, which will result in incorrect decoding of the desired information.

From the above analysis, when the overhearing is imperfect, in order to use WNC for IC, we must remove noise and avoid error propagation at the same time. In the next section, we will propose a WNC-based partial IC strategy to solve these two problems simultaneously.

12.3.2   WNC-based partial interference cancellation strategy

Consider the RAXN as shown in Fig. 12.14; we propose the WNC-based partial IC strategy as follows.

1. There are two time slots in each transmission frame.
2. In the first time slot, s_1 and s_2 broadcast their independent information x_{s_1} and x_{s_2} respectively and simultaneously to the N shared relays and their unintended destinations.
3. The destinations decode the overheard signal and store the decoded information (which may contain errors) in their memory stacks.
4. In the second time slot, the N shared relays normalize the signals they have received in the first time slot and broadcast the normalized signal simultaneously to the two destinations.
5. Each destination exploits physical layer hints to divide its decoded overheard information in the first time slot into clean and faulty parts, and use the clean part to reconstruct and cancel part of the interference from the received signal in the second time slot to retrieve their desired information.

The main difference between this WNC-based partial IC strategy and the conventional WNC strategy has two folds, which include its "activeness" and its novel approach to use imperfect overhearing to improve the exchange performance. The activeness means that the relays actively and intentionally mix the two sources' signals no matter whether the overheard information is perfect or not. Conventional WNC falls back to the hop-by-hop transmission strategy when the overhearing is imperfect (because in this case interference cannot be canceled to retrieve the desired signal), even if there are only a few incorrectly overheard symbols. We will show that the WNC-based partial IC strategy is able to improve the overall system performance by smartly using the available but imperfect overhearing to cancel part of the interference.

12.3.3   Practical considerations

12.3.3.1   Channel side information

We assume that channel side information is only available at the receivers, i.e., *receiver-side channel state information* (CSIR), which is practicable by inserting a negligibly short training sequence into the message sequences. Moreover, we let the relays broadcast their estimated CSIR and normalization factors (used to satisfy the energy constraint) by embedding them into the training sequences with a negligible overhead compared to the original message length.

12.3.3.2   Synchronization

We do not consider the synchronization issue and assume that all the relays are fully synchronized. The synchronization issue arising from the simultaneous relaying operations can be overcome by a distributed relay selection algorithm, which chooses only the best relay to forward its received signal from the sources in the first time slot. Moreover, from (12.32) and (12.33) to be presented later, the performance of simultaneous relaying is dominated by the best two-hop link between the sources and the corresponding destinations. Thus, a distributed relay selection algorithm does not entail a cost on the achievable DMT. However, we only consider simultaneous relaying in this chapter for mathematical simplicity. The same asymptotic DMT performance can be achieved with a suitable distributed relay selection algorithm such as that in [22].

12.3.3.3   Decoding at the destinations

For the source-to-relay and relay-to-destination links that suffer deep fading, it is feasible to use some physical layer error correction codes or even higher-layer protocols to ensure they are error-free. However, for the overheard links, it is infeasible in practice to spend extra resources to ensure their reliability. Thus, when the destinations try to decode their overheard information in the first time slot, the decoded packets may be imperfect with symbol errors. We divide the decoded overheard information at the destinations in the first time slot into two parts: one part with a high probability to be correct (i.e., clean overheard symbols), the other part with a high probability to be incorrect (i.e., faulty overheard symbols), so that we can use the clean overheard symbols to remove part of the interference in the received signal in the second time slot. Possible methods to mark the decoded symbols as clean or faulty include the use of many well known soft decoders or the confidence values calculated from the physical layer signal as shown in [23, 24]. The comparison between different marking methods and their associated error propagation effects are out of the scope of this chapter.

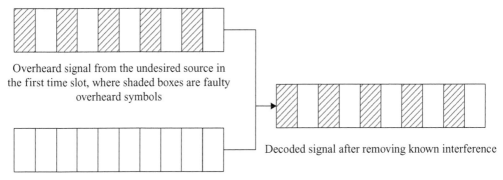

Overheard signal from the undesired source in the first time slot, where shaded boxes are faulty overheard symbols

Decoded signal after removing known interference

Received signal from the relays in the second time slot

Figure 12.15 Decoding at the destinations in the WNC based partial IC strategy.

When each destination receives the signal from the relays in the second time slot, it uses the clean overheard symbols together with the channel side information and the normalization factors to reconstruct part of the interference. Then, it subtracts such reconstructed interference from its received signal in the second time slot at the corresponding positions. This results in dividing the desired signal at the destinations into two parts: one part without interference and the other part with unknown interference. For the second part with unknown interference, we will use traditional decoding method for *medium access control* (MAC) to extract the desired information. Fig. 12.15 illustrates the decoding process at the destinations.

12.3.4   Diversity–multiplexing tradeoff analysis

The main result about the WNC-based partial IC strategy is summarized in the following theorem.

Theorem 12.1   *The achievable DMT lower bound for the proposed WNC-based partial IC strategy is*

$$d(r) > N\left[1 - \frac{2R(2R - R_{t_1} - R_{t_2})}{2R^2 - R_{t_1}^2 - R_{t_2}^2}r\right]. \tag{12.15}$$

Firstly, we analyze the DMT for the part of the signal with known interference. Since there is no difference in processing different symbols, we assume the first part of the received signal at relay R_n is a combination of two super-symbols. In the first time slot, the received signal vector at the N relays is

$$\mathbf{y}_\mathcal{R} = \mathbf{H}_{\mathcal{S},\mathcal{R}} \cdot \mathbf{x}_\mathcal{S} + \mathbf{z}_\mathcal{R}$$

$$= \begin{pmatrix} h_{s_1,R_1} & h_{s_2,R_1} \\ h_{s_1,R_2} & h_{s_2,R_2} \\ & \cdots \\ h_{s_1,R_N} & h_{s_2,R_N} \end{pmatrix} \cdot \begin{pmatrix} x_{s_1} \\ x_{s_2} \end{pmatrix} + \begin{pmatrix} z_{R_1} \\ z_{R_2} \\ \cdots \\ z_{R_N} \end{pmatrix}. \tag{12.16}$$

In the second time slot, the signal transmitted from relay R_n is

$$x_{R_n} = \beta_{R_n} \cdot y_{R_n}, \text{ for } n = 1, 2, \ldots, N. \tag{12.17}$$

Each relay chooses its energy normalization factor β_{R_n} based on its own received signal energy. We use an equal power allocation scheme with the average energy constraint E for each relay in the second time slot of one transmission frame. A more advanced power allocation scheme may enhance the performance in terms of throughput and outage probability in the low *signal to noise ratio* (SNR) regime, but such improvement becomes trivial in the high SNR regime, where a simple equal power allocation scheme is sufficient to achieve the same DMT as that of the optimal power allocation scheme.

The received signal at destination t_1 in the second time slot can be written as

$$y_{t_1} = \mathbf{H}_{\mathcal{R},t_1} \cdot \mathbf{x}_{\mathcal{R}} + z_{t_1} \tag{12.18}$$

where $\mathbf{H}_{\mathcal{R}} = [h_{R_1,t_1}, h_{R_2,t_1}, \ldots, h_{R_N,t_1}]$, $\mathbf{x}_{\mathcal{R}} = [x_{R_1}, x_{R_2}, \ldots, x_{R_N}]^H$, and $[\cdot]^H$ denotes the matrix conjugate transpose. From (12.16) and (12.17), $\mathbf{x}_{\mathcal{R}}$ can also be written as

$$
\begin{aligned}
\mathbf{x}_{\mathcal{R}} &= \boldsymbol{\beta} \cdot \mathbf{y}_{\mathcal{R}} \\
&= \begin{pmatrix} \beta_{R_1} & 0 & \cdots & 0 \\ 0 & \beta_{R_2} & \cdots & 0 \\ & & \cdots & \\ 0 & 0 & \cdots & \beta_{R_N} \end{pmatrix} \cdot \begin{pmatrix} y_{R_1} \\ y_{R_2} \\ \cdots \\ y_{R_N} \end{pmatrix}.
\end{aligned} \tag{12.19}
$$

Substituting (12.16) and (12.19) into (12.18), we get

$$
\begin{aligned}
y_{t_1} &= \mathbf{H}_{\mathcal{R},t_1} \cdot \mathbf{x}_{\mathcal{R}} + z_{t_1} \\
&= \mathbf{H}_{\mathcal{R},t_1} \cdot \boldsymbol{\beta} \cdot \mathbf{y}_{\mathcal{R}} + z_{t_1} \\
&= \mathbf{H}_{\mathcal{R},t_1} \cdot \boldsymbol{\beta} \cdot (\mathbf{H}_{\mathcal{S},\mathcal{R}} \cdot \mathbf{x}_{\mathcal{S}} + \mathbf{n}_{\mathcal{R}}) + z_{t_1} \\
&= [h_{R_1,t_1}, h_{R_2,t_1}, \ldots, h_{R_N,t_1}] \\
&\quad \cdot \begin{pmatrix} \beta_{R_1} & 0 & \cdots & 0 \\ 0 & \beta_{R_2} & \cdots & 0 \\ & & \cdots & \\ 0 & 0 & \cdots & \beta_{R_N} \end{pmatrix} \cdot \begin{pmatrix} h_{S_1,R_1} & h_{S_2,R_1} \\ h_{S_1,R_2} & h_{S_2,R_2} \\ & \cdots \\ h_{S_1,R_N} & h_{S_2,R_N} \end{pmatrix} \\
&\quad \cdot \begin{pmatrix} x_{S_1} \\ x_{S_2} \end{pmatrix} + \tilde{z}_{t_1}.
\end{aligned} \tag{12.20}
$$

Thus,

$$y_{t_1} = \sum_{n=1}^{N} h_{S_1,R_n} \beta_{R_n} h_{R_n,t_1} x_{S_1} + \sum_{n=1}^{N} h_{S_2,R_n} \beta_{R_n} h_{R_n,t_1} x_{S_2} + \sum_{n=1}^{N} h_{R_n,t_1} \beta_{R_n} n_{R_n} + z_{t_1}. \tag{12.21}$$

With CSIR and normalization factors received from the relays and estimated by the destinations, each destination can remove the known interference from its received signal in the second time slot. For destination t_1, $\sum_{n=1}^{N} h_{S_2,R_n} \beta_{R_n} h_{R_n,t_1} x_{S_2}$ is the known interference and thus can be removed from (12.21). Hence, we can write

$$y_{t_1} = \sum_{n=1}^{N} h_{S_1,R_n} \beta_{R_n} h_{R_n,t_1} x_{S_1} + \sum_{n=1}^{N} h_{R_n,t_1} \beta_{R_n} n_{R_n} + z_{t_1}. \qquad (12.22)$$

The accumulated noise at destination t_1 (including the part forwarded from the relays) can be written as

$$\tilde{z}_{t_1} = \sum_{n=1}^{N} h_{R_n,t_1} \beta_{R_n} n_{R_n} + z_{t_1}, \qquad (12.23)$$

where the normalization factor β_{R_n} is chosen to satisfy the energy constraint as follows

$$\begin{aligned}
|\beta_{R_n}|^2 &\leq \frac{E}{E|h_{S_1,R_n}|^2 + E|h_{S_2,R_n}|^2 + \sigma^2} \\
&= \frac{\rho}{\rho|h_{S_1,R_n}|^2 + \rho|h_{S_2,R_n}|^2 + 1},
\end{aligned} \qquad (12.24)$$

where σ^2 is the noise variance, and $\rho = E/\sigma^2$.

Let w_n denote the exponential order of $|\beta_{R_n}|^2$, and $v_{i,n}$ and $u_{n,j}$ denote the exponential orders of $\frac{1}{|h_{S_i,R_n}|^2}$ and $\frac{1}{|h_{R_n,t_j}|^2}$, respectively, for $i, j \in \{1, 2\}$ and $n \in \{1, \ldots, N\}$. Thus, from (12.24), we can easily see that

$$w_n \leq \min\{v_{1,n}, v_{2,n}, 1\}. \qquad (12.25)$$

In order to satisfy (12.25), we choose w_n as follows:

$$w_n = (v_{1,n}, v_{2,n})^-, \qquad (12.26)$$

where $(x)^- = \min\{x, 0\}$. This choice for w_n will ensure that β_{R_n} satisfies the average energy constraint (12.24), and the fact that outage events belonging to set O^+ will make w_n, i.e., the exponential order of β_{R_n}, vanish in all the DMT analytical expressions.

Let $w_{\tilde{z}_{t_1}}$ and $w_{z_{t_1}}$ denote the exponential orders of the variances of $\tilde{z}_{t_1}$ and z_{t_1}, respectively. From (12.23) and (12.26), we have

$$w_{\tilde{z}_{t_1}} = \max_{n=1,2,\ldots,N} \{(-u_{n,1})^+\} + w_{z_{t_1}} = w_{z_{t_1}}. \qquad (12.27)$$

Thus, the DMT of the strategy depends only on the channel matrix, but not on the variance of the accumulated noise. For analytical simplicity, we assume that the accumulated noise is equal to the noise at each destination, which does not affect the DMT analysis. Thus, we can rewrite (12.22) as

$$y_{t_1} = \sum_{n=1}^{N} h_{S_1,R_n} \beta_{R_n} h_{R_n,t_1} x_{S_1} + z_{t_1}, \qquad (12.28)$$

and y_{t_2} can be similarly rewritten as

$$y_{t_2} = \sum_{n=1}^{N} h_{S_2,R_n} \beta_{R_n} h_{R_n,t_2} x_{S_2} + z_{t_2}. \qquad (12.29)$$

From (12.28) and (12.29), we notice that they are very similar to *multiple-input single-output* (MISO) channels, and thus should have similar DMT characteristics. We obtain a lower bound of the DMT by approximating the exponential order of the error probability of a *maximum-likelihood* (ML) decoder with that of the outage probability. From the definition of the outage probability, we know that

$$P_{O_1} = P\left[I(x_{s_1}; y_{t_1}|x_{s_2}) < R_{t_1}\right]$$

$$= P\left[\log_2\left(1 + \rho \sum_{n=1}^{N} |h_{s_1,R_n}|^2 |\beta_{R_n}|^2 |h_{R_n,t_1}|^2\right) < r_{t_1}\log_2\rho\right], \quad (12.30)$$

$$P_{O_2} = P[I(x_{s_2}; y_{t_2}|x_{s_1}) < R_{t_2}]$$

$$= P\left[\log_2\left(1 + \rho \sum_{n=1}^{N} |h_{s_2,R_n}|^2 |\beta_{R_n}|^2 |h_{R_n,t_2}|^2\right) < r_{t_2}\log_2\rho\right]. \quad (12.31)$$

In the high SNR regime, the exponential order of β_{R_n} vanishes and thus we have

$$\lim_{\rho\to\infty} \frac{I(x_{s_1}; y_{t_1}|x_{s_2})}{\log_2\rho} = \lim_{\rho\to\infty} \frac{\log_2\left(1 + \rho \sum_{n=1}^{N} |h_{s_1,R_n}|^2 |\beta_{R_n}|^2 |h_{R_n,t_1}|^2\right)}{\log_2\rho}$$

$$= \max_{n=1,2,\ldots,N}\{1 - \upsilon_{1,n} - u_{n,1}\}^+, \quad (12.32)$$

$$\lim_{\rho\to\infty} \frac{I(x_{s_2}; y_{t_2}|x_{s_1})}{\log_2\rho} = \lim_{\rho\to\infty} \frac{\log_2\left(1 + \rho \sum_{n=1}^{N} |h_{s_2,R_n}|^2 |\beta_{R_n}|^2 |h_{R_n,t_2}|^2\right)}{\log_2\rho}$$

$$= \max_{n=1,2,\ldots,N}\{1 - \upsilon_{2,n} - u_{n,2}\}^+. \quad (12.33)$$

Thus, from (12.30), (12.31), (12.32), and (12.33), the outage events sets O^+ should be defined as

$$O_1^+ = \left\{(\mathbf{v}, \mathbf{u}) \in R^{2N+}| \max_{n=1,2,\ldots,N}\{1 - \upsilon_{1,n} - u_{n,1}\}^+ < r_{t_1}\right\}, \quad (12.34)$$

$$O_2^+ = \left\{(\mathbf{v}, \mathbf{u}) \in R^{2N+}| \max_{n=1,2,\ldots,N}\{1 - \upsilon_{2,n} - u_{n,2}\}^+ < r_{t_2}\right\}. \quad (12.35)$$

From (12.34) and (12.35), we can easily see that, in order for the outage events to happen, the following two constraints should be satisfied:

1. $\upsilon_{1,n} + u_{n,1} > 1 - r_{t_1}, \forall n \in \{1, 2, \ldots, N\}$.
2. $\upsilon_{2,n} + u_{n,2} > 1 - r_{t_2}, \forall n \in \{1, 2, \ldots, N\}$.

Since the outage probability would be dominated by the probability of the outage event with the largest exponential order, i.e., the outage event with the smallest $d_O(r)$, we can write

$$P_{O_1} \doteq \rho^{-d_{O_1}(r_{t_1})} \quad (12.36)$$

for $d_{O_1}(r_{t_1}) = \inf_{(\mathbf{v},\mathbf{u}) \in O^+} \left[\sum_{n=1}^{N} (v_{1,n} + u_{n,1}) \right]$, and

$$P_{O_2} \doteq \rho^{-d_{O_2}(r_{t_2})}$$

for $d_{O_2}(r_{t_2}) = \inf_{(\mathbf{v},\mathbf{u}) \in O^+} \left[\sum_{n=1}^{N} (v_{2,n} + u_{n,2}) \right]$.

Thus, we can lower-bound $d_{O_1}(r_{t_1})$ and $d_{O_2}(r_{t_2})$ as follows:

$$d_{O_1}(r_{t_1}) \geqslant \inf_{(\mathbf{v},\mathbf{u}) \in O^+} \left[\sum_{n=1}^{N} (v_{1,n} + u_{n,1}) \right] > N(1 - r_{t_1}), \qquad (12.37)$$

$$d_{O_2}(r_{t_2}) \geqslant \inf_{(\mathbf{v},\mathbf{u}) \in O^+} \left[\sum_{n=1}^{N} (v_{2,n} + u_{n,2}) \right] > N(1 - r_{t_2}). \qquad (12.38)$$

As $d_{O_1}(r_{t_1})$ and $d_{O_2}(r_{t_2})$ also serve as lower bounds for $d_1(r_{t_1})$ and $d_2(r_{t_2})$, respectively, we can further write

$$d_1(r_{t_1}) > N(1 - r_{t_1}), \qquad (12.39)$$

$$d_2(r_{t_2}) > N(1 - r_{t_2}). \qquad (12.40)$$

Now, we show the DMTs of $N(1 - r_{t_1})$ and $N(1 - r_{t_2})$ are actually also upper bounds for the parts of the received signal with known interference at the two destinations, respectively. Assume, in the first time slot, the two sources transmit their independent signals x_{s_1} (R_{t_1} bits) and x_{s_2} (R_{t_2} bits) reliably to the N relays. In practice, this may not be true due to the wireless fading environment and noise corruption. However, this assumption is sufficient to give a DMT upper bound for the part of the received signal with known interference. In the second time slot, let the N relays fully cooperate through a genie. Since the two destinations cannot cooperate, the best achievable performance is obtained by viewing the transmissions from the relays to the destinations in the second time slot as two channels.

Without any cooperation between the destinations, the DMTs of the two MISO channels are $N(1 - r_{t_1})$ and $N(1 - r_{t_2})$, respectively. Jointly considering two independent MISO channels together does not increase the diversity gain, because the two MISO channels in the second time slot are statistically independent and without any cooperation. Thus, the upper bounds are given by

$$d_1(r_{t_1}) < N(1 - r_{t_1}), \qquad (12.41)$$

$$d_2(r_{t_2}) < N(1 - r_{t_2}). \qquad (12.42)$$

Combining (12.39), (12.40), (12.41), and (12.42), we find that the DMTs of the strategy for the known interference part of the received signal at the two destinations are

$$d_1(r_{t_1}) = N(1 - r_{t_1}), \qquad (12.43)$$

$$d_2(r_{t_2}) = N(1 - r_{t_2}). \qquad (12.44)$$

In order to get the relationship between the diversity gain d and the multiplexing gain r, we need to map points in (12.43) from a coordinate system with r_{t_1} as x-axis to a new coordinate system with r as x-axis. Since $r_{t_1} = \frac{R_{t_1}}{R} r$, points $(0, N)$ and $(1, 0)$ are mapped to points $(0, N)$ and $(\frac{R_{t_1}}{R}, 0)$, respectively. Thus, in the new coordinate system, the DMT (12.43) changes to

$$d_1(r_{t_1}) = N \left(1 - \frac{R}{R_{t_1}} r \right), \tag{12.45}$$

and similarly we have

$$d_2(r_{t_2}) = N \left(1 - \frac{R}{R_{t_2}} r \right). \tag{12.46}$$

Finally, from (12.10) and taking the consumption of two time slots in one cooperation frame into consideration, we get the overall DMT for the known interference part of the received signal at both destinations as

$$d_I(r_{t_1}, r_{t_2}) = N \left[1 - \left(\frac{2R}{R_{t_1} + R_{t_2}} \right) \right], \tag{12.47}$$

where $d_I(r_{t_1}, r_{t_2})$ denotes the diversity gain of the known interference part of the received signal at both destinations.

Secondly, we analyze the DMT for the part of the received signal with unknown interference. For this part of the received signal, since no information is correctly overheard, we cannot remove the interference term from (12.21). Thus, we have no choice but to use traditional decoding methods for MAC.

Note that from the received signals at both destinations, we can at least extract $I(x_{s_1}; y_{t_1}) + I(x_{s_2}; y_{t_2})$ amount of desired information by treating interference as noise. Moreover, in the high SNR regime, for a two-user interference channel, the total achievable multiplexing gain is unity. Because the multiplexing gain for MAC $\{x_{s_1}, x_{s_2}\} \to y_{t_1}$ is also unity, we have $\lim_{\rho \to \infty} \frac{I(x_{s_1}, x_{s_2}; y_{t_1})}{\log_2 \rho} = \lim_{\rho \to \infty} \frac{I(x_{s_1}; y_{t_1}) + I(x_{s_2}; y_{t_2})}{\log_2 \rho}$, and it is sufficient to consider the interference unknown part of the received signal at both destinations as an N-by-1 MISO channel with capacity $I(x_{s_1}, x_{s_2}; y_{t_1}) = I(x_{s_2}; y_{t_1}) + I(x_{s_1}; y_{t_1} | x_{s_2})$ to give a DMT lower bound. In the high SNR regime, the exponential order of β_{R_n} vanishes and thus we have

$$\lim_{\rho \to \infty} \frac{I(x_{s_2}; y_{t_1})}{\log_2 \rho} = \lim_{\rho \to \infty} \frac{\log_2 \left(1 + \frac{\rho \sum_{n=1}^{N} |h_{s_2, R_n}|^2 |\beta_{R_n}|^2 |h_{R_n, t_1}|^2}{\rho \sum_{n=1}^{N} |h_{s_1, R_n}|^2 |\beta_{R_n}|^2 |h_{R_n, t_1}|^2 + 1} \right)}{\log_2 \rho}$$

$$= \left[\max_{n=1,2,\ldots,N} \{1 - v_{2,n} - u_{n,1}\} - \left(\max_{n=1,2,\ldots,N} \{1 - v_{1,n} - u_{n,1}\} \right)^+ \right]^+. \tag{12.48}$$

From the definition of outage probability, we have

$$
\begin{aligned}
P_O &\doteq P\left[I(x_{s_1}; y_{t_1}) + I(x_{s_2}; y_{t_2}|x_{s_1}) < R - R_{t_1} + R - R_{t_2}\right] \\
&= P\left[\log_2\left(1 + \frac{\rho \sum_{n=1}^{N} |h_{s_1,R_n}|^2 |\beta_{R_n}|^2 |h_{R_n,t_1}|^2}{\rho \sum_{n=1}^{N} |h_{s_2,R_n}|^2 |\beta_{R_n}|^2 |h_{R_n,t_1}|^2 + 1}\right) \right.\\
&\quad + \log_2\left(1 + \rho \sum_{n=1}^{N} |h_{s_2,R_n}|^2 |\beta_{R_n}|^2 |h_{R_n,t_2}|^2\right) \\
&\quad \left. < \left(r_{t_1}^c + r_{t_2}^c\right)\log_2 \rho \right],
\end{aligned}
\tag{12.49}
$$

where $r_{t_1}^c$ denotes the multiplexing gain of x_{s_1} at destination t_1 with $(R - R_{t_1})$ amount of information and unknown interference, and $r_{t_2}^c$ is similarly defined.

From (12.33), (12.48), and (12.49), the outage events set O^+ can be defined as

$$
\begin{aligned}
O^+ = \left\{(\mathbf{v}, \mathbf{u}) \in \mathbb{R}^{3N+} \right| &\left[\max_{n=1,2,\ldots,N}\{1 - v_{2,n} - u_{n,1}\}\right. \\
&\left. - \left(\max_{n=1,2,\ldots,N}\{1 - v_{1,n} - u_{n,1}\}\right)^+\right]^+ \\
&\left. + \left[\max_{n=1,2,\ldots,N}\{1 - v_{1,n} - u_{n,1}\}\right]^+ < r_{t_1}^c + r_{t_2}^c\right\}.
\end{aligned}
\tag{12.50}
$$

Thus, in order for the outage events to happen, the following constraints must be satisfied:

$$
v_{2,n} + u_{n,1} > 1 - r_{t_1}^c - r_{t_2}^c, \forall n \in \{1, 2, \ldots, N\}.
\tag{12.51}
$$

Therefore, we can lower-bound the DMT $d_{II}(r_{t_1}^c, r_{t_2}^c)$ of the interference unknown part of the received signal at both destinations as

$$
\begin{aligned}
d_{II}(r_{t_1}^c, r_{t_2}^c) &\geqslant d_O(r_{t_1}^c, r_{t_2}^c) \\
&\geqslant \inf_{(\mathbf{v},\mathbf{u})\in O^+}\left[\sum_{n=1}^{N}(v_{1,n} + u_{n,1} + v_{2,n})\right] \\
&> N\left(1 - r_{t_1}^c - r_{t_2}^c\right).
\end{aligned}
\tag{12.52}
$$

Because $r_{t_1}^c = \frac{R - R_{t_1}}{R}r$ and $r_{t_2}^c = \frac{R - R_{t_2}}{R}r$, following similar steps from (12.43) to (12.45), we can further write

$$
d_{II}\left(r_{t_1}^c, r_{t_2}^c\right) > N\left[1 - \left(\frac{R}{R - R_{t_1}} + \frac{R}{R - R_{t_2}}\right)r\right].
\tag{12.53}
$$

From (12.47), (12.53), and (12.10), we find that the overall achievable DMT for the WNC based partial IC strategy is lower-bounded by

$$
d(r) > N\left[1 - \frac{2R(2R - R_{t_1} - R_{t_2})}{2R^2 - R_{t_1}^2 - R_{t_2}^2}r\right].
\tag{12.54}
$$

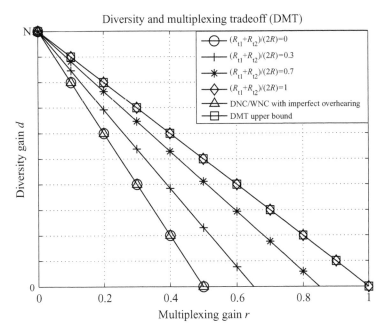

Figure 12.16 DMT of WNC-based partial IC strategy with imperfect overhearing.

As shown in Fig. 12.16, the more information overheard by the destinations, the better DMT the proposed WNC-based partial IC strategy can achieve. In the case of perfect overhearing, it can achieve the DMT upper bound, which is obtained by viewing the channel model as two two-hop fully cooperative MISO channels. In the case of imperfect overhearing, conventional WNC strategy falls back to hop-by-hop transmission strategy, while the WNC-based partial IC strategy can still improve the overall system performance by using the partially correctly overheard information to cancel part of the interference. This indicates that, despite introducing bi-directional interference to both destinations due to the imperfect overhearing, WNC in general improves the overall system throughput and robustness regardless of the quality of the overhearing links, if an appropriate approach is used to explore the imperfect overheard information.

Theorem 12.1 also answers the fundamental question that has motivated the research in this chapter: if WNC uses more redundancy to achieve the same BER as that of the conventional hop-by-hop transmission strategy, which strategy has higher throughput? Or, if the conventional hop-by-hop transmission strategy increases its rate to that of WNC, which one has lower BER? From Fig. 12.16, it can be easily seen that, setting either performance measurement (between the diversity and multiplexing gains that indicate the error and throughput performance, respectively) the same for the two strategies, the other performance measurement of WNC is always strictly better than that of the conventional hop-by-hop transmission strategy. This proves the fundamental superiority of WNC over conventional hop-by-hop transmission even with imperfect overhearing.

12.4   Construction of distributed coding solutions

We have shown in Lemma 12.6 that possible network coding gain comes from the GBNs, which are formed by path intersections of multiple unicast sessions. Thus, to find network coding gain and network coding solution is equivalent to checking whether there are GBNs with network coding gain around the path intersections of multiple unicast sessions.

Lemma 12.7 tells us that the paths of multiple unicast sessions may contain several GBNs, but they must have the same set of intersecting nodes $\mathcal{B}$. Since the network coding gain comes from the intersections of the main routes, i.e., nodes shared by routes of multiple unicast sessions, it is upper bounded by the cardinality of the intersecting node set $|\mathcal{B}|$. Thus, the task of finding network coding gain is equivalent to checking whether the network coding saving exceeds the minimum possible network coding penalty around the intersections, i.e., the minimum possible number of nodes in the opportunistic paths that are not used by the existing paths, $\min(|\mathcal{W} \setminus \mathcal{R}|)$.

The network coding solution can be easily obtained after finding the GBNs with network coding gain. The network coding operations should be performed at the starting node of the intersections in order to mix the messages, and at the ending nodes of the opportunistic routes in order to remove the interference.

Our goal is to design a distributed and easy to implement systematic scheme to crawl all the physical connected links of the network with multiple unicast sessions, using some probe information to find all the coding opportunities. In the following, we develop a powerful yet practical algorithm to detect coding opportunities and construct network coding solutions, without global knowledge of network topology or overwhelming overhead.

The distributed four-way handshaking coding opportunity detection algorithm is given as follows.

1. Let each source node send a probe packet in order to form the intersections among paths of the multiple unicast sessions.
   a. Each probe packet should contain the unicast session's identification and its source and destination addresses.
   b. Each intermediate node forwards its received probe packets using its own local forwarding table and the destination addresses in the received packets.
   c. Each intermediate node only forwards the probe packets that are intended for it.
   d. When an intermediate node detects a new unicast session traveling through it, it stores the unicast flow's identification in its own local buffer.
2. After each destination node receives the probe packet intended for it, it sends a detection packet back to the corresponding source node to detect the intersections.
   a. Each detection packet should contain the unicast session's identification and its source and destination addresses. Moreover, it also contains a designated field, called the intersection field, to record the intersections involving this unicast session.

    b. Each detection packet is sent along the reverse direction of the unicast session's original path.

    c. When a detection packet arrives at an intermediate node, the router checks whether it serves other unicast sessions by searching its local buffer for other unicast sessions' identifications.

       i. If the router serves a new unicast session, which is not present in the detection packet's intersection field, it updates the detection packet by adding an entry in the intersection field containing this intersecting unicast session's identification, current router's address, and number of intersections with this intersecting unicast session (unity in this case).

      ii. If the router serves a unicast flow, which is not new to the detection packet, it updates the detection packet by adding the current router's address to the intersection field and increasing the number of intersections with this intersecting unicast session by unity.

3. After each source node receives the detection packet, it knows exactly the intersecting unicast sessions' identifications, their intersections' addresses, and the number of intersections along their paths. If a source node sees other intersecting sessions along its paths, it sends an opportunistic packet to form the opportunistic paths.

    a. Each opportunistic packet should contain the unicast session's identification, all the intersecting unicast flows' identifications, and their corresponding intersections' addresses. Moreover, it also contains a designated field, called the path field, to record the path the opportunistic packet has traveled.

    b. When an intermediate node receives an opportunistic packet, it adds an entry containing the current router's address and the unicast flows it serves to the path field, and then forwards it to its next hops.

    c. Each intermediate node saves a copy of its received opportunistic packets in its own local buffer.

4. After each destination node receives the opportunistic packets, it sends back a collection packet to detect the network coding penalty.

    a. Each collection packet should contain the unicast session's identification, all the intersecting unicast sessions' identifications, and their intersections' addresses. Moreover, it also contains a designated field, called the penalty field, to record the minimum possible network coding penalty and the corresponding opportunistic paths for each intersecting unicast session.

    b. Each collection packet is sent along the reverse direction of the unicast session's original path.

    c. When an intermediate node receives a collection packet, it checks whether it serves any intersecting unicast session in the collection packet. If it finds any such intersecting unicast session, it adds a freeze flag to the collect packet to indicate that the penalty field information associated with this intersecting unicast flow is not changeable anymore.

    d. For the intersecting unicast sessions it does not serve, the intermediate node checks for each of them whether it has any path from the collection packet's source or the intersecting unicast session's source to itself that does not travel through their intersections along its original path, by examining the saved opportunistic packets in its own buffer.

  i. If it finds one or more such paths, it counts how many nodes in each such path serve neither the received collection packet's unicast session nor the intersecting unicast session.

  ii. It compares the smallest number of nodes counted in paths from the above step with the penalty value in the collection packet, and replaces the value of the penalty and the corresponding opportunistic path in the collection packet with this smallest number and its corresponding path if the smallest number is less than the penalty value in the collection packet. If there is no penalty or corresponding path in the collection packet, it simply adds an entry with this smallest number and its corresponding path to the penalty field (i.e., the initial record).

5. When the starting node of some intersections receives the collection packets for all the intersecting unicast sessions it serves, it calculates the SoT from using network coding. If there is any SoT offered by network coding, it adds a coding flag to the collection packet to indicate the network coding decision for this intersecting unicast session.

6. After each source node receives the collection packets, it checks whether there is a coding flag for each of the intersecting unicast flows. If it finds a coding flag, it sends a short control message along the opportunistic route for this intersecting unicast flow to indicate this coding decision.

Network coding solutions can be obtained in a distributed manner by the four-way handshaking algorithm described above. Network coding operations should be performed at the starting node of the intersections and at the ending nodes of the opportunistic paths.

12.5   Summary and conclusion

In this chapter, we first explain the fundamentals of network coding and its state-of-the-art development. Then, we give coding gain upper bounds with and without practical geometry considerations, as well as illustrative examples to show the performance gains and effectiveness of applying simple network coding solutions to HCNs. After that, we analyze the efficiency and reliability performance of network coding using a simple yet ubiquitous two-way relay-aided X network. Finally, we introduce a low-complexity and distributed coding solution construction method that is ideal for HCNs and other future all IP-based wireless networks.

Network coding is a powerful novel technique that smartly controls the interference from different source nodes to improve the overall network performance. Overlaying network coding techniques upon existing cellular networks provides a simple and economical way to tremendously enhance the network performance without too many hardware replacements. Moreover, it has many other advantages over traditional routing techniques beyond this chapter, such as network error correction and balanced network traffic, although many challenges come together with the promises. Therefore, network coding is no doubt one of the most popular and promising topics in both the communication and information theory societies nowadays.

References

[1] S. Katti, H. Rahul, W. Hu, D. Katabi, M. Médard and J. Crowcroft, XORs in the air: practical wireless network coding. In *Proceedings of the 2006 Conference on Applications, Technologies, Architectures, and Protocols for Computer Communications* (New York: ACM Press, 2006), pp. 243–254.

[2] R. Ahlswede, N. Cai, S.-Y. Li and R. Yeung, Network information flow. *IEEE Transformations on Information Theory*, **46**:4 (2000), 1204–1216.

[3] S.-Y. Li, R. Yeung and N. Cai, Linear network coding. *IEEE Transformations on Information Theory*, **49**:2 (2003), 371–381.

[4] S. Jaggi, P. Sanders, P. Chou, M. Effros, S. Egner, K. Jain and L. Tolhuizen, Polynomial time algorithms for multicast network code construction. *IEEE Transformations on Information Theory*, **51**:6 (2005), 1973–1982.

[5] T. Ho, M. Medard, R. Koetter, D. Karger, M. Effros, J. Shi and B. Leong, A random linear network coding approach to multicast. *IEEE Transformations on Information Theory*, **52**:10 (2006), 4413–4430.

[6] S. Riis, Linear versus non-linear boolean functions in network flow. In *38th Annual Conference on Information Sciences and Systems* (2004).

[7] R. Dougherty, C. Freiling and K. Zeger, Insufficiency of linear coding in network information flow. *IEEE Transformations on Information Theory*, **51**:8 (2005), 2745–2759.

[8] R. Koetter and M. Medard, An algebraic approach to network coding. *IEEE/ACM Transformation on Networking*, **11**:5 (2003), 782–795.

[9] A. R. Lehman and E. Lehman, Complexity classification of network information flow problems. In *Proceedings of the 15th Annual ACM-SIAM Symposium on Discrete Algorithms* (2004), pp. 142–150.

[10] Y. Wu, P. A. Chou and S. Y. Kung, *Information Exchange in Wireless Networks with Network Coding and Physical-Layer Broadcast*, Technical Report MSR-TR-2004-78 (2004).

[11] S. Zhang, S. C. Liew and P. P. Lam, Hot topic: physical-layer network coding. In *Proceedings of the 12th Annual International Conference on Mobile Computing and Networking* (New York: ACM Press, 2006), pp. 358–365.

[12] S. Katti, S. Gollakota and D. Katabi, Embracing wireless interference: analog network coding. In *Proceedings of the 2007 Conference on Applications, Technologies, Architectures, and Protocols for Computer Communications* (New York: ACM Press, 2007), pp. 397–408.

[13] Y. Wu, On constructive multi-source network coding. In *IEEE International Symposium on Information Theory* (2006), pp. 1349–1353.

[14] M. Yang and Y. Yang, A linear inter-session network coding scheme for multicast. *Seventh IEEE International Symposium on Network Computing and Applications* (2008), pp. 177–184.

[15] C.-C. Wang and N. B. Shroff, Beyond the butterfly – a graph-theoretic characterization of the feasibility of network coding with two simple unicast sessions. In *IEEE International Symposium on Information Theory* (2007), pp. 121–125.

[16] C.-C. Wang and N. B. Shroff, Intersession network coding for two simple multicast sessions. In *Proceedings of Annual Allerton Conference on Communications, Control, and Computing* (2007).

[17] N. Ratnakar, D. Traskov and R. Koetter, Approaches to network coding for multiple unicasts. In *International Zurich Seminar on Communications* (2006), pp. 70–73.

[18] D. Traskov, N. Ratnakar, D. Lun, R. Koetter and M. Medard, Network coding for multiple unicasts: an approach based on linear optimization. In *IEEE International Symposium on Information Theory* (2006), pp. 1758–1762.

[19] D. M. Chiu, R. Yeung, J. Huang and B. Fan, Can network coding help in P2P networks? In *Fourth International Symposium on Modeling and Optimization in Mobile, Ad Hoc and Wireless Networks* (2006), pp. 1–5.

[20] S. Katti, D. Katabi, H. Balakrishnan and M. Medard, Symbol-level network coding for wireless mesh networks. *ACM SIGCOMM Computer Communication Review*, **38**:4 (2008), 401–412.

[21] L. Zheng and D. Tse, Diversity and multiplexing: a fundamental tradeoff in multiple-antenna channels. *IEEE Transactions on Information Theory*, **49**:5 (2003), 1073–1096.

[22] Z. Ding, K. Leung, D. Goeckel and D. Towsley, On the study of network coding with diversity. *IEEE Transactions on Wireless Communications*, **8**:3, 1247–1259.

[23] T. L. L. Hanzo and B. Yeap, *Turbo Coding, Turbo Equalisation, and Spacetime Coding: for Transmission over Fading Channels* (Wiley, 2002).

[24] K. Jamieson and H. Balakrishnan, PPR: partial packet recovery for wireless networks. In *Proceedings of the 2007 Conference on Applications, Technologies, Architectures, and Protocols for Computer Communications* (New York: ACM Press, 2007), pp. 409–420.

13 Cognitive radio

Miguel López-Benítez

13.1 Introduction

Cognitive radio (CR) has recently become one of the most intensively studied paradigms in wireless communications. In its broadest sense, a CR can be thought of as an enhanced smart *software defined radio* (SDR). The terms SDR and CR were introduced by J. Mitola in 1992 [1] and 1999 [2], respectively. SDR, sometimes shortened to software radio, is generally a multi-band radio that supports multiple air interfaces and protocols, and is reconfigurable through software running on a *digital signal processor* (DSP), *field-programmable gate array* (FPGA), or general-purpose microprocessor [3]. CR, usually built upon an SDR platform, is a context-aware intelligent radio capable of autonomous reconfiguration by learning from and adapting to the surrounding communication environment [4]. CRs are capable of perceiving and sensing their *radio frequency* (RF) environment, learning about their radio resources, *user equipment* (UE), and application requirements, and adapting their configuration and behavior accordingly. From this definition, two main characteristics of CR can be identified: cognitive capability (ability to capture information and learn from the radio environment) and reconfigurability (which enables the transmitter parameters to be dynamically programmed and modified according to the radio environment).

An important specific application often associated with CR is *dynamic spectrum access* (DSA). DSA, despite being a broader concept [5–7], is commonly understood as the reutilization of licensed RF bands by unlicensed UEs provided that the legitimate licensed UEs are not using the reused frequencies at a given time or in a given region of space. The basic underlying principle of DSA is to permit unlicensed (secondary) UEs to access in an opportunistic and non-interfering manner some licensed bands temporarily/spatially unoccupied by licensed (primary) UEs. The DSA concept is illustrated in Fig. 13.1. Temporal spectrum opportunities arise when the primary system remains inactive (i.e., not transmitting) for a certain time interval. Secondary UEs take profit of these inactivity periods to opportunistically access the spectrum. As illustrated in Fig. 13.1(a), secondary transmissions do not overlap in time with the primary transmissions, and therefore primary and secondary UEs can coexist within the same coverage area. Spatial spectrum opportunities arise when a spectrum band is exploited by the primary system within a bounded area, thus enabling the reuse of the same band by secondary UEs well outside this area. As shown in Fig. 13.1(b), secondary communications can overlap in time with the primary transmissions as long as the reuse distance is large enough to guarantee interference-free operation.

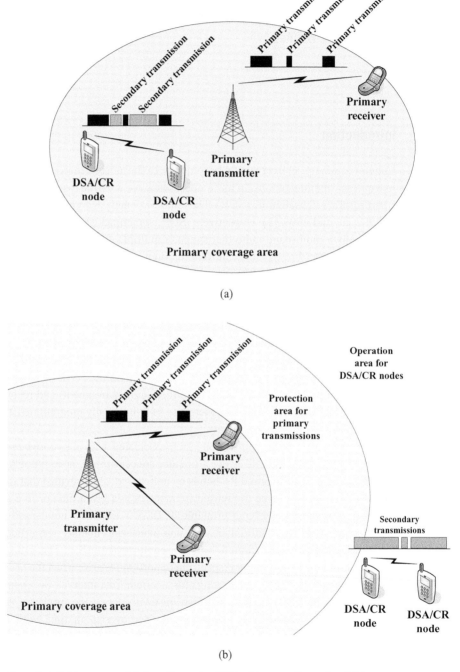

Figure 13.1 DSA concept: (a) time-dimension DSA; (b) space-dimension DSA.

The key enabler of the DSA concept is the CR technology [8]. CRs are capable to sense the spectrum occupancy and opportunistically adapt transmission parameters to utilize empty frequency bands without causing harmful interference to primary networks. A CR is able to reconfigure several parameters such as the communication technology (to adapt to specific communication needs), operating frequency (to take advantage of spectrum gaps detected on different frequency bands), modulation and/or channel coding (to adapt to the application requirements and the instantaneous channel quality conditions), and transmission power (to control interference). Based on the characteristics of the detected spectrum gaps, also referred to as *spectrum holes* or *white spaces*, these parameters can be reconfigured so that the CR is switched to a different spectrum band, the transmitter and receiver parameters are reconfigured, and the appropriate communication protocol parameters and modulation schemes are used.

During recent years, the DSA/CR concept has gained popularity as a promising solution to conciliate the existing conflicts between the ever-increasing spectrum demand and the demonstrated spectrum underutilization resulting from static spectrum regulations and management policies. Work on DSA has in fact become so prominent that DSA is often considered as the defining characteristic of a CR. However, this is a limited view that neglects many of the interesting areas of application for the CR technology. While DSA is certainly an important application of CR, the latter represents a much broader communications paradigm where many aspects of wireless communication systems can be improved via cognition and reconfiguration.

Recently, the DSA/CR technology has also received increasing attention in the context of future *Long Term Evolution* (LTE) communication systems and, in particular, in the context of *heterogeneous cellular networks* (HCNs). DSA/CR has the potential to provide simple solutions to overcome the challenges faced by LTE HCNs and improve the effectiveness and efficiency of already existing methods. In this context, this chapter presents a detailed overview of the DSA/CR technology, and explores the potential application of DSA/CR concepts and techniques in HCNs. First, Section 13.2 provides an exhaustive review of the functionalities and techniques associated with the DSA/CR technology. While conceptually simple, the realization of the DSA/CR concept poses great difficulties in practice, and requires specific technological solutions in order to identify spectrum opportunities, coordinate opportunistic spectrum access among a number of nodes, and guarantee interference-free coexistence with a wide variety of primary systems. Section 13.3 then discusses the potentials of DSA/CR methods and techniques in the context of HCNs. Several practical scenarios are explored, and various implementation alternatives along with the corresponding pros and cons are discussed. Afterwards, Section 13.4 reviews standardization efforts in the context of DSA/CR systems. Finally, Section 13.5 summarizes the chapter and provides some concluding remarks.

13.2   Cognitive radio techniques

The design and conception of any former *radio access technology* (RAT) has relied on the presumption of a set of channels or radio resources known in advance and permanently available to the UEs. CR represents a novel spectrum-aware communication paradigm where the communicating nodes need to obtain information about the radio environment, identify the availability of resources for opportunistic transmission, and enable mechanisms to maintain communication through unreliable and intermittently unavailable channels, while avoiding interference to the legitimate (primary) UEs. This challenging communication environment requires the introduction of new functionalities and techniques in order to guarantee a reliable, efficient, and interference-free spectrum access. The set of specific functions required by a DSA/CR system can be broadly categorized into four groups, namely *spectrum awareness*, *spectrum selection*, *spectrum sharing*, and *spectrum mobility*, which are discussed in the following.

13.2.1   Spectrum awareness

This function is in charge of obtaining relevant information and knowledge of the surrounding radio environment. From the viewpoint of DSA/CR, the most relevant aspect is accurately identifying which portions of the spectrum are available to secondary UEs for opportunistic use (spectrum holes or white spaces). To this end, CRs intelligently track idle frequency bands that are dynamic in both time and space, and detect the presence of primary UEs if they appear while a secondary UE operates in a licensed band. A CR can make use of various techniques to obtain knowledge of the radio environment, including *spectrum sensing*, *geolocation databases*, and *beacon signals*.

13.2.1.1   Spectrum sensing

A simple method to identify spectrum opportunities is to perform a spectrum scan and determine, by means of appropriate signal processing methods, whether a primary signal is present in the scanned band [9]. The spectrum sensing problem can be formulated as a binary hypothesis testing problem where the CR sensor has to determine whether the captured samples correspond to noise samples, and therefore there is no primary signal in the sensed spectrum band, or on the other hand some licensed signal is present. The main advantage of spectrum sensing is that the DSA/CR system does not need to rely on any external system to obtain knowledge of the environment. However, due to the unavoidable presence of noise and other degrading effects, spectrum sensing algorithms may make mistakes. In particular, a busy channel may be declared to be idle (event referred to as *missed detection*), which may result in harmful interference to primary UEs. Similarly, an idle channel may be determined to be busy (event referred to as *false alarm*), which results in missed transmission opportunities and a lower spectrum utilization. Therefore, spectrum sensing algorithms operate under a certain probability of error. In general, the performance of any spectrum sensing method degrades as the *signal to noise ratio* (SNR) decreases. Classical detection theory states that degradation in the detection performance due to reduced SNR can be countered by increasing the sensing time [10, 11]. Nevertheless, there is an SNR limit, referred to as the *SNR wall*, below which a primary signal cannot be reliably detected, irrespective of how long the sensing period is [12, 13]. The existence of SNR walls has been verified experimentally

[14–16] and imposes a fundamental limitation on the detection capabilities of spectrum sensing, and hence on the ability to reliably identify spectrum opportunities.

Signal processing techniques

A significant number of signal detection methods, commonly referred to as *spectrum sensing techniques*, have been developed in the context of DSA/CR to identify the presence of primary signals [17–21]. The existing solutions provide different tradeoffs between the sensing time required to attain a certain detection performance, the algorithm complexity, and computational cost, as well as the signal detection capabilities. Their practical applicability, however, depends on how much information is available about the primary signal. In the most generic case, a DSA/CR UE is not expected to be provided with any prior information about the primary signals that may be present within a certain frequency band. When the secondary receiver cannot gather sufficient information, the energy detection principle [22] can be used due to its ability to work irrespective of the signal to be detected. Energy detection, also referred to as *radiometric detection*, compares the signal energy received in a certain frequency band to a properly set decision threshold. If the signal energy lies above the threshold, the band is declared to be busy. Otherwise the band is supposed to be idle, and can be accessed by DSA/CR UEs. Although a simple and widely applicable method, energy detection is highly susceptible to uncertainty in the noise power, and cannot distinguish a weak primary signal from noise or interference. Despite its practical performance limitations, energy detection has gained popularity as a spectrum sensing technique for DSA/CR due to its general applicability and simplicity, as well as its low computational and implementation costs. Energy detection is often considered as a de facto sensing method in DSA/CR.

Energy detection constitutes a good option when a DSA/CR UE knows nothing about the primary signal or when complexity and computational cost is a major concern. Otherwise, more advanced detection methods that exploit primary signal characteristics (e.g., modulation, coding, pilot signals, preambles, or synchronization sequences) can be employed to attain a better detection performance. When the primary signal is known to the secondary UE, the optimal detector in *additive white Gaussian noise* (AWGN) is the *matched filter*, since it maximizes the received SNR [23, 24]. Matched filters, however, require detailed knowledge of the primary signal, including modulation type and order, pulse shape, or packet format. If this information is not accurate, the matched filter then performs poorly [25]. Matched filters are commonly employed in practice for the coherent detection of pilot signals. Pilot-based coherent detection is capable of distinguishing a primary signal from interference and noise, and therefore works under low SNR. Most existing wireless systems usually transmit pilot signals to help receivers to perform synchronization or channel estimation. For example, analogical *television* (TV) signals have narrow band pilots for audio and video carriers, digital TV signals have also pilots, *code division multiple-access* (CDMA) systems have dedicated spreading codes for pilot and synchronization channels and *orthogonal frequency division multiplexing* (OFDM) systems such as LTE have reference signals for synchronization and channel estimation. A drawback of matched filter detection is that CRs would need a specific receiver and dedicated circuitry to achieve synchrony with each type of primary signal as required for coherent detection [26], which may result in high complexity and implementation costs in HCNs.

An alternative method is cyclostationary feature detection, which was initially proposed in [27, 28]. This detection method exploits the built-in periodicity of modulated signals that results from sine wave carriers, pulse trains, repeating spreading, hopping sequences, or cyclic prefixes. This embedded periodicity results in cyclostationarity, meaning that the average and autocorrelation of the modulated signal exhibit periodicity. By analyzing the cyclic autocorrelation function of the captured signal or its two-dimensional spectrum correlation function, the modulated signal energy can be differentiated from the noise energy, since the modulated signal is cyclostationary with spectral correlation while noise is a wide-sense stationary signal with no correlation. Therefore, cyclostationary feature detectors are also able to discriminate a primary signal from interference or noise, even under low-SNR conditions, and as such can perform better than energy detection. In contrast to pilot-based coherent detection, the same detection procedure is employed to detect any primary signal. However, some characteristics of the primary signal (more concretely, the cyclic frequency of the signal's autocorrelation function) need to be known beforehand. In a multi-RAT environment where some modulated primary signals may be unknown to DSA/CR UEs, an exhaustive search of the cyclic frequency needs to be performed, which implies a huge complexity and computational cost as well as some loss of the capability to differentiate the intended primary signal from cyclostationary interference.

Covariance-based detection is another alternative that determines the presence of a primary signal based on the covariance matrix of the captured signal [29]. This detection method does not require any prior information, but relies on the presumption that the primary signal received at the CR UE is autocorrelated, which can be exploited to distinguish the signal from the (uncorrelated) noise. The performance of this detection method depends on the statistics of the primary signal. In the extreme worst case where the primary signal is completely uncorrelated (i.e., the signal would appear like white noise), a covariance-based detection would fail. However, real signals are usually correlated due to the dispersive nature of real channels. Otherwise correlation can be introduced by oversampling the received signal or using multiple receive antennas [29].

Every spectrum sensing method requires a different amount of knowledge of the primary signal, and is well suited to a particular scenario. Covariance-based detection is in general suitable when the primary signal is known to be correlated, or it is feasible to induce correlations by oversampling the signal or employing multiple receive antennas. Cyclostationary detection is adequate when the cyclic frequency of the primary signal is known. Matched filter detection is appropriate for primary signals whose pilot is known. When nothing is known about the primary signal, the energy detection method can still be employed. It is worth noting that advanced sensing methods in general outperform the conventional energy detection scheme by exploiting specific signal properties, which not only assumes some prior knowledge of the primary signal, but also restricts the field of application to particular primary RATs. The performance improvement is normally obtained at the expense of an increased algorithm complexity and computational cost. Recently, alternative methods based on variants of the energy detection scheme have been proposed, which provide a better detection performance while preserving a similar level of complexity, computational cost, and field of application [30].

Existing spectrum sensing methods are not mutually exclusive, but can be combined and used together to perform a more efficient spectrum sensing of wide frequency bands. In a first stage, a low-complexity energy detector can be employed to search for potential idle subbands. In a second stage, more sophisticated spectrum sensing methods with better detection performances can be applied to selected candidate subbands, and determine whether they are actually available for secondary usage.

The discussed spectrum sensing techniques are mainly intended to provide a binary busy/idle decision on the occupancy state of a channel. While this information may be sufficient for DSA/CR operation, more detailed information can be extracted by means of advanced methods. A good example is the concept of a *blind standard recognition sensor* [31], which can be used to classify the captured primary signal into one of a set of predefined radio standards. The method analyzes the bandwidth of the detected signal, compares the spectral shape to a set of predefined reference shapes, and determines the presence of a frequency-hopping or spread-spectrum signal as well as single/multi-carrier properties. The information obtained in this analysis is fused by means of some logical rules or more advanced methods such as neural networks in order to determine the radio standard to which the captured signal belongs. Without assuming any prior knowledge of the primary signal, this method can determine with significant precision the presence of some of the most common radio standards in current multi-RAT networks such as the *Global System for Mobile Communications* (GSM), *Digital Video Broadcasting* (DVB), *Digital Audio Broadcasting* (DAB), *Local Multipoint Distribution Service* (LMDS), *Digital Enhanced Cordless Telecommunications* (DECT), *Wireless Fidelity* (WiFi), *Worldwide Interoperability for Microwave Access* (WiMAX), or LTE. This information can be exploited by the DSA/CR network to properly adapt to the operating environment and improve the overall spectrum use and system performance.

Spectrum sensing architectures

A CR may employ a single radio for both spectrum sensing and data transmission. In this approach, the CR reserves a certain portion of time to monitor the target primary channel (referred to as *sensing time*), during which the captured samples are processed by applying one of the spectrum sensing methods described above. Based on the sensing result, the CR may decide to transmit or not for a certain period of time (referred to as *transmission time*), after which a new sensing event is scheduled in order to determine whether the channel state has changed and therefore identify new spectrum holes or vacate the channel to avoid interference. The transmission time may be constant (periodic spectrum sensing) or modified according to certain criteria (adaptive spectrum sensing). The main advantage of this architecture is that a single radio is employed to obtain spectrum awareness and transmit data. However, this approach does not allow for a full exploitation of spectrum opportunities for data transmission since some portion of time needs to be reserved for spectrum sensing. Moreover, this approach may result in interference with primary UEs that become active during transmission periods (the sensing periodicity needs to be selected carefully in order to minimize interference).

These drawbacks can be overcome with a dual-radio architecture where one radio is used to transmit data and the other one to perform spectrum sensing. This approach may virtually result in interference-free operation since the data radio immediately stops when the sensing radio detects a primary signal. Moreover, the inefficiencies caused by sensing and transmitting with a single radio are also removed, since the data radio can start exploiting a spectrum hole as soon as it appears and is detected by the sensing radio. However, the use of two simultaneous radios results in an increased complexity and economical cost per CR device as well as a shorter life in battery-powered UEs.

Cooperative sensing

In non-cooperative detection, DSA/CR UEs detect the primary transmitter signal independently based on local observations. On the other hand, cooperative detection solutions rely on the exchange of sensing information from multiple secondary UEs. Cooperative sensing is theoretically more accurate since it helps to solve the *hidden-terminal* problem and mitigates the degrading effects of shadowing and multi-path fading. The hidden-terminal problem occurs when a DSA/CR UE does not detect a distant primary transmitter and decides to transmit, causing harmful interference to a nearby primary receiver. Cooperative sensing can solve this problem, since some DSA/CR UEs in the coverage area of the primary transmitter may detect its presence and exchange sensing information to warn other DSA/CR UEs, which otherwise would need to cope with demanding sensitivity requirements[1] [32]. Cooperative sensing can therefore solve the hidden-terminal problem while relaxing the sensitivity requirements of individual DSA/CR nodes. A DSA/CR UE may also misdetect a primary signal, even within the coverage area of the primary transmitter, because of the shadowing effects of buildings and other surrounding obstacles. Cooperative sensing can also help to solve this situation. Since shadowing can exhibit correlations over relatively large areas (e.g., between locations blocked by the same obstacle), the cooperative sensing gain is limited by the cooperation footprint (i.e., the area over which UEs cooperate) [33]. In essence, a few independent nodes are more robust than many correlated nodes. The degrading effects of multi-path fading can also be overcome by means of cooperation, because multi-path at different radios is essentially uncorrelated (multi-path is correlated over lengths of the order of the wavelength). Thus, the presence of multiple radios can help to reduce the effects of severe multi-path fading at a single radio, since they provide multiple independent realizations of the primary signal. Cooperative sensing can also be used to reduce the detection time of weak primary signals, thereby increasing the frequency agility of the overall network [34, 35].

Cooperative detection is usually conducted in two successive stages, namely sensing (DSA/CR nodes perform spectrum sensing independently) and reporting (spectrum

---

[1] In temporal reuse of spectrum holes, the secondary UEs are located within the coverage area of the primary transmitter, as illustrated in Fig. 13.1(a). Therefore, the detection of the primary system is relatively easy, since DSA/CR nodes only need to have a similar detection sensitivity level as the regular primary receivers. However, DSA/CR nodes exploiting spatial white spaces operate outside the intended coverage area of the primary transmitter, as shown in Fig. 13.1(b), and therefore they need to be notably more sensitive than regular primary receivers.

sensing information is exchanged). Cooperative spectrum sensing can be implemented in a centralized or distributed manner [36]. In centralized methods, sensing information from DSA/CR nodes is gathered at a designated controller (a secondary base station or one of the DSA/CR nodes), which combines the received information, makes a final decision and communicates the result to the involved DSA/CR nodes [37]. In distributed schemes, DSA/CR nodes make individual decisions after sharing sensing information. Distributed cooperation schemes may be easier to implement but may not achieve the detection performance of centralized methods [26].

Different amounts of sensing information can be exchanged. In an extreme case, only one bit of information indicating the observed busy/idle channel state can be reported. The designated controller can then apply various hard-combination methods.

- OR-rule: the channel is declared busy if *at least one* sensing report indicates a busy channel.
- AND-rule: the channel is declared busy if *all* sensing reports indicate a busy channel.
- Majority-rule: the channel is declared busy if *most* sensing reports indicate a busy channel.
- K-out-of-N-rule: the channel is declared busy if *at least K out of N* sensing reports indicate a busy channel.

Alternatively, cooperative DSA/CR UEs can exchange more detailed information, such as the signal energy or power locally observed. The designated controller can then apply various soft-combination methods such as *maximal ratio combining* (MRC) or *equal gain combining* (EGC) [38]. Soft-decision schemes may result in better detection performance but require more bandwidth to exchange sensing information.

Cooperative sensing has some limitations and drawbacks. First of all, cooperation introduces the need for a reliable control channel to exchange sensing information, which may be an issue under the lack of dedicated spectral resources. This may be specially challenging in some particular cases, for example in macro–femto scenarios where there is neither interface for UE-to-UE communication nor X2 interface. Under a limited number of reliable control channels, DSA/CR nodes may need to efficiently share control channels by means of appropriate access methods and protocols. Furthermore, cooperation may also require significant amounts of bandwidth to exchange sensing information and incur in delays and additional energy consumption, which may be an issue in battery-powered UEs. These problems can be alleviated by means of censoring methods to discard unreliable or useless reports, along with data compression techniques to reduce the amount of exchanged information [39, 40]. Bandwidth, delay, and energy consumption can be further reduced by grouping cooperative nodes into hierarchical cluster structures, where the most favorable node in each cluster (in terms of propagation conditions) gathers sensing information and then censors, fuses, and compresses the data before reporting to the upper cluster head [41].

Cooperative sensing can provide significant benefits. However, the feasibility of an available and reliable control channel to exchange the sensing information, the overhead associated with sensing information exchange, and the additional energy consumption in battery-powered UEs represent significant challenges in practice.

13.2.1.2 Geolocation databases

As discussed above, spectrum sensing has some practical limitations and can be somewhat unreliable. An alternative solution, although less flexible and dynamic, is to rely on a centrally managed spectrum usage database. This approach is based on maintaining a frequently updated and centrally located database with information about the regional spectrum usage, including location of primary transmitters, coverage areas, frequencies of operation, transmission powers, radio technologies, operation requirements, etc. When using the database to provide necessary spectrum usage information to the DSA/CR nodes, the location of the DSA/CR nodes is required, hence the term *geolocation database*. Geolocation databases are an alternative method to obtain knowledge of the radio environment, and can be used to identify primary systems in a certain region and ensure that DSA/CR nodes do not attempt to operate within protected areas where secondary transmissions would result in harmful interference levels.

While spectrum sensing methods provide binary information about the busy/idle state of the channels in a frequency band, geolocation databases can provide more detailed and reliable information about the primary system, which can be exploited for a more efficient use of the spectrum. With geolocation databases, DSA/CR nodes do not need to implement computationally complex spectrum sensing methods, but need to implement geolocation methods such as the *Global Positioning System* (GPS) or triangulation, which can be moved to *base stations* (BSs) in infrastructure-based DSA/CR networks in order to reduce UE costs. However, the use of databases involves some practical problems. From a technical point of view, an important problem is the location accuracy. The locations of DSA/CR nodes need to be determined with acceptable accuracy levels to ensure that the information provided by the database corresponds to the real environment where the DSA/CR nodes operate. Existing geolocation alternatives are not perfect and are characterized by certain accuracy errors. Another important technical problem is the selection of the information contained in the database and the structure of the database. Moreover, geolocation databases do not constitute an adequate choice for spectrum bands where the characteristics of the primary system and the spectrum occupancy patterns are highly dynamic. Updating the database after a change of the primary operation conditions and retrieving the updated information from the secondary system is a process that involves bi-directional communication and requires some time. The use of databases in primary systems with highly dynamic characteristics may result in retrieving outdated information that does not reflect the instantaneous circumstance of the primary system. For this reason, the deployment of geolocation databases has been proposed for primary systems that are static or vary over long time scales such as the TV system. The transmitter characteristics (location, frequency, power, etc.) in this band are stable or change very infrequently. From an administrative point of view, the most notable drawback of geolocation databases is the significant economical cost associated to the infrastructure required to deploy and maintain databases. Another controversial issue is who owns and maintains the database (the primary network, the secondary network, a new commercial operator, or a governmental/regulatory organism) and the way the database is managed and accessed depending on the ownership.

13.2.1.3 Beacon signals

Another alternative method to obtain knowledge of the radio environment relies on the use of regional beacon signals that are broadcast in appropriate signaling channels. Beacon signals convey real-time information about the primary systems present in the geographical area where the beacon signal is broadcast, along with various kinds of information on spectrum usage. DSA/CR UEs can then detect the regional beacon signals present at their location and obtain information about the surrounding environment. The use of beacon signals can be envisaged as a network-aided approach based on an active agreement between the secondary network and the original spectrum licensee, whereby the primary network shares real-time information with the secondary network regarding spectrum utilization. This approach would allow secondary systems to conceptually have perfect knowledge of current spectrum usage as well as possibly knowledge about traffic trends and future frequency usage [42]. However, if the primary system is based on a legacy radio technology, enabling such interaction would require additional modifications to the already existing primary network, which might not be possible or economically feasible. Alternatively, beacon signals can also broadcast information obtained from spectrum databases. In this case, beacon signals can be thought of as an alternative means to access spectrum databases.

The essence of the beacon-signal approach has been realized in various particular definitions such as the *spectrum information channel* in [43], the *common spectrum coordination channel* (CSCC) in [44], and the *cognitive pilot channel* (CPC) concept [45–47]. The latter concept has received great attention and constitutes an specially interesting approach in the context of multi-RAT networks. The CPC was originally conceived as a solution for conveying the necessary information from the network side to allow mobile UEs to be aware of the surrounding environment and available RATs in order to facilitate its connection to the network. However, the scope of CPC was extended to become a means of spectrum awareness [46] and an enabler for radio systems coexistence [47]. DSA/CR systems can exploit the information transmitted in the CPC to gain knowledge of the primary radio systems operating within a certain geographical area and their specific characteristics.

The design and deployment of the CPC concept (and beacon signals in general) require some practical aspects to be taken into consideration. One important aspect is the CPC operating frequency. The CPC can be broadcast at a fixed frequency harmonized among primary RATs at a global or regional level, which facilitates the operation of mobile UEs. A simpler solution from the primary operator's viewpoint is to use a set of convenient frequencies that may be neither harmonized nor fixed, which requires UEs to scan frequency bands to search for the CPC signal. Another aspect is the RAT used to convey the CPC, which can be one or more of the RATs deployed making use of the logical channels already provided by the set of available RATs (in-band CPC) or a separate CPC-specific RAT operating in another frequency (out-band CPC). The main advantage of the out-band approach is that any CPC-compliant UE can retrieve the information conveyed in the CPC regardless of the supported RATs. However, this approach requires new infrastructure and a harmonized frequency to transmit the CPC

Table 13.1 Comparison of spectrum awareness methods.

	Spectrum sensing	Geolocation databases	Beacon signals
Infrastructure complexity/ cost	Low	High	Medium/High
UE complexity/cost	High	Medium	Low
Legacy compatibility	High	Medium/Low	Low
Reliability	Low/Medium	High	Medium/High
Spectrum dynamism	High	Low	High
Need for external system/ provider	No	Yes	Yes
Need for additional spectrum	No	No	Yes/No
Specific issues	Time and energy consumption	Positioning system	Standardized solution

channel. On the other hand, the in-band approach reuses the existing infrastructure and does not require frequency harmonization. However, it requires that UEs scan several frequency bands to find the RAT where the CPC is transmitted. Moreover, if the particular RAT is not supported by the UE, the CPC cannot be found. This problem can be solved with a combined CPC approach where an out-band CPC broadcasts a list of primary RATs and the frequencies where the in-band CPC for each RAT can be found, and each in-band CPC transmits detailed information for the corresponding RAT. This approach prevents UEs from having to scan frequency bands and allows the information to be retrieved regardless of the supported RATs. However, a harmonized out-band CPC frequency is still needed. Depending on the considered delivery mode, CPC can be deployed in broadcast or on-demand modes. In broadcast CPC, the information is transmitted periodically and continuously making use of a downlink broadcast channel. In on-demand CPC, the information is transmitted under request when needed by the secondary UE, which requires uplink communication channels to request the CPC information (in addition to the downlink channels), but is more efficient in terms of power consumption and required bit rates [48].

13.2.1.4 Comparison of spectrum awareness methods

The spectrum awareness methods presented in previous sections are not mutually exclusive but can be combined and employed together. For instance, the information retrieved from spectrum databases or beacon signals can be employed to conduct spectrum sensing with some prior knowledge of the sensed spectrum bands instead of continually blindly scanning over large regions of the spectrum. The complementary characteristics of different spectrum awareness methods make it possible to combine them into a more complex approach capable to provide improved spectrum awareness.

Table 13.1 provides a comparison of spectrum awareness methods. Geolocation databases and beacon signals place on the primary system the responsibility of providing spectrum awareness to the secondary system by either updating the information contained at a centralized database or broadcasting the relevant data on regional beacons. Spectrum occupancy information is reliable since it is provided by the primary system, which guarantees minimum interference and an efficient use of spectrum. However, these approaches need a communication means between primary and secondary systems, which may involve some modifications to legacy primary systems and hence lead to compatibility issues. The deployment of geolocation databases and beacon signals implies a significant cost in infrastructure and requires location information along with either ubiquitous connectivity to the spectrum database or standardized channels to broadcast the beacon signals. On the other hand, spectrum sensing solely relies on local observations of the spectrum occupancy without the need of an external system providing spectrum awareness. While spectrum sensing leads to more complex secondary UEs, no modifications to the primary system are required (i.e., it is compatible with legacy systems) and infrastructure costs are notably lower. However, spectrum sensing consumes time and energy and is somewhat unreliable due to practical limitations.

13.2.2   Spectrum selection

The spectrum selection function is in charge of selecting the most suitable spectrum to meet the communication requirements of the secondary UEs in the DSA/CR system. Spectrum selection can be performed at the band level (a specific licensed spectrum band is selected for the whole DSA/CR system) and/or at the channel level (a particular licensed channel within a licensed spectrum band is selected for each DSA/CR UE). Moreover, spectrum selection can be centralized or decentralized. In centralized spectrum selection, a centralized entity selects the spectrum band or channel for each DSA/CR UE. In decentralized spectrum selection, each DSA/CR UE selects individually, based on a set of agreed policies or internal criteria, the spectrum band or channel to be accessed. In the latter case, some contention or coordination mechanisms may be necessary to avoid collisions among secondary UEs (see Section 13.2.3).

The spectrum selection function can be divided into two steps. In the first step, the characteristics of the spectrum holes identified in different spectrum bands are analyzed and characterized based on certain metrics. This step is referred to as *spectrum analysis* or *spectrum characterization*. In the second step, a decision on the spectrum holes to be exploited is made, based not only on the characteristics of the available spectrum but also on internal requirements and policies of the DSA/CR system and possibly external policies too. This step is referred to as *spectrum decision*.

13.2.2.1   Spectrum analysis

In multi-RAT networks where a wide variety of RATs may be present, the spectrum holes identified in various spectrum bands may have quite different characteristics depending on the underlaying radio technology of the primary system. It is therefore necessary to adequately characterize the properties of the identified spectrum holes in order to enable the DSA/CR system to select those that best fit the particular needs and requirements of the DSA/CR UEs. The purpose of the spectrum analysis/characterization function is to quantify the properties of the identified spectrum holes so that the spectrum decision function can more easily determine the most convenient choice. Thus, it is essential to define metrics and parameters that can represent the quality of a particular spectrum hole and its suitability to meet the needs and requirements of the DSA/CR UEs. Two broad groups of parameters can be identified, namely *RF metrics* and *primary activity metrics*.

RF metrics are related to the RF properties of the channels where the spectrum holes are identified. Some examples of relevant RF metrics are discussed in the following

- Bandwidth: One of the probably most relevant properties of a spectrum hole is its bandwidth, as it is one of the aspects determining the maximum data rate that can be accommodated. If the rest of the parameters remain unchanged (e.g., noise level, interference level, channel quality conditions, modulation and coding scheme, etc.), a larger bandwidth implies a higher data rate and hence turns out to be more attractive for high-data-rate applications. The fragmentation of the spectrum holes in a certain band is also an important aspect to be taken into account. In general, large contiguous bandwidths are more attractive, since this simplifies the operation of the DSA/CR system. However, DSA/CR systems based on *multi-carrier modulation* (MCM) methods such as OFDM or *orthogonal frequency division multiple access* (OFDMA) can also exploit opportunities in spectrum bands where the available free bandwidth is fragmented. This can be accomplished by nulling the carrier frequencies where a primary transmission is present. A spectrum band with narrow spectrum holes may not be interesting for a DSA/CR system with other modulation methods, but it may be attractive for MCM systems if the aggregated bandwidth of the spectrum holes is larger than the overall bandwidth available in other spectrum bands or meets the UEs' needs. Bandwidth fragmentation is an important aspect to characterize in order to allow the spectrum decision function to select the most convenient spectrum holes for the DSA/CR UEs.
- Interference: Some spectrum bands are more crowded than others [49, 50]. In general, crowded bands experience higher levels of interference, which may result in reduced data rates. Therefore, spectrum bands with lower interference levels turn out to be more attractive for secondary operation since higher data rates can be expected.
- Emission limits: To guarantee interference-free operation, the DSA/CR system must control the maximum transmission power of the DSA/CR UEs based on the locations of the potential primary victims, their distances from the DSA/CR transmitters, the minimum *signal to interference plus noise ratio* (SINR) required by the primary receivers, etc. Thus, the maximum power level at which the DSA/CR system can transmit depends on the primary system operating in each candidate band. This is a relevant aspect, since it determines the maximum communication distances for the DSA/CR system as well as the expectable SINR levels and data rates. Depending on the intended coverage area for the DSA/CR system, the emission limits in certain spectrum bands may require the DSA/CR network to select other less attractive spectrum bands. To avoid interference to primary systems, the emission limits need to be carefully taken into account in spectrum analysis/characterization. It is worth noting that, while the previous metric (interference) refers to the interference received by the DSA/CR

system from the primary or other secondary systems if the spectrum band is selected, this metric (emission limits) refers to the interference caused by the DSA/CR system to the primary UEs in the candidate band.

- Frequency: The path loss increases with frequency, meaning that for the same transmission power the communication range of a DSA/CR UE decreases as the frequency of operation increases. Moreover, the experienced path loss also has an impact on the transmission power required for a certain target data rate at the receiver, which may be constrained by the interference tolerable by the primary receivers. Therefore, the lower regions of the spectrum may be more appealing for DSA/CR systems targeting larger coverage areas and/or higher data rates.

Primary activity metrics are related to the spectrum occupancy patterns of the primary system in a certain spectrum band. Since there is no guarantee that a spectrum band or channel will be available during the entire duration of a DSA/CR communication, it is important to characterize the activity patterns of primary UEs. An adequate characterization of the primary activity can help to determine the spectrum band where primary and secondary transmissions can coexist with minimum probability of mutual disruptions. Some relevant primary activity metrics are discussed in the following

- Duty cycle: The duty cycle of a spectrum band or channel can be defined as the probability, fraction, or percentage of time that the band or channel is busy (i.e., occupied by a primary transmission). This parameter provides an estimation of the overall time period the spectrum would be available for opportunistic transmissions. Thus, bands or channels with lower duty cycles are more attractive since they will provide more transmission opportunities. The average duty cycle is a common metric to characterize the occupancy level of a band or channel. However, some spectrum bands where channel load is not constant may exhibit a time-dependent pattern of the duty cycle through the day [51]. In such a case, the minimum and maximum duty cycle values may be a convenient method to determine upper and lower bounds for the number of opportunities that can be expected from a certain band or channel.
- Arrival rate of busy/idle periods: This metric determines the average rate at which busy/idle periods appear in the channel and can be useful to estimate how often the DSA/CR system can access or has to vacate the spectrum. This metric may be especially interesting for delay-sensitive traffic that needs to access the spectrum regularly in order to avoid delayed data transmissions. A spectrum band or channel with long busy and long idle periods can have the same average duty cycle as another band or channel with shorter busy and shorter idle periods. However, the latter may be more convenient for delay-sensitive traffic since it would allow more frequent transmissions, thus potentially reducing transmission delays.
- Expected occupancy/vacancy duration: This metric quantifies the length of the busy/idle intervals present in a band or channel, which can be useful to determine the time a DSA/CR UE will have to wait before accessing the spectrum or the time the channel will be available for transmission. This parameter is closely related to the arrival rate of busy/idle periods and can be thought of as an alternative characterization approach. The statistical distributions of busy/idle periods can be a useful metric as well

(e.g., the distribution of idle periods can be employed to determine the probability that a channel is available for a predefined minimum time interval).

The primary activity metrics described above can be used to provide an elementary characterization of the primary UE activity based on past observations of spectrum occupancy. A more sophisticated characterization approach is to make use of advanced spectrum occupancy models [52, 53], and estimate the model parameters based on the spectrum occupancy history. This not only enables a more sophisticated characterization of the primary activity patterns, but can also be used to predict or estimate the future spectrum activity based on past observations of spectrum occupancy.

The RF metrics and primary activity metrics described above represent basic metrics that can be combined in order to provide a more sophisticated characterization of spectrum holes. For example, an effective available bandwidth metric can be defined based on the bandwidth and the duty cycle of a channel as $BW_{eff} = BW(1 - DC)$ where BW is the RF bandwidth of the channel and DC is its duty cycle. This metric can be employed to estimate the effective opportunistic capacity of a channel. For example, based on this metric, a 2 MHz channel with an average occupancy of 20% would be more attractive than a 4 MHz channel with an average occupancy of 80%, since the effective available bandwidth is 1.6 MHz for the former while the value of this parameter is 0.8 MHz for the latter. Other combinations of the presented metrics are also possible, not only among them but also including other aspects such as spatial parameters. For example, the bandwidth of a channel can be multiplied by the coverage area over which the channel is used, thus leading to a space–bandwidth product metric [54]. This metric qualifies the reuse in both spatial and frequency domain. For instance, a 6–8 MHz TV channel occupied by a TV station with a service range of 100 km has different resource consumption compared with the same bandwidth occupied by a *wireless local area network* (WLAN) device with a range of 100 m. The presented metrics can also be employed to estimate other metrics related to the DSA/CR UE *quality of service* (QoS) requirements, such as the channel capacity and attainable data rates as well as expected error rates and transmission delays. The most convenient set of metrics employed to characterize the identified spectrum holes is a problem-specific aspect that depends on the objectives and particular scenario under study.

13.2.2.2 Spectrum decision

After properly characterizing the identified spectrum holes, the spectrum decision function selects an appropriate band or channel and the corresponding parameters such as modulation and coding schemes, transmission power, etc. The spectrum decision is made based not only on the available spectrum holes and their characteristics, as quantified by the previous or other metrics, but also taking into account internal requirements (e.g., QoS requirements) and policies (e.g., the economical cost of accessing licensed spectrum cannot exceed a certain maximum value) of the DSA/CR system and possibly on external policies (e.g., some regions of spectrum such as military bands cannot be accessed or may be accessed under certain conditions).

Spectrum decision criteria in existing literature are frequently based on the estimated channel capacity. A common decision criterion is the maximum-capacity-based spectrum decision (MCSD) [55], which is aimed at maximizing the total network throughput by selecting the spectrum that provides the highest expected capacity. This criterion can be appropriate for best-effort applications where there are no strict QoS requirements. For delay- and jitter-sensitive applications (e.g., real-time traffic) a more convenient approach is the minimum-variance-based spectrum decision (MVSD) [55], which selects the spectrum that provides the minimum capacity variation. This spectrum decision criterion is motivated by the fact that high capacity variations cause delays and jitter. Therefore, a channel where a sustained data rate can be maintained is more convenient for real-time traffic. Alternatively, a spectrum band or channel can be selected based not only on the expected capacity but also on the expected error rates or transmission delays. The maximization of spectrum efficiency can also be included as an objective in the spectrum decision criterion. The spectrum decision may sometimes face conflicting alternatives, specially in multi-RAT networks with a wide variety of systems. For example, a DSA/CR may be forced to choose between a spectrum hole with good radio propagation conditions and a high level of primary activity and a spectrum hole with a lower level of primary activity and poorer radio propagation conditions. The decision in these conflict scenarios would need to be made based on objective functions and constraints, which often leads to complex multi-objective optimization problems.

13.2.3 Spectrum sharing

The purpose of the spectrum sharing function is to provide a fair spectrum access to the coexisting secondary UEs and/or networks by coordinating the access to the available spectrum holes while at the same time guaranteeing an adequate interference-free coexistence between primary and secondary systems. Spectrum sharing among wireless UEs or networks may occur either horizontally or vertically [56]. In horizontal spectrum sharing, all the UEs or networks have the same right to access a particular spectrum band. In vertical spectrum sharing, one wireless system (typically, the owner of the spectrum license) has a higher priority to access the spectrum band and its right to exploit the spectrum prevails over the rest of systems. In DSA/CR systems, vertical spectrum sharing takes place between primary and secondary UEs, while horizontal spectrum sharing takes place among secondary UEs and/or networks. Given the hierarchical access structure of DSA/CR where the primary or licensed UEs are given a higher spectrum access priority, the mechanisms required for spectrum sharing between primary and secondary UEs (vertical spectrum sharing) are different compared to the methods employed for spectrum sharing between secondary UEs and/or networks (horizontal spectrum sharing). The two cases are discussed separately in the following.

13.2.3.1 Vertical spectrum sharing

Three main methods for vertical spectrum sharing can be identified, namely *interweave*, *underlay*, and *overlay* [57]. The interweave approach is in line with the original idea of opportunistic spectrum usage as depicted in Fig. 13.1. The basic idea of this strategy is to identify temporal and spatial spectrum gaps not occupied by primary UEs (i.e., spectrum holes or white spaces) and place secondary transmissions within such gaps. The interweave method follows an interference avoidance strategy where DSA/CR UEs are allowed to use the licensed spectrum when primary UEs are not present either temporally or spatially as illustrated in Fig. 13.1. This approach does not impose restrictions on the transmission power of DSA/CR UEs but rather on when and where they transmit. The main drawback of the interweave scheme is the need to accurately identify the available spectrum holes to avoid interference with the primary system. Depending on the particular characteristics of the primary band to exploit, one or more of the spectrum awareness methods presented in Section 13.2.1 may be required. Interweave CRs need to be very agile in order to switch on and off very quickly over various frequency bands in order to exploit spectrum holes. The interweave approach is probably the most popular method considered in CR. The term DSA is indeed commonly used to refer to the interweave strategy.

The underlay approach consists in ensuring that secondary transmissions remain below the maximum allowable interference temperature at the primary receivers so that their operation is not disturbed. As opposed to the interweave approach, underlay CRs are allowed to transmit anywhere and anytime but under severe constraints on their transmission power so that they operate below the noise floor of the primary UEs. In the general context of CR, such requirements are normally met by spreading secondary transmissions over a wide frequency band by means of the *ultra-wide band* (UWB) technology. The UWB signal is limited by a very strict spectral mask and allows very low power transmissions over a large bandwidth. The actual bandwidth of the spectral mask varies in different countries as defined by their corresponding regulatory bodies, but normally overlaps with several licensed systems. The spectral mask ensures that secondary UWB transmissions do not interfere with the primary systems, as the received UWB power at any primary receiver is typically well below the noise floor. Compared to the interweave approach, the main advantage of underlay is that the activity of primary UEs does not need to be tracked (overlapping transmissions are allowed provided that power constraints are respected). However, the strict power limitations imposed by UWB spectral masks reduce the applicability of the underlay approach to short-range scenarios (typically below 10 meters). Longer transmission ranges (up to 300 meters) are possible, but at very reduced data rates. Another concern with the underlay approach is the real interference impact on the primary receivers. Although a single underlay device may be far from affecting the operation of primary receivers in a significant way, the aggregated interference of a large number of underlay devices might raise the noise floor in an area and thus cause an adverse impact on primary systems. It is worth noting that the underlay approach does not necessarily need to rely on the use of UWB. For example, in an HCN environment, macrocells and femtocells can coexist in an underlay fashion, with the femtocell being power controlled in order to not disturb the macrocell.

Similarly to the underlay approach, the overlay strategy also allows concurrent primary and secondary transmissions. The basic idea of the overlay approach is that a fraction of the transmission power available at the secondary UEs is used for secondary communications and the remainder of the power is employed to assist the primary transmissions by relaying the primary signal from the primary transmitter to the primary receiver. By an adequate selection of the power split, the reduction in the primary receiver's SINR due to the interference caused by the fraction of power that is employed for the secondary transmissions can be compensated by the assistance from the secondary relaying. In other words, the power employed for secondary transmissions results in an increased interference component at the primary receiver and thus in a reduced SINR. However, the fraction of secondary power used to relay the primary signal is carefully selected to increase the signal component at the primary receiver by the same amount so that the effective SINR at the primary receiver is the same regardless of the presence of the secondary transmission. The main advantage of the overlay approach is that secondary transmitters may be allowed to increase the interference temperature level beyond the strict limits imposed in the underlay approach, and the detection of the primary signal can be improved. However, there are many significant drawbacks and practical limitations, mainly the degree of complexity in the secondary transceivers. First, the overlay approach assumes that the secondary system has sufficient information and knowledge of the primary message to produce a compatible signal that complements the primary signal and improves the detection probability at the primary receiver. At the same time, the secondary transmitter needs to be able to communicate with the secondary receiver, which implies the use of techniques such as dirty paper coding or successive interference cancellation depending on whether the secondary transmitter, the secondary receiver, or both have knowledge of the primary signal and channel state. Moreover, a sophisticated power control mechanism is required to decide the power split between secondary transmissions and relayed primary transmissions.

13.2.3.2   Horizontal spectrum sharing

In horizontal spectrum sharing, all the UEs have the same right to access the spectrum, and it is the responsibility of the spectrum sharing function to provide a fair spectrum access to the coexisting secondary UEs and/or networks. The main purpose of horizontal spectrum sharing is to coordinate the access of a group of DSA/CR UEs/networks to the available spectrum holes, determining who and when will access which channel, in order to prevent multiple UEs from colliding in overlapping portions of the spectrum. To some extent, this problem is similar to the *medium access control* (MAC) problem in conventional wireless communication systems. However, existing MAC methods are in general inappropriate for DSA/CR systems due to some particular problems not present in other wireless systems. First, DSA/CR UEs need to coexist not only with one another but also with primary UEs. While collisions among secondary UEs are undesirable but acceptable within some limits, collisions with primary UEs are unacceptable and must be avoided at any cost. Moreover, while conventional wireless systems have a certain amount of spectral resources (channels) to share among the wireless UEs, which in general is fixed and known beforehand, the resources available to a DSA/CR system may change constantly in an unpredictable manner. In the worst possible case, there may be no channels to allocate to the DSA/CR UEs, if all the spectral resources are used by the primary system. Furthermore, the same licensed band may be accessed opportunistically by secondary UEs belonging to different DSA/CR systems or networks without a predefined or established internetwork communication mechanism. The particular characteristics and issues of DSA/CR systems require specific spectrum sharing solutions.

Horizontal spectrum sharing solutions can be broadly categorized according to the *architecture, access behavior,* and *scope/field of application.*

According to the architecture, horizontal spectrum sharing solutions can be classified as *centralized* or *distributed.* In centralized solutions, a central entity controls and decides the UE/network that is granted access to the spectrum based on a set of selected access procedures. The central entity can collect reports and measurements from a set of distributed nodes, and use this information to build up a spectrum allocation map according to a predefined policy, which is then communicated to the DSA/CR UEs/networks. On the other hand, in distributed solutions, each DSA/CR UE or network decides when and how to access the spectrum based on their own observations and a local or global spectrum access policy.

Based on the access behavior, horizontal spectrum sharing solutions can be classified as *cooperative* or *non-cooperative.* Cooperative solutions decide on the spectrum allocation taking into account the potential impact on other DSA/CR UEs. Secondary UEs exchange information (e.g., interference measurements) that is taken into account in the spectrum allocation and access decisions. While centralized solutions can in general be regarded as cooperative, distributed cooperative solutions can be possible as well. On the other hand, in non-cooperative solutions, DSA/CR UEs do not exchange information, and decide on the spectrum allocations in a selfish manner without considering the effect of a particular spectrum allocation on the rest of the DSA/CR UEs. In general, cooperative approaches outperform non-cooperative strategies [58], but the minimal communication requirements of the latter and the resulting lower energy consumption make non-cooperative schemes attractive in some practical scenarios.

According to the scope/field of application, horizontal spectrum sharing solutions can be classified as *internetwork solutions* or *intranetwork solutions.* The aim of internetwork solutions is to enable the coordinated coexistence of multiple DSA/CR systems or networks being deployed in overlapping spectrum and location areas. On the other hand, intranetwork schemes are aimed at coordinating spectrum access among the entities (i.e., UEs) inside the same DSA/CR system or network.

Traditionally, internetwork spectrum sharing has been dictated by the regulatory bodies via static frequency allocations. However, spectrum sharing among multiple DSA/CR systems poses unique challenges that have not been considered before in wireless communication systems and require alternative, more flexible solutions. Existing solutions for internetwork spectrum sharing can be categorized into centralized and distributed schemes.

Centralized internetwork spectrum sharing commonly relies on a central entity that plays the role of a spectrum broker, deciding which spectrum is allocated to each DSA/CR network. A spectrum broker is a central network entity that controls the spectrum sharing among different secondary networks. A spectrum broker can be connected to each network and can serve as a spectrum information manager to enable coexistence of multiple secondary networks. One option is to deploy a *spectrum allocation server* (SAS) equipped with CR capabilities that senses the surrounding environment and acquires spectrum utilization information in its coverage area. Upon request, the SAS allocates the available spectrum to overloaded vendors (i.e., service providers such as wireless operators) [59]. Another option is the definition of a spectrum auction framework where a *spectrum policy server* (SPS) performs the centralized allocation of available spectrum in a specified geographical region [60]. In this scheme, the operators dynamically compete for customers as well as portions of available spectrum. The operators bid for the spectrum, indicating the amount they are willing to pay for a given portion of spectrum and for a given time interval. The SPS then decides the spectrum allocation to each operator, maximizing its profit from the bids received. Through demand responsive pricing, the operators try to come up with convincing service offers for the customers, while trying to maximize their profits. UEs then select which operator to use for a given service type. Spectrum auction frameworks are aimed at achieving conflict-free spectrum allocations that maximize auction revenue and spectrum utilization [61]. The design of such frameworks involves compact bidding languages, pricing models to control tradeoffs between revenue and fairness, and fast auction clearing algorithms to compute revenue-maximizing prices and allocations.

Distributed internetwork spectrum sharing commonly relies on a *common control channel* (CCC) that is used to coordinate distributed spectrum reservations and exchange information among various DSA/CR systems on the current spectrum in use. An example of such a CCC is the CPC presented in Section 13.2.1.3. Although the CPC was not specifically designed for the spectrum sharing problem, it can actually be employed as a mechanism to report the spectrum usage of various wireless systems, including the DSA/CR systems present in a certain geographical area. This information can be employed by the UEs of other DSA/CR systems to select in a distributed manner a spectrum band not used by any other wireless (primary or secondary) system. Other example of CCC is the CSCC etiquette protocol proposed in [44]. The CSCC protocol coordinates radio nodes of different wireless technologies in a proactive way, where a common spectrum coordination channel at the edge of the available spectrum bands is allocated for announcement of radio parameters (e.g., node identifier, center frequency, bandwidth, transmit power, data rate, modulation type, data burst duration, interference margin, and service type). Each node is equipped with a low-bit-rate, narrow-band control radio for listening to announcements and broadcasting its own parameters in the CSCC channel. Radio nodes receiving CSCC control information can then initiate appropriate spectrum sharing policies such as *first-come first-served* (FCFS), priority-based sharing, or dynamic pricing auction, to resolve conflicts in spectrum demand and share the resources more efficiently. The hidden-receiver problem can also be solved because the range of the CSCC can be designed to exceed that of regular service data and DSA/CR UEs can also explicitly announce their presence. It has been shown that a simple CSCC implementation can be used to significantly reduce interference between 802.11b and Bluetooth devices operating in close proximity [44] and enable spectrum coexistence between short-range IEEE 802.11b and wide-area IEEE 802.16a networks [62]. Another example of CCC is the *Distributed QoS-based Dynamic Channel Reservation* (D-QDCR) proposed in [63], which allows BSs of different networks not only to compete among them and reserve spectrum based on data volumes or QoS demands but also to distribute the allocated spectrum to the internal DSA/CR UEs.

Intranetwork spectrum sharing solutions can also be classified into centralized (infrastructure-based) and distributed (ad hoc) schemes. While centralized solutions are mainly based on cooperative schemes [64, 65], distributed methods can be cooperative [66–71] or non-cooperative [72–74]. The number and variety of existing solutions is notably prominent, including methods for both infrastructure-based and ad hoc networks based on random, time-slotted, and hybrid protocols [75, 76]. However, they are frequently based on ideas and principles that have been borrowed from already existing MAC solutions developed for conventional wireless systems such as, for example, the *Request-to-Send* (RTS)/*Clear-to-Send* (CTS) and *Network Allocation Vector* (NAV) concepts of the IEEE 802.11 MAC protocol, busy tones, beacon signals, and clustering methods. A significant number of intranetwork spectrum sharing solutions for DSA/CR systems are based on variations and adaptations of classical MAC methods, which are modified to cope with the particular characteristics and issues of DSA/CR systems.

It is worth noting that inter- and intranetwork spectrum sharing solutions are not necessarily mutually incompatible. Some internetwork solutions can be applied to allocate spectrum at the network level and decide how to share a certain amount of spectrum among several DSA/CR systems, while intranetwork solutions can then be employed to allocate spectrum at the UE level and arbitrate the access of UEs inside the same DSA/CR system.

13.2.4   Spectrum mobility

To avoid harmful interference to licensed UEs, secondary UEs must vacate the channel as soon as a licensed primary UE is detected and move to an alternative spectrum hole in another channel and/or spectrum band. This event results in a transition from a channel/band to a different one and is referred to as *spectrum handoff.* The spectrum mobility function is in charge of triggering spectrum handoffs when appropriate (deciding who and when will switch to another channel/band) and enabling the mechanisms to provide a seamless transition of secondary communications between channels/bands while guaranteeing interference-free operation. The implementation of this functionality is in general challenging, specially in the context of multi-RAT networks.

Spectrum handoffs may be triggered by several events. One of the most intuitive reasons for triggering a spectrum handoff is the detection of a primary UE, which may lead to time- and space-dimension spectrum handoffs. A time-dimension spectrum handoff occurs when a licensed UE starts accessing a radio channel that is currently used by an unlicensed UE. In such a case, the unlicensed secondary UE must vacate the channel and switch to an idle channel/band. A space-dimension spectrum handoff occurs as a result of the DSA/CR UE (physical) mobility[2] when the DSA/CR UE moves to a

---

[2] It is worth noting that the spectrum mobility problem in DSA/CR systems is different from the UE (physical) mobility problem in any mobile communication system. UE mobility always implies a change of location of

geographical region where the current channel is busy (i.e., used by a primary system), thus forcing the secondary UE to switch to another channel/band. While the detection of a primary UE may be the most important reason to trigger a spectrum handoff, the need to preserve (or improve if possible) the QoS experienced by the DSA/CR UEs may also trigger a spectrum handoff, for example in the case of degradation of the radio propagation conditions in the current channel or if a better channel is detected.

Depending on the chronological order of the condition that triggers a spectrum handoff and the time instant the spectrum handoff is actually triggered, spectrum handoff methods can be classified as *reactive* or *proactive*. Reactive spectrum mobility methods wait until the triggering condition occurs (i.e., a primary UE is detected, the QoS experienced by a DSA/CR UE falls below certain threshold, etc.) and then perform a spectrum handoff. Proactive spectrum mobility methods are aimed at anticipating the triggering condition by means of appropriate prediction methods and performing a spectrum handoff before the triggering condition actually happens. Reactive methods lead to a more efficient utilization of spectrum since spectrum opportunities are exploited until they must be released mandatorily. However, reactive methods may lead to some interference with primary UEs (during the time interval elapsed between the appearance of a primary UE and the real vacation of the channel) or a temporary degradation of the QoS experienced by a DSA/CR UE (during the time interval elapsed between the detection of the QoS degradation and the change to a better alternative channel). On the other hand, proactive methods avoid these undesirable situations since the spectrum handoff is triggered in advance. However, the spectrum efficiency resulting from proactive methods may be slightly lower since spectrum opportunities may be released before they actually become unusable. The main challenge of proactive methods lies in the need to predict the triggering event with an acceptable accuracy level in order to achieve interference-free and efficient operation. To this end, the use of accurate spectrum occupancy models [52, 53] can be very helpful in estimating future spectrum occupancy trends and anticipating potential triggering conditions.

When a spectrum handoff is triggered, the parameters at different layers of the protocol stack have to be adjusted. This implies not only the modification of parameters of the physical layer such as transmission power, bandwidth, modulation, and coding schemes in order to match the operation conditions of the new channel or spectrum band, but also the adjustment of parameters at higher layers to ensure a seamless transition process. For instance, a spectrum handoff may lead to a high latency, which may require the *Transmission Control Protocol* (TCP) timer to be frozen at the transport layer when an unlicensed UE switches channel, to avoid any misinterpretation of the delay incurred for the acknowledgment message. Every time a spectrum handoff takes place, the protocol stack is shifted from one mode of operation to another. The objective of the spectrum

the mobile UE, which may require a change of serving BS but not always a change of operation frequency. On the other hand, spectrum mobility always implies a change of operation frequency and is not necessarily associated with a change of location of the mobile DSA/CR UE nor the serving BS. Spectrum mobility may be conditioned by the DSA/CR UE mobility (e.g., when the DSA/CR UE moves to a region where the set of available spectrum holes is different, the channel quality conditions degrade or better alternatives can be found) but the converse is not true.

Table 13.2 Parameters defined by the IEEE 802.22 standard to protect primary UEs (reproduced with permission from IEEE©).

Parameter	Value for wireless microphones	Value for TV receivers
CDT	≤2 s	≤2 s
CMT	2 s	2 s
CCTT	100 ms	100 ms
IDT	−107 dBm (over 200 kHz)	−114 dBm (over 6 MHz)

mobility function is to guarantee that spectrum handoffs are performed smoothly and as fast as required to ensure that DSA/CR UEs perceive minimum performance degradation during such transitions, while at the same time avoiding harmful interference with primary UEs. The objective of the spectrum mobility function is therefore twofold: avoid interference with primary UEs and provide seamless transitions to minimize the impact on the perceived QoS (i.e., the secondary UE communication should not be affected by the spectrum handoff).

To avoid harmful interference, an efficient spectrum mobility solution should be able to effectively meet certain interference constraints, which are highly dependent on the particular characteristics of the primary network. For instance, the IEEE 802.22 standard conceived to operate in the TV broadcast bands (see Section 13.4) specifies some parameters to protect the primary receivers as shown in Table 13.2. Two key parameters are the channel detection time (CDT) and the incumbent detection threshold (IDT). The CDT defines the time period for which a primary receiver can withstand interference before the secondary system detects it. It dictates how quickly the secondary system must be able to detect a primary signal exceeding the IDT. Once the primary signal is detected to be higher than the IDT, two other new parameters have to be considered, namely the channel move time (CMT), i.e. the maximum interval after the primary signal is detected before vacating the channel, and the channel closing transmission time (CCTT), which is the aggregate duration of transmissions by secondary devices during the CMT. Although these parameters and their values can be appropriate to protect a primary TV broadcast network, other protection requirements may be necessary for other types of primary system. DSA/CR systems operating in multi-RAT environments must be aware of the interference tolerated by the various primary systems, and the spectrum mobility function must ensure that appropriate actions are taken to meet such requirements.

To preserve the QoS level experienced by DSA/CR UEs, the spectrum mobility function has to provide a seamless transition between different spectrum bands so that the impact of spectrum handoffs on the perceived QoS is minimized. To this end, the spectrum mobility function must be able to provide alternative channels in case the current channel needs to be vacated. A simple reactive approach is to search for idle channels/bands when a spectrum handoff is required. Since the selection of a new operational frequency may take time, reactive solutions have to be efficient in order to ensure that applications do not suffer from severe performance degradation during spectrum handoffs. An alternative proactive solution is to keep track of the available resources and maintain a list of backup channels/bands so that in the event of a spectrum handoff the DSA/CR communication can be switched immediately to a channel/band from the backup list. Proactive solutions reduce the latency of the spectrum handoff process and minimize the degradation of the QoS experienced by the DSA/CR UEs at the cost of the overhead associated with the periodic updating of the backup list.

13.2.5   Summary of cognitive radio techniques and cross-layer design

Previous sections have provided a detailed overview of the functionalities and techniques associated with the DSA/CR technology, which are summarized in Fig. 13.2. While these functionalities have been presented separately, the practical design of DSA/CR systems should take them into account from a joint perspective, as there are close interactions among them. The most clear interaction flow follows a cyclic sequence, as illustrated in Fig. 13.3. First, the spectrum awareness function detects transmission opportunities for the DSA/CR system. Based on the detected opportunities, the spectrum selection function selects the most suitable spectrum holes. The selected spectrum holes are then shared among the DSA/CR entities under the supervision of the spectrum sharing functionality. If the current spectrum hole becomes unavailable, then the spectrum mobility function initiates a spectrum handoff process, which requires the execution of a new cycle in order to find alternative frequencies, select the most convenient spectrum holes, and resume the communication in the new channel, which may be shared with other DSA/CR entities by means of spectrum sharing techniques.

There are, however, other less evident interactions among the processes executed at different levels of the protocol stack that may have a significant impact on the system performance. For instance, in periodic spectrum sensing a sensing frame is divided into a sensing slot and a data transmission slot. Increasing the duration of the sensing slots improves the probability of detection of the primary signal and hence reduces the probability of harmful interference. However, a longer sensing slot implies a shorter transmission slot, meaning that the selection of this simple parameter at the physical layer may have a direct impact on the throughput experienced at the application level, which is known as *sensing–throughput tradeoff* [77]. This aspect should be designed taking into account interference requirements at the physical layer and QoS requirements at the application layer. Another example of a cross-layer problem is the design of MAC-layer sensing algorithms, a key component of DSA/CR systems in charge of deciding what and when senses which channel. MAC-layer sensing schemes involve the joint consideration of physical and MAC-level aspects and their performance is critical to find a channel available for secondary transmission within a minimum searching delay, which has a direct impact on the QoS experienced at the application level [78–80]. As mentioned in Section 13.2.4, the latency incurred by a spectrum handoff may require to frozen timers at various levels of the protocol stack (e.g., MAC, transport, or application) in order to avoid undesired behaviors, which should be handled by the spectrum mobility function. Another example is the joint cross-layer design of cognitive *radio resource management* (RRM) functions involving various combinations of aspects

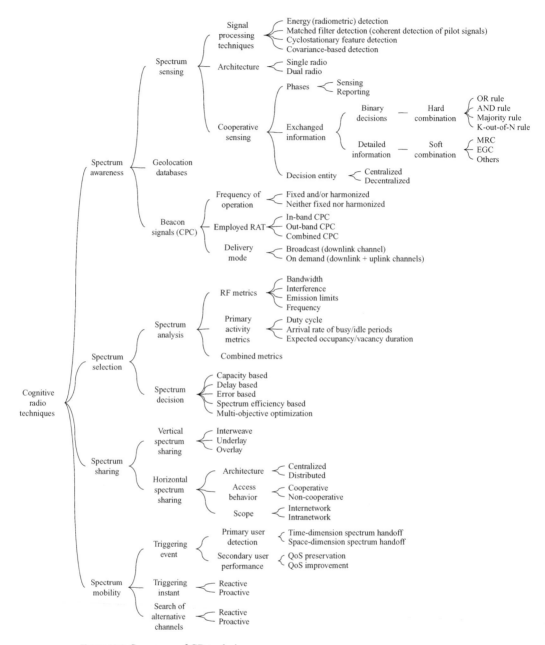

Figure 13.2 Summary of CR techniques.

such as spectrum sensing, admission and load control, scheduling, power control, link adaptation, or routing at the network level.

Since the overall system performance is determined by the combination of the performance of each single layer in the protocol stack and there are close interactions among the functions associated with each layer, it is quite intuitive that the parameters of each

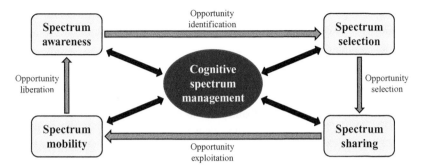

Figure 13.3 Interactions among CR functionalities.

layer should be tuned from a holistic, cross-layer optimization point of view. Cross-layer solutions in wireless communication systems are known to outperform other approaches where the parameters of each layer are optimized separately. The joint optimization of parameters at various layers of the protocol stack can help in overcoming the main challenges for DSA/CR systems, including interference minimization along with the provision of seamless communications and QoS guarantees to DSA/CR UEs.

13.3   Application scenarios for cognitive radio in heterogeneous cellular networks

The DSA/CR concept can be applied to a wide variety of wireless communication scenarios. In initial stages of the technological development, DSA/CR was identified as a new technology for military and emergency networks. Military communication systems operating in foreign countries with dissimilar spectrum allocations can automatically find spectrum bands for opportunistic transmissions without interfering with the local radio communication systems, thus remaining unnoticed. If certain frequencies become noisy or are jammed by the enemy, DSA/CR systems can search for alternative frequency bands for communication, and adapt to the new communication environment. Public safety and emergency networks can also take advantage of the DSA/CR concepts to provide reliable and flexible communication means in exceptional disaster scenarios where the network infrastructure deployed for routine services is partially damaged or totally destroyed. The DSA/CR technology can be used to spontaneously configure new communication links among surviving nodes in alternative frequency bands.

More recently, DSA/CR has been identified not only as a promising solution to improve spectrum usage efficiency in radio communication systems but also as a key technology to improve many aspects of wireless communications, providing intelligence to network entities and enabling UEs to access multiple air interfaces and select the most appropriate alternative under varying communication needs and conditions. As a matter of fact, the standards for many of the classical technologies deployed in multi-RAT networks (e.g., WiFi, WiMAX, and LTE), have recently been modified in order to incorporate coexistence and DSA/CR-like capabilities, as will be discussed in Section 13.4. This section describes various application scenarios for the DSA/CR technology in HCNs.

13.3.1   Rural broadband

Providing Internet connectivity to homes in rural areas by means of a wired infrastructure is prohibitively costly. Broadband *point to multi-point* (P2MP) links between a BS and each home become a more convenient and cost-efficient solution. In this scenario, homes with external directional antennas in the rooftop act as static remote UEs on the UE side while BSs with omnidirectional or sectorized antennas act as the access point providing Internet access. In order to provide broadband Internet access over reasonable coverage distances, the transmission frequencies should ideally be placed in the lower region of the spectrum. However, lower spectrum bands are in general subject to a more intensive usage [49, 50], which makes it difficult to find fixed spectrum bands that are available over large geographical areas. The DSA/CR technology can be employed to identify available frequency bands in the local area of operation of the BS, which may be different among various BSs, thus enabling broadband connectivity over large coverage areas. A DSA/CR system to provide Internet connectivity in this scenario reusing TV white spaces is defined in the IEEE 802.22 standard, which will be described in more detail in Section 13.4.

13.3.2   Dynamic backhaul

Backhaul connections from remote UEs and access networks to a *core network* (CN) can be implemented in a cost-efficient manner by means of wireless connections. The nodes providing the backhaul can be *point to point* (P2P) links acting as relays as well as traffic aggregators combining several links from the access side of the network into one link towards the CN. The backhaul links can be provided using DSA/CR opportunistic access, thus making this solution more cost effective than traditional wireless backhaul solutions based on purchased spectrum licenses. While the aggregated traffic volume at the core side of the backhaul could be fairly constant, the traffic at the access side or through relays/aggregators could be highly variable. A DSA/CR dynamic backhaul could adjust and optimize dynamically the bandwidth consumption at every node of the backhaul network based on the experienced traffic level. Moreover, the DSA/CR technology would allow the backhaul entities to be mobile, reconfiguring operation links and parameters at each node in order to provide an adequate traffic routing from UEs towards the CN. The use of DSA/CR can also be more convenient to search for alternative connections in case of link failures.

13.3.3   Cognitive ad hoc networks

Ad hoc scenarios are in general characterized by highly dynamic properties. Mobile ad hoc nodes communicate in a self-configuring and infrastructureless manner, following a distributed multi-hop architecture and based on dynamic and varying network topologies. Each node can move independently in any direction, changing its links to other mobile ad hoc nodes frequently. In such dynamic scenarios, the DSA/CR concept appears as the natural communication technology, leading to the concept of cognitive ad hoc networks. The DSA/CR technology can be employed to dynamically adapt to the varying communication conditions of the ad hoc nodes, finding appropriate frequencies to establish new links as the nodes move along arbitrary paths. Each node can obtain knowledge of its environment and select the most appropriate channel and operation parameters in order to coexist with the rest of nodes and maintain the network. DSA/CR concepts can be used to adapt the radio transmissions to cope with the rapid variations of the radio environment.

13.3.4 Capacity extension in cellular networks

The success of cellular communication networks has resulted in a steadily increasing demand, not only for the traditional voice service, but also for more recent data multimedia services that are extremely bandwidth consuming. Current cellular systems are facing severe problems to accommodate the ever-increasing traffic demands. The DSA/CR technology can provide a satisfactory solution to this challenging situation by enabling cellular communication systems to utilize spectrum holes found in other spectrum bands in addition to their own licensed spectrum. In this scenario, cellular networks, acting as secondary systems, can take advantage of spectrum opportunities found in other spectrum bands and reuse portions of licensed spectrum in a interference-free manner. This situation would establish a relation of vertical spectrum sharing between the primary system operating in the licensed band and the secondary cellular networks, and a relation of horizontal spectrum sharing among cellular networks opportunistically reusing portions of the same primary spectrum band. Mobile operators can therefore gain additional bandwidth to alleviate the current situation of spectrum congestion in cellular networks [49, 50] and meet the experienced traffic demands. This solution may be specially interesting for future broadband LTE systems, targeting peak data rates up to 300 Mbit/s. Moreover, the OFDMA technology employed in the radio interface of LTE systems is well suited to efficiently reuse fragmented spectrum holes by nulling the frequency carriers where a primary signal is present. DSA/CR techniques can be used to coordinate the opportunistic access of several operators to a licensed band and enhance the capacity and coverage of cellular networks.

13.3.5 Direct UE-to-UE communication in cellular networks

Cellular communication networks are infrastructure-based systems where all the traffic between any pair of mobile UEs goes through the serving BSs. When the end-points of the communication are sufficiently close to establish a direct link between them, a direct UE-to-UE link can be established so that the mobile UEs can communicate with no traffic going through the BSs. This communication approach reduces the traffic load supported by the infrastructure side of the network, which may be of different significance depending on whether the mobile UEs are served by the same or different

BSs.[3] Moreover, a better spectrum efficiency is achieved. While in the classical communication approach two links are required, one between each mobile UE and the BSs, a single link between the mobile UEs is needed in direct UE-to-UE communication. Furthermore, in the former the distance between the mobile UEs and the BSs may be of the order of magnitude of the cell size, while in the latter the distance between two mobile UEs can be much smaller, thus leading to lower transmission powers and reduced interference levels. The DSA/CR technology represents a suitable enabler for this scenario by allowing mobile DSA/CR UEs to detect the proximity of the other UE, find a suitable frequency for the direct link and take appropriate actions to reconfigure their operation parameters.

13.3.6   Coordination and cognitive X2 links

In the context of HCNs, *home evolved NodeBs* (HeNBs) can form an ad hoc network for coordination purposes and the application of *intercell interference coordination* (ICIC) techniques. Notice that the coordination among HeNBs is a challenging task in HCNs due to the lack of a direct interface among HeNBs, similar to the X2 interface among *evolved NodeBs* (eNBs). Coordination among HeNBs would require some kind of communication and signaling overhead through gateways. On the other hand, the establishment of direct links over-the-air among HeNBs based on cognitive techniques would enable a more efficient communication and coordination of the femtocells. Although HeNBs are not mobile, the uncoordinated user-deployed feature may lead to varying network topologies that may change in an unpredictable manner. Cognitive ad hoc techniques can be well suited to this scenario, providing the means for the establishment and management of the network. Cognitive HeNBs would be in charge of acquiring spectrum awareness by means of spectrum sensing. This information could possibly be shared among the nodes of the network (i.e., the HeNBs) based on cooperative sensing methods. The nodes would then select appropriate carrier frequencies for the establishment of individual P2P links, either on the same spectrum band as the macrocell or in external underutilized bands, based on opportunistic spectrum access methods. This cognitive ad hoc network would be able to reconfigure the employed links in the eventual case that the carrier frequency becomes idle in the associated macrocell or external spectrum band, or if one or more of the HeNBs are switched on or off. Cognitive ad hoc networks have been studied extensively [81], and existing solutions could be implemented in HeNBs to enable over-the-air interfemtocell coordination mechanisms without involving additional network elements of the HCN.

13.3.7   Cognitive femtocells

Femtocells are low-power, short-range, consumer-deployed access points connected to the fixed network infrastructure that provide broadband indoor coverage at homes, offices, or public hotspots (e.g., cafés, airports, or shopping centers). Femtocell access

---

[3] Two UEs served by different BSs may be close to each other at the edge of their respective cells.

points (i.e., HeNBs) provide gateway functionalities to indoor UEs by connecting them to the network operator through a broadband wired connection (cable or DSL line) or a wireless backhaul link. Femtocells are deployed by the consumer (i.e., without network planning) and in general operate in the same frequency band as the surrounding macrocell and other neighboring femtocells, thus leading to some potential femtocell-to-macrocell (cross-layer) and femtocell-to-femtocell (co-layer) interference.

Controlling interference between femtocells and macrocells in HCNs is a challenging problem, and several alternatives have been proposed. Interference can be managed by means of static and dynamic frequency reuse schemes [82, 83], where the macrocell and the femtocell share different portions of the spectrum allocated to the macrocell, and coordinated resource allocation schemes [84–86], where both the femtocell and the macrocell use the same spectrum, relying on adequate coordination mechanisms. In the former the spectrum partitioning may not result in an optimum exploitation of the available resources, while in the latter it is very challenging, in practice, to coordinate resource allocations between eNBs and HeNBs due to the lack of a direct interface defined between these entities (the X2 interface enables inter-eNB communication at the macrocell level only). Communication between eNBs and HeNBs might be achieved be means of links through the CN and the corresponding gateways, namely the *System Architecture Evolution Gateway* (SAEGW) and the *home gateway* (HGW). However, the deployment of HeNBs should not affect the CN since it may also be shared with other RATs. Moreover, the aforementioned solutions may not be completely satisfactory in dense deployments due to signaling and delay issues. An alternative appealing solution is to provide HeNBs with cognitive capabilities and formulate macrocells as primary BSs and HeNBs as secondary BSs exploiting spectrum opportunities with minimum interference with the former, which in this scenario plays the role of a primary system composed of a primary BS (i.e., the eNB) and primary mobile UEs (i.e., the macrocell UEs). Cognitive femtocells can exploit their capability to observe the radio environment in real time and adapt dynamically in order to achieve spectrum coexistence at the cross- and co-layer levels in a way that is transparent to the macrocell. In general, cognitive interference management techniques [87–91] result in a more efficient spectrum exploitation compared to spectrum partitioning methods, without the need of direct coordination among macrocells and femtocells. However, the introduction of cognitive femtocells in HCN environments requires some particular considerations.

First of all, one of the most important aspects of this scenario is the need of HeNBs to acquire information on their surrounding radio environment. Several spectrum awareness methods were presented in Section 13.2.1. The concept of a geolocation database might be applied in this scenario in the form of a database of frequency resources, where the location information and expected interference range of nodes (at both the macrocell and femtocell level) is stored and used [92]. However, this approach would require some kind of cross-layer coordination to maintain the database, which poses some issues as discussed above. An alternative is to exploit the signaling channels used by eNBs to inform their associated UEs of sub-channel allocations. These signaling channels can be exploited by HeNBs as beacon signals to gain information on the macrocell spectrum usage. As eNBs will broadcast this information with relatively high power, any HeNB in the macrocell should be able to detect the occupied sub-channels as long as the "DL receiver-sniffing" functionality is implemented in order to receive and process eNB transmissions. An alternative and appealing solution is to allow HeNBs to perform spectrum sensing on the spectrum band allocated to the macrocell in order to identify channels that are not being utilized at the time by the macrocell, and use these channels opportunistically for femtocell communications. Spectrum sensing can provide information about the radio environment in a relatively fast manner, without dependence on the macrocell or interaction with other femtocells. Moreover, as the spectrum sensing functionality would be implemented at HeNBs, it would not require any additional network infrastructure and would not be an issue for the UEs' battery life. Various sensing methods, such as those presented in Section 13.2.1.1, can be used, depending on the desired level of complexity and cost of HeNBs. Moreover, the knowledge of the RAT employed by the primary system (i.e., LTE/OFDMA) can be exploited to make use of detection methods specifically designed for such a signal format [93–95]. As a single HeNB sensing node might not be able to detect an eNB (primary) signal in the presence of channel degrading effects such as shadowing and multi-path fading, an interesting option to be considered is the introduction of cooperative sensing methods. Different HeNBs observe different (i.e., statistically independent) realizations of the same macrocell signals. Therefore, sharing of the sensing information among a set of nearby HeNBs inside the same macrocell can provide detection gains, thus resulting in a more reliable estimation of the frequencies used by the macrocell. Moreover, as the sensing is implemented at HeNBs, the additional energy consumption associated with the sensing cooperation would not be an issue. However, a reliable communication means among HeNBs is required to exchange sensing information. This drawback can be overcome with the establishment of a cognitive ad hoc network, in the macrocell spectrum band or an external band, as discussed in Section 13.3.3.

Based on the gathered spectrum usage information and the analysis of the detected spectrum opportunities (bandwidth, emission limits, occupancy statistics, etc.), each HeNB can then select the carrier frequencies that best suit the needs and requirements of the femtocell UEs. Interference constraints at both the cross- and co-layer levels require appropriate spectrum sharing methods to be employed. In a cognitive femtocell environment, there exists a relation of vertical spectrum sharing between the (primary) macrocell and the (secondary) femtocells. A femtocell can access the macrocell spectrum following an interweave approach by occupying carrier frequencies temporarily unused by the macrocell. While this method does not impose severe transmission power limits, femtocell transmissions are restricted to the inactivity periods of the macrocell, which may be an issue for delay-sensitive services. In contrast, an underlay approach can be adopted, with the femtocell being power-controlled to avoid cross-layer interference to the macrocell [96]. Femtocell transmissions would be allowed at any time, but under strict transmission power limits, which may constrain the achievable data rates. The method providing the highest femtocell capacity and performance depends on the macrocell level of activity and tolerable interference as well as the traffic requirements. Moreover, cognitive femtocells have the same right to opportunistically access the macrocell spectrum and therefore a relation of horizontal spectrum sharing holds among the deployed femtocells. Owing to the unplanned and uncoordinated deployment of HeNBs, distributed solutions appear as the natural approach to manage co-layer interference in cognitive femtocells. Moreover, the establishment of an ad hoc network of HeNBs can be exploited not only to share spectrum sensing information as discussed above, but also implement cooperative horizontal spectrum sharing approaches, which in general outperform non-cooperative schemes.

The previous discussions assume that cognitive femtocells operate in the same spectrum as the macrocell. However, an interesting alternative is the exploitation of the femtocells' cognitive capabilities to access other spectrum bands subject to low utilization levels [97]. This approach would allow us to not only alleviate both the cross- and co-layer interference, but also increase the overall system capacity, as the allocated spectrum would be exclusively used by the macrocell, while femtocells reuse other less crowded spectrum bands. The downside of this alternative is that it may result slightly more challenging, as, for example, the RAT employed in the target band may be unknown (i.e., spectrum sensing methods specifically designed and optimized for particular signal formats cannot be employed), and different femtocells may access spectrum blocks allocated to different primary transmitters (i.e., cooperative sensing methods may turn out to be less useful).

In summary, the implementation of the DSA/CR concept and associated techniques in HeNBs can provide an efficient method for interference management. Cognitive femtocells, despite being deployed uncoordinatedly, can determine the spectrum usage in their surrounding environment at any moment, select suitable frequencies to avoid interference with the macrocell, and share spectrum with other femtocells efficiently. The same logic can be applied to picocells, relay nodes, and other HCN low-power nodes.

13.4   Standardization activities: the future of cognitive radio systems

An important and essential aspect in the commercial feasibility and deployment of the DSA/CR concept is its standardization. The most important standardization initiatives to date have been brought forward by the IEEE. Nevertheless, other international standardization organizations or industry associations such as the *International Telecommunication Union* (ITU), the Wireless Innovation Forum (formerly known as SDR Forum), the *European Telecommunications Standards Institute* (ETSI), and the European association for standardizing information and communication systems (ECMA international) have been or are working in the development of DSA/CR standards as well [98].

The most popular standards on DSA/CR are the IEEE 802.22 standard for *wireless regional area network* (WRAN) using white spaces in the TV frequency spectrum and the IEEE P1900 series of standards in the area of dynamic spectrum management [99] developed by the IEEE *Standards Coordinating Committee* (SCC) 41, formerly known as IEEE P1900 Standards Committee. However, there are several other, lesser known, related activities within IEEE as well. Many other completed IEEE 802 standards already include DSA/CR-like capabilities or related building blocks that evolved from coexistence activities [100].

IEEE 802.22 is the first worldwide standard based on the DSA/CR technology [101–103]. The most prominent target application of IEEE 802.22 is wireless broadband access in rural and remote areas, with performance comparable to those of existing fixed broadband access technologies serving urban and suburban areas. The standard defines the physical (PHY) and MAC layers of a DSA/CR-based air interface [104, 105] for use by license-exempt devices on a non-interfering basis in spectrum allocated to the TV broadcast service. The standard considers the utilization of TV bands for two main reasons, namely the favorable propagation characteristics of the lower-frequency bands allocated to the TV service and the consequent larger coverage areas, and the considerable amount of available TV white space [106–108]. The 802.22 project initially identified the North American frequency range of operation between 54 and 862 MHz, which was extended to between 47 and 910 MHz in order to meet additional international regulatory requirements. Since there is no worldwide uniformity in channelization for TV services, the standard also accommodates the various international TV channel bandwidths of 6, 7, and 8 MHz. The 802.22 system defines a fixed point-to-multipoint (PTM) architecture where each BS communicates with up to 255 associated stations, referred to as *customer premises equipments* (CPEs), by means of time division multiplexing in the *downstream* (DS) direction (BS to CPE) and a demand assigned *time division multiple access* (TDMA) scheme in the *upstream* (US) direction (CPE to BS). The specified BS coverage range is 30 km (with some additions to support 100 km under favorable propagation conditions) and targets peak throughput rates at the edge of the coverage area of 1.5 Mbit/s per UE in DS and 384 kbit/s in US.

While IEEE 802.22 focuses on the development of specific mechanisms for the PHY and MAC layers, the IEEE P1900 series concentrates on the development of architectural concepts and specifications for policy-based network management with dynamic spectrum access in a multi-RAT network composed of incompatible wireless technologies such as Third/Fourth Generation (3G/4G), WiFi, and WiMAX. The series is composed of six standards defining terminology and concepts (IEEE P1900.1), recommended practice for interference and coexistence analysis (IEEE P1900.2), dependability and evaluation of regulatory compliance (IEEE P1900.3), architectural building blocks (IEEE P1900.4 [109]), policy language, and policy architectures (IEEE P1900.5), as well as spectrum sensing aspects (P1900.6).

Other IEEE standardization activities have been initiated to address coexistence issues among various systems or to make amendments to existing standards with the aim of supporting coexistence with license-exempt devices. For instance, several amendments have been made to the PHY and MAC layers of the IEEE 802.11 standard to support channel access and coexistence in TV white spaces. Some examples include the introduction of new functionalities such as sensing of other transmitters (IEEE 802.11af, also known as Wi-Fi 2.0, White-Fi or "Wi-Fi on steroids" [110, 111]), *transmit power control* (TPC) and *dynamic frequency selection* (DFS) (IEEE 802.11h), and some extensions thereof (IEEE 802.11y). Similarly, the IEEE 802.16h amendment develops improved mechanisms to enable coexistence among license-exempt systems based on the IEEE 802.16 standard, and to facilitate the coexistence of such systems with primary UEs

[112]. IEEE 802.19 deals with coexistence issues between unlicensed wireless networks, such as IEEE 802.11/15/16/22. The focus of the IEEE 802.19 standard is on the development of technology-independent methods for supporting coexistence among potentially dissimilar networks that will operate in a common TV white space channel.

In addition to the work performed by the IEEE, some other organizations are working on the definition of standards for DSA/CR systems. ITU-R has published two technical reports [113, 114] on the application of SDR to the *International Mobile Telecommunications* (IMT) 2000 global standard for 3G mobile communications and other land mobile systems. In September 2009 the Wireless Innovation Forum (formerly known as the SDR Forum) initiated a test and measurement project in order to develop a set of use cases, test requirements, guidelines, and methodologies required for secondary opportunistic access to TV white space. The project addresses critical functions such as spectrum sensing, interference avoidance, database performance, and policy conformance. In October 2009, the ETSI Technical Committee for Reconfigurable Radio Systems (RRS) published a series of technical reports [115] summarizing various feasibility studies and examining standardization needs and opportunities. In December 2009, ECMA released the first DSA/CR standard for personal/portable devices operating in TV white space [116, 117]. The standard specifies the PHY and MAC layers of a DSA/CR system with flexible network formation, mechanisms for protection of primary UEs, adaptation to different regulatory requirements, and support for real-time multimedia traffic. The ECMA-392 standard is expected to enable new applications such as in-home high-definition video streaming and interactive TV broadcasting services. More recently, in February 2011, and after the decision of the Federal Communications Commission (FCC) to remove spectrum sensing requirements and use TV white space databases to identify unused TV channels, the *Internet Engineering Task Force* (IETF) joined the standardization efforts by releasing a draft on Protocol to Access White Space (PAWS) databases [118, 119]. The protocol will explore the various aspects of a messaging interface between white space devices and (multiple) white space databases.

Despite the number of initiatives and activities carried out so far, standardization of DSA/CR systems constitutes an exciting challenge still requiring much more effort. Different countries may have different spectrum regulations. While this appears to be reasonable as a result of different social and economic environments, this situation complicates the standardization of DSA/CR systems and the development of worldwide standards. Moreover, there exists a variety of organizations working independently on different standards. Furthermore, the evolving trend towards converged multi-RAT networks poses unique challenges. Although some consolidation is required in this area, the fact is that regulation, standardization, and evolution towards heterogeneity is ongoing, and the final impact remains unknown. How these aspects can be harmonized constitutes a big question yet to be answered in the foreseeable future.

13.5   Summary and conclusions

CR is a revolutionary communications paradigm based on the principles of cognition of the radio environment and reconfiguration to intelligently adapt to new communication conditions. The CR concept is expected to deeply impact the way spectrum will be accessed, shared and exploited in future wireless communication systems. Owing to its advantages and flexibility, DSA/CR has been proposed not only as a promising solution to improve the spectrum usage efficiency, but also as a relevant technology to improve many aspects of wireless communications, providing intelligence to the network entities and enabling the access to multiple spectrum bands and air interfaces, and select the most appropriate communication alternative.

While conceptually simple, the practical implementation of the DSA/CR concept requires effective and efficient solutions to obtain information about the surrounding radio environment, identify the availability of resources for opportunistic transmission, and maintain the communication through unreliable and intermittently unavailable channels, while avoiding interference to the legitimate users of the spectrum. This challenging communication scenario claims new functionalities and techniques that were not required in legacy wireless communication systems. This chapter has provided a detailed overview of the technological solutions associated with the DSA/CR concept, including methods to acquire spectrum awareness, select the most appropriate frequency of operation, and share spectrum opportunities with other DSA/CR UEs and networks while coexisting with primary systems in a non-interfering manner. Some cross-layer issues and design tradeoffs have also been discussed. The potential application of DSA/CR methods and techniques in the context of HCNs has been analyzed, exploring several scenarios of practical interest and discussing various implementation alternatives. Finally, recent standardization efforts aimed at developing new systems based on DSA/CR principles as well as including DSA/CR capabilities in already existing standards have been reviewed. DSA/CR has the potential to provide significant benefits and simple solutions to overcome many of the challenges faced by future HCNs.

References

[1] J. Mitola, Software radios – survey, critical evaluation and future directions. In *Proceedings of the National Telesystems Conference (NTC 1992)* (1992), pp. 13/15–13/23.

[2] J. Mitola and G. Q. Maguire, Cognitive radio: making software radios more personal. *IEEE Personal Communications Magazine*, **6**:4 (1999), 13–18.

[3] J. Mitola, *Software Radio Architecture* (Wiley-Interscience, 2000).

[4] J. Mitola, *Cognitive Radio Architecture* (Wiley-Interscience, 2006).

[5] Q. Zhao and A. Swami, A survey of dynamic spectrum access: signal processing and networking perspectives. In *Proceedings of the IEEE International Conference on Acoustics, Speech and Signal Processing (ICASSP 2007)* (2007), vol. 4, pp. IV/1349–/IV–1352.

[6] Q. Zhao and B. M. Sadler, A survey of dynamic spectrum access. *IEEE Signal Processing Magazine*, **24**:3 (2007), 79–89.

[7] M. M. Buddhikot, Understanding dynamic spectrum access: taxonomy, models and challenges. In *Proceedings of the Second IEEE International Symposium on New Frontiers in Dynamic Spectrum Access Networks (DySPAN 2007)* (2007), pp. 649–663.

[8] K. G. Shin, H. Kim, A. W. Min and A. Kumar, Cognitive radios for dynamic spectrum access: from concept to reality. *IEEE Wireless Communications*, **17**:6 (2010), 64–74.

[9] A. Ghasemi and E. S. Sousa, Spectrum sensing in cognitive radio networks: requirements, challenges and design trade-offs. *IEEE Communications Magazine*, **46**:4 (2008), 32–39.

[10] S. M. Kay, *Fundamentals of Statistical Signal Processing: Detection Theory* (Prentice Hall, 1998), vol. 2.

[11] H. V. Poor, *An Introduction to Signal Detection and Estimation*, 2nd edn. (Springer, 1998), vol. 2.

[12] R. Tandra and A. Sahai, SNR walls for signal detection. *IEEE Journal of Selected Topics in Signal Processing*, **2**:1 (2008), 4–17.

[13] R. Tandra and A. Sahai, SNR walls for feature detectors. In *Proceedings of the Second IEEE International Symposium on New Frontiers in Dynamic Spectrum Access Networks (DySPAN 2007)* (2007), pp. 559–570.

[14] D. Cabric, A. Tkachenko and B. W. Brodersen, Experimental study of spectrum sensing based on energy detection and network cooperation. In *Proceedings of the First International Workshop on Technology and Policy for Accessing Spectrum (TAPAS 2006)* (2006), pp. 1–8.

[15] D. Cabric, A. Tkachenko and R. W. Brodersen, Spectrum sensing measurements of pilot, energy, and collaborative detection. In *Proceedings of the IEEE Military Communications Conference (MILCOM 2006)* (2006), pp. 1–7.

[16] M. López-Benítez, F. Casadevall and C. Martella, Performance of spectrum sensing for cognitive radio based on field measurements of various radio technologies. In *Proceedings of the 16th European Wireless Conference (EW 2010), Special Session on Cognitive Radio* (2010), pp. 1–9.

[17] J. Ma, G. Y. Li and B. H. Juang, Signal processing in cognitive radio. *Proceedings of the IEEE*, **97**:5 (2009), 805–823.

[18] Y. Zeng, Y.-C. Liang, A. T. Hoang and R. Zhang, A review on spectrum sensing for cognitive radio: challenges and solutions. *EURASIP Journal on Wireless Communications and Networking* (2010), 1–15.

[19] T. Yücek and H. Arslan, A survey of spectrum sensing algorithms for cognitive radio applications. *IEEE Communications Surveys and Tutorials*, **11**:1 (2009), 116–130.

[20] D. D. Ariananda, M. K. Lakshmanan and H. Nikookar, A survey on spectrum sensing techniques for cognitive radio. In *Proceedings of the Second International Workshop on Cognitive Radio and Advanced Spectrum Management (CogART 2009)* (2009), pp. 74–79.

[21] D. Noguet *et al.*, *Sensing Techniques for Cognitive Radio – State of the Art and Trends* (2009), IEEE SCC 41 P1900.6 White Paper http://grouper.ieee.org/groups/scc41/6/documents/white_papers/P1900.6_WhitePaper_Sensing_final.pdf.

[22] H. Urkowitz, Energy detection of unknown deterministic signals. *Proceedings of the IEEE*, **55**:4 (1967), 523–531.

[23] R. Price and N. Abramson, Detection theory. *IEEE Transactions on Information Theory*, **7**:3 (1961), 135–139.

[24] J. G. Proakis, *Digital Communications*, 5th edn. (McGraw-Hill, 2008).

[25] A. Sahai, N. Hoven and R. Tandra, Some fundamental limits on cognitive radio. In *Proceedings of the 42nd Allerton Conference on Communications, Control, and Computing (Allerton Conference 2004)* (2004), p. 10.

[26] D. Cabric, S. M. Mishra and R. W. Brodersen, Implementation issues in spectrum sensing for cognitive radios. In *Conference Record of the 38th Asilomar Conference on Signals, Systems and Computers (ACSSC 2004)* (2004), vol. 1, 772–776.

[27] W. A. Gardner, Signal interception: a unifying theoretical framework for feature detection. *IEEE Transactions on Communications*, **36**:8 (1988), 897–906.

[28] W. A. Gardner and C. M. Spooner, Signal interception: performance advantages of cyclic-feature detectors. *IEEE Transactions on Communications*, **40**:1 (1992), 149–159.

[29] Y. Zeng and Y.-C. Liang, Spectrum-sensing algorithms for cognitive radio based on statistical covariances. *IEEE Transactions on Vehicular Technology*, **58**:4 (2009), 1804–1815.

[30] M. López-Benítez, Improved energy detection spectrum sensing for cognitive radio. *IET Communications, Special Issue on Cognitive Communications*, **6**:8 (2012), 785–796.

[31] J. Palicot, C. Moy and R. Hachemani, Multilayer sensors for the sensorial radio bubble. *Physical Communication*, **2**:1–2 (2009), 151–165.

[32] N. Hoven and A. Sahai, Power scaling for cognitive radio. In *Proceedings of the International Conference on Wireless Networks, Communications and Mobile Computing (WirelessComm 2005)* (2005), vol. 1, pp. 250–255.

[33] A. Ghasemi and E. S. Sousa, Collaborative spectrum sensing for opportunistic access in fading environments. In *Proceedings of the First IEEE International Symposium on New Frontiers in Dynamic Spectrum Access Networks (DySPAN 2005)* (2005), pp. 131–136.

[34] G. Ganesan and L. Ye, Cooperative spectrum sensing in cognitive radio, Part I: Two user networks *IEEE Transactions on Wireless Communications*, **6**:6 (2007), 2204–2213.

[35] G. Ganesan and L. Ye, Cooperative spectrum sensing in cognitive radio, Part II: Multiuser networks *IEEE Transactions on Wireless Communications*, **6**:6 (2007), 2214–2222.

[36] W. Wang, L. Zhang, W. Zou and Z. Zhou, On the distributed cooperative spectrum sensing for cognitive radio. In *Proceedings of the International Symposium on Communications and Information Technologies (ISCIT 2007)* (2007), pp. 1496–1501.

[37] S. M. Mishra, A. Sahai and R. W. Brodersen, Cooperative sensing among cognitive radios. In *Proceedings of the IEEE International Conference on Communications (ICC 2006)* (2006), vol. 4, pp. 1658–1663.

[38] J. Ma and Y. Li, Soft combination and detection for cooperative spectrum sensing in cognitive radio networks. In *Proceedings of the IEEE Global Telecommunications Conference (GLOBECOM 2007)* (2007), pp. 3139–3143.

[39] C. Sun, W. Zhang and K. B. Letaief, Cooperative spectrum sensing for cognitive radios under bandwidth constraints. In *Proceedings of the IEEE Wireless Communications and Networking Conference (WCNC 2007)* (2007), pp. 1–5.

[40] C.-H. Lee and W. Wolf, Energy efficient techniques for cooperative spectrum sensing in cognitive radios. In *Proceedings of the Fifth IEEE Consumer Communications and Networking Conference (CCNC 2008)* (2008), pp. 968–972.

[41] C. Sun, W. Zhang and K. B. Letaief, Cluster-based cooperative spectrum sensing in cognitive radio systems. In *Proceedings of the IEEE International Conference on Communications (ICC 2007)* (2007), pp. 2511–2515.

[42] M. J. Marcus, CR: cooperative radio or confrontational radio. In *Proceedings of the Second IEEE International Symposium on New Frontiers in Dynamic Spectrum Access Networks (DySPAN 2007)* (2007), pp. 208–211.

[43] M. M. Buddhikot, P. Kolodzy, S. M. and K. Ryan and J. Evans, DIMSUMnet: new directions in wireless networking using coordinated dynamic spectrum In *Proceedings of the Sixth IEEE International Symposium on a World of Wireless Mobile and Multimedia Networks (WoWMoM 2005)* (2005), pp. 78–85.

[44] D. Raychaudhuri and X. Jing, A spectrum etiquette protocol for efficient coordination of radio devices in unlicensed bands. In *Proceedings of the 14th IEEE International Symposium on Personal, Indoor and Mobile Radio Communications (PIMRC 2003)* (2003), vol. 1, pp. 172–176.

[45] P. Houze, S. B. Jemaa and P. Cordier, Common pilot channel for network selection. In *Proceedings of the IEEE 63rd Vehicular Technology Conference (VTC 2006-Spring)* (2006), vol. 1, pp. 67–71.

[46] O. Sallent, J. Pérez-Romero, R. Agustí and P. Cordier, Cognitive pilot channel enabling spectrum awareness. In *Proceedings of the IEEE International Conference on Communications Workshops (ICC Workshops 2009)* (2009), pp. 1–6.

[47] M. Filo, A. Hossain, A. R. Biswas and R. Piesiewicz, Cognitive pilot channel: enabler for radio systems coexistence. In *Proceedings of the Second International Workshop on Cognitive Radio and Advanced Spectrum Management (CogART 2009)* (2009), pp. 17–23.

[48] J. Pérez-Romero, O. Sallent, R. Agustí and L. Giupponi, A novel on-demand cognitive pilot channel enabling dynamic spectrum allocation. In *Proceedings of the second IEEE International Symposium on New Frontiers in Dynamic Spectrum Access Networks (DySPAN 2007)* (2007) pp. 46–53.

[49] M. López-Benítez, A. Umbert and F. Casadevall, Evaluation of spectrum occupancy in Spain for cognitive radio applications. In *Proceedings of the IEEE 69th Vehicular Technology Conference (VTC 2009 Spring)* (2009), pp. 1–5.

[50] M. López-Benítez, F. Casadevall, A. Umbert, J. Pérez-Romero, R. Hachemani, J. Palicot and C. Moy, Spectral occupation measurements and blind standard recognition sensor for cognitive radio networks. In *Proceedings of the Fourth International Conference on Cognitive Radio Oriented Wireless Networks and Communications (CrownCom 2009)* (2009), pp. 1–9.

[51] M. López-Benítez and F. Casadevall, Empirical time-dimension model of spectrum use based on discrete-time Markov chain with deterministic and stochastic duty cycle models. *IEEE Transactions on Vehicular Technology*, **60**:6 (2011), 2519–2533.

[52] M. López-Benítez, Spectrum usage models for the analysis, design and simulation of cognitive radio networks. Ph.D. dissertation, Department of Signal Theory and Communications, Universitat Politècnica de Catalunya (UPC) (2011).

[53] M. López-Benítez and F. Casadevall, *Spectrum Usage Models for the Analysis, Design and Simulation of Cognitive Radio Networks* (Springer, 2012), ch. 2.

[54] X. Liu and W. Wang, On the characteristics of spectrum-agile communication networks. In *Proceedings of the First IEEE International Symposium on New Frontiers in Dynamic Spectrum Access Networks (DySPAN 2005)* (2005), pp. 214–223.

[55] W.-Y. Lee and I. F. Akyildiz, A spectrum decision framework for cognitive radio networks. *IEEE Transactions on Mobile Computing*, **10**:2 (2011), 161–174.

[56] N. Devroye, P. Mitran and V. Tarokh, Cognitive decomposition of wireless networks. In *Proceedings of the First International Conference on Cognitive Radio Oriented Wireless Networks and Communications (CROWNCOM 2006)* (2006), pp. 1–5.

[57] S. Srinivasa and S. A. Jafar, The throughput potential of cognitive radio: a theoretical perspective. *IEEE Communications Magazine*, **45**:5 (2007), 73–79.

[58] C. Peng, H. Zheng and B. Y. Zhao, Utilization and fairness in spectrum assignment for opportunistic spectrum access. *Mobile Networks and Applications*, **11**:4 (2006), 555–576.

[59] X. Li and S. A. Zekavat, Inter-vendor dynamic spectrum sharing: feasibility study and performance evaluation. In *Proceedings of the Second IEEE International Symposium on New Frontiers in Dynamic Spectrum Access Networks (DySPAN 2007)* (2007), pp. 412–415.

[60] O. Ileri, D. Samardzija and N. B. Mandayam, Demand responsive pricing and competitive spectrum allocation via a spectrum server. In *Proceedings of the First IEEE International*

Symposium on New Frontiers in Dynamic Spectrum Access Networks (DySPAN 2005) (2005), pp. 194–202.

[61] S. Gandhi, C. Buragohain, L. Cao, H. Zheng and S. Suri, A general framework for wireless spectrum auctions. In *Proceedings of the Second IEEE International Symposium on New Frontiers in Dynamic Spectrum Access Networks (DySPAN 2007)* (2007), pp. 22–33.

[62] X. Jing and D. Raychaudhuri, Spectrum co-existence of IEEE 802.11b and 802.16a networks using the CSCC etiquette protocol. In *Proceedings of the First IEEE International Symposium on New Frontiers in Dynamic Spectrum Access Networks (DySPAN 2005)* (2005), pp. 243–250.

[63] G. F. Marias, Spectrum scheduling and brokering based on QoS demands of competing WISPs. In *Proceedings of the First IEEE International Symposium on New Frontiers in Dynamic Spectrum Access Networks (DySPAN 2005)* (2005), pp. 684–687.

[64] C. Raman, R. D. Yates and N. B. Mandayam, Scheduling variable rate links via a spectrum server. In *Proceedings of the First IEEE International Symposium on New Frontiers in Dynamic Spectrum Access Networks (DySPAN 2005)* (2005), pp. 110–118.

[65] V. Brik, E. Rozner, S. Banerjee and P. Bahl, DSAP: a protocol for coordinated spectrum access. In *Proceedings of the First IEEE International Symposium on New Frontiers in Dynamic Spectrum Access Networks (DySPAN 2005)* (2005), pp. 611–614.

[66] L. Cao and H. Zheng, Distributed spectrum allocation via local bargaining. In *Proceedings of the Second Annual IEEE Communications Society Conference on Sensor and Ad Hoc Communications and Networks (SECON 2005)* (2005), pp. 475–486.

[67] J. Zhao, H. Zheng and G.-H. Yang, Distributed coordination in dynamic spectrum allocation networks. In *Proceedings of the First IEEE International Symposium on New Frontiers in Dynamic Spectrum Access Networks (DySPAN 2005)* (2005), pp. 259–268.

[68] L. Ma, X. Han and C. C. Shen, Dynamic open spectrum sharing MAC protocol for wireless ad hoc networks. In *Proceedings of the First IEEE International Symposium on New Frontiers in Dynamic Spectrum Access Networks (DySPAN 2005)* (2005), pp. 203–213.

[69] G. Auer, H. Haas and P. Omiyi, Interference aware medium access for dynamic spectrum sharing. In *Proceedings of the Second IEEE International Symposium on New Frontiers in Dynamic Spectrum Access Networks (DySPAN 2007)* (2007), pp. 399–402.

[70] J. Huang, R. A. Berry and M. L. Honig, Spectrum sharing with distributed interference compensation. In *Proceedings of the First IEEE International Symposium on New Frontiers in Dynamic Spectrum Access Networks (DySPAN 2005)* (2005), pp. 88–93.

[71] C. Cordeiro and K. Challapali, C-MAC: a cognitive MAC protocol for multi-channel wireless networks. In *Proceedings of the Second IEEE International Symposium on New Frontiers in Dynamic Spectrum Access Networks (DySPAN 2007)* (2007), pp. 147–157.

[72] H. Zheng and L. Cao, Device-centric spectrum management. In *Proceedings of the First IEEE International Symposium on New Frontiers in Dynamic Spectrum Access Networks (DySPAN 2005)* (2005), pp. 56–65.

[73] S. Sankaranarayanan, P. Papadimitratos, A. Mishra and S. Hershey, A bandwidth sharing approach to improve licensed spectrum utilization. In *Proceedings of the First IEEE International Symposium on New Frontiers in Dynamic Spectrum Access Networks (DySPAN 2005)* (2005), pp. 279–288.

[74] Q. Zhao, L. Tong and A. Swami, Decentralized cognitive MAC for dynamic spectrum access. In *Proceedings of the First IEEE International Symposium on New Frontiers in Dynamic Spectrum Access Networks (DySPAN 2005)* (2005), pp. 224–232.

[75] C. Cormio and K. R. Chowdhury, A survey on MAC protocols for cognitive radio networks. *Ad Hoc Networks*, **7**:7 (2009), 1315–1329.

[76] T. V. Krishna and A. Das, A survey on MAC protocols in OSA networks. *Computer Networks*, **53**:9 (2009), 1377–1394.

[77] Y.-C. Liang, Y. Zeng, E. C. Y. Peh and A. T. Hoang, Sensing–throughput tradeoff for cognitive radio networks. *IEEE Transactions on Wireless Communications*, **7**:4 (2008), 1326–1337.

[78] L. Yang, L. Cao and H. Zheng, Proactive channel access in dynamic spectrum networks. In *Proceedings of the Second International Conference on Cognitive Radio Oriented Wireless Networks and Communications (CrownCom 2007)* (2007), pp. 487–491.

[79] M. Hamid, A. Mohammed and Z. Yang, On spectrum sharing and dynamic spectrum allocation: MAC layer spectrum sensing in cognitive radio networks. In *Proceedings of the Second International Conference on Communications and Mobile Computing (CMC 2010)* (2010), pp. 183–187.

[80] C. Guo, T. Peng, Y. Qi and W. Wang, Adaptive channel searching scheme for cooperative spectrum sensing in cognitive radio networks. In *Proceedings of the IEEE Wireless Communications and Networking Conference (WCNC 2009)* (2009), pp. 1–6.

[81] I. F. Akyildiz, W. Y. Lee and K. R. Chowdhury, CRAHNs: cognitive radio ad hoc networks. *Ad Hoc Networks*, **7**:5 (2009), 810–836.

[82] D. López-Pérez, G. Roche, A. Valcarce, A. Juttner and J. Zhang, Interference avoidance and dynamic frequency planning for WiMAX femtocells networks. In *Proceedings of the 11th IEEE Singapore International Conference on Communication Systems (ICCS 2008)* (2008), pp. 1579–1584.

[83] D. López-Pérez, A. Juttner and J. Zhang, Optimisation methods for dynamic frequency planning in OFDMA networks. In *Proceedings of the 13th International Telecommunications Network Strategy and Planning Symposium (Networks 2008)* (2008), pp. 1–28.

[84] V. Chandrasekhar and J. G. Andrews, Spectrum allocation in two tier networks. In *Proceedings of the 42nd Asilomar Conference on Signals, Systems and Computers* (2008), pp. 26–29.

[85] Z. Bharucha, A. Saul, G. Auer and H. Haas, Dynamic resource partitioning for downlink femto-to-macro-cell interference avoidance. *EURASIP Journal on Wireless Communications and Networking* (2010), 12.

[86] S. Rangan, Femto–macro cellular interference control with subband scheduling and interference cancelation. In *Proceedings of the IEEE GLOBECOM 2010 Workshops* (2010), pp. 695–700.

[87] Y.-Y. Li, M. Macuha, E. S. Sousa, T. Sato and M. Nanri, Cognitive interference management in 3G femtocells. In *Proceedings of the IEEE 20th International Symposium on Personal, Indoor and Mobile Radio Communications (PIMRC 2009)* (2009), pp. 1118–1122.

[88] L. Zhang, L. Yang and T. Yang, Cognitive interference management for LTE-A femtocells with distributed carrier selection. In *Proceedings of the IEEE 72nd Vehicular Technology Conference Fall (VTC 2010-Fall)* (2010), pp. 1–5.

[89] T. Yang, L. Zhang and L. Yang, Cognitive-based distributed interference management for home-eNB systems with single or multiple antennas. In *Proceedings of the IEEE 21st International Symposium on Personal Indoor and Mobile Radio Communications (PIMRC 2010)* (2010), pp. 1260–1264.

[90] S. Kaimaletu, R. Krishnan, S. Kalyani, N. Akhtar and B. Ramamurthi, Cognitive interference management in heterogeneous femto–macro cell networks. In *Proceedings of the IEEE International Conference on Communications (ICC 2011)* (2011), pp. 1–6.

[91] A. Attar, V. Krishnamurthy and O. N. Gharehshiran, Interference management using cognitive base-stations for UMTS LTE. *IEEE Communications Magazine*, **49**:8 (2011), 152–159.

[92] M. E. Sahin, I. Guvenc, J. Moo-Ryong and H. Arslan, Handling CCI and ICI in OFDMA femtocell networks through frequency scheduling. *IEEE Transactions on Consumer Electronics*, **55**:4 (2009), 1936–1944.

[93] S.-Y. Tu, K.-C. Chen and R. Prasad, Spectrum sensing of OFDMA systems for cognitive radio networks. *IEEE Transactions on Vehicular Technology*, **58**:7 (2009), 3410–3425.

[94] Y. Xin, K. Kim and S. Rangarajan, Multitaper spectrum sensing of OFDMA signals in frequency selective fading environment. In *Proceedings of the Sixth International ICST Conference on Cognitive Radio Oriented Wireless Networks and Communications (CROWNCOM 2011)* (2011), pp. 56–60.

[95] I. Harjula and A. Hekkala, Spectrum sensing in cognitive femto base stations using Welch periodogram. In *Proceedings of the IEEE 22nd International Symposium on Personal Indoor and Mobile Radio Communications (PIMRC 2011)* (2011), pp. 2305–2309.

[96] D. Choi, P. Monajemi, S. Kang and J. Villasenor, Dealing with loud neighbors: the benefits and tradeoffs of adaptive femtocell access. In *Proceedings of the IEEE Global Telecommunications Conference (GLOBECOM 2008)* (2008), pp. 1–5.

[97] Z. Zhao, M. Schellmann, H. Boulaaba and E. Schulz, Interference study for cognitive LTE-femtocell in TV white spaces. In *Proceedings of the 2011 Technical Symposium at ITU Telecom World (ITU TW 2011)* (2011), pp. 153–158.

[98] Y. Zeng, Y.-C. Liang, Z. Lei, S. W. Oh, F. Chin and S. Sun, Worldwide regulatory and standardization activities on cognitive radio. In *Proceedings of the Fourth IEEE International Symposium on New Frontiers in Dynamic Spectrum Access Networks (DySPAN 2010)* (2010), pp. 1–9.

[99] R. V. Prasad, P. Pawelczak, J. A. Hoffmeyer and H. S. Berger, Cognitive functionality in next generation wireless networks: standardization efforts. *IEEE Communications Magazine*, **46**:4 (2008), 72–78.

[100] M. Sherman, A. N. Mody, R. Martinez, C. Rodriguez and R. Reddy, IEEE standards supporting cognitive radio and networks, dynamic spectrum access, and coexistence. *IEEE Communications Magazine*, **46**:7 (2008), 72–79.

[101] C. Cordeiro, K. Challapali, D. Birru and N. S. Shankar, IEEE 802.22: the first worldwide wireless standard based on cognitive radios. In *Proceedings of the First IEEE International Symposium on New Frontiers in Dynamic Spectrum Access Networks (DySPAN 2005)* (2005), pp. 328–337.

[102] C. Cordeiro, K. Challapali, D. Birru and N. S. Shankar, IEEE 802.22: an introduction to the first wireless standard based on cognitive radios. *Journal of Communications*, **1**:1 (2006), 38–47.

[103] C. Stevenson, G. Chouinard, Z. Lei, W. Hu, S. Shellhammer and W. Caldwell, IEEE 802.22: the first cognitive radio wireless regional area network standard. *IEEE Communications Magazine*, **47**:1 (2009), 130–138.

[104] C. Cordeiro, K. Challapali and M. Ghosh, Cognitive PHY and MAC layers for dynamic spectrum access and sharing of TV bands. In *Proceedings of the First International Workshop on Technology and Policy for Accessing Spectrum (TAPAS 2006)* (2006), p. 3.

[105] K. Challapali, C. Cordeiro and D. Birru, Evolution of spectrum-agile cognitive radios: first wireless internet standard and beyond. In *Proceedings of the Second Annual International Workshop on Wireless Internet (WICON 2006)* (2006), p. 27.

[106] B. Scott and M. Calabrese, *Measuring the TV White Space Available for Unlicensed Wireless Broadband*. Free Press and New America Foundation, Technical Report (2005).

[107] M. Mishra and A. Sahai, *How Much White Space is There?* Electrical Engineering and Computer Sciences, University of California at Berkeley, Technical Report UCB/EECS-2009-3 (2009).

[108] M. Nekovee, Quantifying the availability of TV white spaces for cognitive radio operation in the UK. In *Proceedings of the IEEE International Conference on Communications Workshops (ICC Workshops 2009)* (2009), pp. 1–5.

[109] S. Buljore, V. Merat, H. Harada, S. Filin, P. Houze, K. Tsagkaris, V. Ivanov, K. Nolte, T. Farnham and O. Holland, IEEE P1900.4 system overview on architecture and enablers for optimised radio and spectrum resource usage. In *Proceedings of the Third IEEE International Symposium on New Frontiers in Dynamic Spectrum Access Networks (DySPAN 2008)* (2008), pp. 1–8.

[110] A. Stirling, White spaces – the new Wi-Fi? *International Journal of Digital Television*, **1**:1 (2010), 69–83.

[111] S. Deb, V. Srinivasan and R. Maheshwari, Dynamic spectrum access in DTV whitespaces: design rules, architecture and algorithms. In *Proceedings of the 15th Annual International Conference on Mobile Computing and Networking (MobiCom 2009)* (2009), pp. 1–16.

[112] P. Piggin and K. L. Stanwood, Standardizing WiMAX solutions for coexistence in the 3.65 GHz band. In *Proceedings of the Third IEEE International Symposium on New Frontiers in Dynamic Spectrum Access Networks (DySPAN 2008)* (2008), pp. 1–7.

[113] ITU, *Software Defined Radio in IMT-2000, the Future Development of IMT-2000 and Systems Beyond IMT-2000*, International Telecommunication Union, Technical Report ITU-R M.2063 (2005).

[114] ITU, *Software-Defined Radio in the Land Mobile Service*, International Telecommunication Union, Technical Report ITU-R M.2064 (2005).

[115] ETSI, *Reconfigurable Radio Systems (RRS); Summary of Feasibility Studies and Potential Standardization Topics*, European Telecommunications Standards Institute, Technical Report ETSI TR 102 838 v1.1.1 (2009).

[116] J. Wang, M. S. Song, S. Santhiveeran, K. Lim, G. Ko, K. Kim, S. H. Hwang, M. Ghosh, V. Gaddam and K. Challapali, First cognitive radio networking standard for personal/portable devices in TV white spaces. In *Proceedings of the Fourth IEEE International Symposium on New Frontiers in Dynamic Spectrum Access Networks (DySPAN 2010)* (2010), pp. 1–12.

[117] ECMA, *MAC and PHY for Operation in TV White Space*. European Association for Standardizing Information and Communication Systems, Technical Report ECMA-392 (2009).

[118] S. Probasco, G. Bajko and B. Rosen, *Protocol to Access White Space Database: Problem Statement and Requirements* (2011). http://tools.ietf.org/html/draft-patil-paws-problem-stmt-01

[119] T. Derryberry and B. Patil, *Protocol to Access White Space Database: Overview and Use Case Scenarios* (2011). http://tools.ietf.org/html/draft-probasco-paws-overview-usecases-00

14 Energy-efficient architectures and techniques

Weisi Guo, Min Chen and Athanasios V. Vasilakos

14.1 Introduction

Digital information exchange is seen as a key enabler in modern economics, and one of its most challenging aspects is the heterogeneous nature of modern wireless networks. Over 1.4 billion *user equipments* (UEs) are connected to the cellular network with over 3 million *base stations* (BSs) [1]. The volume of data communicated via *information communication technology* (ICT) infrastructures has increased by more than tenfold over the past 5 years. According to [2], the global mobile data traffic is expected to reach 6.3 exabytes per month by 2015, which is more than 26 times the mobile data traffic per month in 2010. A recognized target by the United Nations is to improve both the coverage and the capacity of cellular networks, in order to foster economic growth and reduce the wealth and knowledge gap among countries. It is important to achieve the aforementioned level of wireless connectivity, whilst consuming a low amount of energy and incurring a low cost, due to the growing concern over the environmental damage caused by carbon emissions. The greenhouse effect is mainly caused by excessive emissions of carbon dioxide (CO_2) in the last century. As reported in [3–5], human industrial activities emit twice as much CO_2 as natural processes can absorb at the moment. Globally, ICT infrastructures consume approximately 3% of the world's total energy [5, 6]. In particular, up to 20% of the energy consumption of the ICT industry is attributed to wireless networks [7], the scale of which is still growing explosively [4]. Roughly 70% of the wireless network energy is consumed by the outdoor macrocell BSs. The global cellular network consumes 60 TWh of electricity, which is equivalent to the output of three to four 2000 MW power plants and the consumption level of 20 million medium households in the United Kingdom. The global utility bill for the cellular network operators stands at over $10 billion in 2010.

Recent research has demonstrated that the energy saving potential of the current cellular network is limited to approximately 30%. However, up to 80% energy savings can be achieved if the network is redeployed in a *heterogeneous cellular network* (HCN) configuration. Making cellular networks green would have a tangible positive impact on reducing the carbon footprint of ICT, and also achieve a long-term profitability of mobile service operators. It was found that an energy-efficient HCN can reduce the operation and maintenance cost of the network by 40–60%.

In the face of increasing data demands and uncertain average revenue per user, operators are looking at ways to reduce the running costs in order to improve their competitiveness. Since 2008, the global level of average *capital expenditure* (CAPEX) and *operational expenditure* (OPEX) (around \$500 billion) has fallen by up to 6–8% per year. Many operators are also pledging to reduce carbon emissions; e.g., Vodafone aims to reduce CO_2 emissions by 50% in developed markets by 2020. Nearly half of a typical cellular network's operational and maintenance costs are due to the energy expenditure of BSs.

Recently, both industry and academy are paying substantial attention to research on green cellular networks, which leads to comprehensive efforts in designing new energy-efficient architectures, protocols, and algorithms, targeting various types of wireless network, e.g., cellular networks, mobile ad-hoc networks, sensor networks, etc. In this chapter, we focus on infrastructure-based cellular networks,[1] which serve telecommunication voice calls and Internet data service functionalities to mobile UEs. In particular, the emphasis of this chapter is on HCN architectures and their potential to deliver improved performance at lower energy and cost consumption levels.

Given that over 70% of cellular-network energy expenditure is consumed by outdoor BSs, research into energy efficiency has predominantly focused on cellular *radio access networks* (RANs). The key UE-performance-related challenges faced by the mobile communication industry are the following.

- *Quality of Service* (QoS): For a UE traffic demand that has spatial and temporal variations, how to deploy the network to maximize the user experienced performance?
- *Efficiency*: For a given traffic profile, how to deploy the network and transmit data in the most energy- and cost-efficient way. And how does this efficiency vary with the traffic load?

The relation between performance and energy consumption is illustrated in Fig. 14.1 [10]. For any given performance requirement, there is a saturation point, where energy consumption can no longer be reduced. Research has shown that, for a fixed network deployment, the performance saturation point is dictated by hardware performance, while for a flexible HCN deployment, the performance saturation point is dictated by the achievable capacity of the network architecture and transmission techniques. To demonstrate the complexity of an HCN, Fig. 14.2 summarizes the multi-dimensional optimization challenges in designing a green cellular network [10]. In order to achieve significant performance improvements, the engineering challenge lies in how to integrate cross-layer techniques that involve *systems and architecture*, *transmission techniques*, and *hardware design*.

Green communication techniques in cellular networks have been intensively studied across academia and industry, whilst ICT-related government agencies, public institutions and standardization bodies, such as the *3rd Generation Partnership Project* (3GPP) and the *Institute of Electrical and Electronics Engineers* (IEEE), have also started the regulation and standardization of energy efficiency metrics [11–15]. It is timely and desirable to compile those efforts to offer a comprehensive view on the state-of-the-art green communication techniques.

---

[1] More specific surveys on the latter two can be found in [8] and [9], respectively.

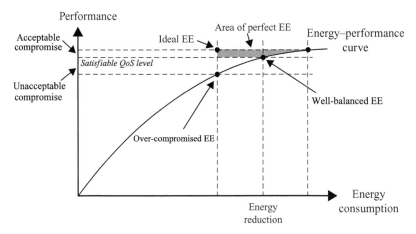

Figure 14.1 Energy–performance curve of a typical cellular network, a network device, a transmission algorithm, or an application service, etc., indicating the tradeoff between energy consumption and capacity performance, where EE is an abbreviation for energy efficiency [10].

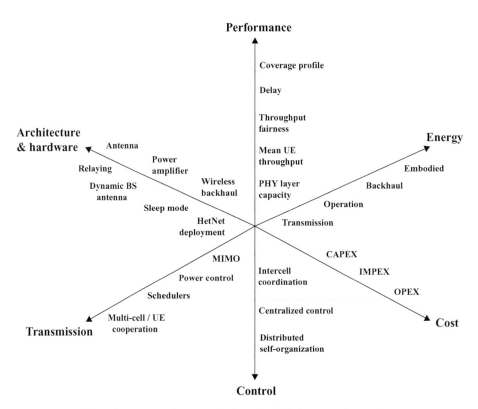

Figure 14.2 Multiple dimensions of green techniques for cellular networks: performance, energy, cost, control, transmission, architecture and hardware.

As previously mentioned, recent research has shown that HCN can significantly improve network performance and reduce energy consumption. In this chapter, the authors present a detailed survey on key state-of-the-art research efforts in HCN deployment and performance. In particular, the chapter examines the fundamental tradeoffs between performance and energy consumption, HCN architectures proposed, dynamic BS designs, advanced transmission schemes, hardware improvements, and cross-layer integration solutions.

14.2   Green cellular projects and metrics

Within the past decade, a growing number of green projects have been funded to facilitate the research, experiment, deployment, and evaluation of improving energy efficiency for cellular networks. In this section, we present a summary of such research projects and a list of energy efficiency metrics used to gauge the performance of green cellular networks.

14.2.1   Green cellular network projects

Governments have funded and developed a series of directives aimed at obligating corporations to become more energy efficient, using greener techniques and thereby reducing CO_2 emissions. A list of key research projects around the world can be found in Table 14.1, which lists their participants and target research themes.

One of the earliest large-scale research projects is the Mobile Virtual Center of Excellence Core 5 *Green Radio* project [7, 16], which is a three-year industry–government funded research program (2009–2012). Other notable pioneers of green cellular network projects include the *GreenTouch* project [10], the *EARTH* project [17, 18], and the *Green IT* project [19]. A common objective set by these projects is to reduce the power consumption of cellular networks from 2 to 1000 fold.

The core research of the aforementioned projects is to investigate the theoretical and practical energy-efficiency limitations of current cellular networks, whilst developing a new generation of energy-efficient cellular architectures, transmission techniques, and hardware equipments. Most of these projects mainly focus on *Long Term Evolution* (LTE) and LTE-Advanced systems, but *High Speed Packet Access* (HSPA) technology has also been considered for immediate impact.

These research projects have shown that a single technique alone can achieve up to 30% energy saving. In order to achieve a more energy-efficient network, an integrated approach to research will be needed in order to combine different cross-layer techniques and exploit their synergies. How this can be achieved will be discussed in the following sections.

Table 14.1 A summary of green cellular network projects.

Project	Organizer	Region	Participants	Targets	Working Emphasis
EARTH	European Commission FP7 IP (3 years / 15 million €)	Europe	European main mobile operators and research organizations	Mobile networks	• energy aware radio and network technology • energy-efficient deployment, architecture, adaptive management • multi-cell cooperation
Green IT	METI & JEITA (Japan)	Japan	Over 100 companies, institutes and organizations	IT	• power efficiency at data centers, networks, displays • policy and mechanisms to encourage green IT • collaboration of industry, academia, and government
GreenTouch	GreenTouch Consortium	Global	Experts from industry and academia	Telecom networks and mobile networks	• reinvention of telecom networks • sustainable data networks • optical, wireless, electronics, routing, architecture, etc.
OPERA-Net	CELTIC / EUREKA (3 years / 5 million €)	Europe	European main mobile operators	Mobile networks	• heterogeneous broadband wireless network • mobile radio access network • link-level power efficiency, amplifier, test bed
GREEN-T	CELTIC (3 years / 6 million €)	Europe	European main mobile operators	Mobile networks (particularly 4G)	• multi-standard wireless mobile devices • cognitive radio and cooperative strategies • QoS guarantee
GreenRadio	MVCE (3 years)	UK	UK universities	Base station and handsets of mobile data service	• power amplifier, power-efficient processing • backhaul redesign, multi-hop routing, relaying • resource allocation, dynamic spectrum access
Cool Silicon	Silicon Saxony Management	Global	Over 60 global ICT companies and institutes	ICT	• micro-/nano-technology • media communication • sensor network
Green Grid	8 Main Contributor Companies	Global	Global ICT Companies	Data centers	• data center energy efficiency (design, measurement, metrics)
GSMAMEE	GSM Association Congress	Global	Over 800 mobile operators and 200 companies	Mobile networks	• benchmarking of mobile energy efficiency networks
Green500 Cool IT	Virginia Tech GreenPeace	US Global	Virginia Tech GreenPeace	Supercomputer IT	• benchmarking of greenest & fastest supercomputers • leaderboard of IT brands on the contributions to the green IT

14.2.2   A taxonomy of green metrics

There have been a variety of energy consumption and efficiency metrics, which can be categorized into two major types.

1. *Energy consumption*: the amount of power or energy consumed, or saved compared to a reference. For a realistic transmission equipment or system, the power consumption is typically composed of transmission-dependent and transmission-independent parts. Therefore, calculating the energy consumed over a period of time needs to take into account both transmission and idle periods.
2. *Energy efficiency*: the amount of power or energy used to transmit a successfully received piece of information. Given that the system may not always be transmitting, this metric can be seen as either an instantaneous transmission efficiency (a.k.a. energy consumption ratio (ECR)), or an averaged efficiency over a period of time.

Generally speaking there are two levels of energy metrics: equipment level and system level [20]. Research into transmission techniques and hardware tends to focus on the equipment-level energy efficiency, whereas architecture- and system-level designers focus on system wide energy consumption. In the future, more effective designs of green metrics can target what percentage of power can be further saved under a technique or standard, or how much CO_2 emission can be converted and generated in relation to delivered QoS and system utilization efficiency.

14.2.3   How green are cellular networks?

14.2.3.1   Multiple access schemes

The current *UMTS Terrestrial Radio Access Network* (UTRAN) cellular system is *spread spectrum multiple access* (SSMA) based, with a typical bandwidth of 5 MHz. The LTE system is *orthogonal frequency division multiple access* (OFDMA) based, with a typical bandwidth of up to 20 MHz. Both the spectral efficiency (throughput per Hz) and overall system throughput have been greatly improved in LTE compared to HSPA [21].

- 2.5- to threefold increase in spectral efficiency, primarily due to the improved modulation and coding schemes.
- 10- to 12-fold increase in system throughput, where part of the increase is due to the fourfold increase in bandwidth.

The resulting energy consumption reduction in the network is approximately 35% for the same number of BSs in the RAN. If fewer BSs are deployed as a result of the improved capacity per BS, up to 60% of energy can be saved by adopting LTE instead of HSPA. Therefore, there is a strong incentive to implement LTE and phase out the HSPA RAN.

14.2.3.2   Power consumption model and data

The basic power consumption of a network node is comprised of a transmission-dependent part (radiohead) and transmission-independent part (overhead), i.e.,

$$P_{\text{total}} = P_{\text{radiohead}} + P_{\text{overhead}}$$

$$= \frac{P_{\text{transmit}}}{\mu} + P_{\text{overhead}}, \qquad (14.1)$$

where μ is the transmission efficiency, which is dominated by the power amplifier efficiency performance. The overhead consists of baseband, cooling elements, and the backhaul of the node, which is generally independent of the transmission part.

The current HSPA cellular networks typically deploy sectorized BSs that consume between 800 and 1500 W per BS [22], which is similar to the power consumed by a household in the developed world. This yields a peak power consumption density of approximately 1000 W per square kilometer in urban areas, and 400 W per square kilometer in rural areas. Typical power consumptions of *macrocell base stations* (MBSs) and different types of *low-power nodes* (LPNs) can be summarized as follows [22, 23].

- Macro base stations for overlay and rural coverage: three to six sectors with 1000–2000 W power consumption each, with 50–70% radiohead and 30–50% overhead.
- Micro base stations for suburban, urban, and large street level hotspot coverage: three sectors with 400–600 W power consumption each, with 40–60% radiohead and 40–60% overhead.
- Pico base stations for localized urban street level hotspot and indoor coverage: one sector with 10–100 W power consumption each, with 10–30% radiohead and 70–90% overhead.
- Relay nodes for rural coverage extension and urban capacity improvement: one sector with 10–25 W power consumption each, with 10–20% radiohead and 80–90% overhead. *Relay nodes* (RNs) typically consume less power than other LPNs due to the lack of a physical backhaul.

The above data shows that, whilst LPNs can provide localized capacity improvements at a much lower power consumption level than BSs (4- to 200-fold), the proportion of their transmission-dependent power consumption is much lower. This is due to the low efficiency of power amplifiers for LPNs.

14.3   Fundamental tradeoffs: capacity, energy, and cost

14.3.1   Introduction

In order to achieve a certain throughput, a cellular network typically expends three valuable resources: spectrum, energy, and cost. From an operator's perspective, there is a desire to reduce the energy consumption, whilst maintaining the required level of capacity, preferably the potential for capacity growth. Therefore, the tradeoff between capacity and energy consumption is of fundamental importance, and has been characterized in [24] and [25] for noise-limited and interference-limited channels, respectively.

Moreover, since reducing the energy consumption has an impact on the OPEX of cellular networks, there is a three-way tradeoff between capacity, energy, and cost, which has been characterized in [26] for an interference-limited multi-BS network.

14.3.2   Fundamental energy saving limits

There are fundamental limits on how much energy can be saved by different deployment and transmission techniques. Energy and cost saving techniques can be generally divided into two categories.

- *Capacity improvement*: increases network capacity and reduces the transmission time of the network. This reduces the radiohead power consumption ($P_{\text{radiohead}}$) of each BS or LPN. The greater the system-level capacity that can be achieved per unit of power consumed, the higher the energy saving. The energy saving is given by [26]

$$\text{energy saving}_{\text{hardware-limited}} \approx \left(1 - \frac{\text{traffic}}{\text{capacity}}\right) \frac{P_{\text{radiohead}}}{P_{\text{total}}}\%, \qquad (14.2)$$

where for a large capacity improvement the bound is limited by the ratio of transmission-dependent power consumption and total power consumption. Therefore, the bound is a *hardware-limited* bound. The fundamental energy saving achievable is approximately 30–60% for BSs and 10–30% for LPNs, depending on the hardware [26].
- *Infrastructure reduction*: maintains the network capacity with a less power-consuming deployment. The greater the system-level capacity that can be achieved per unit of power consumed, the higher the energy saving in re-deployment, i.e.,

$$\text{energy saving}_{\text{capacity-limited}} \approx \left(1 - \frac{\text{traffic}}{\text{capacity}}\right)\%, \qquad (14.3)$$

which is a *capacity-limited* bound. This can reduce the total operational energy consumption of the RAN by up to 100% in theory, but in reality only 50–70% reduction can be achieved with dynamic deployments of LPNs in an HCN configuration, which will be discussed later on in this chapter in greater detail. The disadvantage is that infrastructure-reducing techniques have less flexibility for capacity growth than capacity-improving techniques.

14.3.3   Maximum spectral and energy efficiency

The *Global System for Mobile Communications* (GSM) network is conventionally deployed with a dense frequency reuse pattern to minimize interference at the cost of *spectral efficiency* (SE). This is primarily due to the weak physical layer forward-error-correction codes employed, which requires a high received signal power. The system-level performance of such a network is noise or propagation limited. The SE is defined as the amount of bandwidth used to deliver a bit of information successfully

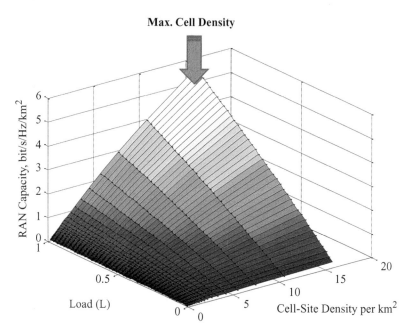

Figure 14.3 HCN spectral efficiency as a function of traffic load and deployment density (reproduced with permission from IEEE©).

(in bits/Hz). In most cellular network configurations, a spectral efficient deployment consists of a dense cluster of LPNs, together with a macro-BS overlay to account for high-mobility UEs [24, 26]. This is shown in Fig. 14.3, where the SE is plotted against the BS deployment density and the traffic load offered, and the maximum SE is achieved at the peak density and load.

The energy efficiency (EE) is defined as the amount of energy required to transmit a bit of information successfully (in bits/J). In a noise-limited network, the optimal EE is typically achieved with a dense cluster of LPNs and a macro-BS overlay to account for high-mobility UEs [24]. In HSPA and LTE, the use of robust convolutional and turbo codes means that the network is relatively interference tolerable and can deploy co-channel configurations to maximize SE. For example, a high EE can be achieved by deploying pico-BSs with an inter-BS distance of 300–500 m [26]. The rationale behind this is that, as the inter-BS distance decreases, the number of effective interferers is reduced, improving the EE. This is shown in Fig. 14.4, where the EE is plotted against the BS deployment density and the traffic load offered, and the maximum EE is achieved at the peak BS density and traffic load.

14.3.4   Maximum cost efficiency

The cost efficiency (CE) is defined as the amount of cost used to deliver a bit of information successfully (in bits/$). The cost of RAN can generally be broken into a CAPEX and an OPEX. More specifically,

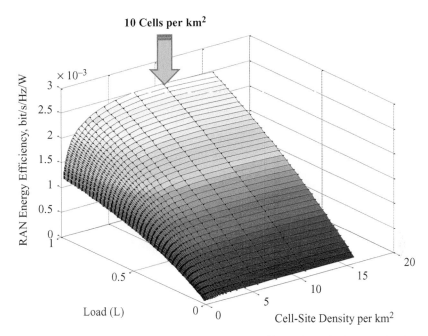

Figure 14.4 HCN energy efficiency as a function of traffic load and deployment density (reproduced with permission from IEEE©).

- CAPEX: installation, upgrade and equipment.
- OPEX: utility bill for energy consumption, backhaul rental, site rental, RAN maintenance.

The CAPEX and annual OPEX of BSs, LPNs, and RNs are approximately given as follows [26]:

- Core network CAPEX: BS ($5000), LPN (free)
- Equipment purchase CAPEX: macro-BS ($50 000), micro-BS ($20 000), pico-BS ($5000), LPN ($300), RN ($600)
- Equipment installation CAPEX: macro-BS ($120 000), micro-BS ($15 000), pico-BS ($3000), LPN (free), RN ($1000)
- Maintenance OPEX: macro-BS ($2500), micro-BS ($1000), pico-BS ($250), LPN (free), RN (free)
- Site rental OPEX: macro-BS ($10 000), micro-BS ($5000), pico-BS ($1000), LPN (free), RN ($300)
- Backhaul rental OPEX: macro-BS ($40 000), micro-BS ($20 000), pico-BS ($10 000), LPN ($1000), RN (free)
- Energy OPEX ($0.2 per kWh): macro-BS ($3500), micro-BS ($2500), pico-BS ($200), LPN ($20), RN ($50)

Typically, the CAPEX is taken out as a loan and the annual cost is dominated by OPEX, of which 15–30% is attributed to energy consumption of the RAN [26].

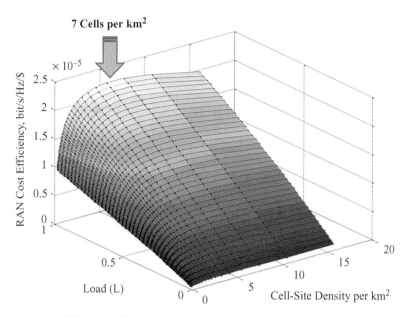

Figure 14.5 HCN cost efficiency as a function of traffic load and deployment density (reproduced with permission from IEEE©).

In an interference-limited network, a good CE can be obtained by deploying micro-BSs with an inter-BS distance of 500–700 m [26]. As shown in Fig. 14.5, where CE is plotted against the BS deployment density and the traffic load offered, and the maximum CE is achieved at a medium BS density and at the peak load. From Fig. 14.3, Fig. 14.4, and Fig. 14.5, we can see that a deployment of micro- and/or pico-BSs with an inter-BS distance of around 500 m would provide close-to-optimal SE, EE, and CE for the network.

Having discussed the theoretical limits on energy and associated cost savings of HCN deployments, the chapter will now examine the specific details of HCN architectures and transmission techniques that can yield a detailed insight into energy savings and capacity improvements.

14.4   Green cellular network architectures

Novel designs of a whole green mobile system would require significant multi-layer optimization or reconstruction, as shown in [6], [27], and [28]. The previously mentioned projects have used a form of *dynamic programming* research, whereby each specific technique is first optimized in isolation and then combined together through a common integration platform, i.e., a multi-layer simulator. In this section, we will look into the different HCN aspects of a RAN and how reconfiguration of the network architecture can save energy.

Table 14.2 A summary of reviewed green techniques for the five major components of an HCN, which are categorized into three areas: processing, communication, and system.

Components	Topics	Category
Data centers	• ON/OFF resource allocation: [27], [43] • Virtualization: [28] • Cooling management: [21], [22]	Processing Processing System
Macrocells	• Dynamic scheduling: [45], [46] • Optimization of cell deployment: [44], [45], [48], [54] • Power saving of power amplifier: [21]	Processing System Communication
Femtocells	• Coverage optimization and power control: [32], [47], [48] • Interference avoidance: [30], [42], [46]	Communication & System Processing
End-hosts	• Energy profiling: [11] • Utilization of multiple radios: [20], [49] • Energy efficient transmissions: [37], [46], [50], [52]	System Communication Communication
Applications & services	• Adaptive power-saving design: [22] • Prediction-based adaptation: [26] • Proxy-based caching • Energy-efficient location-based services	Processing Processing Communication Communication & System

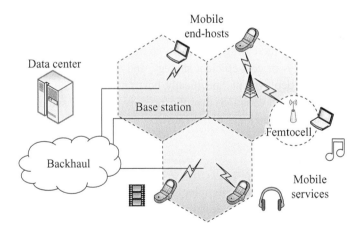

Figure 14.6 A simplified structure of an HCN, which mainly consists of five components: data centers in backhaul, macro BSs, femto BSs, mobile UEs, and mobile services.

As illustrated in Fig. 14.6, an HCN typically consists of five components: outdoor macro-BSs, LPNs, UEs, data centers, and services. Given that approximately 60–75% of the HCN energy is consumed in the outdoor BSs [29], this section examines how this energy consumption level can be reduced from a deployment perspective. In Section 14.5, we will consider the transmission techniques (Table 14.2) used within the five aforementioned components to improve energy efficiency of HCNs.

14.4.1   Homogeneous deployment

Current cellular network deployments, e.g., HSPA, are largely an *irregular homogeneous* network that varies in BS density depending on the traffic and environment.

- Urban deployment: dense deployment of low-power pico BSs, typically two to four pico BSs per square kilometer with a maximum transmit power of 5–10 W each. There might also be an umbrella macro BS overlaid to cover high-mobility UEs. Such an urban deployment typically serves 100–120 active UEs per square kilometer.
- Suburban deployment: less dense deployment of medium-power macro BSs, typically 0.5–2 macro BSs per square kilometer with a maximum transmit power of 10–20 W each. Such a suburban deployment typically serves 40–80 active UEs per square kilometer.
- Rural deployment: sparse deployment of high-power macro BSs, typically 0.2–0.5 macro BSs per square kilometer with a maximum transmit power of 20–40 W each. Such a rural deployment typically serves 20–40 active UEs per square kilometer.

The current strategy behind capacity improvement is to increase the BS density, and there is an approximately linear relationship between network capacity and the number of BSs deployed.

A significant deployment challenge arises due to the asymmetric spatial and temporal distribution of mobile traffic. From a spatial perspective, only 10% of BSs carry more than 50% of the data traffic in both urban and rural environments [30]. From a temporal perspective, the average traffic intensity across regions can vary by up to four- to sixfold during a day. Therefore, increasing BS density to meet increased traffic demands may not be the most energy- or cost-efficient strategy, and thus there is clearly a scope for improvement in RAN deployment. Specifically, how a network architecture can account for the spatial and temporal variations in traffic intensity, whilst consuming the minimum level of energy, is an important question to answer.

14.4.2   Heterogeneous deployment

Heterogeneous deployments typically refer to multi-tier deployments of different categories of network nodes, which are described in more detail in Section 1.2.4.

An example layout of an HCN is shown in Fig. 14.7, where an umbrella macro BS overlaps coverage with multiple micro BSs. Each micro BS is supplemented by a number of LPNs and RNs near the cell edge, where coverage is poor. Fig. 14.8(a) shows a RAN that is modeled as 19 BSs with wrap-around interference [31] (see Chapter 3 for wrap around descriptions). Alternative modeling techniques for HCNs include stochastic geometry [32]. Within this RAN, due to the RNs deployed near the cell edge, as shown in Fig. 14.8(b), a significant improvement in cell-edge performance can be observed in Fig. 14.8(c). However, as shown in Fig. 14.8(d), there is a tradeoff between the achieved capacity and the number of RNs deployed, i.e., too many RNs may erode the mean capacity as a result of excessive mutual interference [33]. In Fig. 14.8(d), the resulting mean capacity map is also shown, whereby the cell-edge performance is improved due

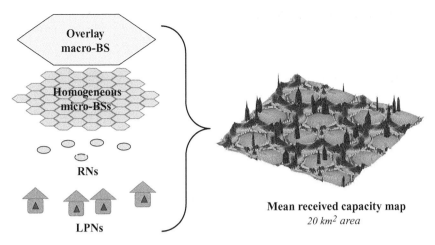

Figure 14.7 HCN architecture: macro BS, micro BS, RNs and LPNs.

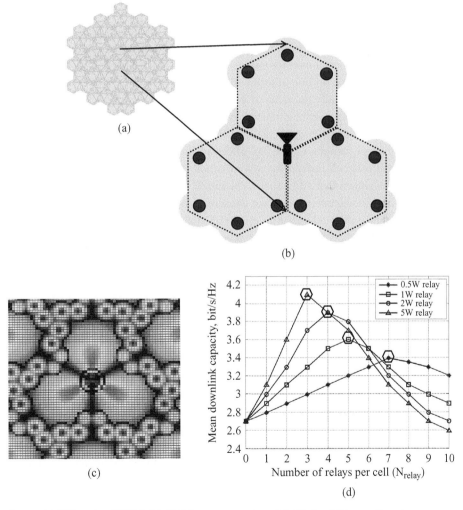

Figure 14.8 RN-assisted HCN: (a) 19 BS wrap-around model, (b) five RN per cell-sector deployment, (c) simulated mean downlink capacity, and (d) optimum number of relays per cell-sector [33].

to the LPNs and RNs deployed. Typically, the energy savings achieved by LPNs and RNs are of a similar level [34].

It is also worth noting that UEs can act as RNs for one another. Cooperative communications use multi-hop communication channels to improve coverage in poor-coverage areas by exploiting macro-diversity. The host-to-host modes supported in *Universal Mobile Telecommunication System* (UMTS) networks will possibly improve the scalability and energy-efficiency of cellular networks [35, 36]. The tethering mode [35] enables the direct data communications among a group of UEs in an ad hoc manner, while the tethering gateway UE will share the packet access with other UEs based on striping transmission schemes. Multi-hop routing and transmission among tethered UEs are also attractive but challenging for improving energy efficiency [36].

14.4.2.1   Fixed deployment

This subsection considers the impact of increased LPN deployments on top of a fixed existing deployment of macro BSs. In order to achieve a similar level of throughput, the primary mechanism for energy reduction is to reduce the transmit power of BSs, thus offloading a portion of traffic to the LPNs.

Traditionally, in areas of poor coverage (e.g., cell edges, alleyways, indoors), BSs have to increase their transmit power to improve UEs' performance. However, for dense network deployments, e.g., in HSPA and LTE networks that deploy BSs in a co-channel manner, actions of one BS affect all its neighboring BSs' co-channel transmissions, and *intercell interference coordination* (ICIC) becomes a key consideration. Research has shown that additional LPN and RN deployment can significantly reduce both the operational energy and cost consumption of a RAN [37–39], where energy savings are significantly enhanced for macro BSs that consume a high level of energy. That is to say, in a fixed deployment scenario:

- High-power macro BSs with a limited number of LPNs/RNs deployed in areas of poor coverage can reduce the energy consumption of the whole network.
- Low-power micro BSs with additional LPNs/RNs deployed in areas of poor coverage may increase the energy consumption of the whole network.

In [33], it has been shown that there is an optimum number of LPNs/RNs deployable for each class of BS considered in terms of energy efficiency. In [40], [38], and [39], the tradeoff between energy savings and capacity is considered from both energy and cost-efficiency perspectives. This tradeoff is illustrated in Fig. 14.9. Results show that the reference HSPA system has a limited achievable throughput, whereas the LTE system can enhance the throughput by approximately six- to eightfold due to increased bandwidth and advanced physical layer modulation and coding schemes. The LTE-Advanced HCN architecture can provide a similar throughput level at a significantly lower energy consumption level. This is achieved by deploying HCN LPNs/RNs that reduce the average distance between UEs and their serving BS, thus reducing the average transmit power of all cells [25].

The total energy saving that can be achieved in a fixed BS deployment scenario is ultimately limited by the ratio between the transmitting radiohead hardware power

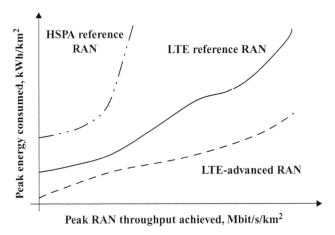

Figure 14.9 Fundamental tradeoff between peak energy and peak throughput of HSPA, LTE, and LTE-Advanced RANs.

consumption and the total power consumption of the BSs, where the total RAN energy saving bound cannot exceed 30–50% for typical macro BSs that are already deployed [26].

14.4.2.2   Re-deployment and dynamic BSs

In order to achieve better energy saving, BSs need to be switched off either on a permanent (redeployment) or temporary (sleep mode) basis. Research has shown that the theoretical energy saving limit in the redeployment case is capacity limited, as opposed to the fixed deployment case [26]. Moreover, it was found that by redeploying the existing BSs and adopting heterogeneous elements such as LPNs and RNs, the energy efficiency of a whole RAN can be improved by 60–80% [34].

Whilst there are long-term operational advantages in re-deployment of BSs, the caveats are the following:

- high CAPEX due to changing BS locations;
- replanning of the network in terms of local path loss models, coverage patterns, and performance dynamics; and
- site rental contracts and permissions.

As a result, research has largely been focused on how to allow existing BSs to be switched on and off dynamically, in order to mimic the effects of redeployment.

A basic mechanism for sleep mode (*passive sleep mode*) is to switch off any BSs that do not have a UE attached [23]. This has the advantage that no UE will be left unprovisioned or require handing over to a neighboring BS. However, the problem is that the sustained probability of a significant number of BSs without a UE attached is fairly low. Thus, the resulting energy saving is only meaningful for a dense deployment of LPNs [41], where the probability of having a significant number of LPNs unused could be larger. The typical energy saving achieved varies from less than 5% for macro

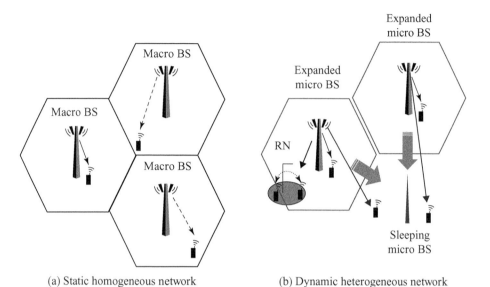

(a) Static homogeneous network
with macro-BSs

(b) Dynamic heterogeneous network
with RN-assisted micro-BSs

Figure 14.10 Different BS deployment strategies: (a) static macro BS deployment; (b) dynamic relay-assisted micro BS deployment.

BSs (0.5–1 per square kilometer), to 20% for micro BSs (2–4 per square kilometer), up to 70% for a dense deployment of pico BSs (7–10 per square kilometer) [34]. This gain can be much lower for femto BSs [42, 43], and depends on efficient idle/active switching modes and supporting architectures.

Compared to passive sleep mode, a more *aggressive sleep mode* algorithm is to switch off BSs with a low traffic load and hand over their UEs to neighboring BSs [44, 45]. An illustration of this technique and a comparison with the conventional macro BS deployment are provided in Fig. 14.10, where a BS is in sleep mode and its nearby UEs are being served by the expanded coverage of neighboring BSs. The major challenge faced by aggressive sleep modes is how to coordinate multiple BSs and manage their contention to sleep while compensating one another. Research into multi-cell coordination has examined centralized and distributed algorithms. Typically, centralized algorithms can achieve up to 50% energy saving at low traffic loads, but require a central controller to have network-wide knowledge of traffic intensity. Although distributed coordination only requires each BS to know its own traffic intensity, it has been shown to achieve a similar level of performance (48% energy saving at low traffic loads) to centralized algorithms. Moreover, the unpredictable behavior of switching off BSs leads to a tradeoff between outage probability and energy saving.

Dynamic BSs can also save energy through dynamic antenna pattern optimization in three different manners.

1. Tilt: downtilt elevation adjustment of BSs' antenna beams to improve the throughput fairness between cell-centre and cell-edge UEs.

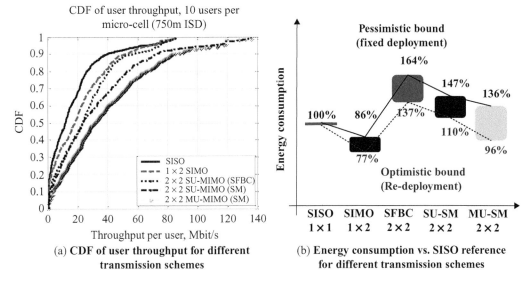

Figure 14.11 Spectral and energy efficiency of MIMO transmission techniques: (a) UE throughput CDF; (b) energy consumption relative to MIMO for fixed and redeployment of RAN.

2. Pan: rotation of the azimuth orientation of cell sectors to balance the traffic load between them.
3. Fan: expand or contract the beamwidth of cell sectors to balance the traffic load between them.

These techniques can be achieved with either mechanical tilting of the BS's antenna array, or novel electrical tilt designs.

14.5   Green cellular transmission techniques

14.5.1   MIMO techniques

Over the past decade, *multiple-input multiple-output* (MIMO) techniques have been incorporated into various wireless network standards, and are currently employed in both outdoor cellular networks and indoor wireless networks. The energy efficiency of MIMO is analyzed in [46]. The results shown in Fig. 14.11 indicate that MIMO transmissions achieve a higher throughput compared to *single-input single-output* (SISO) transmissions. However, compared to SISO BSs, the additional overhead power consumption incurred in MIMO BSs can lead to an overall energy consumption increase, e.g., up to 65% for 2×2 *space frequency block coding* (SFBC) MIMO [22]. A similar story is also true for Multi-User MIMO (MU-MIMO) transmissions, as shown in Fig. 14.11. Considering the fact that additional receiving antennas at the UE do not consume significantly more energy, the most energy-efficient transmission scheme would be *single input multiple output* (SIMO), which can decrease energy consumption by 15% for 1×2 SIMO. The energy saving can be further enhanced by re-deployment of BSs as previously discussed. Re-deploying a small number of BSs with 1×2 SIMO can save up to 33% energy [23].

Given that the capacity demands are met, multiple antennas at the transmitting BS is energy inefficient, due to the linear increase in power consumption with antennas and the diminished return from system-level capacity improvement. The tradeoff is always in favor of configurations with a small number of transmit antennas at high-power BSs and a large number of receive antennas at low-power UEs [25].

14.5.2   Interference reduction

As previously mentioned, cellular networks are increasingly being co-channel deployed, and thus BSs with a full traffic load are mutual interferers with one another. This has transformed RAN performance from what traditionally was noise and propagation limited to interference limited, and has spurred the research into interference reduction, which can be classified into three broad categories.

1. *Interference mitigation*: reduces the level of interference by decreasing the transmit power of nodes. In an HCN that comprises macro BSs and LPNs/RNs, the majority of interference is cross-tier interference. Mitigation of cross-tier interference can be achieved through power control. Research in [39], [47], and [48] investigates how LPNs can adjust transmit power to balance the tradeoff between coverage and energy consumption. Some unsolved issues include the hidden UE problem and how to obtain network knowledge about where interfering nodes are [49]. The most significant energy saving is achieved when UEs are close to their serving BS and the transmit power level of each BS is low.
2. *Interference avoidance*: removes the interference coupling by coordinating transmissions of neighboring cells. Given that not all frequency bands are being utilized simultaneously at low traffic loads, two transmitting nodes can avoid mutual interference by not transmitting on the same sub-band. This can be achieved through intercell coordination, which is typically handled at the scheduler level [50]. Energy saving is more evident when the cells have a lower traffic load.
3. *Interference cancelation*: reduces the level of interference by pre-coding the signal at the transmitter or equalizing the signal at the receiver [51]. This is typically applied to MIMO and multi-BS systems to improve the cell-edge performance, and is known as *coordinated multi-point* (CoMP) transmission and reception. The most significant energy savings occur at the intersection of three adjacent BSs, where the interference is the greatest.

The effective energy saving is very sensitive to conditions such as pathloss and traffic. Each technique generally has a different energy saving *sweet spot*. The typical sweet-spot energy saving is 60–70% in terms of the transmit energy, and 20–30% in terms of the total RAN energy [51].

It has been shown that interference reduction requires knowledge of *channel state information* (CSI) and/or coordination between multiple cells. The latter can be achieved via inter-BS interfaces such as the X2 interface in LTE, but there is a lack of commercial willingness to implement such interfaces between LPNs. Currently, practical interference mitigation and avoidance can only be achieved at the macro BS level.

14.5.3   Scheduling

There are many types of scheduling algorithm [21] that prioritize different performance metrics. An energy-efficient proportional-fair scheduler is proposed in [52], where the transmit energy from the BS to each UE is minimized. Other methods include utilizing free bandwidth and lower-order modulation and coding schemes to reduce the transmit energy consumption. In general, it has been estimated that transmit energy can be reduced by up to 30%, but the total operational energy saving of the network is estimated to be less than 10% [52].

Among all energy-efficient transmission techniques, the most advanced and spectrally efficient techniques are efficient in transmit energy usage, but are not necessarily efficient in total operational energy consumption. Since the idle periods of a BS do not benefit from spectrally efficient transmission techniques, the energy saving achieved by such techniques is typically limited to 10–15% on average.

14.6   Integrated heterogeneous cellular networks

14.6.1   Flexible heterogeneous cellular networks

In this section, a flexible energy-efficient heterogeneous LTE-Advanced network architecture is proposed. For the same traffic load served, its energy efficiency performance is compared to those of the HSPA network and a baseline LTE network. The architecture and transmission techniques of the envisaged green HCN, as illustrated in Fig. 14.7, are summarized as follows.

- *Architecture*: Each micro BS has three to five RNs deployed at the cell edge of each cell sector. LPNs such as femtocells can provide improved access to indoor UEs and hotspot areas. In order to efficiently scale energy consumption with required capacity, each BS employs a dynamic antenna design that can switch from multiple directional cell sectors to a single omnidirectional cell sector. Furthermore, BSs with a low traffic load can be switched off and their UEs can be handed over to the expanded coverage of neighboring BSs. Whether each BS should change its number of active sectors or which operational mode it should use is controlled either by a centralized controller or through distributed decision making. The expected integrated energy saving for an existing fixed deployment of BSs is from 55 to 70%.
- *Techniques*: Use SIMO transmission with interference mitigation among multiple heterogeneous tiers (e.g., between micro BSs and LPNs). Interference avoidance and cancelation can be used among micro BSs, with inter-BS coordination, centralized

control, or game theoretical methods. An energy-aware scheduler can be used to improve the overall energy efficiency of the network, without requiring knowledge of network deployment. The expected integrated energy saving for an existing fixed deployment of BSs is from 25 to 30%.

The rationale behind the above architecture and transmission techniques is that the traffic distribution among BSs is very uneven. Typically, 50% of the data is carried by less than 10% of the BSs. Therefore, an effective energy saving technique is to switch off low-load BSs and expand the coverage of neighboring BSs [44]. Another contributor to energy saving is the HCN architecture of the proposed RAN, where the mean distance from a UE to its serving BS, as well as the mean power consumption per BS, is reduced. The integrated energy saving from both architecture and transmission techniques is estimated to be around 80%. As shown in Fig. 14.9, the proposed green LTE-Advanced architecture can delivery a significant energy saving, particularly at high-throughput regions.

The aforementioned integrated energy saving employs a *dynamic programming* approach, where each technique is optimized in isolation and then integrated under a common framework (e.g., a simulator or a real network). However, the *dynamic programming* approach yields no guarantee on an optimal or near-optimal joint performance. Furthermore, it is difficult to extract or exploit synergies between different techniques. An alternative approach currently under investigation is automation or a *self-organizing network* (SON) at the BS as is discussed in detail in Chapter 6. In the absence of human supervision or knowledge of the optimal network configuration, machine learning allows the BSs to learn what techniques and parameters to adopt under given circumstances.

14.6.2   Self-organizing networks

The introduction of HCN nodes has also increased the network complexity. The optimal network configuration is no longer intuitive nor tractable. Whilst exhaustive simulations can yield insight into which configurations are more desirable for certain scenarios, real-time reconfiguration of RANs requires a different organizational approach. One such approach uses automation or machine learning at the BSs to decide what actions to take. There are two forms of machine learning: *supervised* and *unsupervised*. Supervised learning applies to the case when prior knowledge of the optimal network parameters or behavior is known. Unsupervised learning applies to the case when no prior knowledge exists and performance is measured on the fly by a set of metrics, e.g., energy saving and throughput.

Complex cellular networks usually require unsupervised learning, where the agents (BSs) within the network learn through a combination of observing consequences (e.g., energy saved, throughput achieved) and choosing actions (e.g., sleep mode, CoMP). A set of actions forms a *strategy*, which can be extracted as a jointly optimized integration solution.

A useful approach for devising unsupervised learning algorithms is reinforced learning, a version of which is Q-learning. Given a dynamic traffic environment, a network

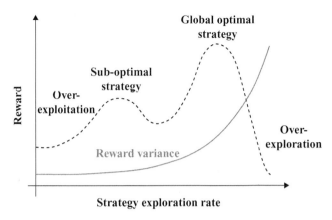

Figure 14.12 Unsupervised machine learning: tradeoff between asymptotic reward (energy saved) and strategy exploration rate.

can choose to *exploit* known strategies that are expected to yield greater reward, or *explore* alternative strategies. Existing research has demonstrated that Q-learning can assist the dynamic optimization of complex wireless networks [53, 54].

In the absence of a full state transition map, there is often a tradeoff between the rate of *exploitation* and the rate of *exploration*. This is demonstrated in Fig. 14.12, where over-exploration of strategies can lead to better understanding of the state transition map, but insufficient resources to effectively exploit the most beneficial strategy, while under-exploration will lead to rapid convergence to a single strategy that may not be optimal. A consequence can be that the SON repeatedly uses a sub-optimal strategy and fails to discover better strategies. A variety of exploration methodologies exist, and a popular one is the *Boltzmann exploration* [55], where the chance of randomly exploring a strategy is weighted by the expectation of reward of all strategies. This is an area of future research, and how much machine learning can further enhance energy savings of HCNs is still to be shown.

14.7   Discussion

14.7.1   Standardization of green cellular networks

Whilst significant research has been conducted on green techniques for cellular networks, standardization is required to explicitly define *what is a green network* and *how green it is*. Currently IEEE is developing specifications for energy-efficient networking techniques, while the *Internet Engineering Task Force* (IETF) is specifying energy consumption monitoring requirements [12] and managed objects [13] to create interoperable standards for green cellular networks. In the meantime, 3GPP has been working on energy saving initiatives and green standards too, targeting current network architectures, e.g., *Technical Specification Group* (TSG) RAN *working group* (WG) 3 for energy saving of UMTS and LTE [14, 15], while the *Worldwide Interoperability for Microwave*

Access (WiMAX) forum is carrying out phase 2 standardization work on femtocell techniques [56]. Given the heterogeneity of emerging cellular networks, future standardization work needs to consider their diverse requirements, flexible methodologies, and vendor-independent attributes.

14.7.2   Pricing in green cellular networks

Another critical issue for greening cellular networks is the lack of an effective pricing scheme. Traditional pricing schemes [57, 58] mainly focus on the service bandwidth and contents. There is still a lack of representative green pricing policies. The compensation-based pricing model in [59] considers the QoS degradation due to energy conservation and resource limitation. The strategies in [60] and [61] combine user pricing, network expenditure, and power control jointly for network-centric and UE-centric radio resource management, respectively.

In the future, researchers must deliver effective pricing strategies for green cellular networks along with QoS guarantee, in a way that encourages networks and UEs to adopt energy-efficient communication strategies.

14.7.3   New energy and materials

While countless efforts have been put into power-saving communication techniques, research activities on discovering new energy sources have been strongly inspired in recent years to push forward the development of green cellular networks [62]. Efficient usage of new power supply resources, such as wind and solar power, can be one promising direction. For instance, a square meter of solar panel can produce 10% of the energy required by a macro BS. It has been announced that about 335 000 cellular BSs will include solar panels by 2013 [62]. Pioneer companies have already developed solar-driven smartphones. However, research activities are still required to increase the efficiency of new energy sources. In addition, utilizing renewable materials and recycling used materials can also contribute to a greener ICT. New methods of energy harvesting for UEs, such as the movement-driven and sound-wave-driven power supply, as well as biological energy sources, may lead to a new level of green cellular networks.

14.8   Conclusion

In this chapter, we have presented a comprehensive survey of the latest green techniques for homogeneous and heterogeneous cellular networks with discussions of their merits and demerits. We have also reviewed recent research projects for green cellular networks and summarized a taxonomy of green metrics. However, there are still many challenges that need to be addressed, especially in emerging green HCN architectures and transmission techniques, as well as how to integrate different cross-layer and cross-tier solutions.

Making cellular networks green would have a tangible positive impact on reducing the carbon footprint of ICT, and also achieve a long-term profitability of mobile service operators. It was found that an energy-efficient HCN can also reduce the operational and maintenance cost of the network by 40–60%. On an architectural level, it was found that the energy saving potential of current cellular networks is limited to approximately 30%, while the energy saving potential of redeploying BSs with the aid of LPNs/RNs in an HCN configuration can increase to 80%. The integration of dynamic BS designs and interference mitigating schemes can yield a further improvement in energy efficiency.

The development and deployment of green cellular networks requires collaboration among cellular service providers, equipment vendors, governments, and the public. Moreover, non-technical factors, such as pricing and marketing strategies, law establishment, service affordability, and UE friendliness, will also play important roles in the success of green HCNs.

Copyright notices

References

[1] G. Fettweis and E. Zimmermann, ICT energy consumption – trends and challenges. In *Proceedings of the 11th International Symposium on Wireless Personal Multimedia Communications* (2008).

[2] CISCO, *Cisco Visual Networking Index: Global Mobile Data Traffic Forecast Update, 2010–2015*, White Paper (2011).

[3] J. Hansen, M. Sato, P. Kharecha, G. Russell, D. Lea and M. Siddall, Climate change and trace gases. *Philosophical Transactions of the Royal Society A: Mathematical, Physical and Engineering Sciences*, **365**:1856 (2007), 1925–1954.

[4] F. Williams, *Green Wireless Communications*, eMobility, technical report (2008).

[5] T. Kelly and S. Head, *ICTS and Climate Change*, ITU-T Technology, technical report (2007).

[6] H. Karl *et al.*, *An Overview of Energy-Efficiency Techniques for Mobile Communication Systems*, report of AG Mobikom WG7, (2003).

[7] S. McLaughlin, Green radio: the key issues. Programme objectives and overview. In *Wireless World Research Forum* (2008).

[8] C. Jones, K. Sivalingam, P. Agrawal and J. Chen A survey of energy efficient network protocols for wireless networks. *Wireless Networks*, **7**:4 (2001), 343–358.

[9] G. Anastasi, M. Conti, M. Di Francesco and A. Passarella, Energy conservation in wireless sensor networks: a survey. *Ad Hoc Networks*, **7**:3 (2009), 537–568.

[10] Y. Yi, *Cellular Greening Via Efficient Bs Control: Topology, On–Off, and Transmssion Power*, Green-Touch, technical report (2011).

[11] M. Haardt, *Future Mobile and Wireless Radio Systems: Challenges in European Research*, The European Commission, technical report (2008).

[12] J. Quittek (ed.), *Requirements for Power Monitoring*. Internet draft draft-quittek-power-monitoring-requirements-00. http://tools.ietf.org/html/draft-quittek-power-monitoring-requirements-00

[13] J. Quittek, *Reference Model for Energy Management*. Internet draft draft-quittek-eman-reference-model-03. http://tools.ietf.org/html/draft-quittek-eman-reference-model-03

[14] D. Knisely, T. Yoshizawa and F. Favichia, Standardization of femtocells in 3GPP. *Communications Magazine, IEEE*, **47**:9 (2009), 68–75.

[15] D. Knisely and F. Favichia, Standardization of femtocells in 3GPP2. *IEEE Communications Magazine*, **47**:9 (2009), 76–82.

[16] *The Mobile Virtual Centre of Excellence, Core 5 Research on Green Radio*. http://www.mobilevce.com/frames.htm?core5research.htm

[17] *Earth Project: Enablers for Energy Efficient Wireless Networks*. https://www.ict-earth.eu

[18] FP7 Consultation Meeting, *Europe Future Mobile and Wireless Radio Systems: Challenges in European Research*, technical report (2008).

[19] *Green IT Initiatives in Japan*. http://www.meti.go.jp/english/policy/GreenITInitiative InJapan.pdf

[20] A. P. Bianzino, A. K. Raju and D. Rossi, Apples-to-apples: a framework analysis for energy-efficiency in networks. *SIGMETRICS Performance Evaluation Review*, **38**:3 (2011), 81–85.

[21] H. Holma and A. Toskala, *LTE for UMTS: OFDMA and SC-FDMA Based Radio Access* (Wiley, 2009).

[22] G. Auer, V. Giannini, I. Godor, P. Skillermark, M. Olsson, M. Imran, D. Sabella, M. Gonzalez, C. Desset and O. Blume, Cellular energy efficiency evaluation framework. In *IEEE 73rd Vehicular Technology Conference (VTC Spring), 2011* (IEEE, 2011), pp. 1–6.

[23] W. Guo and T. O'Farrell, Green cellular network: deployment solutions, sensitivity and tradeoffs. In *Wireless Advanced (WiAd), 2011* (IEEE, 2011), pp. 42–47.

[24] C. Xiong, G. Li, S. Zhang, Y. Chen and S. Xu, Energy and spectral-efficiency tradeoff in downlink OFDMA networks. In *2011 IEEE International Conference on Communications (ICC)* (IEEE, 2011), pp. 1–5.

[25] W. Guo and T. O'Farrell, Power–capacity-tradeoff for low energy interference limited cellular networks. In *IEEE 75th Vehicular Technology Conference (VTC Spring), 2011* (IEEE, 2012).

[26] W. Guo and T. O'Farrell, Capacity–energy–cost tradeoff for small-cell networks. In *IEEE International Workshop on Green Wireless Communications and Networks (GreeNet)* (IEEE, 2012).

[27] L. Correia, D. Zeller, O. Blume, D. Ferling, Y. Jading, I. Godor, G. Auer and L. Van Der Perre, Challenges and enabling technologies for energy aware mobile radio networks. *IEEE Communications Magazine*, **48**:11 (2010), 66–72.

[28] H. Jiang, W. Zhuang and X. Shen, Cross-layer design for resource allocation in 3G wireless networks and beyond. *IEEE Communications Magazine*, **43**:12 (2005), 120–126.

[29] D. Lister, An operator's view on green radio. *Keynote Speech, GreenComm* (2009).

[30] F. Richter, A. Fehske, P. Marsch and G. Fettweis, Traffic demand and energy efficiency in heterogeneous cellular mobile radio networks. In *IEEE 71st Vehicular Technology Conference (VTC 2010-Spring), 2010* (IEEE, 2010), pp. 1–6.

[31] A. Dinnis and J. Thompson, The effects of including wraparound when simulating cellular wireless systems with relaying. In *IEEE 65th Vehicular Technology Conference, 2007. VTC2007-Spring* (IEEE, 2007), pp. 914–918.

[32] S. Wang, W. Guo and T. O'Farrell, Two tier networks with frequency selective surface. In *IEEE International Conference on High Performance Computing and Communications (HPCC)* (IEEE, 2012).

[33] W. Guo and T. O'Farrell, Relay deployment in cellular networks: Planning and optimization. *IEEE Journal on Selected Areas in Communications*, **31** (2012), 1–8.

[34] W. Guo and T. O'Farrell, Small-net vs. relays in a heterogeneous architecture. *Journal of Communications*, **7**:10 (2012), 716–725.

[35] A. Sharma, V. Navda, R. Ramjee, V. Padmanabhan and E. Belding, Cool-tether: energy efficient on-the-fly WiFi hot-spots using mobile phones. In *Proceedings of the Fifth International Conference on Emerging Networking Experiments and Technologies* (ACM, 2009), pp. 109–120.

[36] H. Lei, X. Wang and P. Chong, Opportunistic relay selection in future green multihop cellular networks. In *IEEE 72nd Vehicular Technology Conference Fall (VTC 2010-Fall), 2010* (IEEE, 2010), pp. 1–5.

[37] A. Furuskar, M. Almgren and K. Johansson, An infrastructure cost evaluation of single- and multi-access networks with heterogeneous traffic density. In *IEEE 61st Vehicular Technology Conference, 2005. VTC 2005-Spring. 2005* (IEEE, 2005), vol. 5, pp. 3166–3170.

[38] E. Lang, S. Redana and B. Raaf, Business impact of relay deployment for coverage extension in 3GPP LTE-Advanced. In *2009 IEEE International Conference on Communications Workshops* (IEEE, 2009), pp. 1–5.

[39] Y. Hou and D. Laurenson, Energy efficiency of high QoS heterogeneous wireless communication network. In *IEEE 72nd Vehicular Technology Conference Fall (VTC 2010-Fall), 2010* (IEEE, 2010), pp. 1–5.

[40] T. Quek, W. Cheung and M. Kountouris, Energy efficiency analysis of two-tier heterogeneous networks. In *Wireless Conference 2011 – Sustainable Wireless Technologies (European Wireless), 11th European* (VDE, 2011), pp. 1–5.

[41] I. Ashraf, F. Boccardi and L. Ho, Sleep mode techniques for small cell deployments. *IEEE Communications Magazine*, **49**:8 (2011), 72–79.

[42] V. Chandrasekhar, J. Andrews and A. Gatherer, Femtocell networks: a survey. *IEEE Communications Magazine*, **46**:9 (2008), 59–67.

[43] A. Golaup, M. Mustapha and L. Patanapongpibul, Femtocell access control strategy in UMTS and LTE. *IEEE Communications Magazine*, **47**:9 (2009), 117–123.

[44] Z. Niu, Y. Wu, J. Gong and Z. Yang, Cell zooming for cost-efficient green cellular networks. *IEEE Communications Magazine*, **48**:11 (2010), 74–79.

[45] M. Marsan, L. Chiaraviglio, D. Ciullo and M. Meo, Optimal energy savings in cellular access networks. In *IEEE International Conference on Communications Workshops* (2009), pp. 1–5.

[46] C. Turyagyenda, T. O'Farrell, J. He and P. Loskot, SFBC MIMO energy efficiency improvements of common packet schedulers for the long term evolution downlink. In *IEEE 73rd Vehicular Technology Conference (VTC Spring), 2011* (IEEE, 2011), pp. 1–5.

[47] S. Yeh, S. Talwar, S. Lee and H. Kim, WiMAX femtocells: a perspective on network architecture, capacity, and coverage. *IEEE Communications Magazine*, **46**:10 (2008), 58–65.

[48] H. Claussen, L. Ho and L. Samuel, Self-optimization of coverage for femtocell deployments. In *Wireless Telecommunications Symposium, 2008. WTS 2008* (IEEE, 2008), pp. 278–285.

[49] M. Andrews, V. Capdevielle, A. Feki and P. Gupta, Autonomous spectrum sharing for mixed LTE femto and macro cells deployments. In *INFOCOM IEEE Conference on Computer Communications Workshops, 2010* (IEEE, 2010), pp. 1–5.

[50] C. Turyagyenda, T. O'Farrell and W. Guo, Energy efficient coordinated radio resource management: a two player sequential game modelling for the LTE downlink. *IET Journal on Communication*, **6**:14 (2012), 2239–2249.

[51] T. Le and M. Nakhai, User position aware multicell beamforming for a distributed antenna system. In *2011 IEEE 22nd International Symposium on Personal Indoor and Mobile Radio Communications (PIMRC)* (IEEE, 2011), pp. 1423–1427.

[52] C. Turyagyenda, T. O'Farrell and W. Guo, A novel proportional fair energy packet scheduler for the LTE downlink. In *IEEE 75th Vehicular Technology Conference (VTC Spring), 2011* (IEEE, 2012).

[53] 3GPP, *Self-configuring and Self Optimizing Network (SON) Use Cases and Solutions*. 3GPP Technical Report, TR 36.902 (Release 9, 2009).

[54] R. Razavi, S. Klein and H. Claussen, Self-optimization of capacity and coverage in LTE networks using a fuzzy reinforcement learning approach. In *2011 IEEE 21st International Symposium on Personal Indoor and Mobile Radio Communications (PIMRC)* (IEEE, 2010), pp. 1865–1870.

[55] L. Kaelbling, M. Littman and A. Moore, Reinforcement learning: a survey. *Journal of Artificial Intelligence Research*, **4** (1996), 237–285.

[56] R. Kim, J. Kwak and K. Etemad, WiMAX femtocell: requirements, challenges, and solutions. *IEEE Communications Magazine*, **47**:9 (2009), 84–91.

[57] C. Courcoubetis and R. Weber, *Pricing Communication Networks* (Wiley, 2003), vol. 2.

[58] H. Jung and B. Tuffin, Pricing for heterogeneous services in OFDMA 802.16 systems. In *Sixth International Conference on Wireless On-Demand Network Systems and Services, 2009 (WONS 2009)* (IEEE, 2009), pp. 49–52.

[59] H. Le Cadre, M. Bouhtou and B. Tuffin, A pricing model for a mobile network operator sharing limited resource with a mobile virtual network operator. *Network Economics for Next Generation Networks*, **55** (2009), 24–35.

[60] N. Feng, S. Mau and N. Mandayam, Pricing and power control for joint network-centric and user-centric radio resource management. *IEEE Transactions on Communications*, **52**:9 (2004), 1547–1557.

[61] S. Betz and H. Poor, Energy efficient communications in CDMA networks: a game theoretic analysis considering operating costs. *IEEE Transactions on Signal Processing*, **56**:10 (2008), 5181–5190.

[62] ABIresearch, *Mobile Networks go Green – Minimizing Power Consumption and Leveraging Renewable Energy*, ABIresearch, Technical Report (2008).

Index